Lecture Notes in Artificial Intelligence 1489

Subseries of Lecture Notes in Computer Science
Edited by J. G. Carbonell and J. Siekmann

Lecture Notes in Computer Science
Edited by G. Goos, J. Hartmanis and J. van Leeuwen

Springer
Berlin
Heidelberg
New York
Barcelona
Budapest
Hong Kong
London
Milan
Paris
Singapore
Tokyo

Jürgen Dix   Luís Fariñas del Cerro
Ulrich Furbach (Eds.)

Logics in Artificial Intelligence

European Workshop, JELIA '98
Dagstuhl, Germany, October 12-15, 1998
Proceedings

Series Editors

Jaime G. Carbonell, Carnegie Mellon University, Pittsburgh, PA, USA
Jörg Siekmann, University of Saarland, Saarbrücken, Germany

Volume Editors

Jürgen Dix
Department of Computer Science, University of Maryland
College Park, Maryland 20742, USA
E-mail: dix@cs.umd.edu

Luís Fariñas del Cerro
I.R.I.T. Université Paul Sabatier
118 route de Narbonne, F-31062 Toulouse Cedex, France
E-mail: farinas@irit.fr

Ulrich Furbach
Fachbereich 4: Informatik, Universität Koblenz
Rheinau 1, D-56075 Koblenz, Germany
E-mail: uli@informatik.uni-koblenz.de

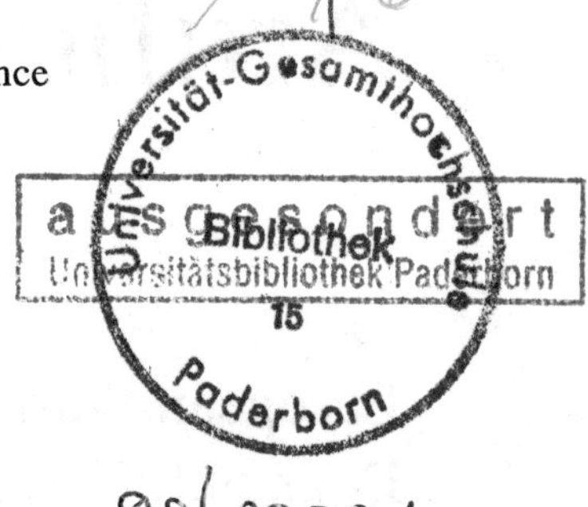

Cataloging-in-Publication Data applied for

Die Deutsche Bibliothek - CIP-Einheitsaufnahme

Logics in artificial intelligence : European workshop ; proceedings / JELIA '98, Dagstuhl, Germany, October 12 - 15, 1998 / Jürgen Dix ... (ed.). - Berlin ; Heidelberg ; New York ; Barcelona ; Budapest ; Hong Kong ; London ; Milan ; Paris ; Santa Clara ; Singapore ; Tokyo : Springer, 1998
(Lecture notes in computer science ; Vol. 1489 : Lecture notes in artificial intelligence)
ISBN 3-540-65141-1

CR Subject Classification (1991): I.2, D.1.6, F.4.1

ISBN 3-540-65141-1 Springer-Verlag Berlin Heidelberg New York

Printed in Germany

Typesetting: Camera ready by author
SPIN 10638847 06/3142 – 5 4 3 2 1 0 Printed on acid-free paper

Preface

This volume contains the papers selected for presentation at the conference and two abstracts from invited speakers. The programme committee selected these 25 papers from 12 countries out of 65 submissions from 17 countries.

The first JELIA meeting was in Roscoff, France, ten years ago. Afterwards, it took place in the Netherlands, Germany, United Kingdom, Portugal, and now again in Germany. The proceedings of the last four meetings appeared in the Springer-Verlag LNCS series, and a selected series of papers of the English and the Portuguese meeting appeared as special issues in the *Journal of Applied Non-Classical Logics* and in the *Journal of Automated Reasoning*, respectively.

The aim of JELIA was and still is to provide a forum for the exchange of ideas and results in the domain of foundations of AI, focusing on rigorous descriptions of some aspects of intelligence. These descriptions are promoted by applications, and produced by logical tools and methods. The papers contained in this volume cover the following topics:

1. *Logic programming*
2. *Epistemic logics*
3. *Theorem proving*
4. *Non-monotonic reasoning*
5. *Non-standard logics*
6. *Knowledge representation*
7. *Higher order logics*

We would like to warmly thank the authors, the invited speakers, the members of the program committee, and the additional reviewers listed below. They all have made these proceedings possible and ensured their quality.

August 1998

Jürgen Dix, Koblenz
Luís Fariñas del Cerro, Toulouse
Ulrich Furbach, Koblenz

Conference Organization

Conference Chair

Jürgen Dix	University of Koblenz, Germany

Program Chairs

Luís Fariñas del Cerro	University of Toulouse, France
Ulrich Furbach	University of Koblenz, Germany

Program Committee

Bruno Buchberger	RISC, Linz, Austria
Ricardo Caferra	LIFIA, Grenoble, France
Robert Demolombe	University of Toulouse, France
Patrice Enjalbert	University of Caen, France
Pascal van Hentenryck	University of Amsterdam, Netherlands
Michael Kohlhase	University of Saarbrücken, Germany
Vladimir Lifschitz	University of Texas, U.S.A.
Kym MacNish	University of Western Australia, Australia
Daniele Mundici	University of Milan, Italy
Ilkka Niemelä	Helsinki University of Technology, Finland
David Pearce	University of Saarbrücken, Germany
Luís Moniz Pereira	Universidade Nova de Lisboa, Portugal
Teodor Przymusinski	University of California at Riverside, U.S.A.
Andrzej Szalas	Warszaw University, Poland
Mary-Anne Williams	University of Newcastle, Australia
Stefan Wrobel	University of Magdeburg, Germany

Organizing Committee (University of Koblenz, Germany)

Chandrabose Aravindan
Peter Baumgartner
Michael Kühn
Frieder Stolzenburg
Bernd Thomas

Sponsors

ESPRIT Compulog Network of Excellence
University of Koblenz
Research Committee 1.2 Inference Systems of the *German Society of Computer Science*

Invited Speakers

Erik Sandewall	Linköping University, Sweden
Bart Selman	Cornell University, U.S.A.
Paul Tarau	Université de Moncton, Canada

Additional Referees

J.J. Alferes
P. Balbiani
P. Baumgartner
B. Beckert
S. Benferhat
P. Besnard
C. Bettini
F. Boniol
T. Boy de la Tour
G. Brewka
A. Burt
M. Castilho
F. Clerin
S. Costantini
J.C. Cunha
C.V. Damasio
S. Demri
M. Dierkes
V. Dugat
C. Dupre
T. Eiter
J. Engelfriet
M.C. Ferreira
S. Fevre
B. Fronhöfer
O. Gasquet
M. Gelfond
T. Gordon
G. Gottlob
P. Gribomont
J. Halpern
H. Herre
A. Herzig
S. Hölldobler
H. Hong
T. Horvath
U. Hustadt
T. Janhunen
I. Jarras
J. Karczmarczuk
S. Kraus
J. Lang
D. Lehmann
R. Li
V. Marek
M. Marin
M. May
N. McCain
M. Milano
Y. Moinard
N. Peltier
M. Protzen
S. Ratschan
H. Rolletschek
D. Sadek
T. Schaub
K. Schlechta
R. Schmidt
D. Seipel
B. Selman
P. Simons
F. Stolzenburg
T. Stratulat
V.S. Subrahmanian
F. Toni
H. Turner
Y.S. Usenko
A. Zanardo
G. Zanon

Previous Workshops

1988 Roscoff	France
1990 Amsterdam	Netherlands
1992 Berlin	Germany
1994 York	United Kingdom
1996 Evora	Portugal

Table of Contents

Logic Programming

Epistemic Logics

Theorem Proving

Non-monotonic Reasoning

Non-standard Logics

Knowledge Representation

Higher Order Logics

Invited Talks

The Well-Founded Semantics Is the Principle of Inductive Definition

Marc Denecker

Department of Computer Science, K.U.Leuven, Belgium
marcd@cs.kuleuven.ac.be

Abstract. Existing formalisations of (transfinite) inductive definitions in constructive mathematics are reviewed and strong correspondences with LP under least model and perfect model semantics become apparent. I point to fundamental restrictions of these existing formalisations and argue that the well-founded semantics (wfs) overcomes these problems and hence, provides a superior formalisation of the principle of inductive definition. The contribution of this study for LP is that it (re-) introduces the knowledge theoretic interpretation of LP as a logic for representing definitional knowledge. I point to fundamental differences between this knowledge theoretic interpretation of LP and the more commonly known interpretations of LP as default theories or auto-epistemic theories. The relevance is that differences in knowledge theoretic interpretation have strong impact on knowledge representation methodology and on extensions of the LP formalism, for example for representing uncertainty.

1 Introduction

With the completion semantics [5], Clark aimed at formalising the meaning of a logic program as a set of definitions. To that aim, he maps a logic program to a set of First Order Logic (FOL) equivalences. Motivated by the research in Nonmonotonic Reasoning, logic programming is currently often seen as a default logic or auto-epistemic logic. In [11], Gelfond proposes a semantics for stratified logic programs based on an auto-epistemic interpretation of the formalism. In [12], Gelfond and Lifschitz motivate the stable semantics for logic programs from the perspective of logic programs as default and auto-epistemic theories.

To compare these readings, consider the program P_0 with unique rule:

$$dead \leftarrow not\ alive$$

P_0 is propositional and hierarchical; all common semantics of LP (completion / perfect [3, 21] / stable [12] / wfs [28]) *agree*; for the above example, the unique model is $\{dead\}$.

In the interpretation of this program as an auto-epistemic theory, P_0 corresponds to the auto-epistemic theory (AEL):

$$AEL(P_0) = \{dead \leftarrow \neg K alive\}$$

which reads as: *one is dead* *if* *it is not believed that* *one is alive.* On the other hand, under completion semantics the meaning of this program is given by the FOL theory:

$$comp(P_0) = \{\neg alive\ ,\ dead \leftrightarrow \neg alive\}$$

These readings show important differences. The completion reading of P_0 states that *alive* is false while the auto-epistemic reading of P_0 gives no information about *alive*; hence *alive* is not known. The completion reading maps implication to equivalence and negation to classical objective negation, while the auto-epistemic reading map negation to a modal operator ($\neg K$) and preserves the implication.

How to explain that, despite this intuitive difference, the stable model -which formalises the default/auto-epistemic reading- corresponds to the model of the completion? The reason is that models in stable semantics and in classical logic play a different role. A stable model is a *belief set*: the set of atoms which are believed, while the model of the completion, as a model of a FOL theory, represents a possible state of the world. Because models in both semantics play a different role, a simple comparison between them does not reveal the different meanings of both semantics.

Actually, a clear and correct model theoretic comparison of the meaning of the auto-epistemic reading and of the completion is possible if done on the basis of the *possible world model* of the auto-epistemic theory and of the set of models of the completion. Both are sets of models; in both sets the role of models is identical: they represent possible states of the world. Such a comparison confirms the intuitive differences between the two readings. The possible world model of the AEL theory $\{dead \leftarrow \neg K alive\}$ is $\{\{dead\}, \{alive, dead\}\}$. This set of models reflects indeed the intuitive meaning of $AEL(P_0)$: *alive* can be true or false, hence nothing is known on *alive*; (therefore) *dead* is always true. Note that the belief set, i.e. the stable model, is the intersection of these possible states. In contrast, the set of models of the completion is the singleton $\{\{dead\}\}$. Interpreted as a possible world model, it represents that *dead* is known to be true, *alive* known to be false.

This observation motivates a closer investigation of the relation between logic programming and inductive definitions. An inductive definition is a form of constructive knowledge. Constructive information *defines* a relation (or a collection of relations) through a constructive process of iterating a recursive recipe. This recipe defines new instances of the relation in terms of the presence (and sometimes the absence) of other tuples of the relation. A broad class of human knowledges in many areas of human expertise, ranging from common sense knowledge situations to mathematics, is of constructive nature. One example is Reiter's formalisation of situation calculus [23]; in this approach, a situation calculus can be understood as an inductive definition on the well-founded poset of situations. Another example is in [26], where we argue that causality information in the context of the ramification problem is a form of constructive information.

Causes, effects and forces propagate in a dynamic system through a constructive process; consequently, the semantics of causality rules is defined by an inductive definition which, by its constructive nature, mirrors the physical process of the effect propagation.

In the context of mathematics, constructive information appears by excellence in inductive definitions. For example, as suggested by the name, the *transitive closure* of a binary relation is naturally perceived as the relation obtained through the construction process of *closing* the relation under the transitivity rule. Not a coincidence, inductive definitions have been studied in constructive mathematics and intuitionistic logic, in particular in the sub-areas of Inductive and Definition logics, Iterated Inductive Definition logics and Fixpoint logics. The main goal of this paper is to review some of this work and to show how inductive definitions are formalised in these areas; this immediately reveals strong relationships with least model and perfect model semantics of logic programming (section 2). I point to fundamental knowledge theoretic problems in these formalisms (section 3) and argue that the logic program formalism under well-founded semantics provides a superior formalisation (section 4). Section 5 considers some implications.

2 Inductive Definitions in mathematics

One can distinguish between positive inductive definitions and definitions by induction on a well-founded set. A prototypical example of a definition by (positive) induction is the one of the transitive closure T_R of a graph R. T_R is defined inductively as follows. T_R contains an arc from x to y if

- R contains an arc from x to y;
- R contains an arc from x to z, and T_R contains an arc from z to y.

It could be formally represented by the rules:

$$\mathcal{D}_{trans} = \begin{cases} tr(X,Y) \leftarrow graph(X,Y) \\ tr(X,Y) \leftarrow graph(X,Z), tr(Z,Y) \end{cases}$$

The intended interpretation of this definition is that the transitive closure is the *least* graph satisfying the implications rather than any graph satisfying the above implications. Alternatively, the transitive closure can be obtained in a constructive way by applying these implications in a bottom up way until saturation. It is commonly known that inductive definitions such as the one of transitive closure cannot be expressed in FOL, and a fortiori, not in the completion semantics[1].

Typical for the above sort of inductive definition is that the induction is positive: i.e. the defined concept depends positively on itself, and hence a unique

[1] A simple counterexample: verify that the unintended interpretation with domain $\{a,b\}$ and $I(graph) = \{(a,a)\}$ and $I(tr) = \{(a,a),(a,b)\}$ satisfies the completion of the implications.

least relation exists. In definitions by (possibly transfinite) induction on a well-founded poset, this is not necessarily the case. In definitions of this kind, a concept is defined for a domain element in terms of strictly smaller elements. An example is the definition of the ordinal powers of a monotonic operator. A simple first order example of such a definition is the definition of even numbers in the well-founded poset $\mathbb{N}, \leq$. One defines that a natural number n is even by induction on $\leq$:

- $n = 0$ is even;
- if n is not even then $n+1$ is even; otherwise $n+1$ is not even.

A formal representation of the definition in the form of implications is:

$$\mathcal{D}^1_{even} = \begin{cases} even(0) \\ even(s(X)) \leftarrow \neg even(X) \end{cases}$$

Now the defined predicate *even* occurs negatively in the body of the rule. Verify that in the natural numbers, this theory has infinitely many minimal models[2].

Its semantics can be described by a constructive process and is also expressed well by the Clark completed definition of the above implications:

$$\forall X.even(X) \leftrightarrow X = 0 \vee \exists Y.X = s(Y) \wedge \neg even(X)$$

A more complex example showing the elements of transfinite induction in a richer context is the concept of *depth* of an element in a well-founded poset $P, \leq$. Define the depth of an element x of P by transfinite induction as the least ordinal which is a strict upper-bound of the depths of elements $y \in P$ such that $y < x$.

Formally, let $F[X,D]$ mean that D is a larger ordinal than the depths of all elements $Y < X$: $F[X,D] \equiv$

$$\forall Y, D_Y.(Y < X \wedge depth(Y, D_Y) \rightarrow D_Y < D)$$

Then, depth is represented by the singleton definition $\mathcal{D}_{depth}$:

$$\begin{aligned} depth(X, D_X) \leftarrow\ & F[X, D_X] \wedge \\ & [\forall D.F[X,D] \rightarrow D_X \leq D] \end{aligned}$$

Construction or Clark completion gives the semantics of this definition. The defined predicate *depth* occurs negatively in the body of the rule, and as a consequence, multiple unintended minimal models may exist[3].

One application of this definition is the definition of depth of a tree. Here the well-founded poset is the set of trees (with values from a given domain D) without infinite branches in a domain; the partial order is the subtree relation. For finitely branching trees, the depth is always a natural number; for infinitely

[2] E.g. $\{even(0), even(2), ..\}$ but also $\{even(0), even(1), even(3), even(5), ..\}$.

[3] E.g. in the context of $\mathbb{N}, \leq$, an unintended minimal model is $\{depth(0,0), depth(0,1), depth(1,2), .., depth(n, n+1), ..\}$.

branching trees, the depth may be an infinite ordinal. E.g. the tree with branches $(0,1,2),(0,2,3,4),(0,3,4,5,6),..,(0,n,..,2n),..$ is a tree with depth ∞.

The above two types of inductive definitions require a different sort of semantics. This raises the question whether a uniform principle of inductive definition can be proposed which is correct for all inductive definitions and hence generalises and integrates completion and minimisation. The first attempt to formalise such a principle was in the context of Iterated Inductive Definitions.

The study of inductive definitions in mathematics has started with Post [19], Spector [24] and Kreisel [15]. Important work in this area includes [9,16,18,2,4]. An offspring of this research is fixpoint logic, currently used in databases [1]. Below is an overview of ideas proposed in the area of Inductive, Iterated Inductive Definitions (IID) and fixpoint logics. The overview is an attempt to give a faithful and comprehensive presentation of the essential ideas in these areas, while I have taken the freedom to reformulate syntax or semantics in order to increase uniformity and comprehensibility.

2.1 Positive Inductive Definitions

Positive Inductive Definitions have been formalised in various ways. In the style of [9], an inductive definition on a given interpretation M is represented as a formula:

$$p(\overline{X}) \leftarrow F[\overline{X},p]$$

where $F[\overline{X},p]$ is a First Order Logic (FOL) formula with only positive occurrences of the defined symbol p but arbitrary occurrences of symbols interpreted in M. In fixpoint logic, the relation p would be denoted $\Gamma_{\Psi}F[\overline{X},\Psi]$ (here p is replaced by a predicate variable Ψ).

[2] studies inductive definitions in a *abstract* representation with an obvious correspondence with definite logic programs. A definition on a domain D of propositional symbols is represented as a possibly infinite set $\mathcal{D}$ of rules $p \leftarrow B$ with $p \in D, B \subseteq D$[4].

[2] gives an overview of three equivalent mathematical principles for describing the semantics of a (Positive) inductive definition. They are equivalent with the way the least model semantics of definite logic programs can be defined [27].

- The model can be defined as the least model of the implications. E.g., in [9], this minimal model semantics is expressed through a circumscription-like axiom (expressing that p must be the least predicate rather than a minimal one).
- The model can be expressed constructively as the least fixpoint of a T_P-like operator associated with the definition. In the presentation of [2], inductive

[4] Definitions represented in the other style can be represented in this abstract way. Given the mathematical structure M and formula $F[\overline{X},p]$, define the domain D as the set of atoms $p(\overline{x})$ with $\overline{x} \in M^n$. Define $\mathcal{D}$ as the set of rules $p(\overline{x}) \leftarrow B$ for each $\overline{x}$ and each set B of p-atoms such that $M \models F[\overline{x},B]$; meaning that F is true for $\overline{x}$ in M when p is interpreted as the set B.

definitions are dually defined as monotonic T_P-like operators. This is the common way in fixpoint logic (hence the name).

- The model can be expressed also as the interpretation in which each atom has a *proof tree.* Also this formalisation has been used in LP in [7]. Because it is less commonly used in LP, I present it here for a slightly extended version of the formalism of [2].

  Let be given a symbol domain D, including a subset $D_o \subseteq D$ which includes the truth values $\mathbf{t}, \mathbf{f}$, an interpretation M interpreting the symbols of D_o such that $M(\mathbf{t}) = \mathbf{t}, M(\mathbf{f}) = \mathbf{f}$. The symbols of D_o are called the open or interpreted symbols. Also given is a definition $\mathcal{D}$ which is a set of rules $p \leftarrow B$ with head $p \in D \setminus D_o$ and body B consisting of atoms of $D \setminus D_o$ and positive or negative literals of D_o[5]. The set $Defined(\mathcal{D}) = D \setminus D_o$ is called the set of defined symbols, the set of open symbols D_o is often denoted $Open(\mathcal{D})$. We assume that each symbol $p \in Defined(\mathcal{D})$ has at least one rule $p \leftarrow B \in \mathcal{D}$ (it may be the rule $p \leftarrow \{\mathbf{f}\}$). Also the body B of a rule is never empty (B may be the singleton $\{\mathbf{t}\}$).

  A $\mathcal{D}$-proof-tree $\mathcal{T}$ of $p \in D$ is a tree of literals of D with p as root such that:
  - all leaves of $\mathcal{T}$ are positive or negative open literals; all non-leaves contain defined atoms;
  - for each non-leaf node p with set of immediate descendants B: $p \leftarrow B \in \mathcal{D}$;
  - $\mathcal{T}$ is loop-free; i.e. contains no infinite branches.

  The model $\overline{M}^{\mathcal{D}}$ of $\mathcal{D}$ given M can be characterised as the set of atoms $p \in D$ which occur in the root of a proof-tree $\mathcal{T}$ such that all leaves are true literals in M. Note that interpreted literals have proof-trees consisting of one node; as a consequence, $\overline{M}^{\mathcal{D}}$ extends M.

2.2 Iterated Inductive Definitions

The logics of Iterated Inductive Definitions are or can be seen as attempts to formalise the mathematical principle of definition by (transfinite) induction on a well-founded order. Iterated Inductive definitions were first introduced in [15] and later studied in [9] and [16]. [2] formulates the intuition of Iterated Inductive Definitions in the following way. Given a mathematical structure M fixing the interpretation of the interpreted predicates and function symbols, a positive inductive definition $\mathcal{D}$ prescribes the interpretation of the defined predicate(s). Once the interpretation of the defined symbols p is fixed, M can be extended with these interpretations, yielding a new interpretation $\overline{M}^{\mathcal{D}}$. On top of this structure, again new predicates may be defined in the similar way as before. The definition of this new predicates may depend negatively on the defined predicates

[5] Allowing positive or negative open literals is an extension to the formalism of [2]. It does not introduce any complexity because the interpretation of these literals is given. This extension will facilitate the leap to inductive definitions with recursion over negation.

p as these are interpreted in $\overline{M}^{\mathcal{D}}$. This principle can be iterated in an arbitrary, even transfinite sequence of positive inductive definitions.

In [2], the abstract definition logic defined there is not explicitly extended with this idea, but given the above intuition, the extension with negation is straightforward. Given a domain D and mathematical structure M, an Iterated Inductive Definition (IID) would be a possibly transfinite sequence $\mathcal{D} = (\mathcal{D}_\alpha)_{\alpha<\alpha_\mathcal{D}}$ of positive inductive definitions such that:

- each defined symbol p is defined in a unique $\mathcal{D}_{\alpha_p}$; we call α_p the stratum of p;
- for each rule $p \leftarrow B \in \mathcal{D}$, for each defined atom $q \in B$, $\alpha_q \leq \alpha_p$; for each defined atom q such that $\neg q \in B$, $\alpha_q < \alpha_p$.

The model $\overline{M}^{\mathcal{D}}$ of a definition can be obtained by transfinitely *iterating* the principle of positive inductive definition over the sequence $(\mathcal{D}_\alpha)_{\alpha<\alpha_\mathcal{D}}$.

There is an obvious correspondence between Iterated Inductive Definitions (IID's) and stratified logic programs under perfect model semantics [3, 20, 21]. Already in 84, [14] defines a semantics for stratified logic programs based on the Iterated Inductive Definition (IID) logic defined in [16]. To my knowledge, this was really the first time that the perfect model semantics for stratified logic programs was defined. Apparently this work stayed largely unnoticed, perhaps because, like the semantics in [16], it is based on sequent calculus, which to some extend increases the mathematical complexity and obscures the simple intuitions underlying this semantics.

Though the intuition of IID's as formulated in [2] is straightforward, it is not easy to see how this idea is implemented in IID logics such as those of [9], [4] and also in [16]. The reason for this seems as follows. The goal of this research was to investigate theoretical expressivity of transfinite forms of IID's. As explained in [4], a definability study makes only sense in a finitely represented logic, while transfinite IID's in the abstract setting above are per definition infinite objects. [9] investigates IID's encoded in an IID-form, a single FOL formula of the form $F[N, X, P]$, and expresses its semantics in a circumscription-like second order formula. The problem is that this encoding is extremely tedious and this blurs the simple intuitions behind this work and the similarities with the perfect model semantics.

Nevertheless, it is interesting -if only from historical perspective- to see how transfinite definitions can be encoded finitely as an IID-form and how a perfect model-like semantics can be expressed in such a notation. Consider the following definition $\mathcal{D}^2_{even}$, constructed for the sole purpose of illustrating the encoding:

$$\mathcal{D}^2_{even} = \begin{cases} (0) & even(0) \leftarrow \mathbf{t} \\ (n+1) & even(n+1) \leftarrow \neg even(n) \\ (n) & even(n) \leftarrow even(n) \\ (\infty) & sw \leftarrow even(n), even(n+1) \\ (\infty+1) & ok \leftarrow \neg sw \end{cases}$$

The symbol sw (which abbreviates *something_wrong*) represents that two subsequent numbers are even, and ok is its negation. This definition can be stratified (the strata of the defined predicates are given). The model obtained after $\infty+2$ iterations is $\{ok, even(2n)|n \in \mathbb{N}\}$.

To encode such an abstract IID, a binary meta-predicate h (of *holds*) is used: $h(\alpha, p)$ means that the stratum of p is α and that p is defined true. The first step in encoding such an abstract IID ($\mathcal{D}_\alpha$) yields a possibly infinite disjunction $F[N, X, P]$. For any rule $p \leftarrow \{.., q, r, \neg s, ..\}$ with $\alpha_q = \alpha_p, \alpha_r < \alpha_p, \alpha_s < \alpha_p$, add one disjunct:

$$N = \alpha_p \wedge X = p \wedge .. \wedge P(q) \wedge h(\alpha_r, r) \wedge \neg h(\alpha_s, s) \wedge ..$$

This disjunct is obtained as a conjunction of $N = \alpha_p \wedge X = p$, corresponding to the head p, a conjunct $P(q)$ for any atom q of the same stratum as the head, and a literal $h(\alpha_r, r)$ and $\neg h(\alpha_s, s)$ for the other literals $r, \neg s \in B$ defined in lower strata[6].

The result is an infinitary formula $F[N, X, P]$. Here, N ranges over the ordinals $\alpha < \alpha_{\mathcal{D}}$, X over atoms and P over sets of atoms. The formula corresponding to $\mathcal{D}^2_{even}$ is the following infinitary disjunction with disjuncts for each $0 \leq n$:

$$\begin{cases} N = 0 \wedge X = even(0) \vee \\ N = n+1 \wedge X = even(n+1) \wedge \neg h(n, even(n)) \vee ... \\ N = n \wedge X = even(n) \wedge P(even(n)) \vee ... \\ N = \infty \wedge X = sw \wedge \; h(n, even(n)) \wedge \\ \qquad\qquad\qquad\qquad h(n+1, even(n+1)) \quad \vee ... \\ N = \infty + 1 \wedge X = ok \wedge \neg h(\infty, p) \end{cases}$$

There is only one step more to go to reduce this formula to an equivalent finite IID-form. But first, we show how to express the semantics of the IID. Two axioms express essentially that at each stratum α, the set $h(\alpha, .) \equiv \{p|h(\alpha, p)$ is true$\}$ satisfies the definition $\mathcal{D}_\alpha$. These axioms express the principle of positive inductive definition: that this set must satisfy the implications of $\mathcal{D}_\alpha$ and that it must be contained in each set satisfying the implications. Below, $F[P(\tau)/h(N, \tau)]$ denotes the formulas obtained by replacing each expression $P(\tau)$ for arbitrary term τ by $h(N, \tau)$.

The first axiom expresses that for each ordinal α and given h for lower strata, $h(\alpha, .)$ satisfies the implications in $\mathcal{D}_\alpha$:

$$\forall N, X.\{h(N, X) \leftarrow F[P(\tau)/h(N, \tau)]\}$$

---

[6] In [9], literals $h(\alpha_q, q)$ are replaced by open formulas $h(\alpha_q, q) \wedge \alpha_q < N$. This open formula represents the restriction of h to strata $< N$ (atoms q at higher strata are false in $h(\alpha_q, q) \wedge \alpha_q < N$). The resulting, more complex axioms can be seen to be equivalent with our axioms for IID-forms obtained from a stratified abstract IID $\mathcal{D}$. The reason for this choice seems to be that the stratification condition, which can be defined nicely for abstract IID's, cannot easily be formulated directly for IID-forms. The more complex axioms determine a unique h predicate even if $F[N, X, P]$ encodes a nonstratifiable or incorrectly stratified definition (but the semantics may be unnatural then); in that case, our simpler axioms do not determine a unique h-predicate due to mutual dependencies between predicates defined at lower and at higher level.

One can verify that if one assigns the values α_p to N and p to X and eliminates false disjuncts, then this complex formula reduces to:

$$h(\alpha_p, p) \leftarrow \begin{cases} .. \\ .. \wedge h(\alpha_q, q) \wedge \neg h(\alpha_r, r) \wedge .. \;\; \vee \\ .. \end{cases}$$

with a disjunct for each $p \leftarrow \{.., q, \neg r, ..\} \in \mathcal{D}$.

The second axiom expresses that for each ordinal α, $h(\alpha, .)$ is contained in each set Ψ which satisfies the implications of $\mathcal{D}_\alpha$. It is a second order axiom, using a set variable Ψ which ranges over sets of atoms and it is a variant of a circumscription axiom:

$$\forall N.\forall \Psi.[\forall X.\Psi(X) \leftarrow F[P(\tau)/\Psi(\tau)]] \rightarrow [\forall X.h(N, X) \rightarrow \Psi(X)]$$

Finally, the infinitary IID-form F should be further encoded by a finite formula. This involves:

- encoding ordinals by a (primitive recursive) well-ordering on natural numbers. E.g. the total order $2 \preceq 3 \preceq .. \preceq 0 \preceq 1$ is a well-ordering encoding the ordinals $0, 1, .., \infty, \infty + 1$.
- encoding atoms by natural numbers: an obvious proposal here is to encode each atom by the natural number encoding the stratum of the atom; i.e. $even(n)$ by $n + 2$, sw by 0 and ok by 1.
- encoding tuples of natural numbers by natural numbers. Details of this are tedious and irrelevant for this paper; we omit them.

In this encoding, an infinite number of disjuncts can be represented in a finite formula using quantification in the natural numbers. The different sets of disjuncts are encoded as follows:

$$\begin{array}{l}
\{N = 0 \wedge X = even(0)\} \\
\qquad \longrightarrow \quad N = 2 \wedge X = 2 \\
\{N = n + 1 \wedge X = even(n + 1) \wedge \neg h(n, even(n)) \mid n \in \mathbb{N}\} \\
\qquad \longrightarrow \quad \exists M.N = M + 1 \wedge 2 \leq M \wedge X = N \wedge \neg h(M, M) \\
\{N = n \wedge X = even(n) \wedge P(even(n)) \mid n \in \mathbb{N}\} \\
\qquad \longrightarrow \quad 2 \leq N \wedge X = N \wedge P(N) \\
\{N = \infty \wedge X = sw \wedge h(n, even(n)) \wedge \;\; h(n + 1, even(n + 1)) \mid n \in \mathbb{N}\} \\
\qquad \longrightarrow \quad N = 0 \wedge X = 0 \wedge \exists M.[2 \leq M \wedge h(M, M) \wedge h(M + 1, M + 1)] \\
\{N = \infty + 1 \wedge X = ok \wedge \neg h(\infty, p)\} \\
\qquad \longrightarrow \quad N = 1 \wedge X = 1 \wedge \neg h(0, 0)
\end{array}$$

The resulting finite IID-form is:

$$\begin{cases}
N = 2 \wedge X = 2 \vee \\
\exists M.[N = M + 1 \wedge 2 \leq M \wedge X = N \wedge \neg h(M, M)] \vee \\
2 \leq N \wedge X = N \wedge P(N) \vee \\
N = 0 \wedge X = 0 \wedge \exists M.[2 \leq M \wedge h(M, M) \wedge h(M + 1, M + 1)] \vee \\
N = 1 \wedge X = 1 \wedge \neg h(0, 0)
\end{cases}$$

2.3 Inflationary Fixed-point Logic

[2] proposes another extension of positive inductive definitions with negation. With an arbitrary formula $F[\overline{X}, P]$ with negative occurrences of P allowed, the resulting T_P-like operator $\mathcal{T}_{F[\overline{X},P]}$ is not monotonic and may not have a least fixpoint. However, the operator $\mathcal{T}^i_{F[\overline{X},P]}(I) = I \cup \mathcal{T}_{F[\overline{X},P]}(I)$ is increasing (though not monotonic) and therefore a fixpoint can be constructed. This idea has been used in fixpoint logic with *inflationary semantics* [1].

Inflationary fixpoint logic is known to be expressive; however, it is not a natural formalisation of inductive definitions over a well-founded set, and therefore, this extension is not relevant in the context of this paper. For example, if we construct a formula $F_{even}[X, even]$ for $\mathcal{D}^1_{even}$ in the same way as for positive inductive definitions, we obtain: $X = 0 \vee \exists Y.X = s(Y) \wedge \neg even(Y)$.
After one application of the inflationary fixpoint operator, the unintended fixpoint $\{even(n)|n \in \mathbb{N}\}$ is obtained.

3 A critique on Iterated Inductive Definitions

The stratified IID formalisms provide a correct treatment of inductive definitions with negation. The IID-forms as defined in e.g. [9] was not intended for use for Knowledge Representation and is absolutely unsuitable for such purpose. But any stratified formalism for inductive definitions with negation will pose certain fundamental problems.
(1) A stratification of a definition does not provide any information about the defined relations. This can be seen from the fact that choosing another stratification for a definition has no impact on its semantics; moreover, there exists ways to construct the semantics of an IID without recurring to a predefined syntactical stratification. It is undesirable that in IID's, a stratification must be chosen and this choice is explicitly reflected in the representation of the definition.

(II) The stratification of an Iterated Inductive Definition is based on a syntactical criterion. As a consequence, a rule set formulated for one alphabet may be stratifiable whereas the corresponding rule set in a *linguistic variant* of the alphabet may be non-stratifiable. The following variant of the definition $\mathcal{D}^1_{even}$ illustrates this. Assume that we use the alphabet: $\{even(n), successor(n,m)|n, m \in \mathbb{N}\}$ with a predicate representation of the concept of successor. In this alphabet, the natural representation of the inductive definition of *even* is the set with for each $n, m \in \mathbb{N}$ the following rules:

$$\mathcal{D}^3_{even} = \begin{cases} successor(n+1, n) \\ even(0) \\ even(n) \leftarrow successor(n,m), \neg even(m) \end{cases}$$

This variant definition cannot be stratified due to the presence of rules $even(m) \leftarrow successor(m,m), \neg even(m)$. A good formalisation should not be as dependent of intuitively innocent linguistic variance.

(III) As a formalisation of inductive definitions on well-founded posets, the requirement of stratified IID's of an explicit stratification is problematic in general. A definition of a concept (like evenness or depth) for x in terms of all $y < x$ is mathematically well-constructed; yet a stratification for such a definition may be in general unknown. As an example, consider the inductive definition of *depth* of an element in a well-founded order or the depth of a tree. The need of an explicit stratification is unnecessary and unnatural.

3.1 WFS: An improved Principle of Inductive Definition

In this section, I argue that the mathematics of (a variant of) the well-founded semantics of logic programming [28] provides an improved formalisation of the principle of inductive definition.

Just like the perfect model, the model $\overline{M}^{\mathcal{D}}$ of a stratified Iterated Inductive Definition $\mathcal{D}$ is obtained by iterating the positive induction principle and constructing a sequence $(M_\alpha)_{\alpha<\alpha_{\mathcal{D}}}$ of interpretations of increasing sub-domains which starts with M and gives gradually better approximations of the model $\overline{M}^{\mathcal{D}}$. Each M_α defines the truth value of all symbols of the sub-alphabet Σ_α and leaves atoms defined at later levels undefined. The role of the stratification in this process is to *delay* the use of some part of the definition until enough information is available to safely apply the positive induction principle on that part of the definition.

The same ideas can be *implemented* in a different way, without relying on an explicit syntactical partitioning of the definition. Instead of using 2-valued interpretations of sub-alphabets, partial interpretations can be used. Here, a partial interpretation is a partial function from the set of atoms D to $\{\mathbf{t}, \mathbf{f}\}$. Equivalently, we use the classical formalisation as a total function from the set of atoms D to $\{\mathbf{t}, \mathbf{u}, \mathbf{f}\}$[7]. The positive induction principle can be conservatively extended for definitions with negation. For a definition $\mathcal{D}$, we define the Positive Induction Operator $\mathcal{PI}_{\mathcal{D}}$ which takes as input a partial interpretation I representing well-defined truth values for a subset of atoms, and derives an extended partial interpretation defining the truth values of other atoms that can be derived by positive induction. Definition of truth values of atoms for which not enough information is available is delayed. The model of a definition is obtained then by a fixpoint construction.

From a knowledge theoretic point of view, the key problem in the above enterprise is the definition of the principle of positive induction in the context of definitions with negation. A formalisation based on proof-trees shows most clearly the structural similarities between positive induction for PID's and for inductive definitions with negation.

[7] This formalisation is mathematically equivalent with the previous one, is more common and leads to more elegant mathematics. Note that in this view, $\mathbf{u}$ plays a similar role as null-values in databases: just as a null value, $\mathbf{u}$ is not a real truth value, it is a place holder for an (as yet) undefined truth value. Below, I return to the issue of interpretation of $\mathbf{u}$.

We formalise the above ideas for a formalism which is the natural extension of the abstract definitions of [2] with negation; at the same time, it is an infinitary version of the propositional LP-formalism. Given is a domain D of propositional symbols. In the new, more general setting, a definition $\mathcal{D}$ consists of rules in which positive and negative open or defined literals may appear in the (nonempty) body. As before, the set of defined symbols that appear in the head of a rule is denoted $Defined(\mathcal{D})$; the set of open or interpreted symbols is denoted as $Open(\mathcal{D})$. Also given is an interpretation M of the open symbols $Open(\mathcal{D})$.

The definition of a $\mathcal{D}$-proof-tree $\mathcal{T}$ as defined in section 2.1 hardly needs to be altered: it is a tree of literals of D such that:

- leaves contain open literals or negative defined literals; non-leaves contain defined atoms $p \in Defined(\mathcal{D})$;
- each non-leaf p has a set of direct descendants B such that $p \leftarrow B \in \mathcal{D}$;
- no infinite branches.

Hence, leaves contain interpreted literals and negations of defined atoms. Note that interpreted atoms have proof-trees consisting of one root node.

Definition 1. *The* Positive Induction Operator $\mathcal{PI}_\mathcal{D}$ *maps partial interpretations I to I' such that* $\forall p \in D$*:*

- $I'(p) = \mathbf{t}$ *if p has a proof-tree with all leaves true in I.*
- $I'(p) = \mathbf{f}$ *if each proof-tree of p has a false leaf in I;*
- $I'(p) = \mathbf{u}$ *otherwise, i.e. no proof-tree of p has only true leaves but there exists at least one without false leaves.*

The Positive Induction Operator is a monotonic operator w.r.t. the precision order $\leq_p$, the point-wise extension of $\mathbf{u} \leq_p \mathbf{f}, \mathbf{u} \leq_p \mathbf{t}$. Monotonic operators w.r.t. $\leq_p$ have a least fixpoint [10]. Hence, each interpretation M of the non-defined symbols can be extended to a unique least fixpoint $\mathcal{PI}_\mathcal{D}\uparrow(M)$.

Definition 2. $\mathcal{PI}_\mathcal{D}\uparrow(M)$ *is the model* $\overline{M}^\mathcal{D}$ *of* $\mathcal{D}$*.*

The structural resemblance between positive induction in PID's and in $\mathcal{PI}_\mathcal{D}$ is apparent. There are some important properties. The first relates this semantics to WFS semantics of logic programs.

Proposition 1. $\mathcal{PI}_\mathcal{D}$ *and the 3-valued stable model operator [22] are identical. The well-founded model of $\mathcal{D}$ is the model* $\overline{M}^\mathcal{D}$ *of* $\mathcal{D}$*.*

Second, this semantics provides a conservative extension of the IID-style semantics, as the WFS is known to generalise least model semantics and perfect model semantics of stratified logic programs.

Third, certain definitions may have partial models (e.g. $\{p \leftarrow \neg p\}$. Note here the changing role of $\mathbf{u}$ during the fixpoint computation and in the fixpoint. When the truth value of an atom is $\mathbf{u}$ at some stage of the fixpoint computation, it means that the truth value of the atom is yet undetermined at this stage. If its truth value is still $\mathbf{u}$ in the fixpoint, it means $\mathcal{D}$ does not allow to constructively define the truth value of p. Hence, undefined atoms in the fixpoint point to ambiguities in the definition.

There seem to be two sensible treatments of ambiguous definitions. They could be considered as *inconsistent*, in a similar sense as in classical logic: ambiguous definitions have no models. In this strict view, Definition 2 is to be refined as:

Definition 3. *If $\mathcal{PI}_{\mathcal{D}}\uparrow(M)$ is 2-valued, then it is the model $\overline{M}^{\mathcal{D}}$ of $\mathcal{D}$; otherwise, $\mathcal{D}$ has no model.*

The result is a 2-valued logic. This is a simple strategy because it avoids potential problems with 3-valued models but it has the disadvantage that no sensible information can be extracted from an ambiguous definition since such a definition entails every formula. This situation is analogous to classical logic.

The more permissive treatment is to allow definitions with partial models. The result is a sort of *paraconsistent* definition logic, i.e. a logic in which definitions with local inconsistencies or local ambiguities do not not entail every formula.

4 Concluding remarks

This paper is a study of the concept of (transfinite) inductive definition. The paper investigates how this concept has been formalised in the past in the ID and IID areas; drawbacks of these formalisations were pointed at and an improved formalisation, inspired by logic programming semantics, is proposed. Strong connections between the formalisations in ID and IID and perfect model semantics but also circumscription semantics have been exposed.

This study is not only relevant as a study of inductive definitions but improves also our understanding of the use of LP for knowledge representation and hence, of the role of LP in Artificial Intelligence. The reading of logic programs as autoepistemic or default theories on the one hand, and as definitions on the other hand, give essentially different perspectives on the meaning of logic programs, on the nature of the negation symbol and the implication symbol in LP.

In general, a knowledge theoretic study as the one in this paper is relevant for developing a knowledge representation methodology. It is (or once was) a widespread view that the advantage of declarative logic for "encoding" knowledge is in its intuitive *linguistic* reading; in the case of this paper: the reading of a set of rules as an inductive definition. This reading of the logic provides the methodological basis for knowledge representation; the tight connection between formal syntax and semantics and a clear intuitive reading facilitates the explicitation of the expert knowledge. Formulas of the theory can be understood by the experts through the linguistic interpretation, without the need of explicitly constructing the formal semantics. Knowledge theoretic studies like the one in this paper, are important to build natural and systematic methodologies for knowledge representation. One aim of this study was to clarify how logic programs can be used for knowledge representation and what sort of knowledge can be represented in it.

A simple illustration of the impact of the linguistic interpretation on knowledge methodology is as follows. The definition that dead means not alive, is naturally expressed in LP under the definition reading by the singleton definition:

$$\{dead \leftarrow \neg alive\}$$

On the other hand, in Extended Logic Programming [13], which is based on the default and AEL view, a correct representation would be:

$$dead \leftarrow \neg alive$$
$$\neg dead \leftarrow alive$$

A knowledge theoretic study is also relevant for the design or extension of a logic. This is also well-illustrated in the case of LP. With respect to knowledge representation, a major problem of LP under the default or auto-epistemic view is that no definite negative information can be represented. This led Gelfond and Lifschitz in [13] to extend the formalism and re-introduce a form of classical negation in Extended Logic Programming.

In the definition view, a logic program entails plenty of definite negative information. As a matter of fact, the problem with standard LP is the strength of its closure mechanism: an atom is assumed false unless it can be proven to be true. As a consequence, representing uncertainty is a serious problem; this problem has received a lot of attention in recent years. In the definition view on standard LP, the problem is because all predicates are defined, have a (possibly empty) definition. Hence, the natural idea is to extend the logic with open predicates which have arbitrary interpretation. In [6], this idea was elaborated in an extension of LP, called Open Logic Programming (OLP). I argued there that OLP provides a knowledge theoretic interpretation of Abductive Logic Programming as a definition logic and that abductive solvers (e.g. SLDNFA [8]) designed for this formalism can be seen as special purpose reasoners on definitions for abduction and deduction[8]. A problem of this work is that it is based on completion semantics; completion is not a good formalisation of induction. To extend this study for the semantics defined in this paper is future work.

The knowledge theoretic interpretation of LP as inductive definitions gives also insight on the relationship with a class of logics outside the area of NMR: definition logics. This class includes fixpoint logics and description logics. In [25], Van Belleghem et al. showed a strong correspondence between OLP-FOL and description logics. To large extend, description logic can be considered as a non-recursive subformalism of OLP-FOL. There is correspondence on the intuitive and semantical level; the differences on the syntactic level are syntactic sugar.

[8] Note that LP and OLP as definition logics do not provide default negation; as I argued in [6], OLP is not a natural formalism to express some sorts of default reasoning problems such as the well-known train crossing example [13] [17]. In order to represent this sort of domains, an autoepistemic modal operator or a default negation operator should be added to definition logic.

The specific syntactic restrictions of description logics have allowed to develop highly efficient reasoning techniques.

Also subject for future work is to substantiate the claim in the introduction, that a broad class of human knowledges in many areas of human expertise, ranging from common sense knowledge situations to mathematics, is of constructive nature, in the sense that (part of) the knowledge is present in the form of a recursive recipe, to be interpreted as defined in this paper. The prominent roles of completion and circumscriptive techniques in NMR and knowledge representation hint at this.

5 Acknowledgements

I thank all colleagues that have commented on this or earlier versions of this paper, in particular Danny De Schreye and Kristof Van Belleghem.

References

1. S. Abiteboul, R. Hull, and V. Vianu. *Foundations of Databases.* Addusin-Wesley Publishing Company, 1995.
2. P. Aczel. An Introduction to Inductive Definitions. In J. Barwise, editor, *Handbook of Mathematical Logic*, pages 739–782. North-Holland Publishing Company, 1977.
3. K.R. Apt, H.A. Blair, and A. Walker. Towards a theory of Declarative Knowledge. In J. Minker, editor, *Foundations of Deductive Databases and Logic Programming.* Morgan Kaufmann, 1988.
4. W. Buchholz, S. Feferman, and W. Pohlers W. Sieg. *Iterated Inductive Definitions and Subsystems of Analysis: Recent Proof-Theoretical Studies.* Springer-Verlag, Lecture Notes in Mathematics 897, 1981.
5. K.L. Clark. Negation as failure. In H. Gallaire and J. Minker, editors, *Logic and Databases*, pages 293–322. Plenum Press, 1978.
6. M. Denecker. A Terminological Interpretation of (Abductive) Logic Programming. In V.W. Marek, A. Nerode, and M. Truszczynski, editors, *International Conference on Logic Programming and Nonmonotonic Reasoning*, pages 15–29. Springer, Lecture notes in Artificial Intelligence 928, 1995.
7. M. Denecker and D. De Schreye. Justification semantics: a unifying framework for the semantics of logic programs. Technical Report 157, Department of Computer Science, K.U.Leuven, 1992.
8. M. Denecker and D. De Schreye. SLDNFA: an abductive procedure for abductive logic programs. *Journal of Logic Programming*, 34(2):111–167, 1997.
9. S. Feferman. Formal theories for transfinite iterations of generalised inductive definitions and some subsystems of analysis. In A. Kino, J. Myhill, and R.E. Vesley, editors, *Intuitionism and Proof theory*, pages 303–326. North Holland, 1970.
10. M. Fitting. A Kripke-Kleene Semantics for Logic Programs. *Journal of Logic Programming*, 2(4):295–312, 1985.
11. M. Gelfond. On Stratified Autoepistemic Theories. In *Proc. of AAAI87*, pages 207–211. Morgan Kaufman, 1987.
12. M. Gelfond and V. Lifschitz. The stable model semantics for logic programming. In *Proc. of the International Joint Conference and Symposium on Logic Programming*, pages 1070–1080. IEEE, 1988.

13. M. Gelfond and V. Lifschitz. Logic Programs with Classical Negation. In D.H.D. Warren and P. Szeredi, editors, *Proc. of the 7th International Conference on Logic Programming 90*, page 579. MIT press, 1990.
14. M. Hagiya and T. Sakurai. Foundation of Logic Programming Based on Inductive Definition. *New Generation Computing*, 2:59–77, 1984.
15. G. Kreisel. Generalized inductive definitions. Technical report, Stanford University, 1963.
16. P. Martin-Löf. Hauptsatz for the intuitionistic theory of iterated inductive definitions. In J.e. Fenstad, editor, *Proceedings of the Second Scandinavian Logic Symposium*, pages 179–216, 1971.
17. J. McCarthy. Applications of Circumscription to Formalizing Common-Sense Knowledge. *Artifical Intelligence*, 28:89–116, 1980.
18. Y. N. Moschovakis. *Elementary Induction on Abstract Structures.* North-Holland Publishing Company, Amsterdam- New York, 1974.
19. E. Post. Formal reduction of the general combinatorial decision problem. *American Journal of Mathematics*, 65:197–215, 1943.
20. H. Przymusinska and T.C. Przymusinski. Weakly perfect model semantics for logic programs. In R.A. Kowalski and K.A. Bowen, editors, *Proc. of the fifth international conference and symposium on logic programming*, pages 1106–1120. the MIT press, 1988.
21. T.C. Przymusinski. On the semantics of Stratified Databases. In J. Minker, editor, *Foundations of Deductive Databases and Logic Programming.* Morgan Kaufman, 1988.
22. T.C. Przymusinski. Well founded semantics coincides with three valued Stable Models. *Fundamenta Informaticae*, 13:445–463, 1990.
23. R. Reiter. The Frame Problem in the Situation Calculus: A simple Solution (Sometimes) and a Completeness Result for Goal Regression. In V. Lifschitz, editor, *Artificial Intelligence and Mathematical Theory of Computation: Papers in Honour of John McCarthy*, pages 359–380. Academic Press, 1991.
24. C. Spector. Inductively defined sets of natural numbers. In *Infinitistic Methods (Proc. 1959 Symposium on Foundation of Mathematis in Warsaw)*, pages 97–102. Pergamon Press, Oxford, 1961.
25. K. Van Belleghem, M. Denecker, and D. De Schreye. A strong correspondence between description logics and open logic programming. In Lee Naish, editor, *Proc. of the International Conference on Logic Programming, 1997*, pages 346–360. MIT-press, 1997.
26. K. Van Belleghem, M. Denecker, and D. Theseider Dupré. Dependencies and ramifications in an event-based language. In *Proc. of the Ninth Dutch Artificial Intelligence Conference, 1997*, 1997.
27. M. van Emden and R.A Kowalski. The semantics of Predicate Logic as a Programming Language. *Journal of the ACM*, 4(4):733–742, 1976.
28. A. Van Gelder, K.A. Ross, and J.S. Schlipf. The Well-Founded Semantics for General Logic Programs. *Journal of the ACM*, 38(3):620–650, 1991.

Combining Introspection and Communication with Rationality and Reactivity in Agents

Pierangelo Dell'Acqua[1], Fariba Sadri[2], and Francesca Toni[2]

[1] Computing Science Department, Uppsala University, Sweden
pier@csd.uu.se
[2] Department of Computing, Imperial College of Science, Technology and Medicine, UK
{fs,ft}@doc.ic.ac.uk

Abstract. We propose a logic-based language for programming agents that can reason about their own beliefs as well as the beliefs of other agents and can communicate with each other. The agents can be reactive, rational/deliberative or hybrid, combining both reactive and rational behaviour. We illustrate the language by means of examples.

1 Introduction

Kowalski & Sadri [12] propose an approach to agents within an extended logic programming framework. In the remainder of the paper we will refer to their agents as KS-agents. KS-agents are *hybrid* in that they exhibit both *rational* (or *deliberative*) and *reactive* behaviour. The reasoning core of KS-agents is a proof procedure that combines forward and backward reasoning. Backward reasoning is used primarily for planning, problem solving and other deliberative activities. Forward reasoning is used primarily for reactivity to the environment, possibly including other agents. The proof procedure is executed within an observe-think-act cycle that allows the agent to be alert to the environment and react to it as well as think and devise plans. Both the proof procedure and the KS-agent architecture can deal with temporal information. The proof procedure (IFF proof procedure [9]) treats both inputs from the environment and agents' actions as *abducibles* (hypotheses).

Barklund et al. [1] and Costantini et al. [6] present an extension to the Reflective Prolog programming paradigm [7, 8] to model agents that are introspective and that communicate with each other. We will refer to this approach as Reflective Prolog with Communication (RPC). In RPC introspection is achieved via the meta-predicate *solve* and communication is achieved via the meta-predicates *tell* and *told*. A communication act is triggered everytime an agent $agent_1$ has a goal of the form $agent_1 : told(agent_2, A)$. This stands both for "$agent_1$ is told by $agent_2$ that A" as well as "$agent_1$ asks $agent_2$ whether A". This goal is solved by $agent_2$ telling $agent_1$ that A, that is, $agent_2 : tell(agent_1, A)$, standing for "$agent_2$ tells $agent_1$ that A". The information is passed from $agent_2$ to $agent_1$ by eventually instantiating A. A main limitation of this approach is that agents cannot tell anything to other agents unless explicitly asked.

The ability to provide agents with some sort of "proactive" communication primitive is widely documented in literature [18–20, 5, 14, 17]. For example, one can model agents that advertise their services so that other agents, possibly with the help of *mediators*, can find agents that provide services for them.

An interesting application of agents is that of a virtual marketplace on the Web where users create autonomous agents to buy and sell goods on their behalf. Chavez & Maes [3], for example, propose a marketplace, where users can create selling and buying agents by giving them a description of the item they want to sell or buy. The main goal of the approach is to help users in the negotiations between buyers and sellers, and to sell the goods better (i.e., at a higher price) than the user would be able to otherwise, by taking advantage of their processing speed and communication bandwidth. Chavez & Maes' agents are (*i*) proactive: "...they try to sell themselves, by going into a marketplace, contacting interested parties (namely, buying agents) and negotiating with them to find the best deal", and (*ii*) autonomous: "...once released into the marketplace, they negotiate and make decisions on their own, without requiring user intervention". Chavez & Maes point out their agents' lack for rationality: "Our experiment demonstrated the need and desire for 'smarter' agents whose decision making processes more closely mimic those of people and which can be directed at a more abstract, motivational level."

In this paper we propose a combination of (a version of) the RPC programming paradigm and KS-agents. In the resulting framework reactive, rational or hybrid agents can reason about their own beliefs as well as the beliefs of other agents and can communicate proactively with each other. In such a framework, the agents' behaviour can be regulated by condition-action rules such as: if I am asked by another, friendly agent about something and I can prove it from my beliefs then I will tell the agent about it.

In the proposed approach, the two primitives for communication in RPC, *tell* and *told*, are treated as abducibles within the cycle of the KS-agent architecture.

The remainder of the paper is structured as follows. In section 2 we give some basic definitions (in particular, we define ordinary and abductive logic programs) and we review the IFF proof procedure. In section 3 we review the KS-agent architecture. In section 4 we review RPC. In sections 5 and 6 we define our approach to introspection and communication in agents, respectively. In section 7 we illustrate the framework by means of an example. In section 8 we conclude and discuss future work.

2 Preliminaries

2.1 Basic definitions

A **logic program** is a set of **clauses** of the form:

$A \leftarrow L_1 \wedge \ldots \wedge L_n \qquad (n \geq 0)$

where every L_i $(1 \leq i \leq n)$ is a literal, A is an atom and all variables are implicitly universally quantified, with scope the entire clause. If $A = p(t)$, with t a vector of terms, then the clause is said to *define* p.

The **completion of a predicate** p [4] defined in a given logic program P by the set of clauses:

$p(t_1) \leftarrow D_1 \quad \dots \quad p(t_k) \leftarrow D_k \qquad (k \geq 1)$

is the **iff-definition**:

$p(x) \leftrightarrow [x = t_1 \wedge D_1] \vee \dots \vee [x = t_k \wedge D_k]$

where x is a vector of variables, all implicitly universally quantified, with scope the entire iff-definition. Any variable in a disjunct D_i which is not in x is implicitly existentially quantified, with scope the disjunct.

If p is not defined in P, then the completion of p is the iff-definition $p(x) \leftrightarrow$ *false*.

The **selective completion** $comp_S(P)$ of a logic program P with respect to a set S of predicates of the language of P is the union of the completions of all predicates in S. Note that the completion of a logic program P [4] is the selective completion of P with respect to all predicates of the language of P, together with Clark's equality theory [4].

An **integrity constraint** is an implication of the form:

$L_1 \wedge \dots \wedge L_n \Rightarrow A \qquad (n \geq 0)$

where $L_1, \dots, L_n$ are literals and A is an atom, possibly *false*. All variables in an integrity constraint are implicitly universally quantified, with scope the entire integrity constraint.

An **abductive logic program** [10] is a triple $\langle P, \mathcal{A}, I \rangle$, where P is a logic program, $\mathcal{A}$ a set of predicates in the language of P, and I a set of integrity constraints. The predicates in $\mathcal{A}$ are referred to as the **abducible predicates** and the atoms built from the abducible predicates are referred to as **abducibles**. $\overline{\mathcal{A}}$ is the set of non-abducible predicates in the language of P.

Without loss of generality (see [10]), we can assume that abducible predicates have no definitions in P. Abducibles can be thought of as hypotheses that can be used to extend the given logic program in order to provide an "explanation" for given queries (or observations). Explanations are required to "satisfy" the integrity constraints. Different notions of explanation and satisfaction have been used in the literature. The simplest notion of satisfaction is consistency of the explanation with the program and the integrity constraints.

In the sequel, constant, function and predicate symbols may be written as any sequence of characters in `typewriter` style. Variables may be written as any sequence of characters in *italic* style starting with a lower-case character. Thus, for example, `buys` may be a predicate symbol, `Tom` may be a constant symbol, while x and $city$ are variables.

Example 1. Let the abductive logic program $\langle P, \mathcal{A}, I \rangle$ be given as follows

$$P = \left\{ \begin{array}{l} \texttt{has}(x,y) \leftarrow \texttt{buys}(x,y) \\ \texttt{has}(x,y) \leftarrow \texttt{steals}(x,y) \\ \texttt{honest(Tom)} \end{array} \right\}$$

$$\mathcal{A} = \left\{ \texttt{buys}, \texttt{steals} \right\}$$

$$I = \left\{ \texttt{honest}(x) \wedge \texttt{steals}(x,y) \Rightarrow \texttt{false} \right\}.$$

Then, given the observation $G = \texttt{has(Tom, computer)}$, the set of abducibles $\{\texttt{buys(Tom, computer)}\}$ is an explanation for G, satisfying I, whereas the set $\{\texttt{steals(Tom, computer)}\}$ is not, because it is inconsistent with I.

2.2 The IFF proof procedure

The IFF proof procedure [9] is a rewriting procedure, consisting of a number of inference rules, each of which replaces a formula by one which is equivalent to it in a theory T of iff-definitions. We assume $T = comp_{\overline{\mathcal{A}}}(P)$, for some given abductive logic program $\langle P, \mathcal{A}, I \rangle$. The basic inference rules are: [1]

1. **Unfolding**: given an atom $\mathrm{p}(t)$ and a definition
   $\mathrm{p}(x) \leftrightarrow D_1 \vee \ldots \vee D_n$ in T,
$\mathrm{p}(t)$ is replaced by $(D_1 \vee \ldots \vee D_n)\theta$, where θ is the substitution $\{x/t\}$.
2. **Propagation**: given an atom $\mathrm{p}(s)$ and an integrity constraint
   $L_1 \wedge \ldots \mathrm{p}(t) \ldots \wedge L_n \Rightarrow A$,
a new integrity constraint $L_1 \wedge \ldots t = s \ldots \wedge L_n \Rightarrow A$ is added.
3. **Logical simplification**:
   $[B \vee C] \wedge E$ is replaced by $[B \wedge E] \vee [C \wedge E]$ **(splitting)**
   $\texttt{not}\, A \wedge B \Rightarrow C$ is replaced by $B \Rightarrow C \vee A$ **(negation elimination)**
   $B \wedge \texttt{false}$ is replaced by $\texttt{false}$, $B \vee \texttt{false}$ is replaced by B, and so on.
4. **Equality rewriting**: applies equality rewrite rules (see [9]) simulating the unification algorithm of [16] and the application of substitutions.

Given an **initial goal** G (a conjunction of literals, whose variables are free), a **derivation** for G is a sequence of formulae $F_1 = G \wedge I, \ldots, F_m$ such that each **derived goal** F_{i+1} is obtained from F_i by applying one of the inference rules, as follows: unfolding - to atoms that are either conjuncts in F_i or conjuncts in bodies of integrity constraints in F_i; propagation - to atoms that are conjuncts in F_i and integrity constraints in F_i; equality rewriting and logical simplification. Every negative literal $\texttt{not}\, A$ as a conjunct in the initial goal as well as in any derived goal is rewritten as an integrity constraint $A \Rightarrow \texttt{false}$.

Every derivation relies upon some control strategy. Some strategies are preferable to others. E.g., splitting should always be postponed as long as possible, because it is an explosive operation.

Let $F_1 = G \wedge I, \ldots, F_n = N \vee Rest$ be a derivation for G such that $N \neq \texttt{false}$, N is some conjunction of literals and integrity constraints, and no inference step can be applied to N which has not already been applied earlier in the derivation. Then, $F_1, \ldots, F_n$ is a **successful derivation**. An **answer extracted from** N is a pair $(\mathcal{D}, \sigma)$ such that

- σ' is a substitution replacing all free and existentially quantified variables in N by variable-free terms in the underlying language and σ' satisfies all equalities and disequalities in N,
- $\mathcal{D}$ is the set of all abducible atoms that are conjuncts in $N\sigma'$ and σ is the restriction of σ' to the variables in G.

[1] The full IFF proof procedure includes two additional inference rules, case analysis and factoring. In this paper, we omit these rules for simplicity.

The full proof procedure is proven sound and complete with respect to a given semantics [9], based on Kunen's three-valued completion semantics.

Example 2. Let $\langle P, \mathcal{A}, I\rangle$ be the abductive logic program in example 1. Then, $comp_{\overline{\mathcal{A}}}(P)$ is

$$\left\{\begin{array}{l}\texttt{has}(x,y) \leftrightarrow \texttt{buys}(x,y) \vee \texttt{steals}(x,y) \\ \texttt{honest}(x) \leftrightarrow x = \texttt{Tom}\end{array}\right\}.$$

The following is a (successful) derivation for the goal $G = \texttt{has}(\texttt{Tom}, \texttt{computer})$:

$F_1 = G \wedge I$

$F_2 = G \wedge [x = \texttt{Tom} \wedge \texttt{steals}(x,y) \Rightarrow \texttt{false}]$ (by unfolding)

$F_3 = G \wedge [\texttt{steals}(\texttt{Tom},y) \Rightarrow \texttt{false}]$ (by equality rewriting)

$F_4 = [\texttt{buys}(\texttt{Tom},\texttt{computer}) \vee \texttt{steals}(\texttt{Tom},\texttt{computer})] \wedge [\texttt{steals}(\texttt{Tom},y) \Rightarrow \texttt{false}]$ (by unfolding)

$F_5 = [\texttt{buys}(\texttt{Tom},\texttt{computer}) \wedge [\texttt{steals}(\texttt{Tom},y) \Rightarrow \texttt{false}]] \vee [\texttt{steals}(\texttt{Tom},\texttt{computer}) \wedge [\texttt{steals}(\texttt{Tom},y) \Rightarrow \texttt{false}]].$ (by splitting)

The answer $(\mathcal{D} = \{\texttt{buys}(\texttt{Tom},\texttt{computer})\}, \sigma = \{\})$ can be extracted from the first disjunct of F_5. An additional successful derivation is $F_1, \ldots, F_9$ with

$F_6 = [\texttt{buys}(\texttt{Tom},\texttt{computer}) \wedge [\texttt{steals}(\texttt{Tom},y) \Rightarrow \texttt{false}]] \vee [\texttt{steals}(\texttt{Tom},\texttt{computer}) \wedge [\texttt{steals}(\texttt{Tom},y) \Rightarrow \texttt{false}] \wedge [y = \texttt{computer} \Rightarrow \texttt{false}]]$ (by propagation)

$F_7 = [\texttt{buys}(\texttt{Tom},\texttt{computer}) \wedge [\texttt{steals}(\texttt{Tom},y) \Rightarrow \texttt{false}]] \vee [\texttt{steals}(\texttt{Tom},\texttt{computer}) \wedge [\texttt{steals}(\texttt{Tom},y) \Rightarrow \texttt{false}] \wedge \texttt{false}]]$ (by equality rewriting)

$F_8 = [\texttt{buys}(\texttt{Tom},\texttt{computer}) \wedge [\texttt{steals}(\texttt{Tom},y) \Rightarrow \texttt{false}]] \vee \texttt{false}$ (by logical simplification)

$F_9 = [\texttt{buys}(\texttt{Tom},\texttt{computer}) \wedge [\texttt{steals}(\texttt{Tom},y) \Rightarrow \texttt{false}]]$ (by logical simplification)

from which the same answer $(\mathcal{D}, \sigma)$ as above can be extracted.

Note that in this example, unfolding and equality rewriting in the conditions of the integrity constraint are performed before any other operation. In general, many of the operations involving the integrity constraints can be done at compile time, and the simplified (more efficient) version of the integrity constraint conjoined to the initial goal.

3 Kowalski-Sadri agents

Every KS-agent can be thought of as an abductive logic program, equipped with an initial goal. The abducibles are *actions* to be executed as well as *observations* to be performed. Updates, observations, requests and queries are treated uniformly as goals. The abductive logic program can be a temporal theory. For example, the event calculus [13] can be written as an abductive logic program:

$$\texttt{holds\_at}(p, t_2) \leftarrow \texttt{happens}(e, t_1) \wedge (t_1 < t_2) \wedge \texttt{initiates}(e, p) \wedge \texttt{not broken}(t_1, p, t_2)$$

$$\texttt{broken}(t_1, p, t_2) \leftarrow \texttt{happens}(e, t) \wedge \texttt{terminates}(e, p) \wedge (t_1 < t < t_2).$$

The first clause expresses that a property p holds at some time t_2 if it is initiated by an event e at some earlier time t_1 and is not broken (i.e. persists) from t_1 to t_2. The second clause expresses that a property p is broken (i.e. does not persist) from a time t_1 to a later time t_2 if an event e that terminates p happens at a time t between t_1 and t_2.

The predicate **happens** is abducible, and can be used to represent both observations, as events that have taken place in the past, or events scheduled to take place in the future. An integrity constraint

$$I_1 \quad \texttt{happens}(e, t) \wedge \texttt{preconditions}(e, t, p) \wedge \texttt{not holds\_at}(p, t) \Rightarrow \texttt{false}$$

expresses that an event e cannot happen at a time t if the preconditions p of e do not hold at time t.

The predicates `preconditions`, `initiates` and `terminates` have application-specific definitions, e.g.

$$\texttt{preconditions}(\texttt{carry\_umbrella}, t, p) \leftarrow p = \texttt{own\_umbrella}$$

$$\texttt{preconditions}(\texttt{carry\_umbrella}, t, p) \leftarrow p = \texttt{borrowed\_umbrella}$$

$$\texttt{initiates}(\texttt{rain}, \texttt{raining})$$

$$\texttt{terminates}(\texttt{sun}, \texttt{raining}).$$

Additional integrity constraints might be given to represent reactive behaviour of intelligent agents, e.g.

$$I_2 \quad \texttt{happens}(\texttt{raining}, t) \Rightarrow \texttt{happens}(\texttt{carry\_umbrella}, t + 1)$$

or to prevent concurrent execution of actions (events)

$$\texttt{happens}(e_1, t) \wedge \texttt{happens}(e_2, t) \Rightarrow e_1 = e_2.$$

The basic "engine" of a KS-agent is the IFF proof procedure, executed via the following cycle:

To cycle at time t,

(*i*) observe any input at time t,
(*ii*) record any such input,
(*iii*) resume the IFF procedure by propagating the inputs,
(*iv*) continue applying the IFF procedure,
using for steps (*iii*) and (*iv*) a total of r units of time,
(*v*) select an atomic action which can be executed at time $t+r+2$,
(*vi*) execute the selected action at time $t+r+2$ and record the result,
(*vii*) cycle at time $t+r+3$.

The cycle starts at time t by observing and recording any inputs from the environment (steps (*i*) and (*ii*)). Steps (*i*) and (*ii*) are assumed to take one unit of time each. Then, the proof procedure is applied for r units of time (steps (*iii*) and (*iv*)). The amount of resources r available in steps (*iii*) and (*iv*) is bounded by some predefined amount n. By decreasing n the agent is more *reactive*, by increasing n the agent is more *rational*. Propagation is applied first (step (*iii*)), in order to allow for an appropriate reaction to the inputs. Afterwards, an action is selected and executed, taking care of recording the result (steps (*v*) and (*vi*)). Steps (*v*) and (*vi*) conjoined are assumed to take one unit of time. Selected actions can be thought of as outputs into the environment, and observations as inputs from the environment. From every agent's viewpoint, the environment contains all other agents.

Selected actions correspond to abducibles in an answer extracted from a disjunct in a derived goal in a derivation. The disjunct represents an *intention*, i.e. a (possibly partial) plan executed in stages. A sensible action selection strategy may select actions from the same disjunct (intention) at different iterations of the cycle. Failure of a selected plan is obtained via logical simplification, after having propagated `false` into the selected disjunct.

Actions that are generated in an intention may have times associated with them. The times may be absolute, for example $\texttt{happens}(\texttt{ring\_bell}, 3)$, or may be within a constrained range, for example $\texttt{happens}(\texttt{step\_forward}, t) \wedge (1 < t < 10)$. In step (*v*), the selected action will either have an absolute time equal to $t+r+2$ or a time range compatible with an execution time at $t+r+2$. In the latter case, recording of the result of the execution instantiates the time of the action.

Integrity constraints provide a mechanism not only for constraining explanations and plans, for example, as in I_1, but also for allowing reactive, condition-action type of behaviour, for example, as in I_2.

4 Reflective Prolog with communication

4.1 Reflective Prolog

Reflective Prolog (RP) [7, 8] is a metalogic programming language that extends the language of Horn clauses [11, 15] to include higher-order-like features.

The language is that of Horn clauses except that terms are defined differently in order to include *names* that are intended to represent at the meta-level the expressions of the language itself. The alphabet of RP differs from the usual alphabet of Horn clauses by making a distinction between variables and *metavariables* and by the presence of *metaconstants* in RP. Metavariables can only be substituted with names of sentences of RP, and metaconstants are intended as names for constants, function and predicate symbols of RP. If c is a constant, a function or a predicate symbol, then we write $'\mathtt{c}$ as a convenient way to represent the metaconstant that names c. Similarly, if $'\mathtt{c}$ is a metaconstant, then its name is $''\mathtt{c}$, and so on.

Furthermore, the alphabet of RP contains the (unary) predicate symbol solve. This allows us to extend at the *meta-level* the intended meaning of predicates (partially) defined at the *object-level*. Clauses defining solve will be referred to as *meta-level clauses*, and clauses defining object-level predicates will be referred to as *object-level clauses*. Metavariables are written as any sequences of characters in *italic* style starting in the upper-case.

Compound terms and atoms are represented at the meta-level as *name terms*. For example, the name of the term $\mathtt{f}(\mathtt{a}, x)$ is the name term $'\mathtt{f}('\mathtt{a}, 'x)$, where $'\mathtt{f}$ and $'\mathtt{a}$ are the metaconstants that name the function symbol f and constant a, respectively, and $'x$ stands for the name of the value of the variable x.

The intended connection between the object-level and the meta-level of RP is obtained by means of the following (inter-level) reflection axioms:

$A \leftarrow \mathtt{solve}('A)$ and $\mathtt{solve}('A) \leftarrow A$, for all atoms A.

The first asserts that whenever an atom of the form $\mathtt{solve}('A)$ is provable at the meta-level, then A is provable at the object-level. These axiom schemata are not explicitly present in any given program but rather they are simulated within the modified SLD-resolution underlying RP.

Suppose, for example, that we want to express the fact that an object obj satisfies all the relations in a given class. If we want to formalise that statement at the object-level, then for every predicate q in class we have to write the clause q(obj). Instead, we may formalise our statement at the meta-level as:

$\mathtt{solve}(P('\mathtt{obj})) \leftarrow \mathtt{belongs\_to}(P, \mathtt{class})$,

where P is a metavariable ranging over the names of predicate symbols.

4.2 Communication

RPC is an extension of Reflective Prolog to accommodate communication between agents [7, 8]. Agents are seen as logic programs. In an agent setting, solve may be seen as representing a given agent's beliefs, for example $\mathtt{agent}_1 : \mathtt{solve}('A)$, for some atom A, stands for "$\mathtt{agent}_1$ believes A". RPC allows for two communication acts, expressed via the two meta-predicates $\mathtt{tell}(X, Y)$ and $\mathtt{told}(X, Y)$. An atom $\mathtt{tell}('\mathtt{agent}_2, 'A)$ in the theory representing $\mathtt{agent}_1$ stands for "$\mathtt{agent}_1$ tells $\mathtt{agent}_2$ that A holds". An atom $\mathtt{told}('\mathtt{agent}_2, 'A)$ in the theory representing $\mathtt{agent}_1$ stands for "$\mathtt{agent}_1$ is told by $\mathtt{agent}_2$ that A holds". The latter atom can also be interpreted as "$\mathtt{agent}_1$ *asks* $\mathtt{agent}_2$ whether A holds", i.e. the predicate told can be used to express queries of agents to other agents.

Communication acts are formalized in RPC by means of (inter-theory) reflection axioms based on the predicate symbols `tell` and `told`:

$[\mathtt{agent}_1 : \mathtt{told}('\mathtt{agent}_2,'A)] \leftarrow [\mathtt{agent}_2 : \mathtt{tell}('\mathtt{agent}_1,'A)]$

for all atoms A, and agent names $'\mathtt{agent}_1$ and $'\mathtt{agent}_2$. The intuitive meaning of each such axiom is that every time an atom of the form $\mathtt{tell}('\mathtt{agent}_1,'A)$ can be derived from a theory $\mathtt{agent}_2$ (which means that $\mathtt{agent}_2$ wants to communicate proposition A to $\mathtt{agent}_1$), the atom $\mathtt{told}('\mathtt{agent}_2,'A)$ is consequently derived in the theory $\mathtt{agent}_1$ (which means that proposition A becomes available to $\mathtt{agent}_1$). These axioms are not explicitly present in any given program but rather they are simulated within the modified SLD-resolution underlying RPC.

These two predicates are intended to model the simplest and most neutral form of communication among agents, with no implication about provability (or truth) of what is communicated, and no commitment about how much of its information an agent communicates and to whom. An agent may communicate to another agent everything it can derive (in its associated theory), or only part of what it can derive, or it may even lie, that is, it communicates something it cannot derive.

The intended connection between `tell` and `told` is that an agent may receive from another agent (by means of `told`) only the information the second agent has explicitly addressed to it (by means of `tell`). Thus, an agent can regulate its interaction with other agents by means of appropriate clauses defining the predicate `tell`.

What use an agent makes of any information given to it by others is entirely up to the agent itself. Thus, the way an agent communicates with others is not hard-wired in the language. Rather, it is possible to define in a program different behaviours for different agents or different behaviours of one agent in different situations. (Several examples of the use of agents in RPC can be found in [1, 6].) For example, the following clauses:

$\mathtt{Bob} : [\mathtt{solve}(X) \leftarrow \mathtt{reliable}(A) \wedge \mathtt{told}(A, X)]$

$\mathtt{Bob} : [\mathtt{solve}(X) \leftarrow \mathtt{told}('\mathtt{John}, '\mathtt{not}\ X)]$

express that an agent `Bob` trusts every agent that is reliable but distrusts `John`. The clause:

$\mathtt{Bob} : [\mathtt{tell}(A, X) \leftarrow \mathtt{agent}(A) \wedge \mathtt{solve}(X)]$

says that the agent `Bob` tells the others whatever he can prove.

RP and its extension RPC rely upon a Prolog-style control strategy and therefore do not allow for agents to tell other agents or ask for information proactively.

5 Adding introspection to KS-agents

In this paper, following the framework of KS-agents, agents are represented as abductive logic programs rather than ordinary logic programs as in RPC. The abductive logic program, in its completed form, is used embedded within a KS-agents' cycle.

In order to accommodate within agents beliefs about different agents, we give solve an additional argument rather than using labels as in RPC. The atom solve($Agent, A$) stands for "$Agent$ believes A". We will assume that solve can only take names of atoms as second argument. By convention, solve(A) will be an abbreviation for solve($Agent, A$) within the program P of $Agent$ itself.

Instead of incorporating the reflection principle linking object- and meta-level within the proof procedure, as in RP, we incorporate it via (a finite set of) clauses to be added to the given (abductive) logic program. (A similar approach is presented in [2].)

Let $\langle P, \mathcal{A}, I \rangle$ be an abductive logic program and let $C = H \leftarrow B$ be a clause in P. Then, we define a reflection principle $\mathcal{RP}$:

$$\mathcal{RP}(C) = \begin{cases} \mathcal{O}(C) & \text{if } C \text{ is an object-level clause} \\ \mathcal{M}(C) & \text{if } C \text{ is a meta-level clause} \end{cases}$$

where

- If $H = \mathtt{p}(t)$, then
$$\mathcal{O}(C) = \{\, \mathtt{solve}(X) \leftarrow X = {}^{\prime}\mathtt{p}({}^{\prime}t) \wedge B \,\}.$$

Note that necessarily $\mathtt{p} \notin \mathcal{A}$, since we assume, without loss of generality, that abducible predicates are not defined in P.

- If $H = \mathtt{solve}({}^{\prime}\mathtt{p}({}^{\prime}t))$ and $\mathtt{p} \notin \mathcal{A}$, then
$$\mathcal{M}(C) = \{\mathtt{p}(x) \leftarrow x = t \wedge B\}.$$

- If $H = \mathtt{solve}(X({}^{\prime}t))$, then
$$\mathcal{M}(C) = \{\, \mathtt{p}(t) \leftarrow X = {}^{\prime}\mathtt{p} \wedge B \mid \text{for every}\, \mathtt{p} \notin \mathcal{A} \text{ and } \mathtt{p} \neq \mathtt{solve} \,\}.$$

- If $H = \mathtt{solve}(X)$, then
$$\mathcal{M}(C) = \{\, \mathtt{p}(x) \leftarrow X = {}^{\prime}\mathtt{p}({}^{\prime}x) \wedge B \mid \text{for every}\, \mathtt{p} \notin \mathcal{A} \text{ and } \mathtt{p} \neq \mathtt{solve} \,\}.$$

$\mathcal{RP}(P)$ is given by the union of $\mathcal{RP}(C)$ for all $C \in P$.

► Let $\langle P, \mathcal{A}, I \rangle$ be an abductive logic program. The **abducibility set** of $\mathcal{A}$, written as $\Gamma(\mathcal{A})$, is the set of clauses:
$$\Gamma(\mathcal{A}) = \{\, \mathtt{solve}(X) \leftarrow X = {}^{\prime}\mathtt{a}({}^{\prime}x) \wedge \mathtt{a}(x) \mid \text{for all } \mathtt{a} \in \mathcal{A} \,\}.$$

The abducibility set of an abductive logic program allows a meta-level representation of provability of abducible predicates.

The intended connection among object-level, meta-level and abducible atoms is captured by the following definition.

► Let $\langle P, \mathcal{A}, I \rangle$ be an abductive logic program.
The **associated program** of P with respect to $\mathcal{A}$, written as $\Delta(P, \mathcal{A})$, is defined as:
$$\Delta(P, \mathcal{A}) = P \cup \mathcal{RP}(P) \cup \Gamma(\mathcal{A}).$$

The **associated integrity constraints** of I with respect to $\mathcal{A}$, written as $\Delta(I, \mathcal{A})$, is defined as:

$$\Delta(I, \mathcal{A}) = I \cup \{\mathtt{solve}(\ulcorner\mathtt{p}(\ulcorner x)) \Rightarrow \mathtt{p}(x) \mid \text{for all } \mathtt{p} \in \mathcal{A}\}.$$

The **meta-abductive logic program** associated with $\langle P, \mathcal{A}, I\rangle$ is:

$$\langle \Delta(P, \mathcal{A}),\ \mathcal{A},\ \Delta(I, \mathcal{A})\rangle.$$

The addition of the new integrity constraints in $\Delta(I, \mathcal{A})$ allows the agent to propagate (and thus compute the consequences of) any new information it receives about abducible predicates in whatever (meta-level or object-level) form, without any need to alter the original set of integrity constraints, I, or the program P.

6 Adding communication to KS-agents

In this paper, we interpret $\mathtt{tell}(X, Y)$ and $\mathtt{told}(X, Y)$ as abducible predicates in meta-abductive logic programs. As for `solve` we can give `tell` and `told` an additional argument instead of introducing labels, to represent communication between agents. For simplicity, we will abbreviate $\mathtt{tell}(\mathit{Agent}, X, Y)$ (resp. `told`) within the program P of *Agent* itself as $\mathtt{tell}(X, Y)$. As with `solve`, we will assume that `tell` and `told` take only names of atoms as their last argument.

Example 3. Let $\mathtt{agent}_1$ be represented by the abductive logic program $\langle P, \mathcal{A}, I\rangle$ with:

$$P = \left\{ \begin{array}{l} \mathtt{solve}(X) \leftarrow \mathtt{told}(A, X) \\ \mathtt{desire}(y) \leftarrow y = \mathtt{car} \\ \mathtt{good\_price}(p, x) \leftarrow p = 0 \end{array} \right\}$$

$$\mathcal{A} = \{\mathtt{tell}, \mathtt{told}, \mathtt{offer}\}$$

$$I\ = \{\mathtt{desire}(x) \wedge \mathtt{told}(A, \ulcorner\mathtt{good\_price}(\ulcorner p, \ulcorner x)) \Rightarrow \mathtt{tell}(A, \ulcorner\mathtt{offer}(\ulcorner p, \ulcorner x))\}.$$

Namely, $\mathtt{agent}_1$ believes anything it is told (by any other agent), and it desires to have a car. The third clause in P says that anything that is free is at a good price. Moreover, if the agent desires something and it is told (by some other agent) of a good price for it, then it makes an offer to the other agent, by telling it. Note that a more accurate representation of the integrity constraint should include time.

The corresponding meta-abductive logic program is

$$\Delta(P,\mathcal{A}) = P \cup \left\{ \begin{array}{l} \texttt{solve}(X) \leftarrow X = \texttt{'desire('}y) \wedge y = \texttt{car} \\ \texttt{solve}(X) \leftarrow X = \texttt{'good\_price('}p,\texttt{'}x) \wedge p = 0 \\ \texttt{desire}(x) \leftarrow X = \texttt{'desire('}x) \wedge \texttt{told}(A,X) \\ \texttt{good\_price}(p,x) \leftarrow X = \texttt{'good\_price('}p,\texttt{'}x) \wedge \texttt{told}(A,X) \\ \texttt{solve}(Y) \leftarrow Y = \texttt{'tell('}A,\texttt{'}X) \wedge \texttt{tell}(A,X) \\ \texttt{solve}(Y) \leftarrow Y = \texttt{'told('}A,\texttt{'}X) \wedge \texttt{told}(A,X) \\ \texttt{solve}(Y) \leftarrow Y = \texttt{'offer('}p,\texttt{'}x) \wedge \texttt{offer}(p,x) \end{array} \right\}$$

$$\mathcal{A} = \{\texttt{tell}, \texttt{told}, \texttt{offer}\}$$

$$\Delta(I,\mathcal{A}) = I \cup \left\{ \begin{array}{l} \texttt{solve('tell('}A,\texttt{'}X)) \Rightarrow \texttt{tell}(A,X) \\ \texttt{solve('told('}A,\texttt{'}X)) \Rightarrow \texttt{told}(A,X) \\ \texttt{solve('offer('}p,\texttt{'}x)) \Rightarrow \texttt{offer}(p,x) \end{array} \right\}.$$

Given an abductive logic program $\langle P, \mathcal{A}, I\rangle$, corresponding to some agent, and the associated meta-abductive logic program $\langle \Delta(P,\mathcal{A}), \mathcal{A}, \Delta(I,\mathcal{A})\rangle$, the agent's cycle applies the IFF procedure with

$$T = comp_{\overline{\mathcal{A}}}(\Delta(P,\mathcal{A})).$$

The equality rewriting rule needs to be modified to take into account equality between names.

Example 4. Let $\langle P, \mathcal{A}, I\rangle$ be the abductive logic program of example 3. Assume that $\texttt{agent}_1$ has the input observation:

$$\texttt{told('agent}_2\texttt{,'good\_price('50,'car))},$$

meaning that $\texttt{agent}_1$ has been told by $\texttt{agent}_2$ of a good price (of 50) for a car. Then, the initial goal G is:

$$\texttt{told('agent}_2\texttt{,'good\_price('50,'car))}.$$

A computed answer for G is

$$\mathcal{D} = \{\texttt{told('agent}_2\texttt{,'good\_price('50,'car))}, \texttt{tell('agent}_2\texttt{,'offer('50,'car))}\}$$
$$\sigma = \{\}.$$

Within the KS-agent architecture, the following requirements are met:

- KS-agents are known by their symbolic names.
- When a KS-agent sends a message, it directs that message to a specific addressee.
- When a KS-agent receives a message, it knows the sender of that message.
- Messages may get lost.

Both $\texttt{tell}$ and $\texttt{told}$ are treated as actions: everytime (the cycle of) an agent $\texttt{agent}_1$ selects a communicative action, i.e., an action of the form

$$\texttt{told}('\texttt{agent}_2, A) \qquad \text{or} \qquad \texttt{tell}('\texttt{agent}_2, A),$$

$\texttt{agent}_1$ will attempt to execute it. If the attempt is successful, the record

$$\texttt{told}('\texttt{agent}_2, A) \qquad \text{or} \qquad \texttt{tell}('\texttt{agent}_2, A)$$

is conjoined to the goals of agent $\texttt{agent}_1$, as an additional input. If the attempt is not successful, the record

$$\texttt{told}('\texttt{agent}_2, A) \Rightarrow \texttt{false} \qquad \text{or} \qquad \texttt{tell}('\texttt{agent}_2, A) \Rightarrow \texttt{false}$$

is conjoined to the goals of agent $\texttt{agent}_1$, as an additional input. In the next section we will formalise an example showing how proactive communication is achieved by executing the proof procedure within cycle.

7 Example

The following example demonstrates the running of the proof procedure within cycle, action selection, proactive communication whereby one agent volunteers information to another, and how during the planning phase such information can help in the choice of intention.

The example is as follows: an agent wishes to register for a conference, let us say Jelia, that is to take place in Paris on the 10th and to make travel arrangements to go to Paris on the 10th (for simplicity we omit the month). So the agent's original goal is the conjunction:

$\texttt{register(Jelia)} \wedge \texttt{travel(Paris, 10)}$.

The agent has the following program, P:

C_1 $\texttt{travel}(\mathit{city}, \mathit{date}) \leftarrow \texttt{go}(\mathit{city}, \mathit{date}, \texttt{train})$

C_2 $\texttt{travel}(\mathit{city}, \mathit{date}) \leftarrow \texttt{go}(\mathit{city}, \mathit{date}, \texttt{plane})$

C_3 $\texttt{early\_registration(Jelia)} \leftarrow \texttt{send\_form}(\texttt{Jelia}, \mathit{date}) \wedge (\mathit{date} < 3)$

C_4 $\texttt{go}(\mathit{city}, \mathit{date}, \mathit{means}) \leftarrow \texttt{book}(\mathit{city}, \mathit{date}, \mathit{means})$

C_5 $\texttt{book}(\mathit{city}, \mathit{date}, \mathit{means}) \leftarrow$
$\quad \texttt{told}('\texttt{ticket\_agent}, '\texttt{available}('\mathit{city}, '\mathit{date}, '\mathit{means})) \wedge$
$\quad \texttt{tell}('\texttt{ticket\_agent}, '\texttt{reserve}('\mathit{city}, '\mathit{date}, '\mathit{means}))$

C_6 $\texttt{solve}(X) \leftarrow \texttt{told}('\texttt{ticket\_agent}, X)$.

Clauses C_1 and C_2 say that one travels to a *city* on a given *date* if one goes there on that *date* by train or by plane. C_3 says that the deadline for early registration for Jelia is the 3rd. C_4 says that one goes to a *city* on a given *date* by some *means* if one makes a booking for that journey. C_5 says that one makes a booking if the ticket_agent confirms availability of ticket and one makes a reservation. Note that in a more thorough representation we would represent the transaction time of when a booking is made (which must be before the time of travel). In that case we will have an extra argument in `tell` and `told` that represents such transaction times. We will ignore this issue here for simplicity.

The agent has the following set of integrity constraints I:

I_1 `register`(*conference*) ⇒ `early_registration`(*conference*)
I_2 `go`(*city*, *date*, *means*) ∧ `strike`(*city*, *date*, *means*) ⇒ `false`.

I_1 states the agent's departmental policy that anyone registering at a conference must take advantage of early registration. I_2 states that one cannot use a mode of transport which is subject to a strike.

Finally, the agent has the following set of abducibles $\mathcal{A}$:
{`send_form`, `tell`, `told`, `register`, `available`, `reserve`, `strike`}.

Now suppose that the cycle of the agent starts at time 1 and that the agent does not observe any input. Thus, the cycle effectively starts at step (*iv*) by applying the IFF proof procedure (using $\langle \Delta(P, \mathcal{A}), \mathcal{A}, \Delta(I, \mathcal{A}) \rangle$ which we do not show here) to the goal obtained by conjoining the original goal and the integrity constraints $\Delta(I, \mathcal{A})$. By repeatedly applying the IFF proof procedure, the goal is transformed (within the initial cycle or within some later iteration, depending on the resource parameter r) into

`register(Jelia)` ∧ `travel(Paris, 10)` ∧ `send_form(Jelia,` *date*`)` ∧ (*date* < 3).

At this point cycle can select (at step (*v*)) the action `send_form(Jelia,` *date*`)` to perform if time is less than 3. If the agent has been too slow and time 3 has already passed the agent has failed its goals. Suppose the agent succeeds. Note that at any time any other agent can send this agent a message. So suppose the ticket agent sends a message that trains are on strike in Paris on the 10th, i.e. at some iteration of cycle at step (*i*) the agent receives the following input

`told('ticket_agent, 'strike('Paris, '10, 'train))`.

In that iteration of cycle this information is propagated (step (*iii*)) and the simplified constraint

`go(Paris, 10, train)` ⇒ `false`

is added to the goal. Meanwhile the sub-goal `travel(Paris, 10)` is unfolded (step (*iv*)) into

`go(Paris, 10, train)` ∨ `go(Paris, 10, plane)`.

The information about the strike will be used to remove the first possibility (i.e the first disjunct) leaving only

`go(Paris, 10, plane)`

which will become the agent's intention. This will be unfolded into the plan

`told('ticket_agent, 'available('Paris, '10, 'plane))` ∧
`tell('ticket_agent, 'reserve('Paris, '10, 'plane))`.

The appropriate actions will be selected during some iterations of cycle.

For lack of space we have ignored the cycle of the `ticket_agent`.

8 Conclusions

We have presented an approach to agents that can reason about their own beliefs as well as beliefs of other agents and that can communicate with each other. The approach results from the combination of the approach to agents in [12] and the approach to meta-reasoning and communication in [6, 1]. We have illustrated the approach by means of a number of examples.

The approach needs to be extended in a number of ways. For simplicity, we have ignored the treatment of time. However, time plays an important role in most agent applications and should be explicitly taken into account.

We have considered only two communication performatives, `tell` and `told`. Existing communication languages, e.g. [5], consider additional performatives, e.g. `deny`, `achieve` and `unachieve`. We are currently investigating whether some of these additional performatives could be defined via communication protocols, as definitions and integrity constraints within our framework.

The primitive `told` is used to express both *active* request for information and *passive* communication. The two roles should be separated out, possibly with the addition of a third predicate `ask`, distinguished from `told`. Then the predicate `told` could be defined in terms of `ask` and `tell`, rather than be an abducible. For example, an agent $\texttt{agent}_1$ is told of X by another agent $\texttt{agent}_2$ if and only if $\texttt{agent}_2$ tells $\texttt{agent}_1$ of X or $\texttt{agent}_1$ actively asks $\texttt{agent}_2$ about X and $\texttt{agent}_2$ gives a positive answer.

We have implicitly assumed that different agents share the same content language. However, this assumption is not essential. Indeed, "translator agents" could be defined, acting as mediators between agents with different content languages. This is possible by virtue of the metalogic features of the language.

Acknowledgments

The authors are grateful to the anonymous referees for their comments. The second and third authors are supported by the UK EPSRC project "Logic-based multi-agent systems".

References

1. J. Barklund, S. Costantini, P. Dell'Acqua, and G. A. Lanzarone. Metareasoning agents for query-answering systems. In Troels Andreasen, Henning Christiansen, and Henrik Legind Larsen, editors, *Flexible Query-Answering Systems*, pages 103–122. Kluwer Academic Publishers, Boston, Mass., 1997.
2. J. Barklund, S. Costantini, P. Dell'Acqua, and G. A. Lanzarone. Reflection Principles in Computational Logic. Submitted to J. of Logic and Computation, 1997.
3. A. Chavez and P. Maes. Kasbah: An agent marketplace for buying and selling goods. In B. Crabtree and N. Jennings, editors, *Proc. 1st Intl. Conf. on the Practical Application of Intelligent Agents and Multi-Agent Technology*, pages 75–90. The Practical Application Company, 1996.
4. K. L. Clark. Negation as failure. In H. Gallaire and J. Minker, editors, *Logic and Data Bases.* Plenum Press, New York, 1978.
5. P. R. Cohen and H. J. Levesque. Communicative actions for artificial agents. In V. Lesser, editor, *Proc. 1st Intl. Conf. on Multiagent Systems*, AAAI Press, pages 65–72. MIT Press, 1995.

6. S. Costantini, P. Dell'Acqua, and G. A. Lanzarone. Reflective agents in metalogic programming. In Alberto Pettorossi, editor, *Meta-Programming in Logic*, LNCS 649, pages 135–147, Berlin, 1992. Springer-Verlag.
7. S. Costantini and G. A. Lanzarone. A metalogic programming language. In G. Levi and M. Martelli, editors, *Proc. 6th Intl. Conf. on Logic Programming*, pages 218–33, Cambridge, Mass., 1989. MIT Press.
8. S. Costantini and G. A. Lanzarone. A metalogical programming approach: Language, semantics and applications. *Int. J. of Experimental and Theoretical Artificial Intelligence*, 6:239–287, 1994.
9. T. H. Fung and R. Kowalski. The IFF proof procedure for abductive logic programming. *J. Logic Programming*, 33(2):151–165, 1997.
10. A. C. Kakas, R. A. Kowalski, and F. Toni. The role of abduction in logic programming. In D. Gabbay, C. Hogger, and A. Robinson, editors, *Handbook of Logic in Artificial Intelligence and Logic Programming*, volume 5, pages 235–324. Oxford University Press, UK, 1998.
11. R. A. Kowalski. Predicate logic as a programming language. In J. L. Rosenfeld, editor, *Information Processing, 1974*, pages 569–574, Amsterdam, 1974. North-Holland.
12. R. A. Kowalski and F. Sadri. Towards a unified agent architecture that combines rationality with reactivity. In Dino Pedreschi and Carlo Zaniolo, editors, *Logic in Databases, Intl. Workshop LID'96*, LNCS 1154, pages 137–149, Berlin, 1996. Springer-Verlag.
13. R. A. Kowalski and M. Sergot. A logic-based calculus of events. *New Generation Computing*, 4:67–95, 1986.
14. Y. Labrou and T. Finin. Semantics and conversations for an agent communication language. In M. N. Huhns and M. P. Singh, editors, *Readings in Agents*, pages 234–242, San Francisco, 1997. Morgan Kaufmann.
15. John W. Lloyd. *Foundations of Logic Programming, Second Edition.* Springer-Verlag, Berlin, 1987.
16. Alberto Martelli and Ugo Montanari. An efficient unification algorithm. *ACM TOPLAS*, 4:258–282, 1982.
17. M. P. Singh. Towards a formal theory of communication for multiagent systems. In *Proc. 12th Intl. Joint Conf. on Artificial Intelligence*, pages 69–74, Sydney, Australia, 1991. Morgan Kaufmann.
18. B. van Linder, W. van der Hoek, and J.-J. Ch. Meyer. Communicating rational agents. In B. Nebel and L. Dreschler-Fischer, editors, *KI-94: Advances in Artificial Intelligence*, LNAI 861, pages 202–213, Berlin, 1994. Springer-Verlag.
19. E. Verharen and F. Dignum. Cooperative information agents and communication. In P. Kandzia and M. Klusch, editors, *Cooperative Information Agents*, LNAI 1202, pages 195–208, Berlin, 1997. Springer-Verlag.
20. G. Wagner. Multi-level security in multiagent systems. In P. Kandzia and M. Klusch, editors, *Cooperative Information Agents*, LNAI 1202, pages 272–285, Berlin, 1997. Springer-Verlag.

Disjunctive Logic Program = Horn Program + Control Program

Wenjin Lu and Ulrich Furbach

Department of Computer Science,University of Koblenz,Germany
lue,uli@informatik.uni-koblenz.de

Abstract. This paper presents an alternative view on propositional disjunctive logic program: Disjunctive program = Control program + Horn program. For this we introduce a program transformation which transforms a disjunctive logic program into a Horn program and a so called control program. The control program consists of only disjunctions of new propositional atoms and controls the "execution" of the Horn program. The relationship between original and transformed programs is established by using circumscription. Based on this relationship a new minimal model reasoning approach is developed. Due to the transformation it is straightforward to incorporate SLD-resolution into the proof procedure.

1 Introduction

Disjunctive logic programs may contain definite or indefinite information which reflects human limitation in understanding the world being modelled. In the absence of negation, that is, for positive disjunctive program, the semantics is defined by so called Generalized Closed World Assumption (GCWA) which is equivalent to minimal model reasoning [Min82]. GCWA allows one to assume an atom to be false if it does not appear in any minimal model of the program.

There are various proof procedures proposed for disjunctive logic programming: Minker and his co-workers used SLI-resolution [LMR92], which is in fact a version of the model elimination procedure; Loveland proposed the near-Horn-Prolog procedures [Lov91] and there are variants of model elimination to deal with disjunctive logic programs[BFS97]. Besides this goal oriented approaches there are as well bottom-up procedures, like SATCHMO[MB88,LRW95] or Hyper tableau[BFN96]. For a unifying view of bottom up and goal oriented approaches see [BF97].

In this paper we propose a novel point of view on propositional disjunctive logic program which is motivated by the following observation.
Given a disjunctive logic program $\mathcal{P}$, for each clause C of the form

$$a_1 \vee ... \vee a_n \leftarrow b_1, ..., b_m$$

we split C into a set $Horn(C)$ of Horn clauses,

$$\begin{aligned} Horn(C) = \{ & a_1 \leftarrow b_1, ..., b_m, \\ & a_2 \leftarrow b_1, ..., b_m, \\ & ...... \\ & a_n \leftarrow b_1, ..., b_m \ \} \end{aligned} \tag{1}$$

Considering the following two sets:

$$Horn(\mathcal{P}) = \prod_{C \in \mathcal{P}} Horn(C) \tag{2}$$

$$\bar{\mathcal{P}}_H = \bigcup_{C \in \mathcal{P}} Horn(C) \tag{3}$$

$Horn(\mathcal{P})$ can be regard as a set of Horn programs, each Horn program in it contains exactly one Horn clause from each $Horn(C)$ $(C \in \mathcal{P})$. It is know that this splitting is complete wrt. minimal models of $\mathcal{P}$ in the sense that for each minimal model M of $\mathcal{P}$, there is a Horn program $\mathcal{P}_M \in Horn(\mathcal{P})$ such that M is the least model of $\mathcal{P}_M$ [IKH92,Lu97]. On the other hand, $\bar{\mathcal{P}}_H$ is a Horn program. Notice that for any $\mathcal{P}_h \in Horn(\mathcal{P})$ $\mathcal{P}_h \subseteq \bar{\mathcal{P}}_H$[1], therefore we can say $\bar{\mathcal{P}}_H$ wrt. minimal models of $\mathcal{P}$ is complete in the sense that for each minimal model M of $\mathcal{P}$, there is a subset $\mathcal{P}_M$ of $\bar{\mathcal{P}}_H$ such that M is the least model of $\mathcal{P}_M$. So under minimal model semantics a disjunctive logic program can be regarded as a Horn program together with a control means which controls the execution of the Horn program (which clauses should be chosen to make a minimal model).

Based on this observation, in this paper we put forward

Disjunctive Program=Control Program + Horn Program

This point of view establishes a relationship between disjunctive logic programming and definite logic programs, the latter has a well-understood declarative semantics and an effective procedural implementation. As a consequence, we can expect to implement minimal model reasoning by using existing techniques (eg. SLD-resolution) developed for definite logic programs.

Other contributions of this paper are as follows:

- A program transformation is introduced which demonstrates the idea presented above. By introducing some new propositional atoms it transforms a disjunctive logic program $\mathcal{P}$ into $\mathcal{P}_C \cup \mathcal{P}_H$, where $\mathcal{P}_C$ is a set of disjunctive facts consisting of only new propositional atoms and $\mathcal{P}_H$ is a Horn program. It turns out that this transformation is sound and complete wrt. minimal models in the sense that M is a minimal model of $\mathcal{P}$ iff there is a minimal model M_C of $\mathcal{P}_C$ such that $M \cup M_C$ is (P, Z)-minimal model of $\mathcal{P}_C \cup \mathcal{P}_H$, where P are the set of all atoms in $\mathcal{P}$ and Z is the set of new propositional atoms introduced by the transformation.

[1] Here for convenience, we abuse an element in $\prod_{C \in \mathcal{P}} Horn(C)$ as a set

- Based on above result, a novel minimal model reasoning procedure is developed which differs from similar work[LMR92,Gin89,Prz89,Ara96] in that it uses SLD-resolution as a basic inference mechanism.

The rest of the paper is organised as follows: after reviewing logic programs and related background knowledge, we provide the transformation in Section 3. Some properties of the transformation are discussed and the soundness and completeness are proved. In Section 4, we discuss how minimal model reasoning can be implemented based on SLD-resolution and develop such a procedure. The paper is finally concluded with some remarks on further work.

2 Preliminaries

Given a first order language L, a *disjunctive logic program* $\mathcal{P}$ is a finite set of *disjunctive program clauses* of the form

$$a_1 \vee \ldots \vee a_n \leftarrow b_1 \wedge \ldots \wedge b_m$$

where every a_i and b_j are atoms with $n \geq 1$ and $m \geq 0$, and all variables are considered to be universally quantified. $a_1 \vee \ldots \vee a_n$ is referred to as the *head* and $b_1 \wedge \ldots \wedge b_m$ as the *body* of the program clause. Usually non-monotonic negations are allowed in the body of a clause, but in this paper we restrict ourselves to positive programs where such negations do not occur. This is not a severe restriction, since minimal model reasoning with positive programs is an important problem to tackle [NNS95,Oli92,IKH92,Gin89,Prz89,Ara96]. When $n = 1$, the program clause is called as a *Horn or definite program clause*, and when $m = 0$, it is called as a *fact*. Note that head of a program clause is not allowed to be empty, and consequently all disjunctive logic programs are consistent. A formula in disjunctive normal form, consisting only of ground atoms, is referred to as a *sentence*. The reader is referred to, for example, [Llo87,LMR92] and references therein, for more information on logic programming and disjunctive logic programs.

The Herbrand base of the language L is usually denoted as HB_L. When no ambiguity arises the subscript L is dropped. Usually, when we consider a program $\mathcal{P}$, we are interested in a Herbrand Base that is restricted to predicate, function and constant symbols that appear in $\mathcal{P}$. If $\mathcal{P}$ has no constant symbols then a dummy constant is assumed. In the sequel we call such Herbrand Base the Herbrand base of $\mathcal{P}$, denote it as $HB_{\mathcal{P}}$ and simply write interpretation and model to denote Herbrand interpretation and Herbrand model resp. of a logic program.

The meaning of a disjunctive logic program is given by the set of its logical consequences. Obviously, negative information can not be handled efficiently in this classical semantics, and hence *generalised closed world assumption* [Min82], referred to as GCWA for short, is usually employed to infer negative information from a disjunctive logic program (see [LMR92] for more discussion and details). GCWA allows one to assume an atom to be false, if it doesn't appear in any minimal model of the program. This has been independently extended for sentences in [YH85,GPP89]. All these versions are not fundamentally different, and for the purpose of this paper, we simply refer to the following definition of closed world assumption in disjunctive logic programming.

In the following we always assume that the given program is instantiated, that is, all clauses in the program are ground and make use of the notion of *minimal models*, *mimimal entailment*, written as $\mathcal{P} \models_{min} \neg\alpha$, and the *immediate consequence operator* $T_{\mathcal{P}}$ for Horn programs.

3 CH-Transformation

In this section we first introduce a program transformation which realizes the idea presented in the introduction. Then we discuss the relationship between original and transformed programs and prove that under minimal model semantics they are equivalent. This result provides a basis for our further discussion.

Definition 1. *Let $\mathcal{P}$ be a disjunctive logic program. For each clause $C \in \mathcal{P}$ of the form:*

$$a_1 \vee a_2 \vee ... \vee a_n \leftarrow b_1, ..., b_m$$

The CH-transformation of C, denoted by $CH(C)$, is a set of clauses defined by:

$$CH(C) = \begin{cases} \{C\} : & n = 1 \\ C_{ch} : & n > 1 \end{cases}$$

where

$$\begin{aligned} C_{ch} = \{ \; & C_1 \vee C_2 \vee ... \vee C_n, \\ & a_1 \leftarrow b_1, ..., b_m, C_1, \\ & a_2 \leftarrow b_1, ..., b_m, C_2, \\ & ...... \\ & a_n \leftarrow b_1, ..., b_m, C_n \; \} \end{aligned}$$

C_1, ..., C_n are new propositional atoms not in $\mathcal{P}$, they are called control variables, the disjunctive fact $C_1 \vee C_2 \vee ... \vee C_n$ is called a control clause. The CH-transformation of $\mathcal{P}$, denoted by $CH(\mathcal{P})$, is defined by

$$CH(\mathcal{P}) = \bigcup_{C \in \mathcal{P}} CH(C)$$

The set of all control variables in $CH(\mathcal{P})$ is denoted by $C_{\mathcal{P}}$

By the definition, the clause in $CH(\mathcal{P})$ is either a Horn clause or a disjunctive fact (control clause). Therefore $CH(\mathcal{P})$ can be represented as follows:

$$CH(\mathcal{P}) = \mathcal{P}_C + \mathcal{P}_H$$

where $\mathcal{P}_C$ is the program consisting of all control clauses in $CH(\mathcal{P})$ and $\mathcal{P}_H$ is the Horn program consisting of all Horn clauses in $CH(\mathcal{P})$. $\mathcal{P}_C$ is called the control program of $\mathcal{P}_H$. The following facts are trivial.

- The disjunctions in $\mathcal{P}_C$ consists of only control variables. No control variable occurs more than once in $\mathcal{P}_C$.
- Each Horn clause in $\mathcal{P}_H$ contains at most one control variable. No control variable occurs two times in $\mathcal{P}_H$.
- Each control variable occurs exactly two times in $CH(\mathcal{P})$, once in $\mathcal{P}_C$ and once in $\mathcal{P}_H$.

For any clause C of the form

$$a_1 \vee a_2 \vee ... \vee a_n \leftarrow b_1, ..., b_m$$

let $Horn(\mathcal{C})$, $Horn(\mathcal{P})$ and $\bar{\mathcal{P}}_H$ be given by formulae 1, 2 and 3 in the introduction resp. For any clause $C \in \mathcal{P}_H$, we denote $\bar{C}$ the clause obtained from C by deleting the control variable in the body of C (if any). Below we state the relationship between $Horn(\mathcal{P})$ and $CH(\mathcal{P})$.

Definition 2. *Let $\mathcal{P}$ be a disjunctive logic program and $C_{\mathcal{P}}$ be the set of all control variables. For any $S \subseteq C_{\mathcal{P}}$, the subprogram of $\mathcal{P}_H$ determined by S, written as $\mathcal{P}_H(S)$, is defined by*

$$\mathcal{P}_H(S) = \{\bar{C} \mid C \in CH(\mathcal{P}) \text{ such that } C \text{ contains either no control variable or one in } S\}$$

Notice the fact that when without considering the control variables, $\bar{\mathcal{P}}_H$ and $\mathcal{P}_H$ are same and for a minimal model M_C of $\mathcal{P}_C$, it contains one and only one control variable from each control clause, then the following proposition is trivial.

Proposition 1. *Let $\mathcal{P}$ be a disjunctive logic program. Then for any $\mathcal{P}_h \in Horn(\mathcal{P})$, there is a minimal model M_C of $\mathcal{P}_C$ such that $\mathcal{P}_H(S) = \mathcal{P}_h$*

Because $Horn(\mathcal{P})$ is complete wrt. minimal model in the sense that for any minimal model M of $\mathcal{P}$, there is a $\mathcal{P}_h \in Horn(\mathcal{P})$ such that M is the least model of $\mathcal{P}_h$. Therefore we have

Theorem 1. *Let $\mathcal{P}$ be a disjunctive logic program. Then for any minimal model M of $\mathcal{P}$, there is a minimal model M_C of $\mathcal{P}_C$ such that M is the least model of $\mathcal{P}_H(M_C)$.*

Example 1. let

$$\begin{aligned}\mathcal{P} = \{\, & p \leftarrow q, r \\ & q \leftarrow r \\ & r \\ & s \leftarrow r_1 \\ & t \leftarrow r_1 \\ & p \vee s \\ & t \vee u \leftarrow r \,\}\end{aligned}$$

then the CH-transformation of $\mathcal{P}$ is as follows.

$$CH(\mathcal{P}) = \mathcal{P}_C \cup \mathcal{P}_H$$

where

$$\mathcal{P}_C = \{ A_1 \vee A_2, \quad B_1 \vee B_2 \}$$

$$\mathcal{P}_H = \{ p \leftarrow q, r \quad q \leftarrow r \quad r \quad s \leftarrow r_1 \quad t \leftarrow r_1 \quad p \leftarrow A_1 \quad s \leftarrow A_2 \quad t \leftarrow r, B_1 \quad u \leftarrow r, B_2 \}$$

and $C_{\mathcal{P}} = \{ A_1, A_2, B_1, B_2 \}$. $M_C = \{ A_1, B_1 \}$ is a minimal model of $\mathcal{P}_C$ and

$$\mathcal{P}_H(M_C) = \{ p \leftarrow q, r \quad q \leftarrow r \quad r \quad s \leftarrow r_1 \quad t \leftarrow r_1 \quad p \quad t \leftarrow r \}$$

Now we discuss the relationship between program $\mathcal{P}$ and $CH(\mathcal{P})$. To do so we need some notations about circumscription [McC80,GPP89,Lif85]

Given first-order theory $\mathcal{T}$, let P and Z be disjoint tuples of predicates from $\mathcal{T}$, then circumscription of P in $\mathcal{T}$ with variable Z is defined as the second-order formula:

$$Circ(T; P; Z) = T(P, Z) \wedge \neg \exists P' Z' (T(P', Z') \wedge P' < P)$$

where $T(P, Z)$ is a theory containing predicate constants P, Z, and P', Z' are tuples of predicate variables similar to P, Z. The set of all predicates other than P, Z from T is denoted by Q, which is called the fixed predicates.

Definition 3. *For any two models M and N of $\mathcal{T}$ we write $M \leq N$ mod(P, Z) if models M and N differ only in how they interpret atoms from P and Z and if the extension of every predicate from P in M is a subset of its extension in*

N. A model M of $\mathcal{T}$ is called (P,Z)-minimal if there is no model N of $\mathcal{T}$ such that $N < M$ mod(P,Z) (ie. such that $M \leq N$ mod(P,Z) but not $N \leq M$ mod(P,Z)).

Note: When $\mathcal{T}$ is a propositional theory, the definition of the relation $\leq$ can be restated as follows: For any two models M and N of $\mathcal{T}$ we write $M \leq N$ *mod*(P,Z) if models M and N differ only in how they interpret atoms from P and Z and $M \cap P \subseteq N \cap P$.

The following theorem is due to [McC80,GPP89,Lif85], which explains the semantics of circumscription.

Theorem 2. *M is a model of $Circ(T;P;Z)$ iff M is a (P,Z)-minimal model of T. In other words, for any formula F, we have $Circ(T;P;Z) \models F$ iff $M \models F$ for every (P,Z)-minimal model M of T.*

Let $\mathcal{P}$ be a disjunctive logic program and $CH(\mathcal{P}) = \mathcal{P}_C \cup \mathcal{P}_H$ be the CH-transformation of $\mathcal{P}$. In the following we always let $P = HB_{\mathcal{P}}$ and for any model M of $CH(P)$, denote $M_C = M \cap C_{\mathcal{P}}$ and $M_H = M \cap P$ and write $M = M_C \cup M_H$.

Lemma 1. *Let $\mathcal{P}$ be a disjunctive logic program and $CH(\mathcal{P}) = \mathcal{P}_C \cup \mathcal{P}_H$. If $M = M_C \cup M_H$ is a model of $CH(\mathcal{P})$, then M_H is a model of $\mathcal{P}$.*

Proof: Let $M = M_C \cup M_H$ be a model of $CH(\mathcal{P})$. If M_H is not a model of $\mathcal{P}$, then there is a clause C of form

$$a_1 \vee a_2 \vee ... \vee a_n \leftarrow b_1, ..., b_m$$

such that $b_1, ..., b_m$ are true in M_H and for any $1 \leq i \leq n$, $a_i \notin M_H$. It must be that $n > 1$, otherwise $C \in CH(\mathcal{P})$, which contradicts that M is a model of $CH(\mathcal{P})$. Let $C_1 \vee C_2 \vee ... \vee C_n$ be the control clause in $CH(C)$; because M is a model of $CH(\mathcal{P})$, there is i $(1 \leq i \leq n)$ such that $C_i \in M$, then $b_1, ..., b_m, C_i$ must be true in M, therefore, $a_i \in M$. But a_i occurs in $\mathcal{P}$, so $a_i \in M_H$. This is a contradiction and we conclude that M_H is a model of $\mathcal{P}$.

Theorem 3. *Let $\mathcal{P}$ be a disjunctive logic program and $CH(\mathcal{P}) = \mathcal{P}_C \cup \mathcal{P}_H$. Then M is a minimal model of $\mathcal{P}$ iff there exists a minimal model M_C of $\mathcal{P}_C$ such that $M_C \cup M$ is an $(P, C_{\mathcal{P}})$-minimal model of $CH(\mathcal{P})$.*

Proof: Let M be a minimal model of $\mathcal{P}$, we prove that there is a M_C such that $M_C \cup M$ is an $(P, C_{\mathcal{P}})$-minimal model of $CH(\mathcal{P})$. For the minimal model M, define M_C as follows: For each clause C of the form

$$a_1 \vee a_2 \vee ... \vee a_n \leftarrow b_1, ..., b_m$$

in $\mathcal{P}$, where $n > 1$, let $C_1 \vee C_2 \vee ... \vee C_n$ be the control clause in $CH(C)$. If $b_1, ..., b_m$ is true in M, then there exists i such that $a_i \in M$, let $C_i \in M_C$. Otherwise let $C_1 \in M_C$. Nothing else is in M_C. It is clear that M_C is a minimal model of $\mathcal{P}_C$.

$M_C \cup M$ is a model of $CH(\mathcal{P})$. If not so, then there is a clause of $a_i \leftarrow b_1, ..., b_m, C_i$ such $b_1, ..., b_m, C_i$ is true in $M_C \cup M$ but a_i is not in M. By the construction of M_C this is impossible because in this case $C_i \in M_C$ if and only if $a_i \in M$.

$M_C \cup M$ is an $(P, C_{\mathcal{P}})$-minimal model of $CH(\mathcal{P})$. If not so, then there is a minimal model $M'_C \cup M'_H$ of $CH(\mathcal{P})$ such that $M'_H \subset M$, By lemma 1, M'_H is a model of $\mathcal{P}$. This contradicts that M is a minimal model of $\mathcal{P}$.

To prove the other direction of the theorem, let M_C be a minimal model of $\mathcal{P}_C$ such that $M_C \cup M$ is a $(P, C_{\mathcal{P}})$-minimal model of $CH(\mathcal{P})$, then by lemma 1, M is a model of $\mathcal{P}$. If M is not a minimal model, then there is a minimal model M' of $\mathcal{P}$ such that $M' \subset M$, for this M', we can define M'_C as above such that $M'_C \cup M'$ is an $(P, C_{\mathcal{P}})$-minimal model of $CH(\mathcal{P})$ but $M' \subset M$. This contradicts that $M_C \cup M$ is an $(P, C_{\mathcal{P}})$-minimal model of $CH(\mathcal{P})$.

By theorem 2 and theorem 3 we finally conclude

Theorem 4. *Let $\mathcal{P}$ be a disjunctive logic program and $CH(\mathcal{P}) = \mathcal{P}_C \cup \mathcal{P}_H$. Then for any formula α which contains no new propositional atoms, we have $\mathcal{P} \models_{min} \alpha$ iff $Circ(CH(\mathcal{P}); P; C_{\mathcal{P}}) \models \alpha$.*

4 Query Answering under Minimal Model Semantics

In this section we discuss how to answer queries under minimal model semantics. Because positive query can be answered by any existing theorem prover, for example PROTEIN [BF94], we concentrate only on negative query, that is, for a given atom q, to evaluate if $\neg q$ is true in all minimal models. The algorithm presented below can be combined with a existing theorem prover to get a sound and complete minimal model reasoning procedure. An example of such an approach can be found in [Ara96]

Our method is based on SLD-resolution due to the CH-transformation. An SLD-resolution proof procedure provides a way to compute answers for a goal from a definite logic program. It starts with a goal clause, $\leftarrow a_1, ..., a_n$, and provides a refutation by deriving an empty goal clause $\Box$. The SLD derivation for definite propositional logic programs can be stated as follows.

Definition 4 (SLD derivation). *Let $\mathcal{P}$ be a definite logic program and G be the goal $\leftarrow a_1, ..., a_m, ..., a_n$. An SLD derivation from $\mathcal{P}$ with top goal G consists of a (finite or infinite) sequence of goals $G_0 (= G), G_1, ...$, such that for all $i \geq 0$, G_{i+1} is obtained from G_i as follows:*

1. *a_m is an atom in G_i and is called the selected atom.*
2. *$a_m \leftarrow b_1, ..., b_q$ is a program clause in $\mathcal{P}$.*
3. *G_{i+1} is the goal $\leftarrow a_1, ..., a_{m-1}, b_1, ..., b_q, a_{m+1}, ..., a_n$*

SLD-derivations can be finite or infinite. An SLD refutation from $\mathcal{P}$ with top goal is a finite SLD derivation of empty set, from $\mathcal{P}$. A finite SLD-derivation can be successful or failed. A successful derivation is one that ends in the empty clause, that is, it is a refutation. A failed SLD-derivation is one that ends in a non-empty goal with the property that no atom in it occurs in the head of clauses in $\mathcal{P}$. SLD resolution is the system that uses SLD derivation as an inference mechanism.

Definition 5. *Let $\mathcal{P}$ be a definite program, G a goal and R a computation rule. Then the SLD-tree for $\mathcal{P} \cup \{G\}$ via R is defined as follows:*

1. *Each node of the tree is a goal(possibly empty).*
2. *The root node is G.*
3. *Let $\leftarrow a_1, ..., a_m, ..., a_k$ $(k \geq 1)$ be a node in the tree and suppose that a_m is the atom selected by R. Then this node has descendent for each input clause $a_m \leftarrow b_1, ..., b_q$. The descendent is*

$$\leftarrow a_1, ..., b_1, ..., b_q, ..., a_k$$

4. *Nodes which are empty clause have no descendents.*

Each branch of the SLD-tree is a derivation of $\mathcal{P} \cup \{G\}$. Branches corresponding to successful derivation are called success branches, branches corresponding to infinite derivation are called infinite branches and branches corresponding to failed derivation are called failure branches.

It has been proved that SLD-resolution is sound and complete wrt. definite programs. In propositional case, it can be stated as follows:

Theorem 5. *Let $\mathcal{P}$ be a definite logic program and $G = \leftarrow a_1, ..., a_n$ be a goal. Then $\mathcal{P} \models a_1, ..., a_n$ iff there is SLD-refutation for $\mathcal{P}$ and top goal G.*

Now we are in a position to present a proof procedure for negative query under minimal model semantics of disjunctive logic program based on CH-transformation and SLD-resolution. The following trivial facts can help us to understand the basic idea behind the method.

Proposition 2. *Let $\mathcal{P}$ be a disjunctive program. Considering CH-transformation of $\mathcal{P}$*

$$CH(\mathcal{P}) = \mathcal{P}_C + \mathcal{P}_H$$

denote $\bar{\mathcal{P}}_H$ the Horn program obtained from $\mathcal{P}_H$ by deleting all control variables. Then

1. *For any $a \in T_{\mathcal{P}_H} \uparrow \omega$, $\mathcal{P} \models_{min} a$.*
2. *For each $a \in T_{\bar{\mathcal{P}}_H} \uparrow \omega$, there is a model M of $\mathcal{P}$ such that a is true in M*
3. *If M is a minimal model of $\mathcal{P}$, then $M \subseteq T_{\bar{\mathcal{P}}_H} \uparrow \omega$*

To prove $\neg q$, assume R to be a computation rule that only selects atoms from $HB_{\mathcal{P}}$, we construct a SLD-tree T for $\mathcal{P}_H \cup \{\leftarrow q\}$. Let $\bar{T}$ be the tree obtained from T by deleting all control variables, then $\bar{T}$ is a SLD-tree for $\bar{\mathcal{P}}_H \cup \{\leftarrow q\}$. For a branch b in T let $\bar{b}$ denote the corresponding branch in $\bar{T}$. A branch b in T may be

1. a success branch. By 1 of Proposition 2, it means q is true in all minimal models of $\mathcal{P}$.
2. an infinite branch. Then $\bar{b}$ is also an infinite branch.
3. a failure branch and ends in a non-empty goal containing atoms in $\mathcal{P}$ (and also possibly some new propositional atoms from $C_\mathcal{P}$). This case corresponds to a failure branch in $\bar{T}$.
4. a failure branch but ends in a non-empty goal consisting of only new propositional atoms. Let the leaf node of the branch b be $\leftarrow N_1, ..., N_m$ $(m > 0)$. In this case b corresponds to a success branch $\bar{b}$ in $\bar{T}$, by 2. of Proposition 2, q is true in some model of $\mathcal{P}$. Notice the fact that N_i $(1 \leq i \leq m)$ indicates that the clause with N_i in its body have been used as input clauses in the derivation corresponding branch b, therefore for any model of $CH(\mathcal{P})$, if it contains N_1,...,N_m, then q must be true in it. Among them if there is a $(P, C_\mathcal{P})$-minimal model of $CH(\mathcal{P})$, then by Theorem 3 we can conclude that there is a minimal model of $\mathcal{P}$ in which q is true.

Hence, the problem of answering a query G is reduced to answering the following problem:

> Let S be a subset of $C_\mathcal{P}$, is there a $(P, C_\mathcal{P})$-minimal model $M = M_C \cup M_H$ of $CH(\mathcal{P})$ such that $S \subseteq M_C$? In other words, is there a model $M = M_C \cup M_H$ of $CH(\mathcal{P})$ such that $S \subseteq M_C$ and M_H is a minimal model of $\mathcal{P}$?

There are many ways to test the minimality of a model [Lu97,Nie96]. The following result is due to [Nie96]

Proposition 3. *[Nie96] Let $\mathcal{P}$ be a set of clauses. An interpretation M is a minimal model of $\mathcal{P}$ iff M is a model of $\mathcal{P}$ and for every atom a, $M \models a$ implies $\mathcal{P} \cup N_\mathcal{P}(M) \models a$, where $N_\mathcal{P}(M) = \{\neg a \mid a$ is an atom appearing in the head of a clause in $\mathcal{P}$ and $M \not\models a\}$.*

Now the problem can be solved as follows:

1. Computing all models containing S of $\mathcal{P}_C$, denote it by $MM(\mathcal{P}_C, S)$ (it can be done by any existing model generation procedure [Lu97,Nie96].
2. For each $M_C \in MM(\mathcal{P}_C, S)$, computing the least model of $\mathcal{P}_H(M_C)$[2] and testing if it is a minimal model of $\mathcal{P}$ using Proposition 3.

Let us summarize the procedure for minimal model reasoning.

Algorithm
Input: A disjunctive logic program $\mathcal{P}$ and an atom q.
Output: $\mathcal{P} \models_{min} \neg q$, if no minimal model contains q, otherwise either $\mathcal{P} \models_{min} q$ or a minimal model in which q is true.

1. Transform $\mathcal{P}$ into $\mathcal{P}_C + \mathcal{P}_H$.

[2] This can be done by Hyper tableau or the immediate consequence operator for Horn programs.

2. Construct a SLD-tree T for $\mathcal{P}_H$ and $\leftarrow q$
3. If there is a success branch in T then return $\mathcal{P} \models_{min} q$, otherwise
4. If there is failure branch ending with subgoal consisting of only control variables and if there is a $(P, C_{\mathcal{P}})$-minimal model $M_C \cup M_{\mathcal{P}}$ such that the subgoal is contained in M_C, then return $M_{\mathcal{P}}$, otherwise
5. return $\mathcal{P} \models_{min} \neg q$.

Before ending this section we illustrate the algorithm with an example.

Example 2. Consider the program $\mathcal{P}$ from example 1. Given goal $\leftarrow p$. A SLD-tree for the Horn program $\mathcal{P}_H$ and goal $\leftarrow p$ is depicted in Fig. 1. In this SLD-tree, there is a success branch, and hence we can conclude $\mathcal{P} \models_{min} p$.

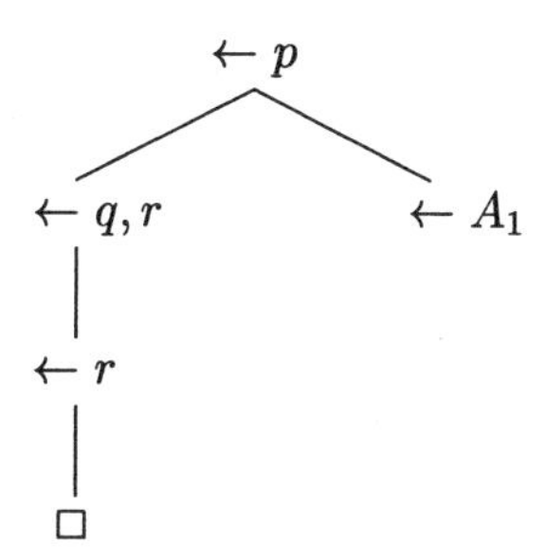

Fig. 1. The SLD-tree for $\mathcal{P}_H$ and $\leftarrow p$

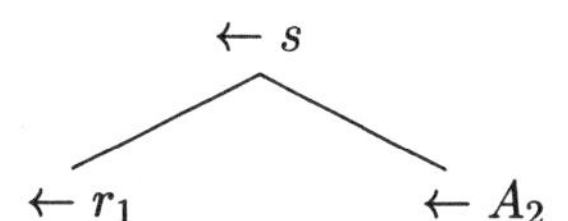

Fig. 2. The SLD-tree for $\mathcal{P}_H$ and $\leftarrow s$

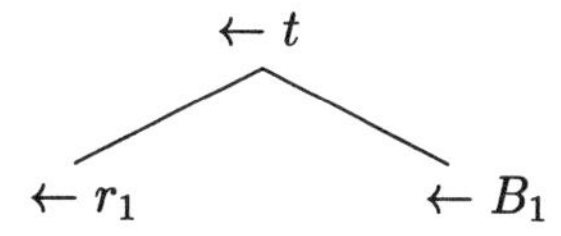

Fig. 3. The SLD-tree for $\mathcal{P}_H$ and $\leftarrow t$

Fig. 2 gives an SLD-tree for the Horn program $\mathcal{P}_H$ and the goal $\leftarrow s$. It contains no success branch but has a branch ending with a subgoal $\leftarrow A_2$. Therefore

a test is needed. There are two minimal models $\{A_2, B_1\}$ and $\{A_2, B_2\}$ of $\mathcal{P}_C$ containing A_2. The models of $CH(\mathcal{P})$ "controlled" by the two models are
$M_1 = T_{\mathcal{P}_H} \uparrow \omega(\{A_2, B_1\}) = \{A_2, B_1, r, p, q, s, t\}$ and
$M_2 = T_{\mathcal{P}_H} \uparrow \omega(\{A_2, B_2\}) = \{A_2, B_2, r, p, q, s, u\}$.
Both M_1 and M_2 are not $(P, C_{\mathcal{P}})$-minimal models, therefore we conclude $\mathcal{P} \models_{min} \neg s$.

Fig. 3 gives an SLD-tree for the Horn program $\mathcal{P}_H$ and goal $\leftarrow t$. It also contains no success branch, but it has a branch ending with a subgoal $\leftarrow B_1$. Therefore a test is needed. There are two minimal models $\{A_1, B_1\}$ and $\{A_2, B_1\}$ of $\mathcal{P}_C$ containing B_1, the models of $CH(\mathcal{P})$ "controlled" by the two models are
$M_1 = T_{\mathcal{P}_H} \uparrow \omega(\{A_1, B_1\}) = \{A_1, B_1, r, p, q, t\}$ and
$M_2 = T_{\mathcal{P}_H} \uparrow \omega(\{A_2, B_1\}) = \{A_2, B_1, r, p, q, s, t\}$
By test, M_1 is a $(P, C_{\mathcal{P}})$-minimal models and by Theorem 3, $M = \{r, p, q, t\}$ is a minimal model of P. So we return M as the output.

5 Conclutions and Further Work

In this paper we proposed a novel point of view on disjunctive logic programming. We gave a transformation which transforms a disjunctive logic program $\mathcal{P}$ into a special disjunctive logic program $CH(\mathcal{P}) = \mathcal{P}_C \cup \mathcal{P}_H$, where $\mathcal{P}_C$ is control program and $\mathcal{P}_H$ is a Horn program. Then the relationship between $\mathcal{P}$ and $CH(\mathcal{P})$ is established. Based on this relation we developed a minimal model reasoning procedure which differs from similar work in that the SLD-resolution can be directly incorporated into it, due to the transformation.

Many interesting topics remain to be done. Currently we are working on the following topics:

- Optimizing the query answering procedure. Many optimizations in the minimal test phase are posible. For example, the minimal model generation method based on E-hyper tableau [Lu97] can be used in the procedure to improve performence.
- Extending the idea for normal logic program under stable model semantics.
- Extending this method for general disjunctive logic program under stable model semantics.
- Extending this method for non-ground case.

Acknowledgements

The author would like to thank all the members of Artificial Intelligence Research Group at the University of Koblenz for their support and useful comments. Special thanks to Chandrabose Aravindan for his invaluable comments and encouragement. The first author is financially supported by the Graduate Scholarship of University of Koblenz.

References

[Ara96] Chandrabose Aravindan. An abductive framework for negation in disjunctive logic programming. In J. J. Alferes, L. M. Pereira, and E. Orlowska, editors, *Proceedings of Joint European workshop on Logics in AI*, number 1126 in Lecture Notes in Artificial Intelligence, pages 252–267. Springer-Verlag, 1996. A related report is available on the web from <http://www.uni-koblenz.de/~arvind/papers/>.

[BF94] Peter Baumgartner and Ulrich Furbach. PROTEIN: A *PRO*ver with a *T*heory *E*xtension *I*nterface. In A. Bundy, editor, *Automated Deduction - CADE-12*, volume 814 of *Lecture Notes in Artificial Intelligence*, pages 769–773. Springer, 1994. Available in the WWW, URL: `http://www.uni-koblenz.de/ag-ki/Systems/PROTEIN/`.

[BF97] Peter Baumgartner and Ulrich Furbach. Calculi for Disjunctive Logic Programming. In Jan Maluszynski, editor, *Logic Programming - Proceedings of the 1997 International Symposium*, Port Jefferson, New York, 1997. The MIT Press.

[BFN96] Peter Baumgartner, Ulrich Furbach, and Ilkka Niemelä. Hyper Tableaux. In *Proc. JELIA 96*, number 1126 in Lecture Notes in Artificial Intelligence. European Workshop on Logic in AI, Springer, 1996.

[BFS97] Peter Baumgartner, Ulrich Furbach, and Frieder Stolzenburg. Computing Answers with Model Elimination. *Artificial Intelligence*, 90(1–2):135–176, 1997.

[Gin89] Matthew L. Ginsberg. A circumscriptive theorem prover. *Artificial Intelligence*, 39:209–230, 1989.

[GPP89] M. Gelfond, H. Przymusinska, and T. Przymusinski. On the relationship between circumscription and negation as failure. *Artificial Intelligence*, 38:75–94, 1989.

[IKH92] K. Inoue, M. Koshimura, and R. Hasegawa. Embedding negation as failure into a model generation theorem prover. In *The 11th International Conference on Automated Deduction*, pages 400–415, Saratoga Springs, NY, USA, June 1992. Springer-Verlag.

[Lif85] V. Lifschitz. Computing circumscription. In *Proceedings of the 9th International Joint Conference on Artificial Intelligence*, pages 121–127, Los Angeles, California, USA, August 1985. Morgan Kaufmann Publishers.

[Llo87] J. W. Lloyd. *Foundations of Logic Programming.* Springer–Verlag, second extended edition, 1987.

[LMR92] Jorge Lobo, Jack Minker, and Arcot Rajasekar. *Foundations of disjunctive logic programming.* MIT Press, 1992.

[Lov91] D. Loveland. Near-Horn Prolog and Beyond. *Journal of Automated Reasoning*, 7:1–26, 1991.

[LRW95] D. Loveland, D. Reed, and D. Wilson. SATCHMORE: SATCHMO with RElevance. *Journal of Automated Reasoning*, 14:325–351, 1995.

[Lu97] Wenjin Lu. Minimal model generation based on e-hyper tableau. In Cristopher Habel Gerhard Brewka and Bernhard Nebel, editors, *Proceedings of KI'97*, number 1303 in Lecture Notes in Artificial Intelligence, pages 99–110. Springer-Verlag, 1997.

[MB88] Rainer Manthey and François Bry. SATCHMO: a theorem prover implemented in Prolog. In Ewing Lusk and Ross Overbeek, editors, *Proceedings of the 9th Conference on Automated Deduction, Argonne, Illinois, May 1988*, volume 310 of *Lecture Notes in Computer Science*, pages 415–434. Springer, 1988.

[McC80] J. McCarthy. Circumscription—a form of non-monotonic reasoning. *Artificial Intelligence*, 13:27–39, 1980.

[Min82] Jack Minker. On indefinite databases and the closed world assumption. In *Lecture Notes in Computer Science 138*, pages 292–308. Springer-Verlag, 1982.

[Nie96] I. Niemelä. A tableau calculus for minimal model reasoning. In *Proceedings of the Fifth Workshop on Theorem Proving with Analytic Tableaux and Related Methods*, pages 278–294, Terrasini, Italy, May 1996. Springer-Verlag.

[NNS95] A. Nerode, R.T. Ng, and V.S. Subrahmanian. Computing circumscriptive databases: I. theory and algorithms. *Information and Computation*, 116:58–80, 1995.

[Oli92] N. Olivetti. A tableaux and sequent calculus for minimal entailment. *Journal of Automated Reasoning*, 9:99–139, 1992.

[Prz89] T.C. Przymusinski. An algorithm to compute circumscription. *Artificial Intelligence*, 38:49–73, 1989.

[YH85] A. Yahya and L. J. Henschen. Deduction in non-horn databases. *Journal of Automated Reasoning*, 1(2):141–160, 1985.

Semantics of Partial-Order Programs

Mauricio Osorio

Departamento de Ingenieria en Sistemas Computacionales,
Universidad de las Americas, Puebla, Mexico
josorio@mail.udlap.mx

Abstract. Partial-order programming is introduced in [JOM95] where it is shown how partial-order clauses help render clear and concise formulations to a different kind of problems, in particular optimization problems. In this paper we present some more examples that we can model using partial-order clauses and we also introduce its Fix-Point semantics. We show that this paradigm and standard logic programming can be naturally integrated in one paradigm. We also discuss WFSCOMP, a new semantics for normal programs, that can be used to give the meaning of general normal+partial-order programs via a translation.

1 Introduction

In this paper, we study the semantics of a logic language whose principal building blocks are *partial-order + normal* program clauses and complete lattice data types. The motivation for our work, however, is very practical in nature: We claim that partial-order clauses and lattices help obtain clear, concise, and efficient formulations of problems requiring the ability to take transitive closures, solve circular constraints, and perform aggregate operations. In a way, our paradigm give us a high level logical notation to represent problems that are difficult to express using only normal clauses.

We present several examples that are naturally expressed in this paradigm.

Example 1.1 (Matrix chain product).
This is a well known example that can be efficiently solved using dynamic programming [Sti87]. Suppose that we are multiplying n matrices $M_1, ...M_n$. This program finds the minimum number of scalar multiplications required to compute the given task.

```
c(I,I) ≤ 0 :- size(N), 1≤I≤ N
c(J,K) ≤ c(J,I)+c(I+1,K) + r(J)*c(I)*c(K) :- J≤I≤ K-1
```

where we encode the size of matrix M_i by `r(I)` and `c(I)`, and we suppose that `c(I)=r(I+1)`. The functions `r,t` have to be provided as part of the code. The *logic* of the matrix chain product problem is very clearly specified in the above program. The program is well defined because the condition `J≤I≤ K-1` ensures that circularity is not present at all. The codomain of function c is the set of Natural numbers. The Natural numbers can be extended to a complete lattice by adding a top element $\top$. The function $+$ has to be extended such that for instance $n + \top = \top$. These details are transparent to the user. Here the bottom-up computation of the fix-point semantics behaves very much like the dynamic programming technique.

Example 1.2 from [JOM95] (Shortest Distance).

```
short(X,Y) ≤ C :- edge(X,Y,C)
short(X,Y) ≤ C+short(Z,Y) :- edge(X,Z,C)
```

Again, we think that the *logic* of the shortest-distance problem is very clearly specified in the above program. The program is well defined even if the extensional database defined by edge defines a directed graph with cycles.

Example 1.3 (The 0-1 knapsack problem).
This is a well known optimization problem that is known to be NP-complete. If however, the capacity, is of order p(n) for some polynomial p(n), the dynamic programming runs in time of order p(n)n, so we have a polynomial algorithm [Sti87].

$\mathtt{kn(I,M)} \geq \mathtt{0}$
$\mathtt{kn(I,M)} \geq \mathtt{kn(I-1,M)}$:- $\mathtt{I} \geq \mathtt{1}$
$\mathtt{kn(I,M)} \geq \mathtt{kn(I-1,M-c(I))+g(i)}$:- $\mathtt{I} > \mathtt{1,c(I)} \leq \mathtt{M}$

Example 1.4 from [JOM95] (Reach).
Sets also have a partial order. We consider finite sets and again we can complete them to ensure a complete lattice. The program works no matter edge defines a direct graph with cycles.

$\mathtt{reach(X)} \geq \{\mathtt{X}\}$
$\mathtt{reach(X)} \geq \mathtt{reach(Y)}$:- $\mathtt{edge(X,Y)}$

We refer the reader to the reference [JOM95] for more examples illustrating the use of set patterns and partial-order clauses. In [JO98a] we present an operational semantics that combines top-down goal reduction, *memo-tables* and a fix-point procedure for solving *functional-constraints.* One contribution of this paper is to observe that our high level notation allows us to use dynamic programming to solve this kind of problems, getting an efficient computation for them.

In [OJ97] we showed how to model these programs by translating them to standard normal clauses. In particular, we showed in [OJ97], that using only the stratified semantics we can capture the meaning of a large class of partial-order programs. This line of research is extended in further detail in [JO98b]. We present here a simple translation that do not require a program to become stratified. However, we have to define a new semantics to capture the intended meaning of the translated programs (since WFS, STABLE and COMP failed). We call this semantics WFSCOMP because it combines WFS and COMP (the two-valued Clark's completion semantics) in a suitable way. This semantics appears interesting in its own right as we will show.

Our paper is organized as follows. Section 1 is the introduction where we give several examples to motivate our approach. In section 2 we give the background, most of this material is taken from [JOM95]. In section 3 we present the fix-point semantics for partial-order programs. In section 4 we discuss the translation to normal programs and the use of WFSCOMP. Finally, we present our conclusions.

2 Background

We present the basic background of partial-order programs as introduced in [JOM95][1]. There are two basic forms of a partial-order clause[2]:

f(terms) $\geq$ *expression* :- *condition*
f(terms) $\leq$ *expression* :- *condition*

condition ::= *goal* | *goal, condition*
goal ::= *p(terms)* | $\neg$ *p(terms)*

where the predicate *p* appearing in *p(terms)* above is an *extensional database predicate*, i.e., one that is defined by ground unit clauses. Terms are made up of constants, variables, and data constructors, while expressions are in addition made up of user-defined functions, i.e., those that appear at the head of the left-hand sides of partial-order clauses. Informally, the declarative meaning of a partial-order clause is that, for all its ground instantiations (i.e., replacing variables by ground terms), the function *f* applied to argument terms is $\geq$ (respectively, $\leq$) the ground term denoted by the expression on the right-hand side, if *condition* is true. In general, multiple partial-order clauses may be used in defining some function *f*. We define the meaning of a ground expression *f(terms)* to be equal to the *least-upper bound* (respectively, *greatest-lower bound*) of the resulting terms defined by the different partial-order clauses for *f*. A program consists of a finite set of partial-order clauses and a finite extensional database.

We now present a model-theoretic semantics for partial-order clauses. For simplicity of presentation, we consider only $\geq$ clauses in this section; the treatment of $\leq$ clauses is symmetric. As noted before, we do not consider the definition of a function using a combination of $\leq$ and $\geq$ clauses. This is why the semantics of $\geq$ clauses can be given in a modular way, without any possibility of interference from $\leq$ clauses, and vice versa. We also consider functions with only one argument but our results carry out straightforward to the general case. Note, however, that this argument can be a general term (which could simulate a multi-argument function using a list), and hence there is no loss of generality by this assumption.

In preparation for the semantics, we first reduce the program respect to the extensional database and then we use the flattened form for all reduced clauses and goals. We use an example to explain the reduction. Let E be as follows:

```
edge(a,b,2)
edge(a,c,3)
```

Then the reduction of program **Shortest distance** respect to `edge` becomes:

`short(a,b)` $\leq$ 2

[1] Definitions and Propositions are taken from [JOM95]
[2] The operational semantics introduced in [JOM95] we also need that where each variable in *expression* also occurs in *terms*

```
short(a,c) ≤ 3
short(a,Y) ≤ 2+short(b,Y)
short(a,Y) ≤ 3+short(c,Y)
```

The idea of flattening has been mentioned in several places in the literature [G*87,Hol89,Jan94,JJ97]. We follow the definition given in [Jan94], and we illustrate it by a simple example.

Example 2.1. Assuming that `f`, `g`, `h`, and `k` are user-defined functions and the remaining function symbols are constructors, the flattened form of a clause

```
f(c(X,Y)) ≥ c1(c2(g(c3(X)), k(d1(h(d2(Y,1)))))).
```

is as follows:

```
f(c(X,Y)) ≥ c1(c2(Y1, Y3)) :- g(c3(X)) = Y1, h(d2(Y,1)) = Y2,
k(d1(Y2)) = Y3.
```

In the above flattened clause, we follow Prolog convention and use the notation `:-` for 'if' and commas for 'and'. All variables are to understood to be universally quantified at the head of the clause, as is customary for definite clauses.
The general form of a flattened clause is

Head :- *Body*

where *Head* is $\mathtt{f}(t) \geq u$, and t and u are terms, and *Body* is of the form $\mathrm{E}_1, \ldots, \mathrm{E}_n$. Each E_i is $\mathtt{f}_i(t_i) = y_i$, where each $\mathtt{f}_i$ is a user-defined function symbol, each y_i is a new variable not present in *Head*, and each t_i is a term that is equivalent to the argument of $\mathtt{f}_i$ in the original, unflattened program clause. Each formula $\mathtt{f}_i(t_i) = y_i$ in *Body* is called a *basic goal*, and a sequence of basic goals is called a *goal sequence*.

The order in which the basic goals are listed on the right-hand side of a flattened clause is the *leftmost-innermost* order for reducing expressions [Man74].

Finally, note that the flattened form of a query is similar to that of *Body*. In order to capture the $\perp$-as-failure assumption, we assume that for every function symbol f in P, the program is augmented by the clause: $\mathtt{f(X)} \geq \perp$.

2.1 Model-Theoretic Semantics

We will work with Herbrand interpretations, where the Herbrand Universe of a program P consists only of ground terms, and is referred to as U_P. The Herbrand Base B_P of a program P consists of ground equality atoms of the form $f(t) = u$, where f is a user-defined function, t is a ground term, and u is a ground term belonging to some complete-lattice domain[3]. Henceforth, we will always use the symbol f to stand for a user-defined (i.e., non-constructor) function symbol.

We develop the model-theoretic semantics without reference to the details of specific lattice domains, such as sets, numbers, etc. This allows our presentation to focus on the essentials of partial-order clauses without digresssing to discuss the axiomatizations (equational theories) of specific data domains. A full treatment of the logical foundations of the set constructors described in section

[3] Since the program is reduced by the extensional database we do not have to include atoms of the form $p(t)$.

1 is given in [Jan94,JJ97], and we refer the reader to these sources for more information. However, in giving examples to illustrate certain points about the semantics, we will need to make use specific data domains. An intuitive understanding of these domains suffices for the examples.

Due to the equational theories for constructors, the predicate = defines an equivalence relation over the Herbrand Universe. But, we can always *contract* a model to a so-called normal model where = defines only an identity relation [Men87] as follows: Take the domain D' of I to be the set of equivalence classes determined by = in the domain U_P of I. Then use Herbrand $\equiv$-interpretations, where $\equiv$ denotes that the domain is a quotient structure. We then should refer to elements in D' by $[t]$, i.e. the equivalence class of the element t, but in order to make the text more readable, we will refer to the $[t]$ elements just as t, keeping in mind that formally we are working with the equivalence classes of t. These details are explained in [Jan94,JJ97].

We assume that every interpretation I includes certain equality and inequality atoms of the form $t_1 = t_2$ and $t_1 < t_2$ according to the fixed intended interpretation of them in the program.

We also assume that, in every interpretation I, f is interpreted as a total function, i.e.,

(i) $(\forall t \in U_P)(\exists u \in U_P)\ f(t) = u \in I$; and

(ii) $f(t) = t_1 \in I$ and $f(t) = t_2 \in I \ \Rightarrow\ t_1 = t_2$.

Definition 1 ([JOM95]).
Let P be a program. An interpretation M is a model of P, denoted by $M \models P$, if for every ground instance, $\mathtt{f}(t) \geq t_1$:- $E_1 \ldots E_k$, *of a $\geq$ clause in P, if $\{E_1 \ldots E_k\} \subseteq M$ then exists an atom $f(t) = u \in M$ and $u \geq t_1$.*

We first briefly motivate our approach to the model-theoretic semantics. Basically, we define the semantics of a function call $f(t)$, where t is a ground term, to be the *glb* (*greatest lower bound*) of all terms defined for $f(t)$ in the different Herbrand models for f (the definition of model is given below). To see we need to take such *glb*s, consider the following trivial program P:

```
f(X) ≥ 1
```

Here, we assume that the result domain for `f` is the lattice of totally-ordered numbers, N: $0 \leq 1 \leq 2 \leq \ldots\top$, for some $\top$. Each model of P interprets `f` as a constant function:

`f(X) = 1`, for all `X` $\in U_P$
`f(X) = 2`, for all `X` $\in U_P$
⋮
`f(X) =` $\top$, for all `X` $\in U_P$

The intended model for function `f`, namely, `f(X) = 1`, for all `X` $\in U_P$, is obtained not by the classical set-intersection ($\cap$) of all models, but instead by the $\sqcap$ of the terms defined for $\mathtt{f}(t)$ in the different models. In the above example, $\sqcap$ is, of course, the *min* operator on numbers.

But not all syntactically well-formed programs have a well-defined meaning. Circularity in function definitions is allowable as long as this occurs through *monotonic* functions. Consider the following program over the boolean lattice:

$a \geq c(b)$
$b \geq c(a)$
$c(\perp) \geq \top$

This program has several models and no one with an intended meaning. Thus, non-monotonic functions are permissible as long as there are no circular definitions through such functions. This motivates our interest in stratified partial-order programs. We begin the discussion of this topic with strongly-stratified programs, defined below, and continue the discussion with cost-monotonic programs in section 2.2.

Definition 2 ([JOM95]).
A program P is strongly-stratified if there exists a mapping function, level : $F \rightarrow \mathcal{N}$*, from the set F of user-defined (i.e., non-constructor) function symbols in P to (a finite subset of) the natural numbers* $\mathcal{N}$ *such that:*
(i) All clauses of the form
$f(term) \geq term$
are permitted
(ii) For a clause of the form
$f(term) \geq g(expr)$
where f and g are user-defined functions, level(f) is greater or equal to level(g) and level(f) is greater than level(h), where h is any user-defined function symbol that occurs in expr.
(iii) no other form of clause is permitted

Note that we have given the above definition using the non-flattened form of program clauses because the definition is easier to understand this way. Although a program can have different level mappings we assume that we select one that has as image a set of consecutive natural numbers that includes 1. For example, in the `reach` program shown before, the function `edge` would be at level one, and the function `reach` would be at level two. The above definition of stratification is, in another sense, very restrictive: it requires a function at any level to be directly defined in terms of other functions at the same level. For instance, the programs in examples 2.5 and 2.6 are not strongly-stratified. We therefore relax this requirement in section 2.3 wherein we introduce *cost-monotonic programs.*

Definition 3 ([JOM95]).
Let P be a set of strongly-stratified program clauses. We define P_k *as those clauses of P for which the user-defined function symbols on the left-hand sides have level* $\leq k$*.*

Definition 4 ([JOM95]).
Given two interpretations I and J for a program P, we define $I \sqsubseteq J$ *if for every* $\mathtt{f}(t) = t_1 \in I$ *there exists* $\mathtt{f}(t) = t_2 \in J$ *such that* $t_1 \leq t_2$ *. We say* $I = J$ *if* $I \sqsubseteq J$ *and* $J \sqsubseteq I$*.*

We will construct the model-theoretic semantics of a strongly-stratified program level by level. Thus, in defining models at some level $j > 1$, all functions from levels $< j$ will have their models uniquely specified. Hence, all interpretations of clauses at some level j will contain the same atoms for every function from a level $< j$. For this reason, we will overload the meaning of the function *level* and use the notation $level(A)$ to refer to the level of the head function symbol of atom A:

Definition 5 ([JOM95]).
For any interpretation I, $I_k := \{\ A : A \in I \wedge level(A) \leq k\}$.

Definition 6 ([JOM95]).
For any two interpretations I and J of a program P,
$I \sqcap J := \{f(t) = u \sqcap u' \ : \ f(t) = u \in I, \ \ f(t) = u' \in J, \ f$ *a function symbol of* $P, \ t \in U_P\ \}$

Definition 7 ([JOM95]).
For any set X of interpretations, $\sqcap X$ is the natural generalization of the previous definition.

Proposition 1 ([JOM95]).
Let X be a set of models for a program P with j levels such that for any $I \in X$ and $J \in X$, $I_{j-1} = J_{j-1}$. Then $\sqcap X$ is also a model.

Definition 8 ([JOM95]).
Given a program P with j levels, we define the model-theoretic semantics of P as:

for $j = 1$, $\mathcal{M}(P_1) := \sqcap\{M : M \models P_1\}$, and
for $j > 1$, $\mathcal{M}(P_j) := \sqcap\{M : M_{j-1} = \mathcal{M}(P_{j-1})$ and $M \models P_j\}$.

Definition 9 ([JOM95]).
Given a program P with j levels and a goal sequence G, we say that substitution θ is a correct answer for G if $\mathcal{M}(P_j) \models G\theta$.

2.2 Cost-Monotonic Programs

The *strongly-stratified language* defined in section 2.1 permits the definition of one function directly in terms of another function at the same level or lower level. However, the *cost-monotonic language* defined below permits the definition of one function in terms of another function at the same level using *monotonic* functions. In the following definitions, as before, we assume functions with one argument.

Definition 10. *A function f is monotonic if $t_1 \leq t_2 \Rightarrow f(t_1) \leq f(t_2)$.*

Definition 11 ([JOM95]).
A program P is cost-monotonic if there exists a mapping function, $level : F \rightarrow$

$\mathcal{N}$, from the set F of user-defined (i.e., non-constructor) function symbols in P to (a finite subset of) the natural numbers $\mathcal{N}$ such that:
(i) Every clause as defined in 2.2 is permitted. A clause of this form is called S-S clause (S-S stands for strongly-stratified).
(ii) For a clause of the form
$f(terms) \geq m(g(expr))$
where m is a monotonic function, $level(f)$ is greater than $level(m)$, $level(f)$ is greater or equal to $level(g)$ and $level(f)$ is greater than $level(h)$, where h is any function symbol that occurs in $expr$. A clause of this form is called a G-S clause (G-S stands for general-stratified).
(iii) no other form of clause is permitted.

In the above definition, note that f and g are not necessarily different. Also, non-monotonic "dependence" occurs only with respect to lower-level functions. We can in fact have a more liberal definition than the one above: First, since a composition of monotonic functions is monotonic, the function m in the above syntax can also be replaced by a composition of monotonic functions. Second, it suffices if the *ground instances* of program clauses are stratified in the above manner. This idea is, of course, analogous to that of *local stratification* [Prz88], except that we are working with functions rather than predicates. It should be clear that the presence of monotonic functions does not call for any alteration of the model-theoretic semantics.

Finally, we would like to note that in general it is not decidable that we can syntactically check whether a function definition is monotonic. For certain domains, such as sets, is possible to detect the monotonicity property in many (but not all) cases by a simple syntactic check.

3 Fix-Point Semantics

The set of interpretations defines a complete lattice, where the bottom interpretation is the one where every function evaluates to $\perp$ and the top interpretation is the one where every function evaluates to $\top$. These interpretations are ordered as given in definition 2.4. We provide a fix-point characterization of the declarative semantics. We will define a $\mathcal{T}_P$ operator (monotonic over a program layer with respect to $\sqsubseteq$) that maps interpretations of a given program layer into interpretations for the same program layer, where the fix-point semantics of lower level subprograms has been already computed. Unless stated otherwise, we assume that every program is cost-monotonic.

Definition 12. *The definition of $\mathcal{T}_P$ is as follows:*
$\mathcal{T}_P(I) := \{f(t) = s$ *: f is a functional symbol, t is ground term, and* $s^I := gts(f(t), I)\}$
where
$gts(e, I) := lub(\{t_1^I : e \geq t_1$ *:- cond* $\in P$ *and cond is true in* $I\}$

Even though the mapping $\mathcal{T}_P$ is, in general, not monotonic, it does have an important property similar to monotonicity for stratified normal programs, as described in [Llo87]. The property is the following.

Lemma 1. *Let P be a program. We define its Fix-Point Semantics, denoted by $\mathcal{F}(P)$, as follows:*
Suppose $P = P_1$, then $\mathcal{T}_P$ is monotonic over the lattice of interpretations for P, and so $\mathcal{F}(P) := LFP(\mathcal{T}_P)$ is well defined. Where LFP denotes the least fix-point.

Suppose $P = P^{\leq}_{k+1}$, $k \geq 1$. Let
$\Lambda_P := \{ I$: I is an interpretation for P, where $I_k = \mathcal{F}(P_k)$
Let $\mathcal{F}(P) := LFP(\mathcal{T}_P)$, ($\mathcal{T}_P$ over Λ_P)

It is not hard to see that $\mathcal{F}$ is well defined.

Theorem 1. *For any program P, $\mathcal{M}(P) = \mathcal{F}(P)$.*

To compute the fix-point semantics, we can use the naive approach to compute the stratified semantics (very much as it is standard in normal programs). That is, we compute the stratified semantics level by level. To compute the stratified semantics at a given level we start with the bottom interpretation and then we iterate $\mathcal{T}_P$ to compute the fix-point semantics. Not always we arrive to the fix-point semantics in a finite number of steps, as the following program shows:
$\mathsf{a} \geq 1 + \mathsf{a}$
Clearly, only at ω number of steps we arrive to the least fix-point, i.e. to $\mathsf{a} = \top$. However, every program that we have tried that comes from a "real" problem was solved in a finite number of steps.

The naive evaluation is a bottom-up strategy which follows directly the Fix-Point semantics that depends on applying the $\mathcal{T}_P$ operator in an iterative way. The seminaive method which uses the same approach as naive evaluation but is applied to generate new tuples, here corresponds to apply the evaluation only to update the tuples. This strategy is nothing else than dynamic programming [Sti87].

3.1 Fix-Point Semantics of Stratified Normal+Partial-Order Programs

We consider here the integration of normal clauses and partial-order clauses. For that, we extend the goals that can occur in *condition* to include equational assertions of the form $f(terms) = X$, but the level of f should be less than the level of the function in the head of the given clause. We also accept *normal clauses* as defined in [Llo87] but now they can also include equational assertions in the body of the clause. Again, the level of a functional symbol in an equational assertion that occurs in the body should be less than the level of the predicate symbol that occurs in the head of the clause. We point out that now our interpretations include functional atoms as well as the usual predicate atoms.

Definition 13. *Given an interpretation I, we define $I^f := \{A(t) = s \in I \mid$ and A is a functional symbol $\}$. And we define I^r to be the complement of I^f respect to I, that is I^r consists of predicate atoms.*

Definition 14. *Given two interpretations I, J, define $I \sqsubseteq_{\subseteq} J$ if $I^f \sqsubseteq J^f$ and $I^r \subseteq J^r$.*

Note that $\sqsubseteq_{\subseteq}$ defines a partial order on the set of interpretations of any given program.

Definition 15. *We say that a model is minimal if it is minimal under the partial order $\sqsubseteq_{\subseteq}$.*

As an example consider the following program:

```
p ≥ { X } :- q(X).
q(1).
q(2) :- q(2).
```

$I := \{\mathtt{q}(1), \mathtt{q}(2), \mathtt{p} = \{1, 2\}\}$ is an interpretation of the program, where $I^f := \{\mathtt{p} = \{1, 2\}\}$ and $I^r := \{\mathtt{q}(1), \mathtt{q}(2)\}$. Moreover, I is a model of the program. Another model of the program is $J = \{\mathtt{q}(1), \mathtt{p} = \{1\}\}$ Note that $J \sqsubseteq_{\subseteq} I$ and J is a minimal model. The program is considered stratified, where the definition of q defines the first level and the definition of p defines the second level. So, to compute the stratified semantics we first compute the minimal model for:

```
q(1).
q(2) :- q(2).
```

which is $\{\mathtt{q}(1)\}$. We do this using the standard monotonic operator as defined in [Llo87]. Then we compute the minimal model of:

```
p ≥ { X } :- q(X).
```

where the semantics of q is known. We do this by using the operator $\mathcal{T}_P$ defined in this section. We obtain then J, as the stratified semantics of this program. We think that any reader with some familiarity with the Fix-Point Semantics of standard logic programs can figurated out the final details of the complete formalization of the Stratified Normal+Partial-Order programs.

4 Semantics based on normal programs

The strategy here is to translate a normal+partial-order program to a standard normal program and then to define the semantics of the translated normal program as the semantics of the original program. We have studied this problem in detail in [OJ97] but we consider here what we claim is the most natural translation. Under this translation, none of the well known semantics gives the intended meaning to our programs. So, we define a new semantics that, as we will show, is interesting itself. A main result is that this semantics gives the intended meaning to our stratified normal+partial-order program. We work in this section with the *normal form* of a program. This form is obtained from the flattened form by

replacing every assertion of the form $f(t) = t1$ by the atom $f_{=}(t, t1)$ and every assertion of the form $f(t) \geq t1$ by $f_{\geq}(t, t1)$.
Except for minor changes, the following four definitions are taken from [OJ97].

Definition 16. *Given a stratified normal+partial-order program P, we define P' to be as follows: Replace each partial-order clause of the form*
E_0 :- *condition*, $E_1, \ldots, E_k, \ldots, E_n$
by the clause
E_0 :- *condition*, $E_1, \ldots, E_k^*, \ldots, E_n$
where E_0 is of the form $f_{\geq}(t_1, X_1)$, E_k is of the form $g_{=}(t_k, X_k)$, E_k^* is of the form $g_{\geq}(t_k, X_k)$ and f and g are (not necessarily different) functions at the same level^P. Note that when a clause is strongly-stratified we have $k = n$.

Definition 17. *Given a program P, we define head(P) to be the set of head functional symbols of P, i.e., the head symbols on the literals of the left-hand sides of the partial-order clauses.*

Definition 18. *Given a program P, a predicate symbol $\mathtt{f}_{\geq}$ which does not occur at all in P, we define $ext_1(\mathtt{f})$ as the following set of clauses:*

$\mathtt{f}_{=}$(Z, S) :- $\mathtt{f}_{\geq}$ (Z, S), $\neg$ $\mathtt{f}_{better}$(Z,S)
$\mathtt{f}_{better}$(Z, S) :- $\mathtt{f}_{\geq}$(Z,S1), S1 > S
$\mathtt{f}_{\geq}$(Z, S) :- $\mathtt{f}_{\geq}$(Z,S1), S1 > S
$\mathtt{f}_{\geq}$ (Z, $\perp$)
$\mathtt{f}_{\geq}$(Z,C) :- f $_{\geq}$(Z,C_1), $\mathtt{f}_{\geq}$(Z,C_2), lub(C_1,C_2,C).

The first two clauses are given in [Van92]. We call the last clause, the *lub* clause, and it is omited when the partial order is total. And lub(C_1, C_2, C) interprets that C is the least upper bound of C_1 and C_2. Symmetric definitions have to be provided for $f_{\leq}$ symbols.

Definition 19. *Given a stratified normal+partial-order program P, we define*
$ext_1(P) := \bigcup_{f \in\ head(P)} ext_1(f)$, *and*
$transl_1'(P) := P' \cup ext_1(P)$,

The following basic result is given in [OJ97].

Proposition 2. *For any stratified normal+partial-order program P, $transl_1'(P)$ is stratified.*

As an example of the translation we use program **Reach** given in example 1.4. The relevant clauses of the translated program are:

reach$_{\geq}$(X,Y) :- scons(X, $\emptyset$, Y)[4]
reach$_{\geq}$(X,Z) :- edge(X,Y), reach$_{\geq}$(Y,Z)
reach$_{\geq}$(X,$\emptyset$)
reach$_{\geq}$(Z, S) :- reach$_{\geq}$(Z,S1), S1 > S
reach$_{\geq}$(Z,S) :- reach$_{\geq}$(Z,S1), $\mathtt{f}_{\geq}$(Z,S2), union(S1,S2,S)

[4] To get rid of the set-constructor that has a variable as an argument in reach$_{\geq}$(X, {X})

$\texttt{reach}_{=}\texttt{(Z, S)} \;\texttt{:-}\; \texttt{reach}_{\geq}\texttt{(Z, S)},\; \neg\, \texttt{reach}_{better}\texttt{(Z,S)}$
$\texttt{reach}_{better}\texttt{(Z, S)} \;\texttt{:-}\; \texttt{reach}_{\geq}\texttt{(Z,S1)},\; \texttt{S1} > \texttt{S}$

The following definition is given in [OJ97]. It can be used as a definition of the declarative semantics of a stratified normal+partial-order program.

Definition 20. *For any stratified normal+partial program P, we define D(P), as the stratified model for* $transl_1'(P)$.

Consider the program P: $\texttt{a} \geq \texttt{1 + a}$.
Note that the model-theoretic semantics of P (as defined in section 2) defines that a=$\top$. On the other hand, the stratified model of $transl_1'$(P) interprets a as a partial function, that is, for every ground term t, $\texttt{a}_{=}\texttt{(t)}$ is false in D(P). This is the only "generic" example where both approaches give a different answer. The following is a basic result of this paper and explains formally the above informal claim.

Proposition 3. *Let P be a cost-monotonic program. If every function is defined total in D(P), then* $D(P) = \mathcal{M}(P)$ *restricted to the common language.*

From a computational point of view, the behavior of D(P) is more realistic than the behavior of $\mathcal{M}(P)$.

The problem with this approach (meaning the use of D(P)) is that it only works with cost-monotonic programs. It has been shown that non cost-monotonic programs sometimes make sense. So, we need a translation that works for a larger class of programs. We consider a more direct translation than $transl_1'$, that we will call $transl_1$. Both translations are very similar and closed variants of them have been studied in [OJ97,Van92].

Definition 21. *Given P, we define* $transl_1(P) := P \cup ext_1(P)$.

For our program **Reach** the new translation is as before, but we replace the clause:

$\texttt{reach}_{\geq}\texttt{(X,Z)} \;\texttt{:-}\; \texttt{edge(X,Y)},\texttt{reach}_{\geq}\texttt{(Y,Z)}$

by the clause:

$\texttt{reach}_{\geq}\texttt{(X,Z)} \;\texttt{:-}\; \texttt{edge(X,Y)},\texttt{reach}_{=}\texttt{(Y,Z)}$

Before we test this (more basic) translation with non cost-monotonic programs, we should do it with cost-monotonic programs. This new translation does not always give a stratified program from a cost-monotonic program. We showed in [OJ97] that neither WFS or STABLE defines the intended semantics for it. The apparently simple program **Reach**, becomes a difficult one to have a well defined semantics after the translation. The problem arises when edge induces a graph with cycles as for instance: $\mathbf{Edge}_2$:={edge(1,2), edge(2,3), edge(3,2)}. The stable semantics fails with **Reach** $\cup$ $\mathbf{Edge}_2$ because it gives no stable models at all. The well-founded semantics defines a non total model to the program. The problem is not as serious as with the stable semantics since it has a partial model consistent with the intended interpretation. But note that the WFS agrees in its true/false assignments with the intended model. Some undefined values for reach(2) are:

$\texttt{reach}(2) \geq \{3\}, \texttt{reach}(2) \geq \{2,3\}, \texttt{reach}(2) = \{2\}, \texttt{reach}(2) = \{2,3\}$.
An interesting point is that the well-founded model agrees with the intended model in the assignments of many false values. For instance, $\texttt{reach}_{\geq}(2, \{1\})$ is false in the partial model. This "decision" pruned all the unacceptable "large" models of the program.

What about the Clark's completion? How does it behaves with respect to our program **Reach**? The Herbrand models of the completion of our example are the following:

1. The intended model, i.e., when $\texttt{reach}_{=}(2, \{2,3\})$.
2. Models where there exists s such that $\texttt{reach}_{=}(2, \texttt{s})$ is true, but excluding the intended model. There is a model of this kind for each s such that s $> \{1,2\}$ is true. Call this class of models C_d.
3. Models where for every s, $\texttt{reach}_{=}(2,\texttt{s})$ is false. Let M be a model of this kind. Then, there exists a non-terminating sequence of ground terms s_0, s_1, ..., s_i, such that $s_i < s_{i+1}$ and $\texttt{reach}_{\geq}(2,s_i)$ is true. Call this class of models C_u. To see that the claim is true, we take $s_0=\perp$. Then $\texttt{reach}_{\geq}(2,s_0)$ is true, and by hypothesis $\texttt{reach}_{=}(2,s_0)$ is false, i.e., $\neg\ \texttt{reach}_{=}(2,s_0)$ is true. By the clause that defines $\texttt{reach}_{=}$, we get that $\texttt{reach}_{better}(2,s_0)$ is true and so , by the converse of the caluse that defines $\texttt{reach}_{better}$, it exists s_1 such that, $s_0 < s_1$, $\texttt{reach}_{\geq}(2,s_1)$ is true. We now can apply the same argument taking s_1 instead of s_0 to show that there exists s_2 such that, $s_1 < s_2$, $\texttt{reach}_{\geq}(2,s_2)$ is true. We can apply the argument forever.

So, we have many more models ($C_u \cup C_d$ in this case) than expected. This is because negation is weak for COMP. Next, we show how to combine WFS and COMP to get the intended model. We remind the reader the following well known fact: Give a program P, M is model of COMP(P) iff M is a supported model of P.

Definition 22. *We define the semantics WFSCOMP(P) as the set of literals that are true in every two-valued supported Herbrand model that extends WFS(P). Any such extension is a model that agrees with the true/false assigments given by WFS. If no such model exists then WFSCOMP(P):=WFS(P).*

The following is another result of this section. It is based on the observation that under the given conditions D(P) is the unique model of COMP(transl_1(P)) that extends WFS(transl_1(P)).

Theorem 2. *For every stratified normal+partial-order program P, such that every function is defined total in D(P), D(P)=WFSCOMP($transl_1$(P)).*

We conjecture that the condition *every function is defined total in D(P)* is in fact no required here. But it is direct to see that the theorem is true under the given condition.

What can we say about the behavior of WFSCOMP for general normal programs?

Let us consider the following example considered elsewhere, which is representative for the problems with *reasoning by cases.* Let P be

$a \leftarrow \neg b$

$b \leftarrow \neg a$

$p \leftarrow a$

$p \leftarrow b$

Several authors have argued that since neither a, nor b can be derived in any semantics based on two-valued models the disjuntion $a \vee b$, thus also p should be true. WFS(P) does not fulfill this point. STABLE as well as COMP derive p. So thus our proposed semantics.

STABLE on the other hand is inconsistent for many programs, this occurs when either STABLE has several stable models but even worst when it lacks of stable models. Consider the following program P:

$b \leftarrow a$

$a \leftarrow b$

$a \leftarrow \neg b$

Then P does no have any stable model. One can argue by reasoning by cases (on b) that a should be a consequence of P. Once accepted this fact we observe that b is also a consequence of the program. So, the intended model of P is $\{a, b\}$. This is what COMP(P) defines, as well as our proposed semantics.

We argue that COMP is very weak to infer negative literals but not so to infer positive literals. On the other hand, WFS is strong enough to infer many negative literal but very weak to infer positive atoms. In general, STABLE derives many literals and so it becomes inconsistent in cases where we argue that there is an intended model. We now see a typical (well known) example where COMP fails to give the intended semantics of a program.

`edge(a,b).`

`edge(c,d).`

`reachable(a).`

`reachable(X)` ← `reachable(Y), edge(Y,X).`

`unreachable(X)` ← ¬`reachable(X).`

Here, `edge(a,b)` means that there is a directed edge from `a` to `b`.

We obviously expect vertices `c,d` to be unreachable, and indeed, Clark's semantics implies it, i.e.,

comp(P)$\models$ `unreachable(c)` and

comp(P)$\models$ `unreachable(d)`

Suppose we add to P the clause `edge(d,c)` and call the resulting program P'. Although we still expect that c and `d` are to be unreachable, the Clark's semantics of P' does not imply that `c` and `d` are unreachable. This example illustrates well why COMP is weak to infer negative literals. Our proposed semantics on the other hand gives the intended model. We consider that WFSCOMP is a good combination of WFS and COMP.

5 Conclusions

We introduced the fix-point semantics of partial-order programs. The seminaive method to evaluate the fix-point corresponds basically to dynamic programming. In this way we see that it is possible to obtain an efficient computation of partial order programs. We saw that this paradigm can be integrated with standard logic programming. We also defined a new declarative semantics of normal programs that in general appears promising. It can be used to give a semantics of normal+partial-order programs, using a direct translation of these class of programs to normal programs. Our general claim is that partial-order programming is useful and should be integrated to logic programming.

References

[G*87] G. Levi, C. Palamidessi, P.G. Bosco, E. Giovannetti, and C. Moiso, "A Complete Semantic Characterization of K-Leaf: A Logic Language with Functions," *Proc. 5th Intl. Conf. on Logic Programming*, Seattle, pp. 993–1005, MIT Press, August 1988.

[Hol89] S. Hölldobler, *Foundations of Equational Logic Programming*, LNAI 353, Springer-Verlag, 1989.

[Jan94] D. Jana, *Semantics of Subset Logic Languages*, Ph.D. dissertation, Department of Computer Science, SUNY-Buffalo, August 1994.

[JJ97] D. Jana and B. Jayaraman, "Set Constructors, Finite Sets, and Logical Semantics," accepted for publication in *Journal of Logic Programming.*

[JO98a] B. Jayaraman and M. Osorio "Theory of Partial-Order Programming", accepted for publication in *Science of computer programming journal.*

[JO98b] B. Jayaraman and M. Osorio "Relating Aggregation and Negation-as-Failure", accepted for publication in *New Generation Computing.*

[JOM95] B. Jayaraman, M. Osorio and K. Moon, "Partial Order Programming (revisited)", *Proc. Algebraic Methodology and Software Technology*, pp. 561-575. Springer-Verlag, July 1995.

[Llo87] J.W. Lloyd, *Foundations of Logic Programming* (2 ed.), Springer-Verlag, 1987.

[Man74] Z. Manna, *A Mathematical Theory of Computation*, McGraw-Hill Publishers, 1974.

[Men87] E. Mendelson, *Introduction to Mathematical Logic*, 3nd ed., Wadsworth, 1987.

[OJ97] M. Osorio and B. Jayaraman, "Aggregation and Well-Founded Semantics$^+$," *Proc. 5th Intl. Workshop on Non-Monotonic Extensions of Logic Programming*, pp. 71-90, LNAI 1216, J. Dix, L. Pereira and T. Przymusinski (eds.), Springer-Verlag, 1997.

[Prz88] T. Przymusinski, "On the Declarative Semantics of Stratified Deductive Databases and Logic Programs," *Proc. Foundations of Deductive Databases and Logic Programming*, J. Minker (ed.), pp. 193-216, Morgan-Kaufmann, 1988.

[Sti87] D.R. Stinson, "An introduction to the Design and Analysis of Algorithms", *Winnipeg, Manitoba, Ca.*, 1987.

[Van92] A. Van Gelder, "The Well-Founded Semantics of Aggregation", *Proc. 9th ACM Symp. on Principles of Database Systems*, 1990, pp. 205-217.

Persistence and Minimality in Epistemic Logic

Wiebe van der Hoek[1], Jan Jaspars[2], and Elias Thijsse[3]

[1] Computer Science, Utrecht University, the Netherlands
wiebe@cs.ruu.nl
[2] Computer Science, University of Amsterdam, the Netherlands
jaspars@wins.uva.nl
[3] Faculty of Arts, Tilburg University, the Netherlands
thysse@kub.nl

Abstract. We give a general approach to characterizing minimal information in a modal context. Our modal treatment can be used for many applications, but is especially relevant under epistemic interpretations of the operator $\Box$. Relative to a modal system **S**, we give three characterizations of minimality of a formula φ and give conditions under which these characterizations are equivalent. We then argue that rather than using bisimulations, it is more appropriate to base information orders on Ehrenfeucht-Fraïssé games to come up with a satisfactory analysis of minimality. Moving to the realm of epistemic logics, we show that for one of these information orders almost all systems trivialize, i.e., either all or no formulas are honest. The other order is much more promising as it permits to minimize wrt positive knowledge. The resulting notion of minimality coincides with well-established accounts of minimal knowledge in **S5**. For **S4** we compare the two orders.

1 Introduction

This paper offers a general account to the issue of *minimal informational content* of modal assertions. This issue is perhaps most prominent in the area of modal *epistemic* logic [6] where researchers have addressed questions like what it means that one agent knows *more* than another one, or whether it makes sense to claim that one 'only knows that φ'. To demonstrate that such questions are non-trivial, note that in an epistemic system with negative introspection, the assumption that one agent can know more in one state than in another yields a contradiction, since in such a case, in the second state, the agent would have knowledge about his ignorance ('I know that I don't know ...') that cannot be shared in the first state. And, in the case of only knowing, it seems defensible to argue that one can honestly claim to only know (that one knows) some atomic fact p —which makes p an honest formula— whereas 'I only know that I know p or that I know q' does not seem to be acceptable.

Studies of 'only knowing' ([6, 14]) and 'all I know' ([11]) have largely been restricted to particular modal systems, such as **S5**, **S4** and **K45**. Recently Halpern [5] has also taken other modal systems such as **K**, **T** and **KD45** into account. Although his approach suggests similar results for e.g. **KD4**, we would like to adopt a more general perspective: *given any modal system, how to characterize the minimal informational content of modal formulas.* Besides *arbitrary* normal systems we prefer to use standard Kripke models, instead of Fagin and Vardi's knowledge structures, and Halpern's tree models.

Our approach is motivated by the question what kind of conclusions can be derived from a modal premise φ, if this premise is understood as minimal information. For instance, if $\varphi = \Box p$ one would derive $\Diamond q$ under this reading of premises (if one only has the information that p, any independent fact q may be possible), whereas in ordinary modal logic $\Diamond q$ is not derived from $\Box p$.[1]

There are three ways to study the question whether a formula φ allows for such a minimal interpretation, and, if so, what can be said about the consequences of φ under this interpretation.

The first approach is a semantical one: Given a formula φ, try to identify models for φ that carry the least information. This approach requires a suitable order between states (i.e., model-world pairs) in order to identify minimal (or, rather, *least*) elements. For the simple (universal) **S5**-models the order coincides with the superset-relation between sets of worlds. Our challenge here is to give a general definition of such an order, which also suits other modal systems.

The second approach is mainly syntactic in nature and presupposes a sublanguage $\mathcal{L}^*$ of 'special' formulas. Given a consistent formula φ, we then try to find a maximally consistent set containing φ with a smallest $\mathcal{L}^*$-part. This approach can be identified as the search for so-called stable expansions, which are related to maximally consistent sets in a straightforward way. Since consistency is defined as a deductive property, and maximally consistent sets pop up as canonical states, there is also a deductive and semantic flavour to this approach.

The third and last approach is purely deductive, and is also known as the disjunction property: φ allows for a minimal interpretation if for any disjunction in $\mathcal{L}^*$ that can be derived from φ, one disjunct is derivable from φ. So, loosely speaking, in choosing between a number of cases, φ forces a decision.

There are several routes to interconnect these approaches. One route is from syntax to semantics. Here the main concern is to find orders on states that preserve the truth of formulas in $\mathcal{L}^*$. Note how such an order models *growth of information*. The reverse route, from semantics to syntax, starts with an order between states, and tries to identify a suitable persistent sublanguage $\mathcal{L}^*$. Our main contribution here is to define orders between models by imposing 'back' (preserving knowledge) and 'forth' (preserving uncertainty) clauses in a way intimately related to Ehrenfeucht-Fraïssé games, giving us a powerful mechanism to fine-tune the order for broad classes of modal logics.

After giving some technical preliminaries in Section 2, in Section 3 we formally link up the three approaches to minimality, and, for two specific orders, we identify the corresponding languages. In Section 4 we evaluate our notions of minimality for epistemic systems. We round off with a conclusion section.

[1] Note that in the partial logic advocated in [8] $\Diamond q$ does not follow from $\Box p$ even when the latter is interpreted as minimal information, because $\Diamond q$ is only true in the presence of positive evidence for q.

2 Preliminaries

Our language $\mathcal{L}$ is just that of modal logic, including the modal operators $\Box$ and its dual $\Diamond$. We assume that formulas φ and ψ in $\mathcal{L}$ are composed from a finite set of propositional atoms $\mathcal{P} = \{p, q, r \ldots\}$, using the modal operators and the classical connectives. A special atom $\bot$ is defined as $(p \wedge \neg p)$, whereas $\top = \neg\bot$.

The function $d : \mathcal{L} \to \mathbb{N}$ calculates the *modal depth* of formulas as follows: $d(p) = d(\top) = d(\bot) = 0$ $(p \in \mathcal{P})$, $d(\neg\varphi) = d(\varphi)$, $d(\varphi \star \psi) = max\{d(\varphi), d(\psi)\}$ for $\star = \wedge, \vee, \to, \leftrightarrow$ and, finally, $d(\Box\varphi) = d(\Diamond\varphi) = 1 + d(\varphi)$. Some properties to be presented are relative to a given subset $\mathcal{L}^* \subseteq \mathcal{L}$. An example of such a sublanguage of $\mathcal{L}$ is $\mathcal{L}_{(n)} = \{\varphi \in \mathcal{L} \mid d(\varphi) \leq n\}$. For a unary operator $\triangle = \neg, \Diamond, \Box$ and language $\mathcal{L}^* \subseteq \mathcal{L}$, $\triangle\mathcal{L}^* = \{\triangle\varphi \mid \varphi \in \mathcal{L}^*\}$. We also use an 'inverse': $\triangle^-\mathcal{L}^*$ denotes $\{\alpha \mid \triangle\alpha \in \mathcal{L}^*\}$.

We use Kripke models $\langle W, R, V\rangle$ as a standard interpretation of the modal language. Instead of Rwv, we also write $v \in R[w]$. Here, the key-notion is the pair $\langle M, w\rangle$ (often written as M, w), also called a *state*, in which each modal formula φ receives its standard interpretation with the typical modal case: $M, w \models \Box\varphi$ iff for all $v \in R[w]$, one has $M, v \models \varphi$. For $\Gamma \subseteq \mathcal{L}$, $M, w \models \Gamma$ means that for all $\gamma \in \Gamma : M, w \models \gamma$. Relative to a given set of models $\mathcal{S}$, consequence is defined by $\Gamma \models_{\mathcal{S}} \varphi$ iff for all $M \in \mathcal{S}$: $M, w \models \Gamma$ implies $M, w \models \varphi$.

A main question in this paper is how truth is preserved by moving from one state to another. Let $\mathcal{L}^* \subseteq \mathcal{L}$ and let S be a set of states. We say that an order $\leq$ on S *preserves* the sublanguage $\mathcal{L}^*$ or that $\mathcal{L}^*$ is *persistent* over $\leq$ in S, iff

$$M, w \leq M', w' \;\Rightarrow\; \text{for all } \varphi \in \mathcal{L}^* : (M, w \models \varphi \Rightarrow M', w' \models \varphi)$$

If the overall converse holds, we say that $\mathcal{L}^*$ *characterizes* $\leq$ on S.

We will discuss several logical systems **S** on top of the minimal modal system **K**, assuming familiarity with the notion of derivability in a modal system **S**. In particular, for a set of premises Γ, we write $\Gamma \vdash_{\mathbf{S}} \varphi$ if there is a derivation of φ (without applications of necessitation to the premises) from Γ in **S**. The formulas φ and ψ are *equivalent in* **S**, or **S**-*equivalent*, if both $\varphi \vdash_{\mathbf{S}} \psi$ and $\psi \vdash_{\mathbf{S}} \varphi$. The logic **S** is called *finitary* for a sublanguage $\mathcal{L}^*$ if it induces finitely many **S**-equivalence classes in $\mathcal{L}^*$. As an immediate consequence of our assumption that $\mathcal{P}$ is finite, we have that every **S** we will consider is *finitary layered*, in the sense that for each $n \in \mathbb{N}$, **S** is finitary for $\mathcal{L}_{(n)}$.

The generalisation that we use to lift $\vdash_{\mathbf{S}}$ from a subset of $2^{\mathcal{L}} \times \mathcal{L}$ to a subset of $2^{\mathcal{L}} \times 2^{\mathcal{L}}$ is more analogous to the one used in sequent calculi (cf. [10]). than to our Hilbert-style presentation of a logical system:

$$\Gamma \vdash_{\mathbf{S}} \Delta \Leftrightarrow \exists \delta_1 \ldots \delta_n \in \Delta : \; \Gamma \vdash_{\mathbf{S}} (\delta_1 \vee \cdots \vee \delta_n)$$

if $\Delta \neq \emptyset$, and $\Gamma \vdash_{\mathbf{S}} \emptyset$ if $\Gamma \vdash_{\mathbf{S}} \bot$.

The minimal system **K** contains the rule of necessitation $(\emptyset \vdash \varphi \Rightarrow \emptyset \vdash \Box\varphi)$ and the modal axiom $K : \Box(\varphi \to \psi) \to (\Box\varphi \to \Box\psi)$ on top of any Hilbert-style axiomatization of propositional logic. Here, we are interested in normal systems **S**, i.e. systems that are obtained by adding modal axioms to **K**. One such extension, **T**, is obtained by adding the axiom $T : \Box\varphi \to \varphi$ to **K**. The system **KD** is named after the axiom that distinguishes it from **K**, which is $D : \Diamond\top$ or, equivalently, $\Box\varphi \to \Diamond\varphi$. In epistemic logic, systems that include axiom 4: $\Box\varphi \to \Box\Box\varphi$ are called *positively introspective,* those that have axiom 5: $\neg\Box\varphi \to \Box\neg\Box\varphi$ are called *negatively introspective.* Two other axioms worth mentioning are B: $\varphi \to \Box\Diamond\varphi$ and G: $\Diamond\Box\varphi \to \Box\Diamond\varphi$. Any combination of the axioms mentioned is called an epistemic logic, here. Typical examples of such systems are **S4** (**T** + 4), **S5** (**S4** + 5) and **S4.2** (**S4** + G).

When $\Box$ is interpreted as belief, the axiom T is often replaced by D, which gives rise to systems that obtain their name directly from the constituting axioms: **KD**, **KD4**, **KD45**, etc.

The set of states verifying **S** (**S**-states for short) is called $\text{STATES}_\mathbf{S}$. For a given formula φ we define $\text{STATES}_\mathbf{S}(\varphi) = \{\langle M, w\rangle \in \text{STATES}_\mathbf{S} \mid M, w \models \varphi\}$.

Given a logic **S** we say that $\varphi \in \mathcal{L}$ satisfies the **S**-*disjunction property* (**S**-DP) over a sublanguage $\mathcal{L}^*$, if φ is **S**-consistent and for every $\psi_1, \psi_2, \ldots \psi_k \in \mathcal{L}^*$:

$$\varphi \vdash_\mathbf{S} (\psi_1 \vee \cdots \vee \psi_k) \;\Rightarrow\; \text{for some } i \leq k : \varphi \vdash_\mathbf{S} \psi_i$$

A set of formulas Γ is **S**-*consistent* if $\Gamma \nvdash_\mathbf{S} \bot$, and *maximally* **S**-*consistent* (**S**-m.c.) if it moreover contains all the formulas φ for which $\Gamma \cup \{\varphi\}$ is consistent. All **S**-m.c. sets together constitute the set of possible worlds $W_\mathbf{S}$ in the canonical model $M_\mathbf{S} = \langle W_\mathbf{S}, R_\mathbf{S}, V_\mathbf{S}\rangle$ for **S**. Thus, maximal consistent sets play a crucial role when proving *(strong) completeness* of **S** wrt any class of models $\mathcal{S}$ that contains $M_\mathbf{S}$: $\Gamma \models_\mathcal{S} \varphi \Rightarrow \Gamma \vdash_\mathbf{S} \varphi$. The converse of this implication is called *(strong) soundness* of **S** wrt $\mathcal{S}$.

Many classes of models have been identified that are sound and complete wrt the systems **S** that were mentioned above (see [3]). Most significant are those classes of models that are determined by a property of their accessibility relation. For instance, for **KD** one takes the serial (i.e., $\forall x \exists y Rxy$) Kripke models, for **T** the accessibility relation is reflexive, in **KD4** it is serial and transitive, in **S5** it is an equivalence relation.

3 Minimal information in modal logic

Let **S** be an arbitrary modal system, $\leq$ a pre-order on $\text{STATES}_\mathbf{S}$. A formula φ is called *honest* with respect to **S** and $\leq$, if there exists a *least* **S**-state verifying $\Box\varphi$. More precisely, φ is **S**-honest (for $\leq$) iff there is an **S**-state M, w such that:

- $M, w \models \Box\varphi$,
- $M', w' \models \Box\varphi \Rightarrow M, w \leq M', w'$ for all $\langle M', w'\rangle \in \text{STATES}_\mathbf{S}$.

In Section 3.1, we will give independent characterizations of honesty.

3.1 General characterizations of minimality

Let $\mathcal{L}^*$ be a sublanguage of $\mathcal{L}$. We consider the following approaches to minimality:

(1) Formula φ has a $\leq$-least verifying **S**-state
(2) φ has an $\mathcal{L}^*$-smallest m.c. expansion
(3) φ has **S**-DP with respect to $\mathcal{L}^*$

The next result, which is visualized in Figure 1, relates the three approaches.

Theorem 1. Let $\mathcal{L}^*$ be *persistent* over $\leq$. Then the minimality approaches (2) and (3) are equivalent, while (1) implies both (2) and (3).

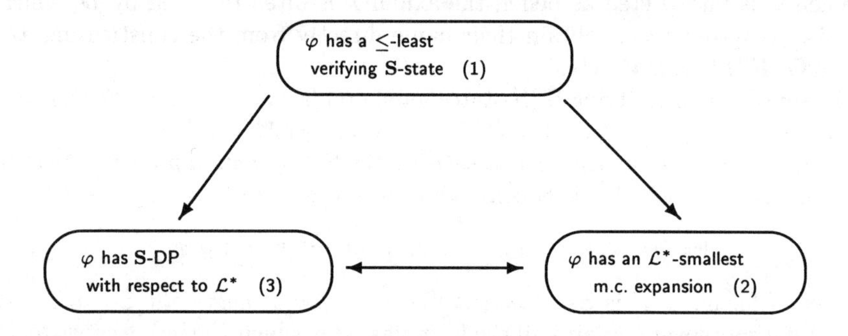

Fig. 1. Relating least states, expansions and disjunction properties wrt $\mathcal{L}^*$

The proof[2] of this and the next theorem invokes the canonical **S**-model and exploits the usual properties of m.c. sets. Without going into details, the equivalence of (2) and (3) is mirrored in the known fact that Σ is an **S**-m.c. set iff both

- Σ is deductively closed, i.e. $\Sigma \vdash_{\mathbf{S}} \psi \Rightarrow \psi \in \Sigma$, and
- Σ has **S**-DP, i.e. Σ is **S**-consistent and for all $\psi_1, \ldots, \psi_k \in \mathcal{L}$: $\Sigma \vdash_{\mathbf{S}} \psi_1 \vee \cdots \vee \psi_k \Rightarrow \Sigma \vdash_{\mathbf{S}} \psi_i$ for some $i \leq k$.

Persistence does not guarantee the implications (3) $\Rightarrow$ (1) and (2) $\Rightarrow$ (1) to hold in general. They can be established by adding the converse of persistence. Let us say that in that case the *minimal information equivalences* hold for $\mathcal{L}^*$ and $\leq$.

Theorem 2. Let $\mathcal{L}^*$ be a persistent sublanguage of $\mathcal{L}$ which also characterizes $\leq$. Then the minimal information equivalences hold for $\mathcal{L}^*$ and $\leq$.

Corollary 1. Let $\mathcal{L}^* \subseteq \mathcal{L}$ be persistent and characterizing for $\leq$. Then the following propositions are equivalent:

- φ is **S**-honest for $\leq$
- $\Box\varphi$ has an $\mathcal{L}^*$-smallest m.c. expansion
- $\Box\varphi$ has **S**-DP over $\mathcal{L}^*$

[2] For full proofs we refer to the forthcoming technical report.

In the literature there is a lot of emphasis on stable sets; in our terminology a stable set is simply the knowledge contained in an m.c. set, or more formally: Σ is stable if $\Sigma = \Box^- \Gamma$ for some m.c. Γ. The second condition for φ being honest can thus be rephrased as:

- φ has a $\Box^- \mathcal{L}^*$-smallest stable expansion[3]

Although this solves the problem of alternative characterization of honesty in an abstract sense, the solution is not entirely satisfactory. Most importantly, it is unclear whether suitable orders exist that enable persistent sublanguages to characterize them. And, if so, we would like to specify them in an independent, insightful way. With this end in view we propose several specific orders in the next subsections.

3.2 Bisimulation and minimality

A first idea which jumps to mind when using preservation results in characterizing minimality, is to employ the notion of bisimulation. Bisimulations are well-known structural descriptions of equivalences between states [2].

Definition 1. Let $M = \langle W, R, V \rangle$ and $M' = \langle W', R', V' \rangle$ be two Kripke models. A *bisimulation* B is a non-empty relation $B \subseteq W \times W'$ such that for all $w \in W$, $w' \in W'$ with Bww':

- $V(w) = V(w')$
- if Rwu for some $u \in W$, then there is a $u' \in W'$ with $R'w'u'$ and Buu' (*forth*)
- if $R'w'u'$ for some $u' \in W'$, then there is a $u \in W$ with Rwu and Buu' (*back*)

If there exists a bisimulation B between the states $\langle M, w \rangle$ and $\langle M', w' \rangle$, we say that the two states bisimulate and we write $\langle M, w \rangle \cong_b \langle M', w' \rangle$

Now, the following well-known result ([2]) guarantees that *any* language $\mathcal{L}^*$ is *invariant* (i.e., two-way persistent) over bisimulations.

Theorem 3. For all $\varphi \in \mathcal{L}$ and all states $\langle M, w \rangle, \langle M', w' \rangle$ one has:
$\langle M, w \rangle \cong_b \langle M', w' \rangle \Rightarrow (M, w \models \varphi \Leftrightarrow M', w' \models \varphi)$

Nevertheless, the existence of a bisimulation is not a necessary condition for modal equivalence of two states. Consider the pair of models in Figure 2, with every world having the same local valuation of propositional variables.

For every natural number n the model M has a branch of length n. M' is similar to M, except that it also has an infinitely long branch. The world 0 verifies the same formulas in both models, but on the other hand, a bisimulation between those two states cannot be given. Similar difficulties arise when we implement weakenings of the notion of bisimulation (e.g., dropping the 'forth'-requirement). It turns out that in the most intuitive adaptations of bisimulation for ordering states on their informative content, the state $\langle M, 0 \rangle$ in the figure above remains a proper extension of $\langle M', 0 \rangle$, while they contain the same information. A refinement of such orders is therefore needed.

[3] Our direct definition of 'stable expansion' generalizes the notion characterized by Moore's fixpoint equation in [13].

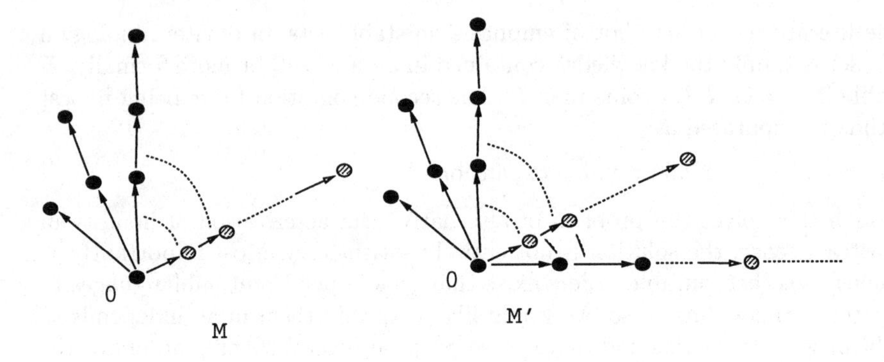

Fig. 2. Two models with the same theory

3.3 Ehrenfeucht-Fraïssé orders

We will now inspect orders inspired by Ehrenfeucht-Fraïssé games.[4] These orders are defined by means of underlying, 'layered' pre-orders. To make the connection, we will present a general lemma that paves the way.

Suppose $\leq^n$ is a pre-order on STATEs for each natural number n ('layer n'). Moreover, let $\leq$ be defined by: ($M = \langle W, R, V\rangle$ and $M' = \langle W', R', V'\rangle$)

$$M, w \leq M', w' \iff \forall n \in \mathbb{N}\ \forall v' \in R'[w']\ \exists v \in R[w] : M, v \leq^n M', v'$$

Finally, let $\mathcal{L}^*$ be a sublanguage and $\mathcal{L}^*_{(n)} = \mathcal{L}^* \cap \mathcal{L}_{(n)}$ be its subset of formulas of modal depth up to n.

Lemma 1 (collecting). If $\mathcal{L}^*_{(n)}$ is persistent and characterizing for $\leq^n$, and $\mathcal{L}^*$ is closed under $\vee$, then $\Box\mathcal{L}^*$ is persistent and characterizing for $\leq$, i.e. $M, w \leq M', w' \iff \forall \varphi \in \Box\mathcal{L}^* : (M, w \models \varphi \Rightarrow M', w' \models \varphi)$.

With this tool we study some important Ehrenfeucht-Fraïssé orders.

General information order. In the first Ehrenfeucht-Fraïssé order the underlying, layered order is an equivalence relation. The relation $\simeq^n$ is defined recursively by:

- $M, w \simeq^0 M', w' \iff V(w) = V'(w')$
- $M, w \simeq^{n+1} M', w' \iff \begin{cases} M, w \simeq^n M', w'\ \& \\ \forall v' \in R'[w']\ \exists v \in R[w] : M, v \simeq^n M', v'\ (\textit{back})\ \& \\ \forall v \in R[w]\ \exists v' \in R'[w'] : M, v \simeq^n M', v'\ (\textit{forth}) \end{cases}$

Then it can be shown that two states are $\simeq^n$-equivalent iff they verify the same formulas up to depth n:

[4] See [4] for the use of Ehrenfeucht-Fraïssé games in first-order predicate logic, in which modal logic can obviously be embedded [2].

$$M, w \simeq^n M', w' \iff \forall \varphi \in \mathcal{L}_{(n)} : (M, w \models \varphi \iff M', w' \models \varphi).$$

From left to right the external equivalence expresses persistence over $\simeq^n$, from right to left it says that $\mathcal{L}_{(n)}$ characterizes $\simeq^n$.

Now we define the general information order $\sqsubseteq$ based on stratified Ehrenfeucht-Fraïssé equivalence:

$$M, w \sqsubseteq M', w' \iff \forall n \in \mathbb{N}\ \forall v' \in R'[w']\ \exists v \in R[w] : M, v \simeq^n M', v'$$

Since $\mathcal{L}_{(n)}$ is persistent and characterizing for $\simeq^n$, lemma 1 shows that $\Box\mathcal{L}$ is both persistent and characterizing for $\sqsubseteq$.

Lemma 2 (modal characterization of $\sqsubseteq$).
$M, w \sqsubseteq M', w' \iff \forall \varphi \in \Box\mathcal{L} : (M, w \models \varphi \Rightarrow M', w' \models \varphi)$.

So, by Theorem 2, we obtain the minimal information equivalences for the general information order.

Theorem 4. *For any* **S**, *the minimal information equivalences hold for* $\sqsubseteq$ *and* $\Box\mathcal{L}$.

By Corollary 1 this implies that φ is honest for $\sqsubseteq$ iff $\Box\varphi$ has an $\Box\mathcal{L}$-smallest m.c. expansion, i.e. φ has a smallest stable expansion. And in terms of disjunction properties, φ is honest iff $\Box\varphi$ has **S**-DP over the full language. As we will show, $\sqsubseteq$ is not a proper order for many epistemic systems. Therefore we introduce yet another Ehrenfeucht-Fraïssé order.

Positive information order. The second order is based on a genuine stratified pre-order. The relation $\preceq^n$ is defined recursively by:

- $M, w \preceq^0 M', w' \iff V(w) = V'(w')$
- $M, w \preceq^{n+1} M', w' \iff \begin{cases} M, w \preceq^n M', w'\ \& \\ \forall v' \in R'[w']\ \exists v \in R[w] : M, v \preceq^n M', v' \quad (\textit{back}) \end{cases}$

Notice the 'forth' direction is typically missing here. Next we define the positive information order $\preceq$ based on the stratified Ehrenfeucht-Fraïssé pre-orders:

$$M, w \preceq M', w' \iff \forall n \in \mathbb{N}\ \forall v' \in R'[w']\ \exists v \in R[w] : M, v \preceq^n M', v'$$

In epistemic terms, the order preserves positive knowledge. We define the relevant sublanguage by:

$$\mathcal{L}^+ = \{\varphi \in \mathcal{L} \mid \varphi \text{ contain neither } \Box \text{ in the scope of } \neg \text{ nor } \Diamond\}$$

So $\Box p \vee \Box q$, $\Box\neg p$ and $\Box p \wedge \neg q$ are members of $\mathcal{L}^+$, but $\neg\Box p$ and $\Diamond p \vee \Box q$ are not. By defining $\Diamond$ as $\neg\Box\neg$, the sublanguage $\mathcal{L}^+$ amounts to the closure under $\wedge$, $\vee$

and $\Box$ of propositional formulas. Recall that $\mathcal{L}^+_{(n)} = \mathcal{L}^+ \cap \mathcal{L}_{(n)}$. One can prove, by induction, the following persistence and characterization result for $\mathcal{L}^+_{(n)}$:

$$M, w \preceq^n M', w' \iff \forall \varphi \in \mathcal{L}^+_{(n)} : (M, w \models \varphi \Rightarrow M', w' \models \varphi) \tag{1}$$

Notice that on the right hand side of (1), we do not have an equivalence now; this is due to the fact that $\mathcal{L}^+_{(n)}$ is not closed under $\neg$ (if $n > 0$).

Again, by the strong persistence property (1) and the collecting Lemma 1, we can easily prove that $\Box\mathcal{L}^+$ is both persistent and characterizing for $\preceq$.

Lemma 3 (modal characterization of $\preceq$).
$M, w \preceq M', w' \iff \forall \varphi \in \Box\mathcal{L}^+ : (M, w \models \varphi \Rightarrow M', w' \models \varphi)$.

Theorem 5. For any **S**, the minimal information equivalences hold for $\preceq$ and $\Box\mathcal{L}^+$.

The latter theorem follows immediately from Lemma 3 and Theorem 2. Using Corollary 1, it implies that φ is honest for $\preceq$ iff $\Box\varphi$ has a $\Box\mathcal{L}^+$-smallest m.c. expansion, i.e., iff φ has an $\mathcal{L}^+$-smallest stable expansion. And in terms of disjunction properties, φ is **S**-honest for $\preceq$ iff $\Box\varphi$ has **S**-DP over $\Box\mathcal{L}^+$.

A simple, yet important case of the positive information order is the submodel relation. $\langle M', w\rangle$ is a submodel of $\langle M, w\rangle$ ($M, w \supseteq M', w$) iff $W' \subseteq W$, $R' = R \cap W' \times W'$ and $V'(u) = V(u)$ for all $u \in W'$. As a consequence of the Łoś theorem in first-order logic, a modal formula is preserved under submodels iff it is equivalent to a 'positive knowledge' formula, i.e. a formula in $\mathcal{L}^+$ (see [12] and [1, thm.2.10]). Since $\Box\mathcal{L}^+ \subseteq \mathcal{L}^+$ this implies, using Lemma 3, that

Corollary 2.

1. $\supseteq$ preserves $\mathcal{L}^+$
2. If $M, w \supseteq M', w$ then $M, w \preceq M', w'$

Since the submodel relation is easily established, this provides a convenient tool for proving that two models are related by the positive information order.

4 Evaluating minimality in epistemic systems

In this section we evaluate the two information orders introduced before in the light of our broad class of epistemic systems, as introduced in Section 2. In fact, we will first prove a negative result for an even larger class. To this purpose we define the notion of a *Geach logic*, see [3]. This is a normal modal system which contains (in addition to **K**) axioms of the form:

$$\Diamond^k\Box^l\varphi \rightarrow \Box^m\Diamond^n\varphi \text{ with } k, l, m, n \in \mathbb{N}, \tag{2}$$

where $\Diamond^k$ is defined recursively: $\Diamond^0\varphi = \varphi$ and $\Diamond^{k+1}\varphi = \Diamond\Diamond^k\varphi$.

The regularity of Geach logics is caused by the general correspondence between an axiom of the form 2 and the class of models in which the accessibility relation is k,l,m,n-confluent:

$$\forall x,y,z \in W : (R^k xy \ \&\ R^m xz) \Rightarrow \exists w \in W : (R^l yw \ \&\ R^n zw). \tag{3}$$

Here R^0 is simply the identity relation, and $R^{k+1} = R \circ R^k$, the relational composition of R and R^k.

Every Geach logic corresponds to a conjunction of relational restrictions as given in (3). One easily verifies that all our epistemic logics are Geach logics.

4.1 General minimality

The general information order specifies that one world is smaller than another world if and only if in the first world less information is true than in the second. It turns out that for most systems this order is not appropriate. In most Geach logics this order trivializes the notion of general honesty: either all formulas are honest, or (nearly) all formulas are dishonest.

Trivial honesty. For weak modal systems such as **K**, **K4**, **KD** and **KD4** it can be proved by a simple model-theoretic technique that all formulae φ such that $\Box\varphi$ is consistent are honest with respect to the general information order. This technique is called simple amalgamation. For two **S**-states a simple amalgamation is constructed by adding one world from which all worlds are accessible which are accessible from the original two states. The construction is depicted in Figure 3. For every formula φ we obtain:

$$(M,w \models \Box\varphi \ \&\ M',w' \models \Box\varphi) \Leftrightarrow M^*,w^* \models \Box\varphi. \tag{4}$$

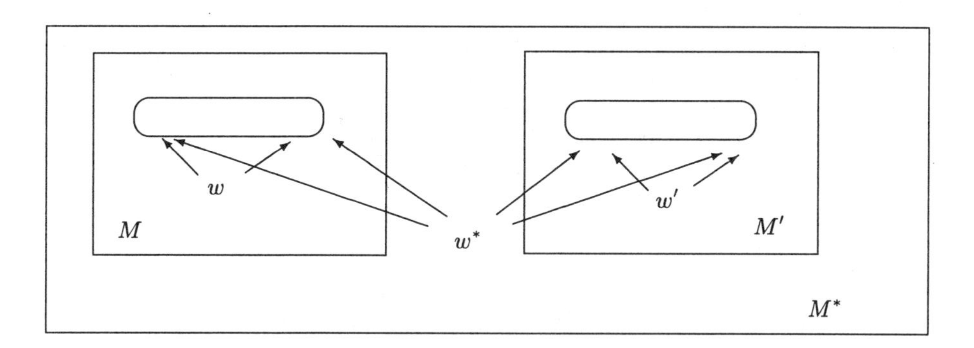

Fig. 3. A simple amalgamation of the states $\langle M,w\rangle$ and $\langle M',w'\rangle$.

Let us suppose that the class of **S**-models is sound and complete wrt **S** and that this class is closed under amalgamation —which is the case for each of these weak modal systems. This proves the disjunction property of any consistent $\Box\varphi$ over $\Box\mathcal{L}$, by using contraposition: Suppose that $\Box\varphi \not\vdash_{\mathbf{S}} \Box\psi_1$, $\Box\varphi \not\vdash_{\mathbf{S}} \Box\psi_2$, then we can find **S**-states with $M, w \models \Box\varphi \wedge \neg\Box\psi_1$ and $M', w' \models \Box\varphi \wedge \neg\Box\psi_2$. By (4) we know that $M^*, w^* \models \Box\varphi$, but obviously, also $M^*, w^* \models \neg(\Box\psi_1 \vee \Box\psi_2)$. In other words, $\Box\varphi \not\vdash_{\mathbf{S}} \Box\psi_1 \vee \Box\psi_2$.

This argument shows that in every system **S** with only axioms of the form $\Box^l\varphi \to \Box^m\Diamond^n\varphi$ with $l, m+n > 0$ trivial honesty occurs, since the semantic conditions for such Geach logics are preserved under simple amalgamation.

Trivial dishonesty. On the other hand, for many Geach logics the notion of full honesty deflates seriously. Such systems often contain theorems of the form $\Box\varphi_1 \vee \Box\varphi_2$ where $\Box\varphi_1$ and $\Box\varphi_2$ are not theorems (*). In particular, systems which incorporate Geach axioms with $k, m > 0$ have the property (*).[5] The disjunction property is then easily violated. In fact, one can prove the following.

Theorem 6. For any epistemic logic **S** except for **T**, **S4**, and the weak systems **K**, **KD**, **K4**, and **KD4**, no **S**-consistent formula is honest.

Systems without consistent honest formulas are **K45**, **KD45** and **S5**, but also those with a milder form of negative introspection such as **S4.2**.

Remnants of full honesty. From the previous paragraphs it follows that the only epistemic systems which do not suffer from trivial honesty or trivial dishonesty are **T** and **S4**. A simple propositional variable p (a fact) is both **T**- and **S4**-honest with respect to the general information order, while $\Box p \vee \Box q$ is dishonest in these two systems.

A semantic test for full honesty in such systems as **T** and **S4** is provided by the notion of *rootability* [9], which is defined on the basis of *reflexive amalgamation.* A reflexive amalgamation of two states is defined in the same way as a simple amalgamation with the only exception that the 'root world' w^* is taken to be reflexive: $R^*w^*w^*$. A formula φ is called rootable if for every pair of $\Box\varphi$-states, a reflexive amalgamation can be found such that $\Box\varphi$ holds in the root world. Rootability implies honesty, but not the other way around. Nevertheless, for every system which is closed under reflexive amalgamation, such as **S4** and **T**, every positive knowledge formula is honest with respect to the general information order if and only if it is rootable. So, for example, p is rootable.

4.2 Positive minimality

As we have seen the systems **T** and **S4** permit non-trivial honest and dishonest formulas. However, there are good reasons to question the feasibility of the notion of full honesty. To begin with, full honesty cannot serve as a good general notion of honesty, since in many systems this notion trivializes, as we have seen in the previous section. Moreover, for epistemic purposes it seems intuitively more sound to exclude formulas which represent ignorance, i.e., formulas of the form $\neg\Box\varphi$, when it comes to minimizing knowledge.

[5] Another set of Geach axioms satisfying (*) are B and the like ($k, l = 0, m > 0$).

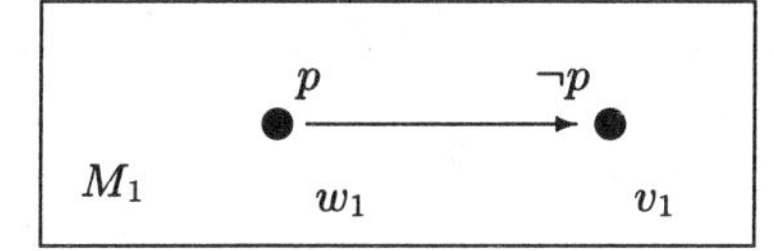

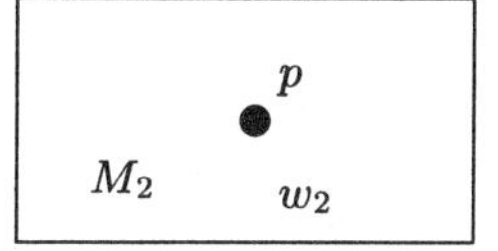

Fig. 4. Two **S4**-models.

To understand the problem, consider the two models above (in all figures, we omit reflexive arrows). Intuitively, one would say that the agent knows more in state $\langle M_2, w_2 \rangle$ than in $\langle M_1, w_1 \rangle$, since the agent considers less possibilities in $\langle M_2, w_2 \rangle$. However, $\langle M_1, w_1 \rangle$ is not smaller than $\langle M_2, w_2 \rangle$ in terms of the general information order. In the first configuration the agent knows that he does not know that p, while in the second he does not know that he does not know that p, since he knows that p. This shows that the general information order on possible worlds does not fit in with our intuition that 'more knowledge' corresponds to 'less uncertainty'.

For the system **S5**, the restriction to positive minimality turns out to be equivalent with the original analysis of honesty in [6]. In fact a more restricted version of minimality is given in [6], viz. with respect to the language $\Box\mathcal{L}_{(0)}$ (factual knowledge). However, it can be shown that in the system **S5** the disjunction property with respect to this restricted language is equivalent to the disjunction property with respect to the language of positive knowledge formulas.

For some modal systems such as **S4**, in which neither general nor positive minimality trivializes, it is interesting to compare the two orders. First let us note that for arbitrary normal systems general honesty implies positive honesty (this easily follows from the disjunction properties):

Theorem 7. *For any normal system* **S**, *if* φ *is* **S**-*honest wrt* $\sqsubseteq$ *then* φ *is also* **S**-*honest wrt* $\preceq$.

However, a similar transfer between different modal systems (and one kind of honesty) is not easily obtained. It may therefore be illuminating to contrast general and positive honesty for **S4** with positive honesty for **S5**. Table 1 displays formulas which are honest ($\surd$) or dishonest ($-$) in the indicated sense.

From Theorem 6, we know that there are no (consistent) formulas that are generally honest in **S5**. Also, Theorem 7 explains why there are no witnesses for the cases 3 and 4 in the table. Cases 5 and 6 show that Theorem 7 cannot be strengthened to an 'if and only if' statement. Finally, note that there is no relationship between positive honesty in **S5** and either general or positive honesty in **S4**.

For illustrative purposes, let us prove both a positive and a negative entry in Table 1. To start with, let us consider $\varphi = \Box p \vee \Box\Diamond q$. In order to demonstrate that $\Box\varphi$ has the **S4**-DP with respect to $\Box\mathcal{L}^+$, let $\alpha, \beta \in \mathcal{L}^+$. To prove that $\Box\varphi \vdash_{\mathbf{S4}} \Box\alpha \vee \Box\beta$ implies that either $\Box\varphi \vdash_{\mathbf{S4}} \Box\alpha$, or $\Box\varphi \vdash_{\mathbf{S4}} \Box\beta$, we argue by contraposition. Hence assume that $\Box\varphi \nvdash_{\mathbf{S4}} \Box\alpha$, and $\Box\varphi \nvdash_{\mathbf{S4}} \Box\beta$. Using completeness, we find two **S4**-models $M = \langle W, R, V \rangle$ and $M' = \langle W', R', V' \rangle$,

case	*formula*	**S4gen**	**S4pos**	**S5pos**
1	$p \vee q$	√	√	√
2	$\Box p \vee q$	√	√	−
3	none	√	−	√
4	none	√	−	−
5	$\Box p \vee \Box\Diamond q$	−	√	√
6	$\Box p \vee \Box\Diamond\Box q$	−	√	−
7	$(\Box(\Box p \vee q) \wedge \neg\Box q) \vee \Box(p \vee r)$	−	−	√
8	$\Box p \vee \Box q$	−	−	−

Table 1. Several (dis-)honest formulas for **S4** and **S5**

with

$$M, w \models \Box\varphi \wedge \neg\Box\alpha, \text{ and } M', w' \models \Box\varphi \wedge \neg\Box\beta$$

Thus, there must be v and v' such that $M, v \models \varphi \wedge \neg\alpha$, and $M', v' \models \varphi \wedge \neg\beta$ (see Figure 5). Now, we build a new model M^* out of M and M' as follows.

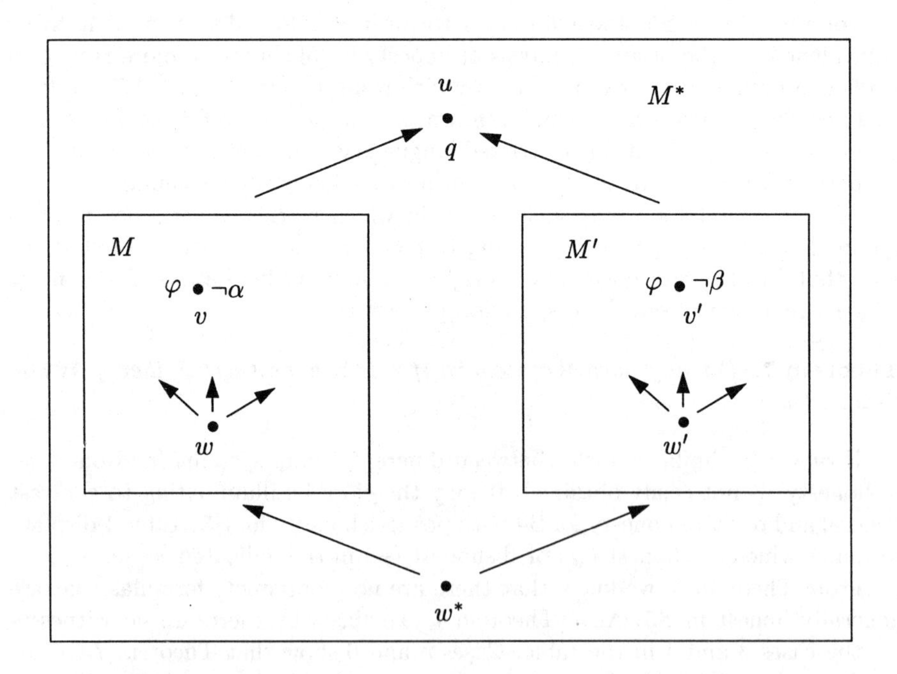

Fig. 5. Adding a common ceiling and root to two **S4**-models

To the reflexive amalgamation of the two models, we add a 'common ceiling' u in which q is true, i.e. let

$$\begin{aligned} W^* &= W \cup W' \cup \{w^*, u\}, \\ R^* &= R \cup R' \cup (W^* \times \{u\}) \cup (\{w^*\} \times W^*), \end{aligned}$$

and V^* equals V on W, V' on W', $V^*(q,u) = 1$ and V^* on w^* is arbitrary. M^* is an **S4**-model, since it is both reflexive and transitive. Because $M^*, u \models q$, we have that that $M^*, w^* \models \Box\varphi$. Finally $M^*, w^* \not\models \Box\alpha \vee \Box\beta$, for suppose that $M^*, w^* \models \Box\alpha$, then in particular $M^*, v \models \alpha$. Since M is carefully constructed to be a submodel of M^*, by the corollary of the Łoś theorem: $M, v \models \alpha$, contradictory to assumption. Thus $M^*, w^* \not\models \Box\alpha$. By a similar argument, using the fact that M' is also a submodel of M^*, $M^*, w^* \not\models \Box\beta$. In all, $M^*, w^* \not\models \Box\alpha \vee \Box\beta$, hence by soundness $\Box\varphi \not\vdash_{\mathbf{S4}} \Box\alpha \vee \Box\beta$. Since $\Box\varphi$ is obviously **S4**-consistent, this completes the proof that φ is **S4**-honest wrt the positive information order.

To illustrate the derivation of a negative result in Table 4.2, let us consider the first '−' at line 7. Let $\varphi = (\Box(\Box p \vee q) \wedge \neg\Box q) \vee \Box(p \vee r)$. We will disprove the **S4**-DP for φ. First of all, it is easily verified that $\Box\varphi \vdash_{\mathbf{S4}} (\Box p \vee q) \vee \Box(p \vee r)$. However, the two models M and M' of Figure 6 illustrate that $\Box\varphi \not\models_{\mathbf{S4}} (\Box p \vee q)$ and that $\Box\varphi \not\models_{\mathbf{S4}} \Box(p \vee r)$, respectively.

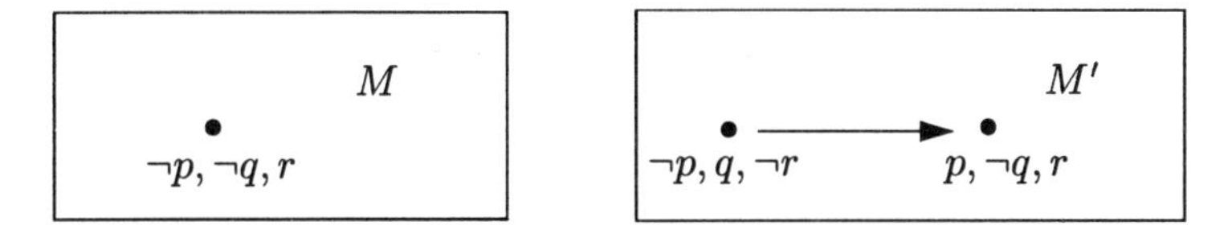

Fig. 6. Two counter models

5 Conclusion

Until now research in the notion of minimal information was mainly devoted to particular modal logics such as **S4** and **S5**. Most often authors also used non-standard semantics and specialized techniques to model minimal information. In this paper we offered a general approach to the representation of minimal information on the basis of standard Kripke models for arbitrary normal modal logics.

The key idea is the use of preservation results for modal logic, and can be sketched as follows. A formula expresses minimal information iff there is a least verifying model; this presupposes an order wrt which the model is minimal. Next determine which formulas are preserved by this order, i.e. which formulas remain true when moving to a 'greater' (more informative) model. If this sublanguage also characterizes the order, this is a suitable information order. Such an order and the sublanguage it preserves then provides precise syntactic, deductive and semantic criteria for minimality which can be used as equivalent tests for representation of minimal information by single, so-called honest, formulas.

In the context of epistemic logic this analysis led to the choice of the positive Ehrenfeucht-Fraïssé order on models for the semantic analysis of minimality and the sublanguage of positive knowledge formulas for the syntactic and the deductive description of minimality. The conclusion is that this description offers an adequate analysis of minimality for epistemic systems.

One may argue that for this analysis trivial honesty (i.e. all consistent formulas are honest) still occurs on the level of weak modal logics such as **K**, **KD**, **K4** and **KD4**, and that therefore, our choice is not adequate for those systems. In epistemic logic however these systems are simply too weak since in these systems a set as, for example, $\{\Box(\Box p \vee \Box\neg p), \neg\Box p, \neg\Box\neg p\}$ is consistent, which seems unacceptable when $\Box$ represents some epistemic attitude (an argument against Hintikka's **KD4** axiomatization of belief [7]). It is implausible that an agent may have the information that he has the information whether p, while on the other hand, he doubts whether p is the case.

Needless to say the generality of our approach may inspire further research in this area. One obvious topic is the application of our results to particular systems, such as extensions of **S4** like **S4.2** and **S4F**. Other lines of future research may be the adaptation of our techniques to multiple agent systems and the move to partial semantics.

References

1. H. Andréka, J. van Benthem & I. Németi, 'Back and Forth Between Modal Logic and Classical Logic', *Journal of the IGPL*, Vol.3, No. 5, pp. 685–720, 1995.
2. J. van Benthem, *Language in Action. Categories, Lambdas and Dynamic Logic*, North-Holland, Amsterdam, 1991.
3. B.F. Chellas, *Modal Logic. An Introduction*, Cambridge University Press, 1980.
4. H.-D. Ebbinghaus & J. Flum, *Finite model theory*, Springer–Verlag, Berlin, 1995.
5. J.Y. Halpern, 'Theory of Knowledge and Ignorance for Many Agents', in *Journal of Logic and Computation*, **7** No. 1, pp. 79–108, 1997.
6. J.Y. Halpern & Y. Moses, 'Towards a theory of knowledge and ignorance', in Kr. Apt (ed.) *Logics and Models of Concurrent Systems*, Springer–Verlag, Berlin, 1985.
7. J. Hintikka, *Knowledge and Belief: An introduction to the Logic of the Two Notions*, Cornell University Press, Ithaca N.Y, 1962.
8. W. van der Hoek, J.O.M. Jaspars, & E.G.C. Thijsse, 'Honesty in Partial Logic'. *Studia Logica*, **56** (3), 323–360, 1996.
9. J.O.M. Jaspars, 'A generalization of stability and its application to circumscription of positive introspective knowledge', *Proceedings of the Ninth Workshop on Computer Science Logic* (CSL'90), Berlin: Springer–Verlag 1991.
10. J. Jaspars & E. Thijsse, 'Fundamentals of Partial Modal Logic', in P. Doherty (ed.) *Partiality, Modality, Nonmonotonicity* (pp.111-141), Stanford: CSLI Publications, Studies in Logic, Language and Information, 1996
11. H.J. Levesque, 'All I know: a study in auto-epistemic logic', in *Artificial Intelligence*, **42**(3), pp. 263–309, 1990.
12. J. Łoś, 'Quelques remarques, théorèmes et problèmes sur les classes définissables d'algèbres', in Th. Skolem et al. (eds.), *Mathematical Interpretation of Formal Systems*, North-Holland, Amsterdam, 1955
13. R.C. Moore,'Semantical considerations on non-monotonic logic', *Artificial Intelligence* 25, pp. 75–94, 1985
14. M. Vardi, 'A model-theoretic analysis of monotonic knowledge', IJCAI85, pp. 509–512, 1985.

Prohairetic Deontic Logic (PDL)

Leendert W.N. van der Torre[1] and Yao-Hua Tan[2]

[1] IRIT, Paul Sabatier University, Toulouse, France
torre@irit.fr
[2] EURIDIS, Erasmus University Rotterdam, Netherlands
ytan@fac.fbk.eur.nl

Abstract. In this paper we introduce Prohairetic Deontic Logic (PDL), a preference-based dyadic deontic logic. An obligation 'α should be (done) if β is (done)' is true if (1) no $\neg\alpha\wedge\beta$ state is as preferable as an $\alpha\wedge\beta$ state and (2) the preferred β states are α states. We show that the different elements of this mixed representation solve different problems of deontic logic. The first part of the definition is used to formalize contrary-to-duty reasoning, that for example occurs in Chisholm's and Forrester's notorious deontic paradoxes. The second part is used to make dilemmas inconsistent. PDL shares the intuitive semantics of preference-based deontic logics without introducing additional semantic machinery such as bi-ordering semantics or ceteris paribus preferences.

1 Introduction

Deontic logic is a modal logic, in which absolute and conditional obligations are represented by the modal formulas $O\alpha$ and $O(\alpha|\beta)$, where the latter is read as 'α ought to be (done) if β is (done).' It can be used for the formal specification and validation of a wide variety of topics in computer science (for an overview and further references see [44]). For example, deontic logic can be used to formally specify soft constraints in planning and scheduling problems as norms. The advantage is that norms can be violated without creating an inconsistency in the formal specification, in contrast to violations of hard constraints. With the increasing popularity and sophistication of applications of deontic logic the fundamental problems of deontic logic become more pressing.

From the early days, when deontic logic was still a purely philosophical enterprise, it is known that it suffers from certain paradoxes. The majority of the paradoxes evaporates under close scrutiny yet some of them – most notably the contrary-to-duty paradoxes – persist. The conceptual issue of these paradoxes is how to proceed once a norm has been violated. Clearly, this issue is of great practical relevance, because in most applications norms are violated frequently. Usually it is stipulated in the fine print of a contract what has to be done if a term in the contract is violated. If the violation is not too serious, or was not intended by the violating party, the contracting parties usually do not want to consider this as a breach of contracts, but simply as a disruption in the execution of the contract that has to be repaired. Hence, the contrary-to-duty paradoxes are important benchmark examples of deontic logic, and deontic logics incapable of dealing with them are considered insufficient tools to analyze deontic reasoning. In this paper we restrict ourselves to the area of *preference-based* deontic logic [12,23,19,10,15,17,4], i.e. deontic logics based on a preference relation in the semantics, because they have proven to be the most appropriate to solve the paradoxes. Still, preference-based deontic logic has the following three problems.

Strong preference problem. Preferences for α_1 and α_2 conflict for $\alpha_1 \wedge \neg\alpha_2$ and $\neg\alpha_1 \wedge \alpha_2$.

Contrary-to-duty problem. A contrary-to-duty obligation is an obligation that is only in force in a sub-ideal situation. For example, the obligation to apologize for a broken promise is only in force in the sub-ideal situation where the obligation to keep promises is violated. Reasoning structures like 'α_1 should be (done), but if $\neg\alpha_1$ is (done) then α_2 should be (done)' must be formalized without running into the notorious contrary-to-duty paradoxes of deontic logic like Chisholm's and Forrester's paradoxes [6,9].

Dilemma problem. Most deontic logicians take the perspective that deontic logic formalizes the reasoning of an authority issuing norms, and such an authority does not intentionally create dilemmas. Consequently, dilemmas should be inconsistent. However, some deontic logicians, most notably in computer science, try to model obligations in practical reasoning, and in daily life dilemmas exist (see e.g. [40,4]). Consequently, according to this alternative perspective dilemmas should be consistent. In this paper we follow the traditional and mainstream perspective. The three formulas $O\alpha \wedge O\neg\alpha$, $O(\alpha_1 \wedge \alpha_2) \wedge O\neg\alpha_1$ and $O(\alpha_1 \wedge \alpha_2|\beta_1) \wedge O(\neg\alpha_1|\beta_1 \wedge \beta_2)$ represent dilemmas and they should therefore be inconsistent. However, the formula $O(\alpha|\beta_1) \wedge O(\neg\alpha|\beta_2)$ does not represent a dilemma and should be consistent.

In this paper we propose Prohairetic (i.e. Preference-based) Deontic Logic (PDL), a logic of non-defeasible obligations in which dilemmas are inconsistent. The basic idea of this logic is that an obligation 'α should be (done) if β is (done)' is true if (1) no $\neg\alpha \wedge \beta$ state is as preferable as an $\alpha \wedge \beta$ state and (2) the preferred β states are α states. The first part of the definition is used to formalize contrary-to-duty reasoning, and the second part is used to make dilemmas inconsistent. PDL shares the intuitive semantics of preference-based deontic logics, and solves the strong preference problem without introducing additional semantic machinery like bi-ordering semantics or ceteris paribus preferences. Moreover, PDL shares the intuitive formalization of contrary-to-duty reasoning of dyadic deontic logic [12,23]. Finally, PDL solves the dilemma problem by making the right set of formulas inconsistent.

This paper is organized as follows. We first discuss the strong preference problem (Section 2) and the contrary-to-duty and dilemma problems (Section 3). We then give an axiomatization of PDL in a modal preference logic (Section 4), and finally we reconsider the three problems in PDL (Section 5).

2 Obligations and preferences

It has been suggested [20, 19, 10, 15, 17, 4] that a unary deontic operator O might be defined in a logic of preference by $O\alpha =_{def} \alpha \succ \neg\alpha$. In the following discussion we assume possible worlds models with a deontic preference ordering on the worlds, i.e. Kripke models $\langle W, \leq, V \rangle$ which consist of a set of worlds W, a binary reflexive and transitive accessibility relation $\leq$ and a valuation function V. The problem is how to lift the preferences between worlds to preferences between sets of worlds or propositions, written as $\alpha_1 \succ \alpha_2$. It is well-known from the preference logic literature [41] that the preference $\alpha \succ \neg\alpha$ cannot be defined by the set of preferences of all α worlds to each $\neg\alpha$ world, because two unrelated obligations like 'be polite' Op and 'be helpful' Oh would conflict when considering 'being polite and unhelpful' $p \wedge \neg h$ and 'being impolite and helpful' $\neg p \wedge h$. Proof-theoretically, if a preference relation $\succ$ has left and right strengthening, then the two preferences $p \succ \neg p$ and $h \succ \neg h$ derive $(p \wedge \neg h) \succ (\neg p \wedge h)$ and $(\neg p \wedge h) \succ (p \wedge \neg h)$. The two derived preferences seem contradictory.

The conflict can be resolved with additional information. For example, politeness may be less important than helpfulness, such that $(p \wedge \neg h) \succ (\neg p \wedge h)$ is less important than $(\neg p \wedge h) \succ (p \wedge \neg h)$. This relative importance of obligations can only be formalized in a defeasible deontic logic, in which obligations can be overridden by other obligations (for an analysis of the conceptual distinctions, see [33, 35]), because $(p \wedge \neg h) \succ (\neg p \wedge h)$ is overridden by $(\neg p \wedge h) \succ (p \wedge \neg h)$, and Op is not in force when only $(p \wedge \neg h) \vee (\neg p \wedge h)$ worlds are considered. However, in this paper we only consider non-defeasible or non-overridable obligations. For such logics, the following three solutions have been considered.

Bi-ordering. Jackson [19] and Goble [10] introduce a second ordering representing degrees of 'closeness' of worlds to solve the strong preference problem. They define the preference $\alpha \succ \neg\alpha$ by the set of preferences of *the closest* α worlds to *the closest* $\neg\alpha$ worlds. The underlying idea is that in certain contexts the way things are in some worlds can be ignored – perhaps they are too remote from the actual world, or outside an agent's control. For example, the obligations Op and Oh are consistent when 'polite and unhelpful' $p \wedge \neg h$ and 'impolite and helpful' $\neg p \wedge h$ are not among the closest p, $\neg p$, h and $\neg h$ worlds.[1]

[1] This solution of the strong preference problem introduces an *irrelevance* problem, because the preferences no longer have left and right strengthening. For example, the preference $(p \wedge h) \succ (\neg p \wedge h)$ cannot even be derived from 'be polite' $p \succ \neg p$, because $p \wedge h$ or $\neg p \wedge h$ may not be among the closest p or $\neg p$ worlds. There is another interpretation of the closeness ordering. 'The closest' could also be interpreted as 'the most normal' as used in the preferential semantics of logics of defeasible reasoning. The 'multi preference' semantics is a formalization of *defeasible* deontic logic [33, 35], sometimes called the logic of prima facie obligations. However, it is not clear that closeness is an intuitive concept for non-defeasible obligations.

Ceteris paribus. Hansson [15, 14, 13] defines $\alpha \succ \neg\alpha$ by a 'ceteris paribus' preference of α to $\neg\alpha$, see [4] for a discussion.[2] That is, for each pair of α and $\neg\alpha$ worlds that are identical except for the evaluation of α, the α world is preferred to the $\neg\alpha$ world. The obligation 'be polite' Op prefers 'polite and helpful' $p \wedge h$ to 'impolite and helpful' $\neg p \wedge h$, and 'polite and unhelpful' $p \wedge \neg h$ to 'impolite and unhelpful' $\neg p \wedge \neg h$, but it does not say anything about $p \wedge h$ and $\neg p \wedge \neg h$, and neither about $p \wedge \neg h$ and $\neg p \wedge h$. Likewise, the obligation 'be helpful' Oh prefers 'polite and helpful' $p \wedge h$ to 'polite and unhelpful' $p \wedge \neg h$, and 'impolite and helpful' $\neg p \wedge h$ to 'impolite and unhelpful' $\neg p \wedge \neg h$. These preferences can be combined in a single preference ordering for $Op \wedge Oh$ that prefers $p \wedge h$ worlds to all other worlds, and that prefers all worlds to $\neg p \wedge \neg h$ worlds.[3]

Consistent dilemmas. Finally, a preference $\alpha \succ \neg\alpha$ can be defined by 'every $\neg\alpha$ world is not as preferable as any α world' (or, maybe more intuitively, $\alpha \not\succ \neg\alpha$ is defined as 'there is an $\neg\alpha$ world w_1 which is at least as preferable as an α world w_2'). The definition is equivalent to the problematic 'all α worlds are preferred to each $\neg\alpha$ world' if the underlying preference ordering on worlds $\leq$ is strongly connected, i.e. if for each pair of worlds w_1 and w_2 in a model M we have either $w_1 \leq w_2$ or $w_2 \leq w_1$. However, the two obligations Op and Oh do not conflict when considering $p \wedge \neg h$ and $\neg p \wedge h$ *when we allow for incomparable worlds*, following [40]. In contrast to the other solutions of the strong preference problem, dilemmas like $Op \wedge O\neg p$ are consistent. We say that the logic does not have the no-dilemma assumption. The preference relation $\succ$ has left and right strengthening, and $p \succ \neg p$ and $h \succ \neg h$ imply $(p \wedge \neg h) \succ (\neg p \wedge h)$ and $(\neg p \wedge h) \succ (p \wedge \neg h)$. However, the latter two preferences are not logically inconsistent. The $\neg p \wedge h$ and $p \wedge \neg h$ worlds are incomparable.[4]

---

[2] Moreover, Hansson defines obligations by the property of negativity. According to this principle, what is worse than something wrong is itself wrong. See [15, 4] for a discussion on this assumption.

[3] In [16] Hansson rejects the use of ceteris paribus preferences for obligations (in contrast to, for example, desires). Moreover, ceteris paribus preferences introduce an *independence* problem. At first sight, it seems that a 'ceteris paribus' preference $\alpha \succ \neg\alpha$ is a set of preferences of all $\alpha \wedge \beta$ worlds to each $\neg\alpha \wedge \beta$ world for all circumstances β such that $\alpha \wedge \beta$ and $\neg\alpha \wedge \beta$ are complete descriptions (represented by worlds). However, consider the preference $p \succ \neg p$ and circumstances $p \leftrightarrow \neg h$. The preference $p \succ \neg p$ would derive $(p \wedge (p \leftrightarrow \neg h)) \succ (\neg p \wedge (p \leftrightarrow \neg h))$, which is logically equivalent to the problematic $(p \wedge \neg h) \succ (\neg p \wedge h)$. The exclusion of circumstances like $p \leftrightarrow \neg h$ is the independence problem. Only for 'independent' β there is a preference of $\alpha \wedge \beta$ over $\neg\alpha \wedge \beta$ (see e.g. [29] for an ad hoc solution of the problem).

[4] It was already argued by von Wright [41] that this latter property is highly implausible for preferences. On the other hand, this solution is simpler than the first two solutions of the strong preference problem, because it does not use additional semantic machinery such as the second ordering or the ceteris paribus preferences. Moreover, it does not have an irrelevance or an independence problem.

Prohairetic Deontic Logic (PDL) proposed in this paper is an extension of the third approach. To formalize the no-dilemma assumption, we write $M \models I\alpha$ for 'the ideal worlds in M satisfy α.' Hence, if we ignore infinite descending chains[5] then we can define that $M \models I\alpha$ if and only if $Pref \subseteq |\alpha|$ where *Pref* stands for the set of most preferred (ideal) worlds of M, and $|\alpha|$ stands for the set of all worlds satisfying α. Obligations are defined as a combination of a strong preference and an ideal preference.

$$O\alpha =_{def} (\alpha \succ \neg\alpha) \land I\alpha$$

The formula $Op \land O\neg p$ is inconsistent, because the formula $Ip \land I\neg p$ is inconsistent. The two obligations 'be polite' Op and 'be helpful' Oh are formalized by (1) p worlds are preferred to or incomparable with $\neg p$ worlds, (2) h worlds are preferred to or incomparable with $\neg h$ worlds, and (3) the ideal worlds are $p \land h$ worlds.

In the following section we argue that this solution is not only simpler than the first two solutions of the strong preference problem, but it also gives a more intuitive solution to the contrary-to-duty problem.

3 Dyadic obligations and contrary-to-duty preferences

The contrary-to-duty problem is the major problem of monadic deontic logic, as shown by the notorious Good Samaritan [2], Chisholm [6] and Forrester [9] paradoxes. The formalization of these paradoxes should be consistent. For example, the formalization of the Forrester paradox in monadic deontic logic is 'Smith should not kill Jones' ($O\neg k$), 'if Smith kills Jones, then he should do it gently' ($k \to Og$) and 'Smith kills Jones' (k). From the three formulas $O\neg k \land Og$ can be derived. The derived formula should be consistent, even if we have 'gentle killing implies killing,' i.e. $\vdash g \to k$, see e.g. [11]. However, this formalization of the Forrester paradox does not do justice to the fact that only in *very* few cases we *seem* to have that $O\neg\alpha \land O(\alpha \land \beta)$ is not a dilemma, and should be consistent. The consistency of $O\neg k \land Og$ is a solution that seems like overkill. Deontic logicians therefore tried to formalize contrary-to-duty reasoning by introducing temporal and preferential notions [37].

B. Hansson [12] and Lewis [23] argued that the contrary-to-duty problem can be solved by introducing dyadic obligations. A dyadic obligation $O(\alpha|\beta)$ is read as 'α ought to be (done) if β is (done).' They define a dyadic obligation by $O_{HL}(\alpha|\beta) =_{def} I(\alpha|\beta)$, where we write $I(\alpha|\beta)$ for 'the ideal β worlds satisfy α.' Hence, if we again ignore infinite descending chains, then we define $M \models I(\alpha|\beta)$ if and only if $Pref(\beta) \subseteq |\alpha|$, where $Pref(\beta)$ stands for the preferred β worlds of M. The introduction of the dyadic representation was inspired by the standard way

[5] The problems caused by infinite descending chains are illustrated by the following example. Assume a model that consists of one infinite descending chain of $\neg\alpha$ worlds. It seems obvious that the model should not satisfy $I\alpha$. However, the most preferred worlds (which do not exist!) satisfy α. See [22, 3] for a discussion.

of representing conditional probability, that is, by $Pr(\alpha|\beta)$ which stands for 'the probability that α is the case given β.' In a dyadic deontic logic the Forrester paradox can be formalized by 'Smith should not kill Jones' $O(\neg k|\top)$, 'if Smith kills Jones, then he should do it gently' $O(g|k)$ and 'Smith kills Jones' k. In this formalization $\top$ stands for any tautology, e.g. $p \vee \neg p$. The obligation $O(g|k)$ is a contrary-to-duty (CTD) obligation of $O(\neg k|\top)$, because an obligation $O(\alpha|\beta)$ is a CTD obligation of the primary obligation $O(\alpha_1|\beta_1)$ if and only if $\alpha_1 \wedge \beta$ is inconsistent. In dyadic deontic logic, the formula $O(\neg k|\top) \wedge O(g|k)$ is consistent, whereas the formula $O(\neg k|\top) \wedge O(g|\top)$ is inconsistent when we have $\vdash g \rightarrow k$.

The Hansson-Lewis dyadic deontic logics have been criticized (see e.g. [24]), because they do not have factual detachment, represented by the formula **FD**: $O(\alpha|\beta) \wedge \beta \rightarrow O\alpha$, i.e. the derivation of absolute obligations from dyadic ones. However, there are good reasons not to accept **FD**. If the logic would have **FD**, then they would reinstate the Forrester paradox, because we would derive $O\neg k \wedge Og$ from $O(\neg k|\top) \wedge O(g|k) \wedge k$. To explicate the difference with dyadic obligations which do have factual detachment and therefore cannot represent the Forrester paradox, we prefer to call the Hansson-Lewis obligations contextual obligations, see also [34]. Instead of **FD** we may have $O(\alpha \,|\, \beta \wedge O(\alpha \,|\, \beta))$ as a theorem, see [43].

Dyadic obligations formalize contrary-to-duty reasoning, without making dilemmas like $O\alpha_1 \wedge O(\neg\alpha_1 \wedge \alpha_2)$ consistent. However, the dyadic representation also introduces a new instance of the dilemma problem, represented by the formula $O(\alpha \,|\, \beta_1) \wedge O(\neg\alpha \,|\, \beta_1 \wedge \beta_2)$. An example is Prakken and Sergot's considerate assassin example, that consists of the two obligations 'Smith should not offer Jones a cigarette' $O(\neg c|\top)$ and 'Smith should offer Jones a cigarette, if he kills him' $O(c\,|\,k)$. Prakken and Sergot [26] argue that the two sentences of the considerate assassin example represent a dilemma, because the obligation $O(c|k)$ is not a CTD obligation of $O(\neg c|\top)$. Hence, $O(\neg c|\top) \wedge O(c|k)$ should be inconsistent, even when there is another premise 'Smith should not kill Jones' $O(\neg k|\top)$.

B.Hansson-Lewis dyadic deontic logics do not give a satisfactory solution for the dilemma problem, because $O_{HL}(\neg c|\top) \wedge O_{HL}(c|k)$ is consistent. In Prohairetic Deontic Logic, dyadic obligations are defined in a similar spirit as the absolute obligations in the previous section.

$$O(\alpha|\beta) =_{def} ((\alpha \wedge \beta) \succ (\neg\alpha \wedge \beta)) \wedge I(\alpha|\beta)$$

The set of obligations $S = \{O(\neg c \mid \top), O(c \mid k)\}$ is inconsistent, because the formula $(\neg c \succ c) \wedge I(c|k)$ is inconsistent, as is shown in Section 5.

4 Axiomatization

Prohairetic Deontic Logic (PDL) is defined in a modal preference logic. The standard Kripke models $M = \langle W, \leq, V \rangle$ of PDL contain a binary accessibility relation $\leq$, that is interpreted as a (reflexive and transitive) *deontic preference ordering*. The advantages of our formalization in a modal framework are twofold.

First, if a dyadic operator is given by a definition in an underlying logic, then we get an axiomatization for free! We do not have to look for a sound and complete set of inference rules and axiom schemata, because we simply take the axiomatization of the underlying logic together with the new definition. In other words, the problem of finding a sound and complete axiomatization is replaced by the problem of finding a definition of a dyadic obligation in terms of a monadic modal preference logic. The second advantage of a modal framework in which all operators are defined, is that $I(\alpha \mid \beta)$ and $\alpha_1 \succ \alpha_2$ can be defined separately. In this section we axiomatize PDL in the following three steps in terms of a monadic modal logic and a deontic betterness relation. See [10, 14] for an analogous stepwise construction of 'good' in terms of 'better' and [43] for a stepwise construction of minimizing conditionals analogous to $O_{HL}(\alpha \mid \beta)$ in terms of a 'betterness' relation.

Ideality (deontic preference) ordering. We start with two monadic modal operators $\Box$ and $\overset{\leftrightarrow}{\Box}$. The formula $\Box\alpha$ can be read as 'α is true in all worlds at least as good (as the actual world)' or '$\neg\alpha$ is necessarily worse,' and $\overset{\leftrightarrow}{\Box}\alpha$ can be read as 'α is true in all worlds.'

$$M, w \models \Box\alpha \text{ iff } \forall w' \in W \text{ if } w' \leq w, \text{ then } M, w' \models \alpha$$
$$M, w \models \overset{\leftrightarrow}{\Box}\alpha \text{ iff } \forall w' \in W \; M, w' \models \alpha$$

The $\Box$ operator will be treated as an S4 modality and the $\overset{\leftrightarrow}{\Box}$ operator as an S5 modality. As is well-known, the standard system S4 is characterized by a partial pre-ordering: the axiom **T**: $\Box\alpha \rightarrow \alpha$ characterizes reflexivity and the axiom **4**: $\Box\alpha \rightarrow \Box\Box\alpha$ characterizes transitivity [18, 5]. Moreover, the standard system S5 is characterized by S4 plus the axiom **5**:$\neg \overset{\leftrightarrow}{\Box}\alpha \rightarrow \overset{\leftrightarrow}{\Box}\neg \overset{\leftrightarrow}{\Box}\alpha$. The relation between the modal operators is given by $\overset{\leftrightarrow}{\Box}\alpha \rightarrow \Box\alpha$. This is analogous to the well known relation between the modal operators for knowledge K and belief B given by $Kp \rightarrow Bp$.

Deontic betterness relation. A binary betterness relation $\alpha_1 \succ \alpha_2$, to be read as 'α_1 is deontically preferred to (better than) α_2,' is defined in terms of the monadic operators. The following betterness relation obeys von Wright's expansion principle [41], because a preference of α_1 over α_2 only compares the two formulas $\alpha_1 \wedge \neg\alpha_2$ and $\neg\alpha_1 \wedge \alpha_2$.

$$\alpha_1 \succ \alpha_2 =_{def} \overset{\leftrightarrow}{\Box}(\alpha_1 \wedge \neg\alpha_2 \rightarrow \Box\neg(\alpha_2 \wedge \neg\alpha_1))$$

We have $M, w \models \alpha_1 \succ \alpha_2$ if we have $w_2 \not\leq w_1$ for all worlds $w_1, w_2 \in W$ such that $M, w_1 \models \alpha_1$ and $M, w_2 \models \alpha_2$, where we write as usual $\Diamond\alpha =_{def} \neg\Box\neg\alpha$. The betterness relation $\succ$ is quite weak. For example, it is not anti-symmetric (i.e. $\neg(\alpha_2 \succ \alpha_1)$ cannot be derived from $\alpha_1 \succ \alpha_2$) and it is not transitive (i.e. $\alpha_1 \succ \alpha_3$ cannot be derived from $\alpha_1 \succ \alpha_2$ and $\alpha_2 \succ \alpha_3$). It is easily checked that the lack of these properties is the result of the fact that we do not have totally connected orderings.

Obligatory. What is obligatory is defined in terms of deontic betterness.

$$I(\alpha|\beta) =_{def} \overleftrightarrow{\Box}(\beta \rightarrow \Diamond(\beta \wedge \Box(\beta \rightarrow \alpha)))$$
$$O(\alpha|\beta) =_{def} (\alpha \wedge \beta \succ \neg\alpha \wedge \beta) \wedge I(\alpha|\beta)$$

We have $M, w \models I(\alpha \mid \beta)$ if the preferred β worlds are α worlds, and α eventually becomes true in all infinite descending chains of β worlds [21, 3]. Finally, we have $M, w \models O(\alpha|\beta)$ if we have $w_2 \not\leq w_1$ for all $w_1, w_2 \in W$ such that $M, w_1 \models \alpha \wedge \beta$ and $M, w_2 \models \neg\alpha \wedge \beta$, and $M, w \models I(\alpha|\beta)$.

The logic PDL is defined by defining these three layers in a modal preference logic.

Definition 1 (PDL). *The bimodal language $\mathcal{L}$ is formed from a denumerable set of propositional variables together with the connectives $\neg$, $\rightarrow$, and the two normal modal connectives $\Box$ and $\overleftrightarrow{\Box}$. Dual 'possibility' connectives $\Diamond$ and $\overleftrightarrow{\Diamond}$ are defined as usual by $\Diamond\alpha =_{def} \neg\Box\neg\alpha$ and $\overleftrightarrow{\Diamond}\alpha =_{def} \neg\overleftrightarrow{\Box}\neg\alpha$.*

The logic PDL is the smallest $S \subset \mathcal{L}$ such that S contains classical logic and the following axiom schemata, and is closed under the following rules of inference,

K	$\Box(\alpha \rightarrow \beta) \rightarrow (\Box\alpha \rightarrow \Box\beta)$	**K′**	$\overleftrightarrow{\Box}(\alpha \rightarrow \beta) \rightarrow (\overleftrightarrow{\Box}\alpha \rightarrow \overleftrightarrow{\Box}\beta)$
T	$\Box\alpha \rightarrow \alpha$	**T'**	$\overleftrightarrow{\Box}\alpha \rightarrow \alpha$
4	$\Box\alpha \rightarrow \Box\Box\alpha$	**4'**	$\overleftrightarrow{\Box}\alpha \rightarrow \overleftrightarrow{\Box}\overleftrightarrow{\Box}\alpha$
R	$\overleftrightarrow{\Box}\alpha \rightarrow \Box\alpha$	**5'**	$\neg\overleftrightarrow{\Box}\alpha \rightarrow \overleftrightarrow{\Box}\neg\overleftrightarrow{\Box}\alpha$

Nes *From α infer $\overleftrightarrow{\Box}\alpha$*
MP *From $\alpha \rightarrow \beta$ and α infer β*

extended with the following three definitions.

$$\alpha_1 \succ \alpha_2 =_{def} \overleftrightarrow{\Box}(\alpha_1 \wedge \neg\alpha_2 \rightarrow \Box\neg(\alpha_2 \wedge \neg\alpha_1))$$
$$I(\alpha|\beta) =_{def} \overleftrightarrow{\Box}(\beta \rightarrow \Diamond(\beta \wedge \Box(\beta \rightarrow \alpha)))$$
$$O(\alpha|\beta) =_{def} (\alpha \wedge \beta \succ \neg\alpha \wedge \beta) \wedge I(\alpha|\beta)$$

Definition 2 (PDL Semantics). *Kripke models $M = \langle W, \leq, V\rangle$ for PDL consist of W, a set of worlds, $\leq$, a binary transitive and reflexive accessibility relation, and V, a valuation of the propositional atoms in the worlds. The partial pre-ordering $\leq$ expresses preferences: $w_1 \leq w_2$ iff w_1 is at least as preferable as w_2. The modal connective $\Box$ refers to accessible worlds and the modal connective $\overleftrightarrow{\Box}$ to all worlds.*

$$M, w \models \Box\alpha \text{ iff } \forall w' \in W \text{ if } w' \leq w, \text{ then } M, w' \models \alpha$$
$$M, w \models \overleftrightarrow{\Box}\alpha \text{ iff } \forall w' \in W\ M, w' \models \alpha$$

The following proposition shows that, as a consequence of the definition in a standard bimodal logic, the soundness and completeness of PDL are trivial.

Proposition 1 (Soundness and completeness of PDL). *Let* $\vdash_{\text{PDL}}$ *and* $\models_{\text{PDL}}$ *stand for derivability and logical entailment in the logic PDL. We have* $\vdash_{PDL} \alpha$ *if and only if* $\models_{PDL} \alpha$.

Proof. Follows directly from standard modal soundness and completeness proofs, see e.g. [18, 5, 28].

We now consider several properties of the dyadic obligations. First, the logic PDL has the following theorem, which is valid for any preference-based deontic logic defined by $O(\alpha|\beta) =_{def} \alpha \wedge \beta \succ \neg\alpha \wedge \beta$ (for any betterness relation $\succ$) or by $O(\alpha|\beta) =_{def} I(\alpha|\beta)$.

R $O(\alpha|\beta_1 \wedge \beta_2) \leftrightarrow O(\alpha \wedge \beta_1|\beta_1 \wedge \beta_2)$,

Second, the logic PDL does not have closure under logical implication. This is a typical property of preference-based deontic logics. For example, the preference-based deontic logics discussed in [19, 15, 10, 4, 17] do not have closure under logical implication either. The following theorem *Weakening of the Consequent* **WC** is *not* valid in PDL.

WC $O(\alpha_1|\beta) \rightarrow O(\alpha_1 \vee \alpha_2|\beta)$

The third property we consider is the following *disjunction rule* **OR**, related to *Reasoning-By-Cases* and Savage's sure-thing principle. It is *not* valid either.

OR $(O(\alpha|\beta_1) \wedge O(\alpha|\beta_2)) \rightarrow O(\alpha|\beta_1 \vee \beta_2)$

The fourth property we consider is so-called *Restricted Strengthening of the Antecedent* **RSA**, expressed by the following theorem of the logic. In can easily be shown that $O(\alpha|\beta_1 \wedge \beta_2)$ can only be derived in PDL from $O(\alpha|\beta_1)$ when we have $I(\alpha|\beta_1 \wedge \beta_2)$ as well.

RSA $(O(\alpha|\beta_1) \wedge I(\alpha|\beta_1 \wedge \beta_2)) \rightarrow O(\alpha|\beta_1 \wedge \beta_2)$

We can add strengthening of the antecedent with the following notion of preferential entailment, that prefers maximally connected models. We say that a model is more connected if its binary relation contains more elements (in our terminology $\{(w_1, w_2), (w_2, w_1)\}$ is therefore more connected than $\{(w_1, w_2)\}$).

Definition 3 (Preferential entailment). *Let the two possible worlds models* $M_1 = \langle W, \leq_1, V\rangle$ *and* $M_2 = \langle W, \leq_2, V\rangle$ *be two PDL models.* M_1 *is at least as connected as* M_2*, written as* $M_1 \sqsubseteq M_2$*, iff for all* $w_1, w_2 \in W$ *if* $w_1 \leq_2 w_2$*, then* $w_1 \leq_1 w_2$*.* M_1 *is more connected than* M_2*, written as* $M_1 \sqsubset M_2$*, iff* $M_1 \sqsubseteq M_2$ *and* $M_2 \not\sqsubseteq M_1$*. The formula* ϕ *is preferentially entailed by* T*, written as* $T \models_{\sqsubset} \phi$*, iff* $M \models \phi$ *for all maximally connected models* M *of* T*.*

The maximally connected models of a set of obligations are unique (for a given W and V) if the transitivity axiom **4**: $\Box\alpha \rightarrow \Box\Box\alpha$ is omitted from the axiomatization. The unique maximally connected model of the set of obligations $S = \{O(\alpha_i \mid \beta_i) \mid 1 \leq i \leq n\}$ has the accessibility relation $\{w_1 \leq w_2 \mid$ there is no $O(\alpha|\beta) \in S$ such that $M, w_1 \models \neg\alpha \wedge \beta$ and $M, w_2 \models \alpha \wedge \beta\}$. However, if axiom **4** is omitted, then 'preferred' in $I(\alpha \mid \beta)$ no longer has a natural meaning. Finally, if only full models are considered in the semantics, i.e. models that contain a world for each possible interpretation, then we can derive for example $\emptyset \models_{\sqsubset} \neg O(p|\top)$, because there cannot be a model with only p worlds. In the following section preferential entailment is illustrated by several examples.

5 The three problems reconsidered

In the introduction of this paper we mentioned three problems: the strong preference problem, the contrary-to-duty problem and the dilemma problem. In this section we show how Prohairetic Deontic Logic solves the three problems. The strong preference problem is that preferences for α_1 and α_2 conflict for $\alpha_1 \wedge \neg\alpha_2$ and $\neg\alpha_1 \wedge \alpha_2$. The following example illustrates that the problem is solved by the dynamics of preferential entailment. It also illustrates why the logic is non-monotonic.

Example 1 (Polite and helpful, continued). Consider the three sets of obligations $S = \emptyset$, $S' = \{O(p|\top)\}$ and $S'' = \{O(p|\top), O(h|\top)\}$. The three unique maximally connected models of S, S' and S'' are represented in Figure 1. With no premises,

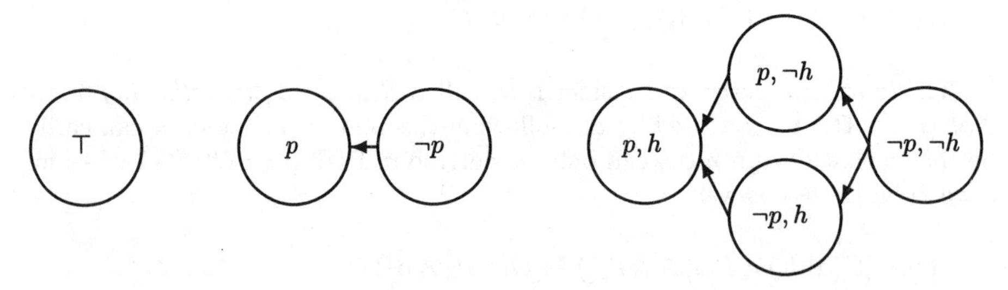

Fig. 1. Unique maximally connected models of $\emptyset$, $\{O(p|\top)\}$ and $\{O(p|\top), O(h|\top)\}$.

all worlds are equally ideal. By addition of the premise $O(p|\top)$, the p worlds are strictly preferred over $\neg p$ worlds. Moreover, by addition of the second premise $O(h|\top)$, the h worlds are strictly preferred over $\neg h$ worlds, and the $p \wedge \neg h$ and $\neg p \wedge h$ worlds become incomparable. Hence, the strong preference problem is solved by representing conflicts with incomparable worlds. This solution uses preferential entailment, a technique from non-monotonic reasoning, because for the preferred models we have that all incomparable worlds refer to some conflict. We have $S' \models_{\sqsubset} O(p|\neg(p \wedge h))$ and $S'' \not\models_{\sqsubset} O(p|\neg(p \wedge h))$. By addition of a formula we loose conclusions.

The solution of the contrary-to-duty problem is based on the dyadic representation. The solution is illustrated by the representation of Forrester's [9] and Chisholm's [6] paradoxes in Prohairetic Deontic Logic. The paradoxes were originally formulated in monadic deontic logic.

Example 2 (Forrester's paradox). Consider $S = \{O(\neg k \mid \top), O(g \mid k), k\}$, where k can be read as 'Smith kills Jones' and g as 'Smith kills him gently,' and g logically implies k. The unique maximally connected model of S is represented in Figure 2. The actual world is any of the k worlds. The formalization of S is unproblematic and the semantics reflect the three states that seem to be implied by the paradox. We have $S \models O(\neg k \vee g \mid \top)$ as a consequence of the theorem $(O(\alpha \mid \neg\beta) \wedge O(\beta \mid \gamma)) \rightarrow O(\alpha \vee \beta \mid \gamma)$, which expresses that $k \wedge \neg g$ is the worst state that should be avoided.

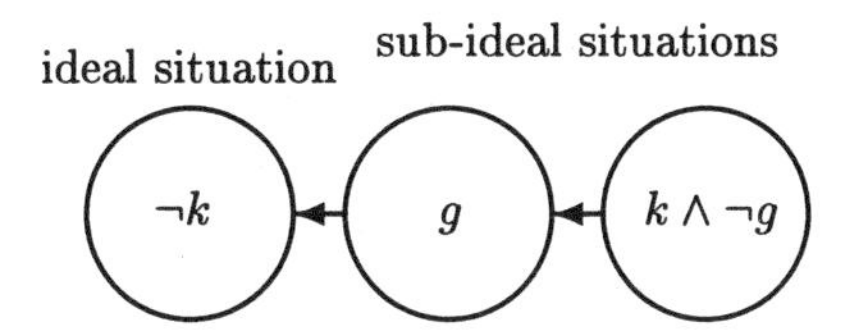

Fig. 2. Unique maximally connected model of $\{O(\neg k|\top), O(g|k), k\}$

Example 3 (Chisholm's paradox). Consider $S = \{O(a|\top), O(t|a), O(\neg t|\neg a), \neg a\}$, where a can be read as 'a certain man going to the assistance of his neighbors' and t as 'telling the neighbors that he will come.' The unique maximally connected model of S is represented in Figure 3. The crucial question of Chisholm's

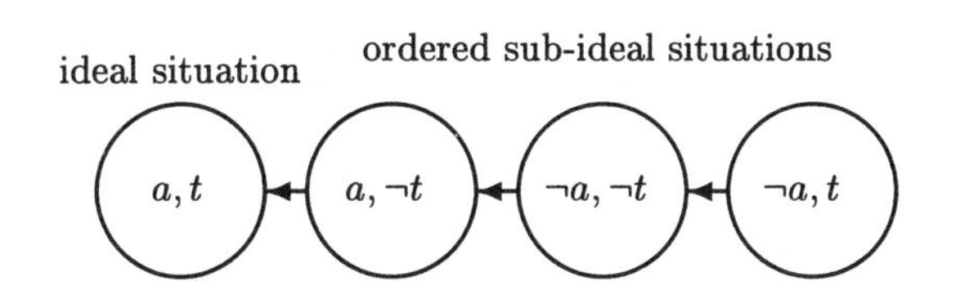

Fig. 3. Unique maximally connected model of $\{O(a|\top), O(t|a), O(\neg t|\neg a), \neg a\}$

paradox is whether there is an obligation that 'the man should tell the neighbors that he will come' $O(t \mid \top)$. This obligation is counterintuitive, given that a is false. This obligation is derived from S by any deontic logic that has so-called deontic detachment (also called deontic transitivity), represented by the following formula $\mathbf{DD}^0$.

$\mathbf{DD}^0$ $(O(\alpha|\beta) \wedge O(\beta|\gamma)) \rightarrow O(\alpha|\gamma)$

However, $\mathbf{DD}^0$ is not valid in Prohairetic Deontic Logic. The obligation 'the man should tell his neighbors that he will come' $O(t|\top)$ cannot be derived in PDL from S. The following weaker version of $\mathbf{DD}^0$ is valid in PDL.

$$\mathbf{RDD}\ (O(\alpha|\beta \wedge \gamma) \wedge O(\beta|\gamma) \wedge I(\alpha \wedge \beta|\gamma)) \to O(\alpha \wedge \beta|\gamma)$$

The obligation that 'the man should go to his neighbors and tell his neighbors that he will come' $O(a \wedge t|\top)$ can (preferentially) be derived from S.

Finally, we show that Prohairetic Deontic Logic solves the dilemma problem, because it makes the considerate assassin set in Example 4 inconsistent, without making the window set in Example 5 inconsistent.

Example 4 (Considerate assassin). Consider $S = \{O(\neg c|\top), O(c|k)\}$, where c can be read as 'Smith offers Jones a cigarette' and k can be read as 'Smith kills Jones.' The set S is inconsistent with $\overset{\leftrightarrow}{\Diamond}(k \wedge \neg c)$, as can be verified as follows. Assume there is a model of S. The obligation $O(c\,|\,k)$ implies $I(c\,|\,k)$, which means that for every world w_1 such that $M, w_1 \models \neg c \wedge k$ there is a world w_2 such that $M, w_2 \models c \wedge k$ and $w_2 < w_1$ (i.e. $w_2 \leq w_1$ and $w_1 \not\leq w_2$). However, the obligation $O(\neg c|\top)$ implies $\neg c \succ c$, which means that for all worlds w_1 such that $M, w_1 \models \neg c \wedge k$ there is not a world w_2 such that $M, w_2 \models c \wedge k$ and $w_2 \leq w_1$. These two conditions are contradictory (if there is such a world w_1).

Moreover, consider $S' = \{O(\neg d|\top), O(d \wedge p|d), O(\neg p|\top)\}$, where d can be read as 'there is a dog' and p as 'there is a poodle.' Prakken and Sergot [27] argue that S' should be inconsistent, based on its analogy with S. For similar reasons as the inconsistency of S above, the set S' is inconsistent in PDL.

Example 5 (Window). Consider $S = \{O(c\,|\,r), O(\neg c\,|\,s)\}$, where c can be read as 'the window is closed,' r as 'it starts raining' and s as 'the sun is shining.' It is argued by von Wright [42] that S does not represent a dilemma and that it should therefore be consistent, see also [1]. In PDL the set S is consistent, and a maximally connected model M of S is given in Figure 4. The ideal worlds satisfy $r \to c$ and $s \to \neg c$, and the sub-ideal worlds either $\neg c \wedge r$ or $c \wedge s$. We have $M \not\models O(c|r \wedge s)$ and thus $S \not\models_{\sqsubset} O(c|r \wedge s)$.

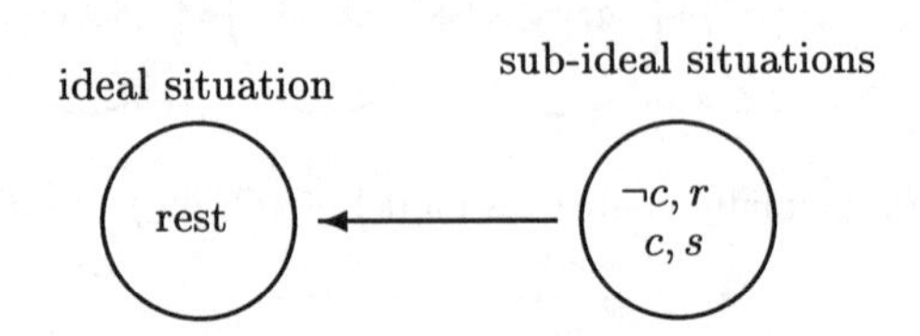

Fig. 4. Preferred model of $\{O(c|r), O(\neg c|s)\}$

Note that there are many maximally connected models. For example, the $c \wedge r$ worlds can be the only preferred worlds, when the $\neg c \wedge s$ worlds are equivalent with the $\neg c \wedge r$ worlds. Alternatively, the $\neg c \wedge s$ worlds can be the only most preferred worlds.

6 Related research

Besides the deontic logics already discussed in this paper, there is a preference-based deontic logic proposed by Brown, Mantha and Wakayama [4]. We write the obligations in their logic as O_{BMW}. At first sight it seems that the logic is closely related to Prohairetic Deontic Logic, because $O_{BMW}\alpha$ also has a mixed representation. However, a further inspection of the definitions reveals that their proposal is quite different from ours. Obligations are defined by

$$O_{BMW}\alpha =_{def} P_f\alpha \wedge A_m\alpha = \overleftarrow{\Box}\neg\alpha \wedge \Diamond\alpha$$

where $P_f\alpha$ is read as 'α is preferred,' $A_m\alpha$ is read as 'α is admissible,' and $\overleftarrow{\Box}\alpha$ is read as 'α is true in all inaccessible worlds.' Hence, $O_{BMW}\alpha$ means 'the truth of α takes us to a world at least as good as the current world and there exists a world at least as good as the current world where α is true' [4, p.200]. The first distinction is that in the logic, dilemmas are consistent (which they consider an advantage, following [40]). Secondly, the motivation for the mixed representation is different. Whereas we introduced the mixed representation to solve both the contrary-to-duty problem (for which we use $\alpha \succ \neg\alpha$) and the dilemma problem (for which we use $I\alpha$), they use the mixed representation to block the derivation of the theorem $O_{BMW}(\alpha_1 \vee \alpha_2) \rightarrow O_{BMW}\alpha_1$, which they consider as 'somewhat unreasonable, since the premise is weaker than the conclusion.' However, it is easily checked that the logic validates $O_{BMW}(\alpha_1 \vee \alpha_2) \wedge A_m\alpha_1 \rightarrow O_{BMW}\alpha_1$. Hence, under certain circumstances stronger obligations can be derived. This counterintuitive formula is not a theorem of PDL.[6]

7 Conclusions

In this paper, we introduced Prohairetic Deontic Logic. We showed that it gives a satisfactory solution to the strong preference problem, the contrary-to-duty problem and the dilemma problem. We now study the use of PDL for legal expert systems and to specify intelligent agents for the Internet [7, 8] like the drafting, negotiation and processing of trade contracts in electronic commerce, and the relevance for logics of desires and goals as these are developed in qualitative decision theory [25, 36]. Moreover, in [38] an implementation is envisaged for a sublanguage of PDL.

Acknowledgements

Thanks to David Makinson for comments on an earlier version of this paper.

[6] The logic PDL is the basis of the logics developed in [32]. A previous version appeared as a part of the two-phase system [30]. Moreover, the logic PDL has been presented at the AAAI 1997 spring symposium on Qualitative Approaches to Deliberation and Reasoning. Finally, the dynamic interpretation of PDL with preferential entailment is related to more recent work on dynamic update semantics for obligations [38, 39] though the formal construction is radically different and other issues have been discussed (in particular the different notions of defeasibility, following [33, 35]).

References

1. C.E. Alchourrón. Philosophical foundations of deontic logic and the logic of defeasible conditionals. In J.-J. Meyer and R. Wieringa, editors, *Deontic Logic in Computer Science: Normative System Specification*, pages 43–84. John Wiley & Sons, 1993.
2. L. Åqvist. Good Samaritans, contrary-to-duty imperatives, and epistemic obligations. *Noûs*, 1:361–379, 1967.
3. C. Boutilier. Conditional logics of normality: a modal approach. *Artificial Intelligence*, 68:87–154, 1994.
4. A.L. Brown, S. Mantha, and T. Wakayama. Exploiting the normative aspect of preference: a deontic logic without actions. *Annals of Mathematics and Artificial Intelligence*, 9:167–203, 1993.
5. B.F. Chellas. *Modal Logic: An Introduction.* Cambridge University Press, 1980.
6. R.M. Chisholm. Contrary-to-duty imperatives and deontic logic. *Analysis*, 24:33–36, 1963.
7. R. Conte and R. Falcone. ICMAS'96: Norms, obligations, and conventions. *AI Magazine*, 18,4:145–147, 1997.
8. B.S. Firozabadhi and L.W.N. van der Torre. Towards an analysis of control systems. In H. Prade, editor, *Proceedings of the ECAI'98*, pages 317–318, 1998.
9. J.W. Forrester. Gentle murder, or the adverbial Samaritan. *Journal of Philosophy*, 81:193–197, 1984.
10. L. Goble. A logic of good, would and should, part 2. *Journal of Philosophical Logic*, 19:253–276, 1990.
11. L. Goble. Murder most gentle: the paradox deepens. *Philosophical Studies*, 64:217–227, 1991.
12. B. Hansson. An analysis of some deontic logics. In R. Hilpinen, editor, *Deontic Logic: Introductionary and Systematic Readings*, pages 121–147. D. Reidel Publishing Company, Dordrecht, Holland, 1971. Reprint from *Noûs*, 1969.
13. S.O. Hansson. A new semantical approach to the logic of preference. *Erkenntnis*, 31:1–42, 1989.
14. S.O. Hansson. Defining "good" and "bad" in terms of "better". *Notre Dame Journal of Formal Logic*, 31:136–149, 1990.
15. S.O. Hansson. Preference-based deontic logic (PDL). *Journal of Philosophical Logic*, 19:75–93, 1990.
16. S.O. Hansson. Situationist deontic logic. *Journal of Philosophical Logic*, 26:423–448, 1997.
17. Z. Huang and M. Masuch. The logic of permission and obligation in the framework of ALX3: how to avoid the paradoxes of deontic logic. *Logique et Analyse*, 149, 1997.
18. H.G. Hughes and M.J. Creswell. *A Companion to Modal Logic.* Methuen, London, 1984.
19. F. Jackson. On the semantics and logic of obligation. *Mind*, 94:177–196, 1985.
20. R.E. Jennings. Can there be a natural deontic logic? *Synthese*, 65:257–274, 1985.
21. P. Lamarre. S4 as the conditional logic of nonmonotonicity. In *Proceedings of the KR'91*, pages 357–367, 1991.
22. D. Lewis. *Counterfactuals.* Blackwell, Oxford, 1973.
23. D. Lewis. Semantic analysis for dyadic deontic logic. In S. Stunland, editor, *Logical Theory and Semantical Analysis*, pages 1–14. D. Reidel Publishing Company, Dordrecht, Holland, 1974.

24. B. Loewer and M. Belzer. Dyadic deontic detachment. *Synthese*, 54:295–318, 1983.
25. J. Pearl. From conditional oughts to qualitative decision theory. In *Proceedings of the 9th Conference on Uncertainty in Artificial Intelligence (UAI'93)*, pages 12–20, 1993.
26. H. Prakken and M.J. Sergot. Contrary-to-duty obligations. *Studia Logica*, 57:91–115, 1996.
27. H. Prakken and M.J. Sergot. Dyadic deontic logic and contrary-to-duty obligations. In D. Nute, editor, *Defeasible Deontic Logic*, pages 223-262. Kluwer, 1997.
28. Y. Moses R. Fagin, J.Y. Halpern and M.Y. Vardi. *Reasoning About Knowledge*. MIT press, 1995.
29. S.-W. Tan and J. Pearl. Specification and evaluation of preferences under uncertainty. In *Proceedings of the KR'94*, pages 530–539, 1994.
30. Y.-H. Tan and L.W.N. van der Torre. How to combine ordering and minimizing in a deontic logic based on preferences. In *Deontic Logic, Agency and Normative Systems. Proceedings of the ΔEON'96*, Workshops in Computing, pages 216–232. Springer Verlag, 1996.
31. L.W.N. van der Torre. Violated obligations in a defeasible deontic logic. In *Proceedings of the ECAI'94*, pages 371–375, 1994.
32. L.W.N. van der Torre. *Reasoning About Obligations: Defeasibility in Preference-based Deontic Logics*. PhD thesis, Erasmus University Rotterdam, 1997.
33. L.W.N. van der Torre and Y.-H. Tan. Cancelling and overshadowing: two types of defeasibility in defeasible deontic logic. In *Proceedings of the IJCAI'95*, pages 1525–1532, 1995.
34. L.W.N. van der Torre and Y.-H. Tan. Contextual deontic logic. In *Proceedings of the First International Conference on Modeling and Using Context (CONTEXT'97)*, pages 1–12, Rio de Janeiro, Brazil, 1997.
35. L.W.N. van der Torre and Y.-H. Tan. The many faces of defeasibility in defeasible deontic logic. In D. Nute, editor, *Defeasible Deontic Logic*, pages 79–121. Kluwer, 1997.
36. L.W.N. van der Torre and Y.-H. Tan. Diagnosis and decision making in normative reasoning. *Artificial Intelligence and Law*, 1998.
37. L.W.N. van der Torre and Y.-H. Tan. The temporal analysis of Chisholm's paradox. In *Proceedings of the AAAI'98*, 1998.
38. L.W.N. van der Torre and Y.-H. Tan. An update semantics for deontic reasoning. In P. McNamara and H. Prakken, editors, *Norms, Logics and Information Systems. New Studies on Deontic Logic and Computer Science*. IOS Press, 1998.
39. L.W.N. van der Torre and Y.-H. Tan. An update semantics for prima facie obligations. In H. Prade, editor, *Proceedings of the ECAI'98*, pages 38–42, 1998.
40. B.C. van Fraassen. Values and the heart command. *Journal of Philosophy*, 70:5–19, 1973.
41. G.H. von Wright. *The Logic of Preference*. Edinburgh University Press, 1963.
42. G.H. von Wright. A new system of deontic logic. In R. Hilpinen, editor, *Deontic Logic: Introductory and Systematic Readings*, pages 105–120. D. Reidel Publishing company, Dordrecht, Holland, 1971.
43. E. Weydert. Hyperrational conditionals. monotonic reasoning about nested default conditionals. In *Foundations of Knowledge Representation and Reasoning*, LNAI 810, pages 310–332. Springer, 1994.
44. R.J. Wieringa and J.-J.Ch. Meyer. Applications of deontic logic in computer science: A concise overview. In *Deontic Logic in Computer Science*, pages 17–40. John Wiley & Sons, Chichester, England, 1993.

Phased Labeled Logics of Conditional Goals

Leendert W.N. van der Torre

IRIT, Paul Sabatier University, Toulouse, France
torre@irit.fr

Abstract. In this paper we introduce phased labeled logics of conditional goals. Labels are used to impose restrictions on the proof theory of the logic. The restriction discussed in this paper is that a proof rule can be blocked in a derivation due to the fact that another proof rule has been applied earlier in the derivation. We call a set of proof rules that can be applied in any order a phase in the proof theory. We propose a one-phase logic of goals containing four proof rules, and we show that it is equivalent to a four-phase logic of goals in which each phase contains exactly one proof rule. The proof theory of the four-phase logic of goals is much more efficient, because other orderings no longer have to be considered.

1 Introduction

In the usual approaches to planning in AI, a planning agent is provided with a description of some state of affairs, a *goal state*, and charged with the task of discovering (or performing) some sequence of actions to achieve that goal. Recently several *logics* for conditional or context-sensitive goals and desires have been proposed [3, 2, 10, 1, 12, 11, 8, 7] in the context of *qualitative* decision theory. In [15] we introduced a version of a labeled deductive system [4] to reason about goals. Labeled goals $G(\alpha|\beta)_L$ can *roughly* be read as 'preferably α if β, against the background of L.' The label keeps track of the context in which the goal is derived. It has some desirable properties not found in other proposals. First, the logic can reason about conflicting goals. This is important, because goals only impose partial preferences, i.e. preferences given some objective and given some context. Objectives can conflict and, as a consequence, goals with overlapping contexts can conflict. Second, the labeled logics are stronger than previous proposals in the sense that they validate strengthening of the antecedent and transitivity. It has been shown in [15] that these proof rules can *only* be combined with the desirable proof rule weakening of the consequent if additional machinery like labels is introduced in the logic. Otherwise counterintuitive conclusions follow.

In the phased labeled logics of goals (PLLG) introduced in this paper we show how to phase derivations. To impose phasing restrictions on the derivations, the labeled logics of goals are extended in two ways. First, a phase is associated with each PLLG-proof rule by an explicitly given phasing function. Second, the label of a PLLG-goal not only contains a set of sets of propositional formulas, the fulfillments (F) of the premises from which the goal is derived, but it also contains an integer, the phase (p) in which the goal is derived. There are two restrictions on the use of proof rules in PLLG. First, as in other labeled logics of goals, a consistency check on F is used to restrict derivations such that all premises can be fulfilled. The fulfillment of a premise $G(\alpha|\beta)$ is the propositional formula $\alpha \wedge \beta$ [19] and the fulfillment of a derived goal depends on the premises from which it is derived. Later in this paper we show how this consistency check restricts derivations to a kind of constructive proofs, in the sense that no goals are derived from conflicts. Second, and this is new, the phase p is used to select the proof rules that still may be applied.

In this paper we focus on a one-phase labeled logic of goals (LLG) and a four-phase labeled logic of goals (4LLG). Both contain the four proof rules strengthening of the antecedent, a type of transitivity, weakening of the consequent and the disjunction rule for the antecedent. Moreover, we show that the conjunction rule for the consequent follows from these four rules. In LLG these rules can be applied in any order, but in 4LLG there is only one order in which they can be applied. Nevertheless, we prove that every formula that can be derived in LLG can also be derived in 4LLG. Consequently, in the proof theory of LLG we can restrict ourselves to one specific order of the proof rules, which makes the logic much more efficient. Finally, we also prove that in 4LLG the consistency check on the label can be replaced by a consistency check on the antecedent and consequent.

2 Phased labeled logic of goals (PLLG)

Phased labeled logics of goals are versions of a labeled deductive system as it was introduced by Gabbay in [4]. Roughly speaking, the label L of a goal $G(\alpha|\beta)_L$ consists of a record of the fulfillments (F) of the premises that are used in the derivation of $G(\alpha|\beta)$, and the phase (p) in which it is derived. Where there is no application of reasoning by cases, F can be taken to be a set of boolean formulas, that grows by joining sets as premises are combined. But in general, to cover the parallel tracks created through reasoning by cases, we need to consider sets of sets of boolean formulas [9].

Definition 1 (Language). *Let $\mathcal{L}$ be a propositional base logic. The language of* PLLG *consists of the labeled dyadic goals $G(\alpha|\beta)_L$, with α and β sentences of $\mathcal{L}$, and L a pair (F,p) that consists of a set of sets of sentences of $\mathcal{L}$ (fulfillments) and an integer (the phase). We write $\models$ for entailment in $\mathcal{L}$.*

Each formula $G(\alpha|\beta)_L$ occurring as a premise has a label that consists of its own (propositionally consistent) fulfillment and phase 0.

Definition 2 (Premise). *A formula $G(\alpha|\beta)_{(\{\{\alpha\wedge\beta\}\},0)}$, where $\alpha \wedge \beta$ is consistent in $\mathcal{L}$, is called a premise of* PLLG.

The phase of a goal is determined by the proof rule used to derive the goal, and the set of fulfillments is the union (OR) or the product (SA, TRANS) of the labels of the premises used in this inference rule, where the product is defined by $\{S_1, \ldots, S_n\} \times \{T_1, \ldots, T_m\} = \{S_1 \cup T_1, \ldots, S_1 \cup T_m, \ldots, S_n \cup T_m\}$. The labels are used to check that fulfillments are consistent and that the phase of reasoning is non-decreasing. In a normative context the consistency check realizes a variant of the *Kantian principle* that '*ought implies can.*'

Definition 3 (PLLG). *Let ρ be a phasing function that associates with each proof rule below an integer called its phase. The phased labeled logic of goals* PLLG *for ρ consists of the inference rules below, extended with the following two conditions $R = R_F + R_p$.*

R_F: *$G(\alpha|\beta)_{(F,p)}$ may only be derived if each $F_i \in F$ is consistent: it must always be possible to fulfill a derived goal and each of the goals it is derived from, though not necessarily all of them at the same time.*

R_p: *$G(\alpha|\beta)_{(F,p)}$ may only be derived if $p \geq p_i$ for all goals $G(\alpha_i|\beta_i)_{(F_i,p_i)}$ it is derived from.*

The inference rules of PLLG *are replacements by logical equivalents (for antecedent and consequent) and the following four rules.*

$$\text{SA}_R : \frac{G(\alpha \mid \beta_1)_{(F,p)}, R}{G(\alpha \mid \beta_1 \wedge \beta_2)_{(F \times \{\beta_2\}, \rho(\text{SA}))}}$$

$$\text{TRANS}_R : \frac{G(\alpha \mid \beta \wedge \gamma)_{(F_1,p_1)}, G(\beta \mid \gamma)_{(F_2,p_2)}, R}{G(\alpha \wedge \beta \mid \gamma)_{(F_1 \times F_2, \rho(\text{TRANS}))}}$$

$$\text{WC}_R : \frac{G(\alpha_1 \mid \beta)_{(F,p)}, R}{G(\alpha_1 \vee \alpha_2 \mid \beta)_{(F, \rho(\text{WC}))}}$$

$$\text{OR}_R : \frac{G(\alpha \mid \beta_1)_{(F_1,p_1)}, G(\alpha \mid \beta_2)_{(F_2,p_2)}, R}{G(\alpha \mid \beta_1 \vee \beta_2)_{(F_1 \cup F_2, \rho(\text{OR}))}}$$

We say $\{G(\alpha_i|\beta_i) \mid 1 \leq i \leq n\} \vdash_{\text{PLLG}} G(\alpha|\beta)$ if there is a labeled goal $G(\alpha|\beta)_L$ that can be derived from the set of goals $\{G(\alpha_i|\beta_i)_{(\{\{\alpha_i \wedge \beta_i\}\},0)} \mid 1 \leq i \leq n\}$.

The unusual transitivity rule (TRANS$_R$) implies, under certain circumstances, the standard transitivity rule as well as the conjunction rule. First, if we have $\rho(\text{SA}) \leq \rho(\text{TRANS}) \leq \rho(\text{WC})$, then we have the following derivation.

$$\dfrac{\dfrac{\dfrac{G(\alpha|\beta)_{(F_1,p_1)}}{G(\alpha|\beta \wedge \gamma)_{(F_1 \times \{\gamma\}, \rho(\text{SA}))}}\text{SA} \quad G(\beta|\gamma)_{(F_2,p_2)}}{G(\alpha \wedge \beta|\gamma)_{(F_1 \times \{\gamma\} \times F_2, \rho(\text{TRANS}))}}\text{TRANS}}{G(\alpha|\gamma)_{(F_1 \times \{\gamma\} \times F_2, \rho(\text{WC}))}}\text{WC}$$

Consequently, the following standard transitivity rule is implied by PLLG if we have $\rho(\text{SA}) = \rho(\text{TRANS}) = \rho(\text{WC})$, i.e. if SA, TRANS and WC are in the same phase, because F_2 implies γ (see also Proposition 2).

$$\text{TRANS}'_R : \frac{G(\alpha \mid \beta)_{(F_1,p_1)}, G(\beta \mid \gamma)_{(F_2,p_2)}, R}{G(\alpha \mid \gamma)_{(F_1 \times F_2, \rho(\text{TRANS}))}}$$

Second, concerning the conjunction rule, if $\rho(\text{SA}) \leq \rho(\text{TRANS})$, then we can first strengthen $G(\alpha_1|\beta)$ to $G(\alpha_1|\beta \wedge \alpha_2)$, and then apply TRANS as follows to derive $G(\alpha_1 \wedge \alpha_2|\beta)$.

$$\dfrac{\dfrac{G(\alpha_1|\beta)_{(F_1,p_1)}}{G(\alpha_1|\beta \wedge \alpha_2)_{(F_1\times\{\alpha_2\},\rho(\text{SA}))}}\text{SA} \qquad G(\alpha_2|\beta)_{(F_2,p_2)}}{G(\alpha_1 \wedge \alpha_2|\beta)_{(F_1\times\{\alpha_2\}\times F_2,\rho(\text{TRANS}))}}\text{TRANS}$$

Consequently, the following conjunction rule is implied by the logic PLLG if we have $\rho(\text{SA}) = \rho(\text{TRANS})$, i.e. if SA and TRANS are in the same phase, because F_2 implies α_2 (Proposition 2).

$$\text{AND}_R : \frac{G(\alpha_1 \mid \beta)_{(F_1,p_1)}, G(\alpha_2 \mid \beta)_{(F_2,p_2)}, R}{G(\alpha_1 \wedge \alpha_2 \mid \beta)_{(F_1\times F_2,\rho(\text{TRANS}))}}$$

Labeled logics of goals can reason about conflicting goals, and they can combine several proof rules without deriving counterintuitive consequences. It has been shown in [15] that this can be achieved by using the fulfillments without using the phases. First, the logics can reason about conflicting goals, because we have $G(p), G(\neg p) \not\vdash_{\text{PLLG}} G(p \wedge \neg p)$ and $G(p), G(\neg p) \not\vdash_{\text{PLLG}} G(q)$, where $G(\alpha)$ is short for $G(\alpha|\top)$, $\top$ stands for any tautology like $p \vee \neg p$ and q is not logically implied by p or $\neg p$. In particular, the fulfillments in the labels are used to block the second derivation step in the following counterintuitive derivation [15], in which a blocked derivation step is represented by a dashed line. Moreover, from the results presented later in this paper follows that this counterintuitive derivation can also be blocked by giving WC a higher phase than AND.

$$\dfrac{\dfrac{G(p)}{G(p \vee q)}\text{WC} \qquad G(\neg p)}{\dfrac{G(q \wedge \neg p)}{G(q)}\text{WC}}\text{AND (dashed)}$$

Second, it is easily checked how the fulfillments in the labels are used to combine strengthening of the antecedent (SA) with weakening of the consequent (WC) without validating the following counterintuitive derivation [15]. Moreover, this counterintuitive derivation can also be blocked by giving WC a higher phase than SA, together with a consistency check on the conjunction of antecedent and consequent.

$$\dfrac{\dfrac{G(c|\top)}{G(c \vee t|\top)}\text{WC}}{G(c \vee t|\neg c)}\text{SA (dashed)}$$

3 Labeled logic of goals (LLG): some examples

In the this section we illustrate the use of the consistency check on F in the labeled logics of goals by a variant of the logic proposed in [15], and in the following sections we study the use of different phases in the proof theory. The logic LLG is the PLLG that consists of only one phase.

Definition 4 (LLG). *The logic* LLG *is the* PLLG *with the phasing function* ρ *defined by* $\rho(\text{SA}) = 1$, $\rho(\text{TRANS}) = 1$, $\rho(\text{WC}) = 1$, $\rho(\text{OR}) = 1$.

The logic LLG derives the proof rules TRANS′ and AND discussed in the previous section, because SA, TRANS and WC (respectively SA and TRANS) are in the same phase. The following example illustrates the labeled logic of conditional goals.

Example 1. Consider the set $S = \{G(a \vee p|\top), G(\neg a|\top)\}$ as premise set, where a can be read as 'buying apples' and p as 'buying pears' (taken from [13]). We have $S \not\vdash_{\text{LLG}} G(p\,|\,a)$, as desired. Below it is shown how two derivations of the counterintuitive $G(p|a)_L$ are blocked. The non-derived goal is counterintuitive, because when a is true (its antecedent) then the first premise is fulfilled and the second is violated. This pattern holds irrespective of whether p is true or false. Buying pears does not 'improve' the situation, once apples are bought. Hence, once a is assumed, there is no longer any reason to infer p.

$$\dfrac{\dfrac{\dfrac{G(a \vee p|\top)_{(\{\{a\vee p\}\},0)} \quad G(\neg a|\top)_{(\{\{\neg a\}\},0)}}{G(\neg a \wedge p|\top)_{(\{\{a\vee p,\neg a\}\},1)}}\,\text{AND}}{G(\neg a \wedge p|a)_{(\{\{a\vee p,\neg a,a\}\},1)}}\,\text{SA}}{G(p|a)_{(\{\{a\vee p,\neg a,a\}\},1)}}\,\text{WC}
\qquad
\dfrac{\dfrac{\dfrac{G(a \vee p|\top)_{(\{\{a\vee p\}\},0)} \quad G(\neg a|\top)_{(\{\{\neg a\}\},0)}}{G(\neg a \wedge p|\top)_{(\{\{a\vee p,\neg a\}\},1)}}\,\text{AND}}{G(p|\top)_{(\{\{a\vee p,\neg a\}\},1)}}\,\text{WC}}{G(p|a)_{(\{\{a\vee p,\neg a,a\}\},1)}}\,\text{SA}$$

The following example illustrates how the transitivity rule formalizes that conditional rules can be applied one after the other. Derivations go 'as far as possible.'

Example 2. Consider the set of goals $S = \{G(a\,|\,b), G(b\,|\,c), G(c\,|\,\top), G(\neg a\,|\,\top)\}$. There is a conflict for a, because we have $S \vdash_{\text{LLG}} G(a\,|\,\top)$ and $S \vdash_{\text{LLG}} G(\neg a\,|\,\top)$. There is not a conflict for b, because we have $S \vdash_{\text{LLG}} G(b|\top)$ and $S \not\vdash_{\text{LLG}} G(\neg b|\top)$. Hence, derivation chains go as far as possible and there is no weak contraposition as, for example, in conditional entailment [5], see also the discussion in [9]. Moreover, consider the set of goals $S' = \{G(a|b), G(\neg c|b \vee c)\}$. We have $S \vdash_{\text{LLG}} G(a|b \vee c)$, as desired. The following derivation illustrates how this more complex form of transitivity is supported.

$$\dfrac{\dfrac{\dfrac{G(a|b)_{(\{\{a\wedge b\}\},0)}}{G(a|b \wedge \neg c)_{(\{\{a\wedge b,\neg c\}\},1)}}\,\text{SA} \qquad G(\neg c|b \vee c)_{(\{\{b\wedge\neg c\}\},0)}}{G(a \wedge \neg c|b \vee c)_{(\{\{a\wedge b,\neg c,b\wedge\neg c\}\},1)}}\,\text{TRANS}}{G(a|b \vee c)_{(\{\{a\wedge b,\neg c,b\wedge\neg c\}\},1)}}\,\text{WC}$$

The third example illustrates how the disjunction rule supports reasoning by cases.

Example 3. Consider the set $S = \{G(a \wedge c|b), G(a \wedge \neg c|\neg b)\}$ (taken from [9]). We have $S \vdash_{\text{LLG}} G(a|\top)$, as desired. The following derivation illustrates this complex type of reasoning by cases.

$$\dfrac{\dfrac{G(a \wedge c|b)_{(\{\{a\wedge c\wedge b\}\},0)}}{G(a|b)_{(\{\{a\wedge c\wedge b\}\},1)}}\text{WC} \quad \dfrac{G(a \wedge \neg c|\neg b)_{(\{\{a\wedge\neg c\wedge\neg b\}\},0)}}{G(a|\neg b)_{(\{\{a\wedge\neg c\wedge\neg b\}\},1)}}\text{WC}}{G(a|\top)_{(\{\{a\wedge c\wedge b\},\{a\wedge\neg c\wedge\neg b\}\}),1)}}\text{OR}$$

On the other hand, consider the set $S = \{G(a|b), G(a|\neg b)\}$. In [14,21] it is argued that $G(a|a \leftrightarrow b)$ is counterintuitive and should therefore not be derived. We have $S \not\vdash_{\text{LLG}} G(a\,|\,a \leftrightarrow b)$, and we have the blocked derivation below. The non-derived goal is counterintuitive, because when $a \leftrightarrow b$ is true (its antecedent) then the first premise is fulfilled when its antecedent is true and the second is violated when its antecedent is true. This pattern holds irrespective of whether a is true or false. Hence, once $a \leftrightarrow b$ is assumed, there is no longer any reason to infer a.

$$\dfrac{\dfrac{G(a|b)_{(\{\{a\wedge b\}\},0)} \quad G(a|\neg b)_{(\{\{a\wedge\neg b\}\},0)}}{G(a|\top)_{(\{\{a\wedge b\},\{a\wedge\neg b\}\},1)}}\text{OR}}{G(a|a \leftrightarrow b)_{(\{a\wedge b, a\leftrightarrow b\},\{a\wedge\neg b, a\leftrightarrow b\}\},1)}}\text{SA (dashed)}$$

In the following section we study the additional expressive power of PLLG over LLG by introducing different phases in the proof theory.

4 Four-phase labeled logic of goals (4LLG)

In this section we discuss a labeled deontic logic which completely orders the derivations of LLG in the following order: SA, TRANS, WC, OR. We call the resulting logic four-phase labeled logic of goals 4LLG.

Definition 5 (4LLG). *The logic* 4LLG *is the* PLLG *with the phasing function* ρ *defined by* $\rho(\text{SA}) = 1$, $\rho(\text{TRANS}) = 2$, $\rho(\text{WC}) = 3$, $\rho(\text{OR}) = 4$.

In Theorem 1 below we show that for each LLG derivation there is an equivalent 4LLG derivation. We first prove three propositions.

Proposition 1. *Consider any potential derivation of* PLLG*, satisfying the condition* R_p *but not necessarily* R_F*. Then the following two conditions are equivalent:*

1. *The final derivation satisfies condition* R_F*,*
2. *The derivation satisfies* R_F *everywhere.*

Proof *The labels are, in a suitable sense, cumulative. Every element of every label in the derivation is classically implied by some element of the label of the final conclusion.*

Proposition 2. *For each goal $G(\alpha \mid \beta)_{(F,p)}$ derived in* PLLG *we have for each $F_i \in F$ that $F_i \models \alpha \wedge \beta$.*

Proof *By induction on the structure of the proof tree. The property trivially holds for the premises, and it is easily seen that the proof rules retain the property.*

Proposition 3. *We can replace two subsequent steps of an* LLG *derivation by an equivalent* 4LLG *derivation.*

Proof *The replacements are given below. For each replacement the original* LLG *derivation as well as its* 4LLG *replacement are given. From Proposition 2 follows that the replacement does not violate the consistency check. For example, consider the reversing e.1. of* OR_4 *and* TRANS_2. *Call the fulfillments of the three premises $G(\alpha|\beta_1 \wedge \gamma)$, $G(\alpha|\beta_2 \wedge \gamma)$ and $G(\beta_1 \vee \beta_2|\gamma)$ respectively F_1, F_2 and F_3. From the* LLG*-derivation follows that each element of $(F_1 \cup F_2) \times F_3$ is consistent, and therefore all elements of $F_1 \times F_3$ and $F_2 \times F_3$ are consistent. Moreover, from Proposition 2 follows for each $F_{1,i} \in F_1$ that $F_{1,i} \models \beta_1$ and for each $F_{2,i} \in F_2$ that $F_{2,i} \models \beta_2$. Consequently, for the* 4LLG*-derivation we have that the labels of the replacements $F_1 \times \{\beta_1 \vee \neg\beta_2\} \times F_3$ and $F_2 \times \{\beta_2 \vee \neg\beta_1\} \times F_3$ are equivalent to $F_1 \times F_3$ and $F_2 \times F_3$, and all elements of them are therefore consistent. The other proofs are analogous.*

$$\dfrac{\dfrac{G(\alpha|\beta \wedge \gamma_1) \quad G(\beta|\gamma_1)}{G(\alpha \wedge \beta|\gamma_1)}\,\text{TRANS}}{G(\alpha \wedge \beta|\gamma_1 \wedge \gamma_2)}\,\text{SA} \qquad \dfrac{\dfrac{G(\alpha|\beta \wedge \gamma_1)}{G(\alpha|\beta \wedge \gamma_1 \wedge \gamma_2)}\,\text{SA} \quad \dfrac{G(\beta|\gamma_1)}{G(\beta|\gamma_1 \wedge \gamma_2)}\,\text{SA}}{G(\alpha \wedge \beta|\gamma_1 \wedge \gamma_2)}\,\text{TRANS}$$

a. Reversing the order of TRANS_2 and SA_1

$$\dfrac{\dfrac{G(\alpha_1|\beta_1)}{G(\alpha_1 \vee \alpha_2|\beta_1)}\,\text{WC}}{G(\alpha_1 \vee \alpha_2|\beta_1 \wedge \beta_2)}\,\text{SA} \qquad \dfrac{\dfrac{G(\alpha_1|\beta_1)}{G(\alpha_1|\beta_1 \wedge \beta_2)}\,\text{SA}}{G(\alpha_1 \vee \alpha_2|\beta_1 \wedge \beta_2)}\,\text{WC}$$

b. Reversing the order of WC_3 and SA_1

$$\dfrac{\dfrac{G(\alpha_1|\beta \wedge \gamma)}{G(\alpha_1 \vee \alpha_2|\beta \wedge \gamma)}\,\text{WC} \quad G(\beta|\gamma)}{G((\alpha_1 \vee \alpha_2) \wedge \beta|\gamma)}\,\text{TRANS} \qquad \dfrac{\dfrac{G(\alpha_1|\beta \wedge \gamma) \quad G(\beta|\gamma)}{G(\alpha_1 \wedge \beta|\gamma)}\,\text{TRANS}}{G((\alpha_1 \vee \alpha_2) \wedge \beta|\gamma)}\,\text{WC}$$

c.1. Reversing the order of WC_3 and TRANS_2

$$\dfrac{G(\alpha|(\beta_1 \vee \beta_2) \wedge \gamma) \quad \dfrac{G(\beta_1|\gamma)}{G(\beta_1 \vee \beta_2|\gamma)}\,\text{WC}}{G((\alpha \wedge (\beta_1 \vee \beta_2)|\gamma)}\,\text{TRANS} \qquad \dfrac{\dfrac{\dfrac{G(\alpha|(\beta_1 \vee \beta_2) \wedge \gamma)}{G(\alpha|\beta_1 \wedge \gamma)}\,\text{SA} \quad G(\beta_1|\gamma)}{G(\alpha \wedge \beta_1|\gamma)}\,\text{TRANS}}{G((\alpha \wedge (\beta_1 \vee \beta_2)|\gamma)}\,\text{WC}$$

c.2. Reversing the order of WC_3 and TRANS_2

$$\frac{\dfrac{G(\alpha|\beta_1)\quad G(\alpha|\beta_2)}{G(\alpha|\beta_1 \vee \beta_2)}\,\text{OR}}{G(\alpha|(\beta_1 \vee \beta_2) \wedge \beta_3)}\,\text{SA} \qquad \frac{\dfrac{G(\alpha|\beta_1)}{G(\alpha|\beta_1 \wedge \beta_3)}\,\text{SA}\quad \dfrac{G(\alpha|\beta_2)}{G(\alpha|\beta_2 \wedge \beta_3)}\,\text{SA}}{G(\alpha|(\beta_1 \wedge \beta_3) \vee (\beta_2 \wedge \beta_3))}\,\text{OR}$$

d. Reversing the order of OR_4 and SA_1

$$\frac{\dfrac{G(\alpha|\beta_1 \wedge \gamma)\quad G(\alpha|\beta_2 \wedge \gamma)}{G(\alpha|(\beta_1 \vee \beta_2) \wedge \gamma)}\,\text{OR}\quad G(\beta_1 \vee \beta_2|\gamma)}{G(\alpha \wedge (\beta_1 \vee \beta_2)|\gamma)}\,\text{TRANS}$$

$$\frac{\dfrac{G(\alpha|\beta_1 \wedge \gamma)\quad \dfrac{G(\beta_1 \vee \beta_2|\gamma)}{G(\beta_1 \vee \beta_2|\gamma \wedge (\beta_1 \vee \neg\beta_2))}\,\text{SA}}{G(\alpha \wedge (\beta_1 \vee \beta_2)|\gamma \wedge (\beta_1 \vee \neg\beta_2))}\,\text{TRANS}\quad \dfrac{G(\alpha|\beta_2 \wedge \gamma)\quad \dfrac{G(\beta_1 \vee \beta_2|\gamma)}{G(\beta_1 \vee \beta_2|\gamma \wedge (\beta_2 \vee \neg\beta_1))}\,\text{SA}}{G(\alpha \wedge (\beta_1 \vee \beta_2)|\gamma \wedge (\beta_2 \vee \neg\beta_1))}\,\text{TRANS}}{G(\alpha \wedge (\beta_1 \vee \beta_2)|\gamma)}\,\text{OR}$$

e.1. Reversing the order of OR_4 and TRANS_2

$$\frac{G(\alpha|\beta \wedge (\gamma_1 \vee \gamma_2))\quad \dfrac{G(\beta|\gamma_1)\quad G(\beta|\gamma_2)}{G(\beta|\gamma_1 \vee \gamma_2)}\,\text{OR}}{G(\alpha \wedge \beta|\gamma_1 \vee \gamma_2)}\,\text{TRANS}$$

$$\frac{\dfrac{\dfrac{G(\alpha|\beta \wedge (\gamma_1 \vee \gamma_2))}{G(\alpha|\beta \wedge \gamma_1)}\,\text{SA}\quad G(\beta|\gamma_1)}{G(\alpha \wedge \beta|\gamma_1)}\,\text{TRANS}\quad \dfrac{\dfrac{G(\alpha|\beta \wedge (\gamma_1 \vee \gamma_2))}{G(\alpha|\beta \wedge \gamma_2)}\,\text{SA}\quad G(\beta|\gamma_2)}{G(\alpha \wedge \beta|\gamma_2)}\,\text{TRANS}}{G(\alpha \wedge \beta|\gamma_1 \vee \gamma_2)}\,\text{OR}$$

e.2. Reversing the order of OR_4 and TRANS_2

$$\frac{\dfrac{G(\alpha_1|\beta_1)\quad G(\alpha_1|\beta_2)}{G(\alpha_1|\beta_1 \vee \beta_2)}\,\text{OR}}{G(\alpha_1 \vee \alpha_2|\beta_1 \vee \beta_2)}\,\text{WC} \qquad \frac{\dfrac{G(\alpha_1|\beta_1)}{G(\alpha_1 \vee \alpha_2|\beta_1)}\,\text{WC}\quad \dfrac{G(\alpha_1|\beta_2)}{G(\alpha_1 \vee \alpha_2|\beta_2)}\,\text{WC}}{G(\alpha_1 \vee \alpha_2|\beta_1 \vee \beta_2)}\,\text{OR}$$

f. Reversing the order of OR_4 and WC_3

Theorem 1 (Equivalence LLG and 4LLG). *Let S be a set of conditional goals. We have $S \vdash_{\text{LLG}} G(\alpha|\beta)$ if and only if $S \vdash_{\text{4LLG}} G(\alpha|\beta)$.*

Proof $\Leftarrow$ *Every* 4LLG *derivation is a* LLG *derivation.* $\Rightarrow$ *We can take any* LLG *derivation and construct an equivalent* 4LLG *derivation, by iteratively replacing two subsequent steps in the wrong order by several steps in the right order, see Proposition 3. If the proof tree is finite, then after a finite number of steps, all derivation steps are ordered, because no set of replacements cycles (and can be used to construct infinite proof trees).*

Proposition 4. 4LLG *is the only four-phase logic of goals for which Theorem 1 holds.*

Proof *Proposition 3 does not hold for any other four phase logic. Counterexamples for reversing the order of each two subsequent steps of* 4LLG *are given below. For* SA *and* TRANS, *the premises* $G(p|\top)$ *and* $G(q|\top)$ *cannot be combined unless one is first strengthened. For* TRANS *and* WC, *if the premise* $G(q|\top)$ *is weakened then it can no longer be used to detach* $G(p|q)$. *For* WC *and* OR, *the latter can only be applied if consequents are equivalent.*

$$\dfrac{\dfrac{G(p \mid \top)}{G(p \mid q)}\text{SA} \quad G(q \mid \top)}{G(p \wedge q \mid \top)}\text{TRANS} \qquad \dfrac{\dfrac{G(p \mid q) \quad G(q \mid \top)}{G(p \wedge q \mid \top)}\text{TRANS}}{G(p \mid \top)}\text{WC} \qquad \dfrac{\dfrac{G(p_1 \mid q_1)}{G(p_1 \vee p_2 \mid q_1)}\text{WC} \quad G(p_1 \vee p_2 \mid q_2)}{G(p_1 \vee p_2 \mid q_1 \vee q_2)}\text{OR}$$

It is easy to see that the following proof rule OR′ is implied by OR and WC in LLG. If we replace OR in LLG by the more general OR′, then we can use it as a phase-3 rule.

$$\text{OR}'_R : \frac{G(\alpha_1 \mid \beta_1)_{(F_1,p_1)}, G(\alpha_2 \mid \beta_2)_{(F_2,p_2)}, R}{G(\alpha_1 \vee \alpha_2 \mid \beta_1 \vee \beta_2)_{(F_1 \cup F_2, \rho(\text{OR}))}}$$

Moreover, Theorem 2 shows that in 4LLG the consistency check on the label can be replaced by a consistency check on the conjunction of the antecedent and consequent of the goal.

Theorem 2. *Consider any potential derivation of* 4LLG, *satisfying the condition* R_p *but not necessarily* R_F. *Then the following four conditions are equivalent:*

1. *The derivation satisfies condition* R_F *throughout phase 1,*
2. *The derivation satisfies* R_F *everywhere,*
3. *Each consequent is consistent with its antecedent throughout phase 1,*
4. *Each consequent is consistent with its antecedent everywhere.*

Proof *Clearly* (2) ⇒ (1) *and* (4) ⇒ (3). *Through phase 1, for each formula the conjunction of antecedent and consequent is equivalent to the unique element of its label. Hence* (1) ⇔ (3). *In phase 2 the conjunction of antecedent and consequent is also equivalent to the unique element of its label, which is equivalent to the label of the first premise of each derivation step. In phase 3 and 4 the rules preserve the consistency of the conjunction of antecedent and consequent, and they also preserve the property that each element of the label is consistent. From this we have* (3) ⇒ (4) *and* (1) ⇒ (2). *Putting this together gives us* (1) ⇔ (2) ⇔ (3) ⇔ (4) *and we are done.*

In the following section we illustrate that the first two phases of the four-phase logic 4LLG, i.e. SA and TRANS, can be combined in one phase without invalidating Theorem 2. Moreover, the latter two phases of 4LLG, i.e. WC and OR, can be combined similarly. The two-phase logic 2LLG first combines goals in derivation chains or arguments (SA and TRANS), and then combines arguments (WC and OR) with reasoning by cases.

5 Two-phase labeled logic of goals 2LLG

The logic 2LLG is the PLLG that combines the first two phases and the last two phases of 4LLG.

Definition 6 (2LLG). 2LLG *is the* PLLG *with the phasing function* ρ *defined by* $\rho(\text{SA}) = 1$, $\rho(\text{TRANS}) = 1$, $\rho(\text{WC}) = 2$, $\rho(\text{OR}) = 2$.

It is easy to see that 2LLG is equivalent to LLG and 4LLG in the sense of Theorem 1, because its phasing is in between the phasing of LLG and 4LLG. Moreover, Theorem 3 below is analogous to Theorem 2 for the two-phase logic 2LLG.

Theorem 3. *Consider any potential derivation of* 2LLG, *satisfying the condition* R_p *but not necessarily* R_F. *Then the following four conditions are equivalent:*

1. *The derivation satisfies condition* R_F *throughout phase 1,*
2. *The derivation satisfies* R_F *everywhere,*
3. *Each consequent is consistent with its antecedent throughout phase 1,*
4. *Each consequent is consistent with its antecedent everywhere.*

Proof *Analogous to the proof of Theorem 2, because in phase 1 the conjunction of antecedent and consequent is also equivalent to the unique element of its label, and in phase 2 the rules preserve the consistency of the conjunction of antecedent and consequent, as well as the property that each element of the label is consistent.*

We can construct other two-phase logics of goals, for example the one in which the first phase consists of only SA, but the following proposition shows that in those logics we cannot restrict ourselves to a consistency check on the conjunction of the antecedent and conjunction.

Proposition 5. *The logic* 2LLG *is the only two-phase* PLLG *that validates versions of Theorem 1 and 2, in which* 4LLG *is replaced by the two-phase logic.*

Proof *From the proof of Proposition 4 follows that* 2LLG *is the only two-phase* PLLG *for which Theorem 1 holds and in which* WC *cannot occur before* TRANS. *The following derivation shows that if* WC *can occur before* TRANS, *then we need the labels for the consistency check.*

$$\dfrac{G(p|q\wedge r)_{(\{\{p\wedge q\wedge r\}\},0)} \quad \dfrac{G(\neg p\wedge q|r)_{(\{\{\neg p\wedge q\wedge r\}\},0)}}{G(q|r)_{(\{\{\neg p\wedge q\wedge r\}\},\rho(\text{WC}))}}\ \text{WC}}{G(p\wedge q|r)_{(\{\{p\wedge q\wedge r,\neg p\wedge q\wedge r\}\},\rho(\text{TRANS}))}}\ \text{TRANS}$$

The surprising theorems follow from the consistency check in PLLG, which is stronger than it seems at first sight. In the following section we illustrate that the consistency check restricts derivations to a type of constructive proofs in the sense that no conclusions may be drawn from conflicts.

6 Conflicting goals

In this section we discuss three derivations related to conflicting goals. They are not valid in PLLG, because their usefulness depends on the application area of the logic of goals. In natural language different types of goals are used through each other, with possible confusion. An agent can either commit himself to a goal, because it maximizes his expected utility, or the goal may be imposed upon him by some authority, e.g. his boss or owner. In the latter case, the agent can simply adopt the goal as a desirable state, or he may interpret it as a reflection of the authority's maximal expected utility. Most formalisms for goals are developed to be used by intelligent robots, whose goals are the commands given by his owner. In the robot case, the logical properties of goals resemble the logical properties of obligations. As a consequence of the analogy between goals and obligations, the following unrestricted conjunction rule may be accepted in the logic of goals.

$$\text{AND}': \frac{G(\alpha_1 \mid \beta)_{(F_1,p_1)}, G(\alpha_2 \mid \beta)_{(F_2,p_2)}}{G(\alpha_1 \wedge \alpha_2 \mid \beta)_{(F_1 \times F_2, \max(p_1,p_2))}}$$

Moreover, the deontic 'ought implies can' axiom $\neg G(\bot|\alpha)$ may be accepted. For a discussion on the unrestricted conjunction rule and the associated problems, as well as a development of logics satisfying this proof rule, we refer to the deontic logic literature (see e.g. [14, 20] for a discussion and references). However, even if the unrestricted conjunction rule and the deontic 'ought implies can' axiom are not accepted in the logic of goals, then there is another interesting issue of conflicting goals. It concerns the following derivation, which we call *Forbidden Conflict* (FC) [17]. In this rule, as well as in the derivations in the remainder of this section, we leave the labels unspecified.

$$\frac{G(\neg a|\top) \quad G(a|b)}{G(\neg b|\top)} \ \text{FC}$$

In contrast to the constructive proofs of PLLG, this proof rule formalizes a conflict averse strategy. If b is the case then a conflict arises; therefore the agent desires that b is false. The following continuation of Example 2 illustrates that derivations no longer go as far as possible, but instead we have weak contraposition.

Example 4. Assume the proof rules TRANS, WC and FC together with the four goals $S = \{G(a|b), G(b|c), G(c|\top), G(\neg a|\top)\}$. The goal $G(b|\top)$ can be derived from $G(b|c)$ and $G(c|\top)$ by TRANS and WC, and the goal $G(\neg b|\top)$ from $G(a|b)$ and $G(\neg a|\top)$ by FC. Hence, there is a conflict for b, because $G(b|\top)$ as well as $G(\neg b|\top)$ can be derived. At first sight, this seems undesirable (see also [9]).

The following derivation is the third and last one we discuss related to conflicting goals.

$$\frac{G(a|b) \quad G(a|\neg b) \quad G(a \rightarrow b|\top)}{G(a \wedge b|\top)}$$

It formalizes a more complex conflict averse strategy. From the two goals $G(a|\neg b)$ and $G(a \to b|\top)$ follows that if $\neg b$ is the case then a conflict arises: a as well as $a \to b$ cannot both be the case as well. The goal $G(a \wedge b|\top)$ can thus be derived as follows.

$$\dfrac{G(a|b) \qquad \dfrac{G(a|\neg b) \quad G(a \to b|\top)}{G(b|\top)}\,\text{FC}}{G(a \wedge b|\top)}\,\text{TRANS}$$

It follows from Proposition 2 that in PLLG this derivation is blocked, because all consistent extensions of the fulfillment of the premise $G(a|\neg b)$ do not derive $a \wedge b$. The following derivation illustrates that the goal can be derived in 4LLG without fulfillment check with the unrestricted conjunction rule AND$'$.

$$\dfrac{\dfrac{G(a|b) \quad G(a|\neg b)}{G(a|\top)}\,\text{OR} \qquad G(a \to b|\top)}{G(a \wedge b|\top)}\,\text{AND}$$

The consequences of conflict averse strategies for PLLG are yet unclear, and left for further research.

7 Related research

Several authors have observed the relation between goals and desires in qualitative decision theory and obligations in deontic logic [10, 1, 8] and we have discussed the relation between decision theory, diagnosis theory and deontic logic in [19]. Labeled deontic logic was introduced in [16, 17], though its properties have not been studied. It does not make a dilemma like $O(p \mid \top) \wedge O(\neg p \mid \top)$ inconsistent, in contrast to traditional deontic logics like for example so-called standard deontic logic (SDL), and we therefore now interpret it as a logic of goals. Makinson [9] extended labeled deontic logic with the disjunction rule to cover reasoning by cases and an unrestricted conjunction rule. Two-phase deontic logic has been proposed in [13], but phased deontic logic has not been related to labeled deontic logic (although it has been suggested in [9]).

In [15] we introduced the *one-phase* labeled logic of goals LLG_0 based on labeled formulas $G(\alpha \mid \beta)_{(F,V)}$ and two consistency checks. In LLG_0, a formula $G(\alpha|\beta)_{(\{\alpha\wedge\beta\},\{\neg\alpha\wedge\beta\})}$ is called a premise, and the label of a goal derived by an inference rule is the union of the labels of the premises used in this inference rule. LLG_0 based on a violation check and a fulfillment check consists of inference rules, extended with a condition $R = R_V + R_F$.

R_V: $G(\alpha \mid \beta)_{(F,V)}$ may only be derived if $\alpha \wedge \beta \not\models \gamma$ for all $\gamma \in V$: fulfilling a derived goal should not imply a violation of one of the goals it is derived from;

R_F: $G(\alpha|\beta)_{(F,V)}$ may only be derived if $\alpha \wedge \beta \not\models \neg\gamma$ for all $\gamma \in F$: it must always be possible to fulfill a derived goal and each of the goals it is derived from, though not necessarily all of them at the same time.

This logic is stronger than the logics proposed in this paper, because it validates the following counterintuitive derivation. Notice that we cannot strengthen the condition of this logic, such that for example it is always possible that all the premises can be fulfilled at the same time, because then the intuitive first derivation of Example 3 is blocked.

$$\dfrac{\dfrac{\dfrac{\dfrac{G(p|r) \quad G(q|r)}{G(p \wedge q|r)}\text{AND} \qquad G(p \wedge q|\neg r)}{G(p \wedge q|\top)}\text{OR}}{G((p \wedge q) \vee s|\top)}\text{WC}}{G((p \wedge q) \vee s|\neg(p \wedge q \wedge r))}\text{SA}$$

Alternatively, and perhaps more intuitively, we take the logic PLLG and replace its SA rule by the following rule SA$*$, and replace R_F by R_F* with a consistency check on the conjunction of the fulfillments in the label and the antecedent of the goal, as in LLG$_0$. We call the resulting logics PLLG$*$.

$$\text{SA}*_R : \frac{G(\alpha \mid \beta_1)_{(F,p)}, R}{G(\alpha \mid \beta_1 \wedge \beta_2)_{(F,\rho(\text{SA}))}}$$

R_F*: $G(\alpha|\beta)_{(F,p)}$ may only be derived if $\{\beta\} \cup F_i$ is consistent for each $F_i \in F$.

It is easy to see that 4LLG* is equivalent to 4LLG. Moreover, we conjecture that Theorem 1 still holds, i.e. that LLG* is equivalent to 4LLG*, and we conjecture that PLLG* is equivalent to PLLG. However, the following example illustrates that the proof of Theorem 1 no longer holds, because some LLG$*$ derivations have intermediate results (here $G(p|q_1)$) which cannot be used in 4LLG*.

$$\dfrac{\dfrac{\dfrac{G(p|\top)}{G(p|q_1)}\text{SA} \qquad G(p|q_2)}{G(p|q_1 \vee q_2)}\text{OR}}{G(p|\neg q_1 \wedge q_2)}\text{SA}$$

Makinson [9] proposes a one-phase labeled logic based on R_F* in which the premises are represented by $G(\alpha|\beta)_{\{\{\alpha\}\}}$, i.e. in which only the consequents are represented in the label. Obviously, Theorem 1 no longer holds. Typical properties are that the disjunction rule always holds, such that the second derivation in Example 3 is not blocked, and the last derivation of Section 6 is also valid. Moreover, the logic has been extended with an unrestricted conjunction rule.

Most other logics of goals that have been proposed have a built in mechanism such that more specific and conflicting goals override more general ones, see e.g. [1, 12, 11]. However, specificity is only one possible rule to decide conflicts, which may be overridden itself (as in legal reasoning, where more general later rules override more specific earlier rules). Moreover, these logics make conflicts like $G(p) \wedge G(\neg p)$ *inconsistent*, which is not in line with the idea that goals can refer to different objectives. Finally, the non-monotonic mechanisms do not formalize the deliberating robber satisfactorily [18].

8 Conclusions and further research

Labeled logic is a powerful formalism to analyze and construct proofs. In this paper we discussed four proof rules in the framework of labeled deduction and we showed how labeled logics can combine the proof rules without deriving counterintuitive consequences. In particular, we showed how derivations can be partitioned into several phases. Surprisingly, the phasing of 4LLG does not restrict the set of derivable conclusions of LLG. Consequently, the phasing of 4LLG can be considered as a useful heuristic to make the proof theory of LLG more efficient, because only a small subset of all proofs of LLG have to be considered to proof the (in)validity of a formula.

Presently we are looking for ways to define a semantics for PLLG, such that also negations and disjunctions of goals are defined. One approach has been given in [9]. A possible worlds semantics can be defined along the lines of the two-phase deontic logic in [13, 21]. Theorem 2 and 3 show that in 4LLG and 2LLG we can get rid of the fulfillments in the label by checking the consistency of the conjunction of the antecedent and consequent of the dyadic operators. Moreover, we can also get rid of the integer in the label by introducing different operators for each phase. For example, for 2LLG we can define the two operators G_1 (with proof rules SA and TRANS) and G_2 (with proof rules WC and OR). The premises are goals $G_1(\alpha_i \mid \beta_i)$, the conclusion is a goal $G_2(\alpha \mid \beta)$, and the two phases are linked to each other with the following new proof rule.

$$\frac{G_1(\alpha|\beta)}{G_2(\alpha|\beta)}$$

To fully grasp the logical properties of goals, we think it is important to not only consider the proof theory, but also to consider the semantic relation between goals, desires, utilities and preferences in a decision-theoretic setting (see e.g. [10, 8, 22]). In particular, goals serve as computationally useful heuristic approximations of the relative preferences over the possible results of a plan [3], and goals are used to communicate desires in a compact and efficient way [6]. We think this is not a rivaling approach to the approach taken in this paper, but a complementary one.

Acknowledgement

Thanks to Patrick van der Laag, David Makinson and Emil Weydert for discussions on the issues raised in this paper, and comments on earlier versions of it. Many of the issues and examples discussed in this paper were raised during a discussion on labeled deontic logic with David Makinson.

References

1. C. Boutilier. Toward a logic for qualitative decision theory. In *Proceedings of the KR'94*, pages 75–86, 1994.

2. J. Doyle, Y. Shoham, and M.P. Wellman. The logic of relative desires. In *Sixth International Symposium on Methodologies for Intelligent Systems*, Charlotte, North Carolina, 1991.
3. J. Doyle and M.P. Wellman. Preferential semantics for goals. In *Proceedings of the AAAI'91*, pages 698–703, 1991.
4. D. Gabbay. *Labelled Deductive Systems*, volume 1. Oxford University Press, 1996.
5. H. Geffner and J. Pearl. Conditional entailment: bridging two approaches to default reasoning. *Artificial Intelligence*, 53:209–244, 1992.
6. P. Haddawy and S. Hanks. Representations for decision-theoretic planning: Utility functions for dead-line goals. In *Proceedings of the KR'92*, pages 71–82, 1992.
7. Z. Huang and J. Bell. Dynamic goal hierarchies. In *Intelligent Agent Systems: Theoretical and Practical Issues*, LNAI 1027, pages 88–103. Springer, 1997.
8. J. Lang. Conditional desires and utilities - an alternative approach to qualitative decision theory. In *Proceedings of the ECAI'96*, pages 318–322, 1996.
9. D. Makinson. On a fundamental problem of deontic reasoning. In P. McNamara and H. Prakken, editors, *Norms, Logics and Information Systems. New Studies on Deontic Logic and Computer Science.* IOS Press, 1998.
10. J. Pearl. From conditional oughts to qualitative decision theory. In *Proceedings of the UAI'93*, pages 12–20, 1993.
11. S.-W. Tan and J. Pearl. Qualitative decision theory. In *Proceedings of the AAAI'94*, 1994.
12. S.-W. Tan and J. Pearl. Specification and evaluation of preferences under uncertainty. In *Proceedings of the KR'94*, pages 530–539, 1994.
13. Y.-H. Tan and L.W.N. van der Torre. How to combine ordering and minimizing in a deontic logic based on preferences. In *Deontic Logic, Agency and Normative Systems. Proceedings of the ΔEON'96*, Workshops in Computing, pages 216–232. Springer, 1996.
14. L.W.N. van der Torre. *Reasoning about Obligations: Defeasibility in Preference-based Deontic Logic.* PhD thesis, Erasmus University Rotterdam, 1997.
15. L.W.N. van der Torre. Labeled logics of conditional goals. In *Proceedings of the ECAI'98*, pages 368–369. John Wiley & Sons, 1998.
16. L.W.N. van der Torre and Y.-H. Tan. Cancelling and overshadowing: two types of defeasibility in defeasible deontic logic. In *Proceedings of the IJCAI'95*, pages 1525–1532. Morgan Kaufman, 1995.
17. L.W.N. van der Torre and Y.-H. Tan. The many faces of defeasibility in defeasible deontic logic. In D. Nute, editor, *Defeasible Deontic Logic*, pages 79–121. Kluwer, 1997.
18. L.W.N. van der Torre and Y.-H. Tan. Deliberate robbery, or the calculating Samaritan. In *Proceedings of the ECAI'98 Workshop on Practical Reasoning and Rationality (PRR'98)*, 1998.
19. L.W.N. van der Torre and Y.-H. Tan. Diagnosis and decision making in normative reasoning. *Artificial Intelligence and Law*, 1998.
20. L.W.N. van der Torre and Y.-H. Tan. Prohairetic Deontic Logic (PDL). In *this volume*, 1998.
21. L.W.N. van der Torre and Y.-H. Tan. Two-phase deontic logic. 1998. Submitted.
22. L.W.N. van der Torre and E. Weydert. Goals, desires, utilities and preferences. In *Proceedings of the ECAI'98 Workshop on Decision Theory meets Artificial Intelligence*, 1998.

Analysis of Distributed-Search Contraction-Based Strategies

Maria Paola Bonacina *

Dept. of Computer Science, University of Iowa, USA
bonacina@cs.uiowa.edu

Abstract. We present a model of parallel search in theorem proving for forward-reasoning strategies, with contraction and distributed search. We extend to parallel search the bounded-search-spaces approach to the measurement of infinite search spaces, capturing both the advantages of parallelization, e.g., the subdivision of work, and its disadvantages, e.g., the cost of communication, in terms of search space. These tools are applied to compare the search space of a distributed-search contraction-based strategy with that of the corresponding sequential strategy.

1 Introduction

The difficulty of fully-automated theorem proving has led to investigate ways of enhancing theorem-proving strategies with parallelism. We distinguish among *parallelism at the term level* (i.e., parallelizing the inner algorithms of the strategy), *parallelism at the clause level* (i.e., parallel inferences within a single search) and *parallelism at the search level* or parallel search [8]. This paper considers *parallel search*: deductive processes search in parallel the space of the problem, and the parallel search succeeds as soon as one of the processes succeeds. A parallel-search strategy may subdivide the search space among the processes (*distributed search*), or assign to the processes different search plans (*multi-search*), or combine both principles. The processes *communicate* to merge their results, and preserve completeness (e.g., if the search space is subdivided or unfair search plans are used). This paper concentrates on distributed search.

Parallel search applies to theorem-proving strategies in general. This paper studies *forward-reasoning* and in particular *contraction-based* strategies (e.g., [13, 19, 9, 4]). There has been much interest in parallelizing these strategies (e.g., [12, 11, 5, 15, 6] and [8, 21] for earlier references), because they behave well sequentially (e.g., Otter [17], RRL [14], Reveal [2], EQP [18] are based on these strategies, and thanks to them succeeded in solving challenge problems, e.g., [2, 1, 18]). However, the parallelization of contraction-based strategies is difficult, primarily because of *backward contraction*: the clauses used as premises are subject to being deleted, the database is highly dynamic, and eager backward contraction (inter-reduce before expanding) diminishes the degree of concurrency of the inferences. The qualitative analysis in [8] showed how these factors affect adversely approaches based on parallelism at the term or clause level.

* Supported in part by NSF grants CCR-94-08667 and CCR-97-01508.

This paper begins a quantitative analysis of *distributed-search contraction-based strategies*, which poses several problems. We need to represent *subdivision* of the search space and *communication*, and measure their advantage and cost, respectively. Since the subdivision may not be effective, the search processes *overlap*: one needs to capture also the disadvantage of duplicated search. These problems are made worse by a fundamental difficulty: the search spaces in theorem proving are *infinite*. Therefore, we cannot analyze subdivision and overlap in terms of total size of the search space. Neither can we rely on the classical complexity measure of time to capture the cost of communication, because theorem-proving strategies are semidecision procedures that may not halt, so that "time" is not defined. Finally, the problems are not well-defined. Since parallel theorem proving is a young field, and an analysis of this kind was not attempted before, there are no standard formal definitions for many of the concepts involved. In recent work [10], we proposed an approach to the analysis of strategies, comprising *a model for the representation of search*, a notion of *complexity of search* in infinite spaces, and measures of this complexity, termed *bounded search spaces*. In this paper we build on this previous work to address the problems listed above.

The first part of the paper (Sections 2 and 3) develops **a framework of definitions for parallel theorem proving** by forward-reasoning distributed-search strategies. Three important properties of parallel search plans (*monotonicity of the subdivision*, *fairness* and *eager contraction*) are identified, and sufficient conditions to satisfy them are given. We point out that it is not obvious that a parallelization of a contraction-based strategy is contraction-based. On the contrary, this issue is critical in the parallelization of forward reasoning. **A model of parallel search** is presented in Section 4. A strategy with contraction not only visits, but also *modifies* the search space [10]. In distributed search also *subdivision* and *communication* modify the search space, and many processes are active in parallel. Our solution is based on distinguishing the *search space* and the *search process*, and yet representing them together in a *parallel marked search graph*. The structure of the graph represents the search space of all the possible inferences, while the *marking* represents the search process, including contraction, subdivision and communication.

Once we have a model of the search, we turn to **measuring benefits and costs of parallelization in terms of search complexity** (Section 5). The methodology of [10] is based on the observation that for infinite search spaces it is not sufficient to measure the generated search space. It is necessary to measure also the effects of the actions of the strategy on the infinite space that lies ahead. An exemplary case is that of contraction, where the deletion of an existing clause may prevent the generation of others. Our approach is to enrich the search space with a notion of *distance*, and consider the *bounded search space* made of the clauses whose distance from the input is within a given bound. The infinite search space is reduced to an infinite succession of *finite* bounded search spaces. Since the bounded search spaces are finite, they can be compared (in a multiset ordering), eliminating the obstacle of the impossibility of comparing infinite spaces. The second fundamental property of the bounded search spaces is that they change with the steps performed by the strategy. In the sequential case, the only factor which modifies the bounded search spaces is contraction, which makes clauses unreachable (infinite distance). In the parallel case, there are also subdivision and communication: subdivision makes the bounded search spaces for the parallel processes *smaller* (advantage of parallelism), while comunication *undoes* in part the effect of the subdivision (disadvantage of parallelism). The *parallel bounded search spaces* for the parallel derivation as a whole measure also the cost of duplicated search due to overlapping processes.

Section 6 applies these tools to **the analysis of distributed-search contraction-based strategies**. In distributed search, eager contraction depends on communication (e.g., to bring to a process a needed simplifier). We discover two related patterns of behaviour, called *late contraction* and *contraction undone*, where eager contraction fails. It follows that search paths that eager contraction would prune are not pruned. While in a sequential derivation the bounded search spaces decrease monotonically due to contraction, in a parallel derivation they may oscillate *non-monotonically*, because they reflect the conflict of subdivision and communication, and the conflict of contraction and communication. However, the incidence of late contraction and contraction undone decreases as the speed of communication increases, and at the limit, if communication takes no time, they disappear. For the overlap, we give sufficient conditions to minimize it relying on local eager contraction, independent of communication.

The last task is to compare a sequential contraction-based strategy $\mathcal{C}$ with its parallelization $\mathcal{C}'$. We prove that if $\mathcal{C}'$ has instantaneous communication and minimizes the overlap, its parallel bounded search spaces are smaller than those of $\mathcal{C}$. On one hand, this result represents a limit that concrete strategies may approximate. For instance, this theorem justifies formally the intuition about improving performance by devising subdivision criteria that reduce the overlap (e.g., [6]). On the other hand, since the hypothesis of instantaneous communication is needed, it represents a negative result on the parallelizability of contraction-based strategies, which contributes to explain the difficulty with obtaining generalized performance improvements by parallel theorem proving.

This kind of analysis is largely new, especially for parallel strategies. Most studies of complexity in deduction analyze the length of propositional proofs as part of the $NP \neq co\text{–}NP$ quest (e.g., [22]), or work with Herbrand complexity and proof length to obtain lower bounds for sets of clauses (e.g., [16]). The study in [20] analyzes measures of duplication in the search spaces of theorem-proving strategies. The full version of this paper, with the proofs and more references, can be found in [7].

2 Parallel theorem-proving strategies

In the *inference system* of a *theorem-proving strategy*, *expansion rules* (e.g., resolution) generate clauses, while *contraction rules* (e.g., subsumption and simplification) delete or reduce clauses according to a well-founded ordering $\succ$ (e.g., the multiset extension of a complete simplification ordering on atoms). An inference rule can be seen as a function which takes a tuple of premises and returns a set of clauses to be added and a set of clauses to be deleted:

Definition 21 *Let Θ be a signature, $\mathcal{L}_\Theta$ the language of clauses on Θ, and $\mathcal{P}(\mathcal{L}_\Theta)$ its powerset. An inference rule f^n of arity n is a function $f^n\colon \mathcal{L}_\Theta^n \rightarrow \mathcal{P}(\mathcal{L}_\Theta) \times \mathcal{P}(\mathcal{L}_\Theta)$. (If f^n does not apply to $\bar{x}$, $f^n(\bar{x}) = (\emptyset, \emptyset)$.)*

Given $\bar{x} = (\varphi_1 \ldots \varphi_n)$, let X be the multiset $\{\varphi_1 \ldots \varphi_n\}$, and $\pi_1(x,y) = x$ and $\pi_2(x,y) = y$ the projection functions:

Definition 22 *Given a well-founded ordering $(\mathcal{L}_\Theta, \succ)$, f^n is an* expansion rule *if $\forall \bar{x} \in \mathcal{L}_\Theta^n$, $\pi_2(f^n(\bar{x})) = \emptyset$. It is a* contraction rule *w. r. t. $\succ$ if either $\pi_1(f^n(\bar{x})) = \pi_2(f^n(\bar{x})) = \emptyset$, or $\pi_2(f^n(\bar{x})) \neq \emptyset$ and $X - \pi_2(f^n(\bar{x})) \cup \pi_1(f^n(\bar{x})) \prec_{mul} \pi_2(f^n(\bar{x}))$.*

The *closure* of a set of clauses S with respect to an inference system I is the set $S_I^* = \bigcup_{k \geq 0} I^k(S)$, where $I^0(S) = S$, $I^k(S) = I(I^{k-1}(S))$ for $k \geq 1$ and $I(S) = S \cup \{\varphi \rceil\ \varphi \in \pi_1(f(\varphi_1 \ldots \varphi_n)),\ f \in I,\ \varphi_1 \ldots \varphi_n \in S\}$.

Clauses deleted by contraction are *redundant*, and inferences that use redundant clauses (without deleting them) are also redundant (e.g., [9, 4]). Using the notion of *redundancy criterion* [3], $R(S)$ denotes the set of clauses redundant with respect to S according to R. A redundancy criterion R and a set of contraction rules I_R *correspond* if whatever is deleted by I_R is redundant according to R ($\pi_2(f^n(\bar{x})) \subseteq R(X - \pi_2(f^n(\bar{x})) \cup \pi_1(f^n(\bar{x}))))$, and if $\varphi \in R(S) \cap S$, I_R can delete φ without adding other clauses to make it redundant ($\pi_1(f^n(\bar{x})) = \emptyset$ and $\pi_2(f^n(\bar{x})) = \{\varphi\}$). I_R and R are based on the same ordering. Let $I = I_E \cup I_R$, distinguishing expansion and contraction rules in I.

Next, a parallel strategy has a system M of *communication operators*, such as *receive*: $\mathcal{L}_\Theta^* \rightarrow \mathcal{P}(\mathcal{L}_\Theta) \times \mathcal{P}(\mathcal{L}_\Theta)$, and *send*: $\mathcal{L}_\Theta^* \rightarrow \mathcal{P}(\mathcal{L}_\Theta) \times \mathcal{P}(\mathcal{L}_\Theta)$, where $receive(\bar{x}) = (\bar{x}, \emptyset)$ (adds received clauses to the database of the receiver), and $send(\bar{x}) = (\emptyset, \emptyset)$ (sending something does not modify the database of the sender).

The other component of a strategy is the *search plan*, which chooses inference rule and premises at each stage of a derivation $S_0 \vdash \ldots S_i \vdash \ldots$, where S_i is the state of the derivation after i steps, usually the multiset of existing clauses. We use *States* for the set of states and *States** for sequences of states. In distributed search, the search plan also controls *communication* and *subdivision*. Since S_I^* is infinite and unknown, the subdivision is built *dynamically*: at stage i the search plan subdivides the inferences that can be done in S_i. For each process p_k, an inference is either *allowed* (assigned to p_k), or *forbidden* (assigned to others):

Definition 23 *A* parallel search plan *is a 4-tuple $\Sigma = \langle \zeta, \xi, \alpha, \omega \rangle$:*

1. *The* rule-selecting function *ζ: States* $\times \mathbb{N} \times \mathbb{N} \rightarrow I \cup M$ takes as arguments the partial history of the derivation, the number of processes and the identifier of the process executing the selection, and returns an inference rule or a communication operator.*
2. *The* premise-selecting function *ξ: States* $\times \mathbb{N} \times \mathbb{N} \times (I \cup M) \rightarrow \mathcal{L}_\Theta^*$ also takes in input the selection of ζ, and satisfies $\xi((S_0 \ldots S_i), n, k, f^m) \in S_i^m$.*

3. *The* subdivision function α: *States** $\times \mathbb{N} \times \mathbb{N} \times (I \cup M) \times \mathcal{L}^*_\Theta \to$ *Bool also takes as argument the selection of* ξ, *and returns true (process allowed to perform the step), or* $false$ *(forbidden), or* $\perp$ *(undefined).*
4. *The* termination-detecting function ω: *States* $\to$ *Bool returns true if and only if the given state contains the empty clause.*

Definition 24 *Given a theorem-proving problem* S, *the* parallel derivation *generated by a strategy* $\mathcal{C} = \langle I, M, \Sigma\rangle$, *with* $\Sigma = \langle \zeta, \xi, \alpha, \omega\rangle$, *for processes* $p_0 \ldots p_{n-1}$ *is made of* n *asynchronous* local derivations $S = S^k_0 \vdash_{\mathcal{C}} \ldots S^k_i \vdash_{\mathcal{C}} \ldots$, *s. t.* $\forall k$, $0 \leq k \leq n-1$, $\forall i \geq 0$, *if* $\omega(S^k_i) = false$, $\zeta((S^k_0 \ldots S^k_i), n, k) = f$, *either* $f = receive$ *and* $\bar{x}$ *is received, or* $f \neq receive$ *and* $\xi((S^k_0 \ldots S^k_i), n, k, f) = \bar{x}$, *and* $\alpha((S^k_0 \ldots S^k_i), n, k, f, \bar{x}) = true$, *then* $S^k_{i+1} = S^k_i \cup \pi_1(f(\bar{x})) - \pi_2(f(\bar{x}))$.

A sequential search plan $\Sigma = \langle \zeta, \xi, \omega\rangle$ has ζ: *States** $\to I$, ξ: *States** $\times I \to \mathcal{L}^*_\Theta$ and ω. A parallel search plan $\Sigma' = \langle \zeta', \xi', \alpha, \omega\rangle$ is a *parallelization by subdivision* of Σ, if: whenever $\zeta'((S_0 \ldots S_i), n, k) \in I$, $\zeta'((S_0 \ldots S_i), n, k) = \zeta(S_0 \ldots S_i)$; and $\xi'((S_0 \ldots S_i), n, k, f) = \xi((S_0 \ldots S_i), f)$. $\mathcal{C}' = \langle I, M, \Sigma'\rangle$ is a parallelization of $\mathcal{C} = \langle I, \Sigma\rangle$, if Σ' is a parallelization of Σ.

3 Monotonicity, fairness and eager contraction

If ζ and ξ can select a certain f and $\bar{x}$, α needs to be defined on their selection, and it should not "forget" its decisions when clauses are deleted:

Definition 31 *A subdivision function* α *is* total on generated clauses *if for all* $S_0 \vdash \ldots S_i \vdash \ldots$, k, n, f^m *and* $\bar{x} \in (\bigcup^i_{j=0} S_j)^m$, $\alpha((S_0 \ldots S_i), n, k, f^m, \bar{x}) \neq \perp$.

Since it is undesirable that permission changes after it has been decided, α is required to be *monotonic* w. r. t. $\perp < false$ and $\perp < true$:

Definition 32 *A subdivision function* α *is* monotonic *if for all* $S_0 \vdash \ldots S_i \vdash \ldots$, n, k, f, $\bar{x}$, *and* $i \geq 0$, $\alpha((S_0 \ldots S_i), n, k, f, \bar{x}) \leq \alpha((S_0 \ldots S_{i+1}), n, k, f, \bar{x})$.

A strategy is *complete* if it succeeds whenever S_0 is inconsistent. Completeness is made of *refutational completeness* of the inference system, and *fairness* of the search plan [9]. A sufficient condition for fairness is (e.g., [3]):

Definition 33 *A derivation* $S_0 \vdash \ldots S_i \vdash \ldots$ *is* uniformly fair *w. r. t.* I *and* R *if* $I(S_\infty - R(S_\infty)) \subseteq \bigcup_{j\geq 0} S_j$, *where* $S_\infty = \bigcup_{j\geq 0} \bigcap_{i\geq j} S_i$ *(persistent clauses).*

In distributed search a process only needs to be fair with respect to what is allowed. Since α is monotonic, allowed ($\exists j\ \alpha((S^k_0 \ldots S^k_j), n, k, f, \bar{x}) = true$) is equivalent to persistently allowed ($\exists i\ \forall j \geq i\ \alpha((S^k_0 \ldots S^k_j), n, k, f, \bar{x}) = true$).

Definition 34 *A local derivation* $S^k_0 \vdash \ldots S^k_i \vdash \ldots$ *is* uniformly fair *w.r.t.* I, R *and* α, *if* $\varphi \in \pi_1(f^m(\bar{x}))$, $\bar{x} \in (S^k_\infty - R(S^k_\infty))^m$ *and* $\exists i \geq 0\ \forall j \geq i$ $\alpha((S^k_0 \ldots S^k_j), n, k, f^m, \bar{x}) = true$ *imply* $\varphi \in \bigcup_{j\geq 0} S^k_j$.

The search plan needs to ensure that for every inference from a tuple of persistent (in $S_\infty = \bigcup_{k=0}^{n-1} S_\infty^k$) non-redundant premises, there is a p_k which is allowed (*fairness of subdivision*) and has all the premises (*fairness of communication*):

Definition 35 *A parallel derivation* $S_0^k \vdash_C \ldots S_i^k \vdash_C \ldots$, *for* $k \in [0, n-1]$, *by* $\Sigma = \langle \zeta, \xi, \alpha, \omega \rangle$, *is* uniformly fair *w. r. t.* I *and* R, *if:*

1. $\forall p_k$, $S_0^k \vdash_C \ldots S_i^k \vdash_C \ldots$ *is uniformly fair w. r. t.* I, R *and* α (local fairness).
2. $\forall f^m \in I$, $\forall \bar{x} \in (S_\infty - R(S_\infty))^m$, *s. t.* $\pi_1(f^m(\bar{x})) \neq \emptyset$, $\exists\ p_k$ *s. t.* $\bar{x} \in (S_\infty^k - R(S_\infty))^m$ *(*fairness of communication*), and* $\exists i \geq 0$, *s. t.* $\forall j \geq i$, $\alpha((S_0^k \ldots S_j^k), n, k, f^m, \bar{x}) = true$ (fairness of subdivision).

If Σ is uniformly fair, its parallelization Σ' inherits local fairness from Σ.

Theorem 31 *If a parallel derivation* $S_0^k \vdash_C \ldots \vdash_C S_i^k \vdash_C \ldots$, *for* $k \in [0, n-1]$, *is uniformly fair with respect to* I *and* R, *then* $I(S_\infty - R(S_\infty)) \subseteq \bigcup_{k=0}^{n-1} \bigcup_{j \geq 0} S_j^k$.

A sequential strategy is *contraction-based* if it features contraction rules and an *eager-contraction* search plan:

Definition 36 *A derivation* $S_0 \vdash_I \ldots S_i \vdash_I \ldots$ *has* eager contraction, *if for all* $i \geq 0$ *and* $\varphi \in S_i$, *if there are* $f^m \in I_R$ *and* $\bar{x} \in S_i^m$, *such that* $\pi_2(f^m(\bar{x})) = \{\varphi\}$, *then* $\exists l \geq i$ *such that* $S_l \vdash S_{l+1}$ *deletes* φ, *and* $\forall j, i \leq j \leq l$, $S_j \vdash S_{j+1}$ *is not an expansion inference, unless the derivation succeeds sooner.*

The parallelization of eager contraction is difficult. First, each local derivation needs to have *local eager contraction*, which is defined as in Def. 36 with expansion replaced by expansion or communication. Second, the strategy should ensure that p_k delete eagerly a clause reducible by a premise generated by p_h. This depends on communication:

Definition 37 *Let* $\varphi \in S_\infty - R(S_\infty)$ *and* i *be the first stage s. t.* $\varphi \in \bigcup_{h=0}^{n-1} S_i^h$. *A parallel derivation has* propagation of clauses up to redundancy *if* $\forall p_h\ \exists j\ \varphi \in S_j^h$, instantaneous propagation of clauses up to redundancy *if* $\forall p_h\ \varphi \in S_i^h$.

Definition 38 *A parallel derivation has* distributed global contraction, *if for all* p_k, $i \geq 0$, *and* $\varphi \in S_i^k$, *if there are* $f^m \in I_R$ *and* $\bar{x} \in (\bigcup_{h=0}^{n-1} S_i^h)^m$ *such that* $\pi_2(f^m(\bar{x})) = \{\varphi\}$, *then* $\exists l \geq i$ *such that* $S_l^k \vdash S_{l+1}^k$ *deletes* φ, *unless* p_k *halts sooner. It has* global eager contraction *if, in addition,* $\forall j, i \leq j \leq l$, $S_j^k \vdash S_{j+1}^k$ *is neither an expansion nor a communication step.*

Global eager contraction generalizes eager contraction to parallel derivations, while distributed global contraction is a weaker requirement which guarantees contraction, but not priority over expansion. In the presence of communication delays, p_k may use as expansion premise a clause which is reducible with respect to $\bigcup_{h=0}^{n-1} S^h$ before it becomes reducible with respect to S^k. Thus, we have:

Lemma 31 *Local eager contraction and propagation of clauses up to redundancy (instantaneous propagation of clauses) imply distributed global contraction (global eager contraction, respectively).*

A parallel strategy is *contraction-based* if it has contraction rules and its search plan has local eager contraction and distributed global contraction.

Theorem 32 *Let $\mathcal{C} = \langle I, \Sigma \rangle$ be a contraction-based strategy and $\mathcal{C}'$ with $\Sigma' = \langle \zeta', \xi', \alpha, \omega' \rangle$ a parallelization by subdivision of $\mathcal{C}$. If Σ' propagates clauses up to redundancy, and for all $f \in I_R$, i, n, k and $\bar{x}$, $\alpha((S_0^k \dots S_i^k), n, k, f, \bar{x}) = true$, $\mathcal{C}'$ is also contraction-based.*

The requirement that all contractions are allowed to all is strong, especially if contraction is not only deletion but also deduction (e.g., simplification). Assume that α forbids contractions, e.g., $f(\bar{x}) = (\{\psi_1, \dots, \psi_m\}, \{\varphi_1, \dots, \varphi_p\})$. Consider a strategy that lets a process delete the φ_j, but not generate the ψ_i, when ζ selects f, ξ selects $\bar{x}$, and α forbids the step. It is sufficient that at least one process generates and propagates the ψ_i to preserve completeness:

Theorem 33 *Let $\mathcal{C}$ and $\mathcal{C}'$ be as in Theorem 32. If Σ' propagates clauses up to redundancy, and whenever $\zeta((S_0^k \dots S_i^k), n, k) = f \in I_R$, $\xi((S_0^k \dots S_i^k), n, k, f) = \bar{x}$, $\alpha((S_0^k \dots S_i^k), n, k, f, \bar{x}) = false$, it is $S_{i+1}^k = S_i^k - \pi_2(f(\bar{x}))$, $\mathcal{C}'$ is also contraction-based.*

4 Search graphs for parallel search

Given a theorem-proving problem S and an inference system I, the *search space* induced by S and I is represented by the hypergraph $G(S_I^*) = (V, E, l, h)$, where V is the set of vertices, l is a vertex-labelling function $l: V \rightarrow \mathcal{L}_\Theta / \doteq$ (from vertices to equivalence classes of variants, so that all variants are associated to a unique vertex), h is an arc-labelling function $h: E \rightarrow I$, and if $f^m(\varphi_1, \dots, \varphi_m) = (\{\psi_1, \dots, \psi_s\}, \{\gamma_1, \dots, \gamma_p\})$ for $f^m \in I$, E contains a hyperarc $e = (v_1 \dots v_n; w_1 \dots w_p; u_1 \dots u_s)$ where $h(e) = f^m$, $n + p = m$, and

- $\forall j\ 1 \leq j \leq n\ l(v_j) = \varphi_j$ and $\varphi_j \notin \{\gamma_1, \dots, \gamma_p\}$ (premises not deleted),
- $\forall j\ 1 \leq j \leq p\ l(w_j) = \gamma_j$ (deleted premises), and $\forall j\ 1 \leq j \leq s\ l(u_j) = \psi_j$ (generated clauses).

W.l.o.g. we consider hyperarcs in the form $(v_1 \dots v_n; w; u)$. Contraction inferences that purely delete clauses are represented as replacement by *true* (a dummy clause such that $\forall \varphi\ \varphi \succ true$), and a special vertex T is labelled by *true*.

The search graph $G(S_I^*) = (V, E, l, h)$ represents the static structure of the search space. The dynamics of the search during a derivation is described by *marking functions* for vertices and arcs:

Definition 41 *A* parallel marked search-graph $(V, E, l, h, \bar{s}, \bar{c})$ *for* $p_0, \ldots p_{n-1}$ *has an n-tuple* $\bar{s}$ *of* vertex-marking functions $s^k: V \to Z$ *such that*

$$s^k(v) = \begin{cases} m & \text{if } m \text{ variants } (m > 0) \text{ of } l(v) \text{ are present for process } p_k, \\ -1 & \text{if all variants of } l(v) \text{ have been deleted by process } p_k, \\ 0 & \text{otherwise;} \end{cases}$$

and an n-tuple $\bar{c}$ *of* arc-marking functions $c^k: E \to \mathbb{N} \times Bool$ *such that* $\pi_1(c^k(e))$ *is the number of times* p_k *executed* e *or received a clause generated by* e, *and* $\pi_2(c^k(e)) = true/false$ *if* p_k *is allowed/forbidden to execute* e.

Hyperarc $e = (v_1 \ldots v_n; w; u)$ is *enabled* for p_k if $s^k(v_j) > 0$ for $1 \leq j \leq n$, $s^k(w) > 0$, and $\pi_2(c^k(e)) = true$.

Definition 42 *A parallel derivation induces n* successions of vertex-marking functions $\{s_i^k\}_{i \geq 0}$, *one per process. For all* $v \in V$, $s_0^k(v) = 0$, *and* $\forall i \geq 0$:

- *If at stage i* p_k *executes an enabled hyperarc* $e = (v_1, \ldots, v_n; w; u)$:

$$s_{i+1}^k(v) = \begin{cases} s_i^k(v) - 1 & \text{if } v = w \ (-1 \text{ if } s_i^k(v) = 1), \\ s_i^k(v) + 1 & \text{if } v = u \ (+1 \text{ if } s_i^k(v) = -1), \\ s_i^k(v) & \text{otherwise.} \end{cases}$$

- *If at stage i* p_k *receives* $\bar{x} = (\varphi_1, \ldots \varphi_n)$, *where* $\varphi_j = l(v_j)$ *for* $1 \leq j \leq n$:

$$s_{i+1}^k(v) = \begin{cases} s_i^k(v) + 1 & \text{if } v \in \{v_1, \ldots v_n\} \ (+1 \text{ if } s_i^k(v) = -1), \\ s_i^k(v) & \text{otherwise.} \end{cases}$$

- *If at stage i* p_k *sends* $\bar{x}$, $s_{i+1}^k(v) = s_i^k(v)$.

Note that $s_0^k(v) = 0$ also for input clauses: the steps of reading or receiving input clauses are included in the derivation (read steps can be modelled as expansion steps), because the subdivision function applies to input clauses.

Definition 43 *A parallel derivation induces n* successions of arc-marking functions $\{c_i^k\}_{i \geq 0}$: $\forall a \in E$, $\pi_1(c_0^k(a)) = 0$ *and* $\pi_2(c_0^k(a)) = true$, *and* $\forall i \geq 0$:

$$\pi_1(c_{i+1}^k(a)) = \begin{cases} \pi_1(c_i^k(a)) + 1 & \text{if } p_k \text{ executes } a \text{ or receives a clause generated by } a, \\ \pi_1(c_i^k(a)) & \text{otherwise;} \end{cases}$$

$$\pi_2(c_{i+1}^k(a)) = \begin{cases} \alpha((S_0 \ldots S_{i+1}), n, k, f, \bar{x}) & \text{if } \alpha((S_0 \ldots S_{i+1}), n, k, f, \bar{x}) \neq \perp, \\ true & \text{otherwise (arcs allowed by default),} \end{cases}$$

where $h(a) = f$ *and* $\bar{x}$ *is the tuple of premises of hyperarc a.*

5 Measures of search complexity

Let $G = (V, E, l, h)$ be a search graph. For all $v \in V$, if v has no incoming hyperarcs, the *ancestor-graph* of v is the graph made of v itself; if $e = (v_1 \dots v_n; v)$ is a hyperarc and $t_1 \dots t_n$ are ancestor-graphs of $v_1 \dots v_n$, the graph with root v connected by e to $t_1 \dots t_n$ is an *ancestor-graph* of v, denoted by $(v; e; (t_1, \dots, t_n))$. (To simplify the notation, we include the deleted premise in $v_1 \dots v_n$.) The set of the ancestor-graphs of v in G is denoted by $at_G(v)$ (or $at_G(\varphi)$). From now on, $G = (V, E, l, h, \bar{s}, \bar{c})$ is the parallel marked search graph searched by $p_0, \dots p_{n-1}$, $G^k = (V, E, l, h, s^k, c^k)$ is the marked search graph for p_k and $G_i^k = (V, E, l, h, s_i^k, c_i^k)$ is the marked search graph for p_k at stage i of a derivation.

If a *relevant* ancestor of φ in $t \in at_G(\varphi)$ is deleted by contraction, it becomes impossible to reach φ by traversing t:

Definition 51 *Let* $t = (v; e; (t_1 \dots t_n)) \in at_G(v)$ *with* $e = (v_1 \dots v_n; v)$. *A vertex* $w \in t$, $w \neq v$, *is* relevant *to* v *in* t *for* p_k *(*$w \in Rev_{G^k}(t)$*) if either* $w \in \{v_1 \dots v_n\}$ *and* $\pi_1(c^k(e)) = 0$, *or* $\exists i\ 1 \leq i \leq n$ *s. t.* w *is relevant to* v_i *in* t_i *for* p_k.

If $\pi_1(c^k(e)) \neq 0$ because p_k executed e, deleting ψ is irrelevant for φ, since ψ has been already used to generate φ. If $\pi_1(c^k(e)) \neq 0$ because p_k received a variant of φ generated by e, deleting ψ is irrelevant for φ, since φ came from the outside.

Given $t \in at_G(\varphi)$, the *p(ast)-distance* measures the portion of t that p_k has visited; the *f(uture)-distance* measures the portion of t that p_k needs to visit to reach φ by traversing t; the *g(lobal)-distance* is their sum:

Definition 52 *For all clauses* φ, *ancestor-graphs* $t \in at_G(\varphi)$ *and processes* p_k:

- *The* p-distance of φ on t for p_k *is* $pdist_{G^k}(t) = |\ \{w \mid w \in t, s^k(w) \neq 0\}\ |$.
- *The* f-distance of φ on t for p_k *is*

$$fdist_{G^k}(t) = \begin{cases} \infty & \textit{if } s^k(\varphi) < 0, \textit{ or} \\ & \exists w \in Rev_{G^k}(t),\ s^k(w) < 0, \\ |\ \{w \mid w \in t, s^k(w) = 0\}\ | & \textit{otherwise.} \end{cases}$$

- *The* g-distance of φ on t for p_k *is* $gdist_{G^k}(t) = pdist_{G^k}(t) + fdist_{G^k}(t)$.

The f-distance of φ in G for p_k *is* $fdist_{G^k}(\varphi) = min\{fdist_{G^k}(t) \mid t \in at_G(\varphi)\}$.
The g-distance of φ in G for p_k *is* $gdist_{G^k}(\varphi) = min\{gdist_{G^k}(t) \mid t \in at_G(\varphi)\}$.

While infinite distance captures the effect of contraction, we consider next subdivision:

Definition 53 *An ancestor-graph* t *is* forbidden *for process* p_k *if there exists an arc* e *in* t *such that* $\pi_1(c^k(e)) = 0$ *and* $\pi_2(c^k(e)) = false$. *It is* allowed *otherwise.*

If p_k receives the clause that e generates, it is $\pi_1(c^k(e)) \neq 0$, and t is no longer forbidden, because it is no longer true that forbidding e prevents p_k from exploring t. This may happen only as a consequence of communication, because

e cannot be executed. Thus, subdivision forbids ancestor-graphs, reducing the search effort of each process, but communication may undo it. Indeed, a strategy employs communication to preserve fairness in presence of a subdivision. For fairness, it is sufficient that every non-redundant path is allowed to one process. If more than one process is allowed, their searches may overlap:

Definition 54 *p_k and p_h* overlap *on ancestor-graph t if t is allowed for both.*

An overlap represents a potential duplication of search effort, hence a waste. A goal of a parallel search plan is to preserve fairness while minimizing the overlap.

Definition 55 *The* bounded search space *within distance j for p_k is the multiset of clauses $space(G^k, j) = \sum_{v \in V, v \neq \top} mul_{G^k}(v, j) \cdot l(v)$, where each $l(v)$ has multiplicity $mul_{G^k}(v, j) = |\{t \mid t \in at_G(v), t \text{ allowed for } p_k, 0 < gdist_{G^k}(t) \leq j\}|$.*

The sequential bounded search spaces $space(G, j)$ are defined in the same way, with multiplicity $mul_G(v, j) = |\{t \mid t \in at_G(v), 0 < gdist_G(t) \leq j\}|$.

Definition 56 *The* parallel bounded search space *within distance j is the multiset of clauses $pspace(G, j) = \sum_{v \in V, v \neq \top} pmul_G(v, j) \cdot l(v)$, where $l(v)$ has multiplicity $pmul_G(v, j) = \lfloor gmul_G(v, j)/n \rfloor$ with $gmul_G(v, j) = \sum_{k=0}^{n-1} mul_{G^k}(v, j)$.*

If p_h and p_k overlap on an ancestor-graph t, t is counted in both $mul_{G^h}(v, j)$ and $mul_{G^k}(v, j)$ and twice in $gmul_G(v, j)$, reflecting the overlap.

Theorem 51 *If α is the constant function true (no subdivision), and no communication occurs, then for all p_k, $i \geq 0$ and $j > 0$, $space(G_i^k, j) = space(G_i, j)$, and $pspace(G_i, j) = space(G_i, j)$.*

6 The analysis

In a sequential derivation, if a clause φ is deleted at stage i and regenerated via another ancestor-graph at some stage $j > i$, a contraction-based strategy will delete it again, and will do so before using φ to generate other clauses [10]. It follows that when φ is re-deleted, it is still relevant on all ancestor-graphs where it was relevant at stage i. Therefore, it is possible to make the following approximation: if $fdist_{G_i^k}(t)$ is infinite, $fdist_{G_j^k}(t)$ can be regarded as infinite for all $j > i$. In a parallel derivation, the situation is more complex. First, we need to consider not only the possibility that a process p_k regenerates φ via another ancestor-graph, but also the possibility that p_k receives another variant of φ from another process, which may not be aware that φ is redundant. Second, we need to consider not only deleted clauses, but also clauses such that $fdist_{G_i^k}(\varphi) = \infty$ because a relevant ancestor has been deleted on every $t \in at_G(\varphi)$. In a sequential derivation, it is impossible for φ to appear at some $j > i$, because it cannot be generated, but in a parallel derivation, φ may still be received from another process at some $j > i$. Thus, we prove a more general result:

Lemma 61 *In a derivation with local eager contraction, for all p_k, i, and φ, if $fdist_{G_i^k}(\varphi) = \infty$ (regardless of whether φ is deleted or made unreachable) and $s_j^k(\varphi) > 0$ for some $j > i$ (regardless of whether φ is received or regenerated), there exists a $q > j$, such that $s_q^k(\varphi) < 0$ (hence $fdist_{G_q^k}(\varphi) = \infty$), and p_k does not use φ to generate other clauses at any stage l, $j \leq l < q$.*

Therefore, we can make the approximation that $fdist_{G_i^k}(\varphi) = \infty$ implies $\forall j > i$ $fdist_{G_j^k}(\varphi) = \infty$. This takes care of clauses that the strategy finds redundant. Consider a non-redundant clause φ and an ancestor-graph t of φ such that $fdist_{G_i^k}(t) = \infty$ because $s_i^k(\psi) = -1$ for a relevant ancestor ψ in t. For simplicity, let ψ be a parent of φ, with arc e from ψ to φ. Assume that p_h has not deleted ψ, executes e, and sends to p_k a φ generated by e. The arrival of φ at some stage $r > i$ makes ψ irrelevant ($\pi_1(c_r^k(e) = 1)$, so that $fdist_{G_r^k}(t) \neq \infty$. There is irrelevance of contraction at p_h, because the clause(s) that contract ψ do not arrive at p_h fast enough to delete ψ before it is used to generate φ. When p_h finally deletes ψ, this deletion is irrelevant to t, because p_h has already executed e: we call this phenomenon *late contraction*. There is irrelevance of contraction at p_k, because the arrival of φ from p_h makes the deletion of ψ irrelevant: we call this phenomenon *contraction undone*. Distributed global contraction guarantees that p_h will delete ψ eventually, so that $s_j^h(\psi) = -1$ for some $j > i$. It is sufficient that p_h executes e and generates φ at a stage $l < j$, and φ arrives at p_k at a stage $r > i$, for this situation to occur. Thus, distributed global contraction is not sufficient to prevent late contraction and contraction undone.

The following theorems integrate all our observations on subdivision, contraction and communication:

1. *If $S_i^k \vdash S_{i+1}^k$ generates ψ, then $\forall j > 0$, $space(G_{i+1}^k, j) \preceq_{mul} space(G_i^k, j)$.* When ψ is generated, the subdivision function α may become defined on a tuple of premises $\bar{x}$ including ψ. If α decides that an arc e with premises $\bar{x}$ is forbidden, ancestor-graphs including e become forbidden, so that the bounded search spaces become smaller.
2. *If $S_i^k \vdash S_{i+1}^k$ replaces ψ by ψ', then $\forall j > 0$, $space(G_{i+1}^k, j) \preceq_{mul} space(G_i^k, j)$.* A contraction step replacing ψ by ψ' prunes those ancestor-graphs whose distance becomes infinite because of the deletion of ψ, and those ancestor-graphs which become forbidden as a consequence of the generation of ψ'.
3. *If $S_i^k \vdash S_{i+1}^k$ sends ψ, then $\forall j > 0$, $space(G_{i+1}^k, j) = space(G_i^k, j)$. If $S_i^k \vdash S_{i+1}^k$ receives ψ, $\forall j > 0$, $\exists\, l \leq i$, $space(G_{i+1}^k, j) \preceq_{mul} space(G_l^k, j)$.* When p_k receives ψ, there may be three kinds of consequences: allowed ancestor-graphs may become forbidden (subdivision), reducing the multiplicity of some clauses; forbidden ancestor-graphs may become allowed (subdivision undone) and relevant deleted ancestors may become irrelevant (contraction undone), increasing the multiplicity of some clauses. However, since communication cannot expand the bounded search spaces, but only undo previous reductions, the resulting bounded search spaces are limited by the bounded search spaces at some previous stage.

These theorems show that the bounded search spaces capture all relevant phenomena: pruning by contraction, subdivision and cost of communication. While in a sequential derivation the bounded search spaces may either remain the same (expansion) or decrease (contraction), in a parallel derivation the bounded search spaces of a process may oscillate *non-monotonically* because of communication. The faster is communication, however, the lesser is the incidence of late contraction and contraction undone; at the limit, if the strategy has instantaneous propagation of clauses up to redundancy, they disappear:

Lemma 62 *In a derivation with local eager contraction and instantaneous propagation of clauses up to redundancy, let e be an arc of $t \in at_G(\varphi)$ which uses ψ and generates $\psi' \in S_\infty - R(S_\infty)$. If $s_i^k(\psi) = -1$ and $\psi \in Rev_{G_i^k}(t)$ for some p_k:*

1. *$\forall p_h$, $\forall j$, $s_j^h(\psi) = -1$ implies $\psi \in Rev_{G_j^h}(t)$ (what is relevant to one process is relevant to all: no late contraction).*
2. *$\forall j \geq i$, $\psi \in Rev_{G_j^k}(t)$ (what is relevant at a stage remains relevant at all following stages: no contraction undone).*

The approximation $fdist_{G_i^k}(t) = \infty \Rightarrow \forall j > i \; fdist_{G_j^k}(t) = \infty$ can be made:

Theorem 61 *In a derivation with local eager contraction and instantaneous propagation of clauses up to redundancy, if $fdist_{G_i^k}(t) = \infty$ and $fdist_{G_j^k}(t) \neq \infty$ for some $0 < i < j$, there exists a $q > j$ such that $fdist_{G_q^k}(t) = \infty$.*

Next, we turn our attention to the overlap. We observe that two overlapping processes may generate variants of the same clause. The following property prevents different processes from generating variants of the same clause:

Definition 61 *A subdivision function α has* no clause-duplication *if for all vertices $u \neq \top$, for any two hyperarcs into u, e_1 with inference rule f and premises $\bar{x}$, and e_2 with inference rule g and premises $\bar{y}$, $\forall i \geq 0$, if $\alpha((S_0, \ldots S_i), n, k, f, \bar{x}) = true$ and $\alpha((S_0, \ldots S_i), n, h, g, \bar{y}) = true$, then $k = h$.*

This property is compatible with fairness, for which one allowed process is sufficient. We show next that the combination of no clause-duplication and local eager contraction minimizes the overlap. There are two kinds of overlap: one caused by the subdivision function itself when it allows the same arc to more than one process, and one caused by communication (e.g., $\pi_2(c^k(e)) = true$ and $\pi_2(c^h(e)) = false$ but $\pi_1(c^h(e)) \neq 0$). No clause-duplication avoids the first kind of overlap by definition. For the second one, assume that p_k is the only process authorized to generate all variants of φ. By local eager contraction, if p_k generates more than one variant of φ, all but one are deleted before being sent to any other process. Thus, p_k may send to another process only one variant, and the same variant to all processes, so that:

Lemma 63 *In a derivation with local eager contraction and no clause-duplication, for any clause φ, if p_h is the only process allowed to generate φ, $\exists r$ such that $\forall k \neq h$, $\forall i \geq r$, $\forall j > 0$, $mul_{G_i^k}(\varphi, j) \leq 1$ (i.e, communication may make* at most one *forbidden ancestor-graph allowed).*

We have all the elements to compare a contraction-based, uniformly fair strategy $\mathcal{C} = \langle I, \Sigma \rangle$ with a parallelization by subdivision $\mathcal{C}' = \langle I, M, \Sigma' \rangle$, which is uniformly fair and contraction-based. Since $\mathcal{C}$ and $\mathcal{C}'$ have the same inference system, the initial search space is the same (i.e., $G_0^k = G_0$ and $space(G_0^k, j) = space(G_0, j)$ for all k and j). We compare first the behaviour on redundant clauses, next on ancestor-graphs including redundant inferences, and then on the remaining ancestor-graphs. We begin by proving three preliminary lemmas:

1. *If $\varphi \in S_i$ for some i, then $\exists p_k$ $\exists j$ such that either $\varphi \in S_j^k$ or $\varphi \in R(S_j^k)$.*
2. *If $\varphi \in R(S_i)$ for some i, then $\forall p_k$ $\exists j$ such that $\varphi \in R(S_j^k)$.*
3. *If $fdist_{G_i}(\varphi) = \infty$ for some i, then $\forall p_k$ $\exists j$ s. t. $\forall l \geq j$, either $fdist_{G_l^k}(\varphi) = \infty$, or all $t \in at_G(\varphi)$ are forbidden for p_k at stage l.*

These allow us to show that all redundant clauses eliminated by $\mathcal{C}$ will be excluded by $\mathcal{C}'$ as well:

Theorem 62 *If $fdist_{G_i}(\varphi) = \infty$ for some i, then there exists an r such that for all $i \geq r$ and $j > 0$, $pmul_{G_i}(\varphi, j) = 0$.*

To show that all ancestor-graphs pruned by $\mathcal{C}$ are pruned by $\mathcal{C}'$, we need to use Lemma 62 to prevent late contraction and contraction undone, and this can be done only under the hypothesis of instantaneous propagation of clauses:

Lemma 64 *Assume that $\mathcal{C}'$ has instantaneous propagation of clauses up to redundancy. If $fdist_{G_i}(t) = \infty$ for some i, then for all p_k there exists a j such that for all $l \geq j$, either $fdist_{G_l^k}(t) = \infty$, or t is forbidden for p_k at stage l.*

The final lemma covers ancestor-graphs that are not pruned. Thus, we need a hypothesis on subdivision, and we assume that $\mathcal{C}'$ has no clause-duplication:

Lemma 65 *Assume that $\mathcal{C}'$ has instantaneous propagation of clauses up to redundancy and no clause-duplication. If $fdist_{G_i}(\varphi) \neq \infty$ for all i, there exists an r such that for all $i \geq r$ and $j > 0$, $pmul_{G_i}(\varphi, j) \leq mul_{G_i}(\varphi, j)$.*

Theorem 63 *If $\mathcal{C}'$ has instantaneous propagation of clauses up to redundancy and no clause-duplication, $\forall j$ $\exists m$ s. t. $\forall i \geq m$ $pspace(G_i, j) \preceq_{mul} space(G_i, j)$.*

Intuitively, a value j of the bound may represent the search depth required to find a proof. If the problem is hard enough that the sequential strategy does not succeed before stage m, the parallel strategy faces a smaller bounded search space beyond m, and therefore may succeed sooner.

Theorem 63 is a limit theorem, in a sense similar to other theoretical results obtained under an ideal assumption. On one hand, it explains the nature of the problem, by indicating in the overlap and the communication-contraction node its essential aspects. On the other hand, it represents a limit that concrete strategies may approximate by improving overlap control and communication.

7 Discussion

If it had been possible to prove that a contraction-based parallelization has smaller bounded search spaces without assuming instantaneous communication, there would have been a ground to expect a generalized success of distributed search, at least to the extent to which smaller bounded search search spaces mean shorter search. However, we found that this type of result does not hold. Therefore, a distributed-search contraction-based strategy may do better than its sequential counterpart, but it is not guaranteed to. When adopting distributed search, one expects that communication will have a cost, and contraction may be delayed. The trade-off is to accept these disadvantages in order to avoid synchronization (a method where parallel processes have to synchronize on every inference in order to enforce eager contraction would be hopeless). Also, one may conjecture that the advantage of subdivision will offset the cost of communication in terms of delayed contraction. Our analysis showed that this conjecture does not hold on the bounded search spaces. In summary, this analysis contributes to explain why the parallelization of efficient forward-reasoning strategies has been an elusive target. Furthermore, the explanation is analytical, rather than based solely on empirical observations.

So little is known about complexity in theorem proving, and strategy analysis, however, that these findings should be regarded as a beginning, not a conclusion. In this paper we have tried essentially to determine whether distributed search may make the search space *smaller* by doing at least as much contraction as the sequential process and adding the effect of the subdivision. Accordingly, we have compared bounded search spaces by comparing the multiplicities of each clause. It remains the question of whether distributed search may take advantage by performing steps in different order, especially contraction steps, hence producing *different* search spaces. Thus, a first direction for further research may be to find other ways to compare the bounded search spaces, which may shed light on other aspects, and possibly other advantages, of distributed search. Another direction for future work is to apply the bounded search spaces to analyze *multi-search contraction-based strategies.* These issues may be connected to, or even require, the continuation of the analysis of sequential contraction-based strategies. In [10], we compared strategies with the same search plan and inference systems different in contraction power. The complementary problem of analyzing sequential strategies with the same inference system but different search plans still needs to be addressed. Finally, we have considered only forward-reasoning strategies; another line of research is to extend our methodology to subgoal-reduction strategies, such as those based on model elimination.

References

1. S. Anantharaman and M. P. Bonacina. An application of automated equational reasoning to many-valued logic. In M. Okada and S. Kaplan, editors, *CTRS-90*, volume 516 of *LNCS*, pages 156–161. Springer Verlag, 1990.

2. S. Anantharaman and J. Hsiang. Automated proofs of the Moufang identities in alternative rings. *J. of Automated Reasoning*, 6(1):76–109, 1990.
3. L. Bachmair and H. Ganzinger. Non-clausal resolution and superposition with selection and redundancy criteria. In A. Voronkov, editor, *LPAR-92*, volume 624 of *LNAI*, pages 273–284. Springer Verlag, 1992.
4. L. Bachmair and H. Ganzinger. A theory of resolution. Technical Report MPI-I-97-2-005, Max Planck Institut für Informatik, 1997.
5. M. P. Bonacina. On the reconstruction of proofs in distributed theorem proving: a modified Clause-Diffusion method. *J. of Symbolic Computation*, 21:507–522, 1996.
6. M. P. Bonacina. Experiments with subdivision of search in distributed theorem proving. In M. Hitz and E. Kaltofen, editors, *PASCO-97*, pages 88–100. ACM Press, 1997.
7. M. P. Bonacina. Distributed contraction-based strategies: model and analysis. Technical Report 98-02, Dept. of Computer Science, University of Iowa, 1998.
8. M. P. Bonacina and J. Hsiang. Parallelization of deduction strategies: an analytical study. *J. of Automated Reasoning*, 13:1–33, 1994.
9. M. P. Bonacina and J. Hsiang. Towards a foundation of completion procedures as semidecision procedures. *Theoretical Computer Science*, 146:199–242, 1995.
10. M. P. Bonacina and J. Hsiang. On the modelling of search in theorem proving – Towards a theory of strategy analysis. *Information and Computation*, forthcoming, 1998.
11. R. Bündgen, M. Göbel, and W. Küchlin. Strategy-compliant multi-threaded term completion. *J. of Symbolic Computation*, 21:475–506, 1996.
12. J. Denzinger and S. Schulz. Recording and analyzing knowledge-based distributed deduction processes. *J. of Symbolic Computation*, 21:523–541, 1996.
13. N. Dershowitz and J.-P. Jouannaud. Rewrite systems. In J. van Leeuwen, editor, *Handbook of Theoretical Computer Science*, volume B, pages 243–320. Elsevier, Amsterdam, 1990.
14. D. Kapur and H. Zhang. An overview of RRL: rewrite rule laboratory. In N. Dershowitz, editor, *3rd RTA*, volume 355 of *LNCS*, pages 513–529. Springer Verlag, 1989.
15. C. Kirchner, C. Lynch, and C. Scharff. Fine-grained concurrent completion. In H. Ganzinger, editor, *7th RTA*, volume 1103 of *LNCS*, pages 3–17. Springer Verlag, 1996.
16. A. Leitsch. *The Resolution Calculus*. Springer, Berlin, 1997.
17. W. McCune. Otter 3.0 reference manual and guide. Technical Report 94/6, MCS Div., Argonne Nat. Lab., 1994.
18. W. McCune. Solution of the Robbins problem. *J. of Automated Reasoning*, 19(3):263–276, 1997.
19. D. A. Plaisted. Equational reasoning and term rewriting systems. In D. Gabbay and J. Siekmann, editors, *Handbook of Logic in Artificial Intelligence and Logic Programming*, pages 273–364. Oxford University Press, New York, 1993.
20. D. A. Plaisted and Y. Zhu. *The Efficiency of Theorem Proving Strategies*. Friedr. Vieweg & Sohns, 1997.
21. C. B. Suttner and J. Schumann. Parallel automated theorem proving. In L. Kanal, V. Kumar, H. Kitano, and C. B. Suttner, editors, *Parallel Processing for Artificial Intelligence*. Elsevier, Amsterdam, 1994.
22. A. Urquhart. The complexity of propositional proofs. *Bulletin of Symbolic Logic*, 1:425–467, 1995.

A Deduction Method Complete for Refutation and Finite Satisfiability

François Bry and Sunna Torge

Computer science Institute, University of Munich, Germany
{bry,torge}@informatik.uni-muenchen.de

Abstract. Database and Artificial Intelligence applications are briefly discussed and it is argued that they need deduction methods that are not only refutation complete but also complete for finite satisfiability. A novel deduction method is introduced for such applications. Instead of relying on Skolemization, as most refutation methods do, the proposed method processes existential quantifiers in a special manner which makes it complete not only for refutation, but also for finite satisfiability. A main contribution of this paper is the proof of these results.
Keywords: Artificial Intelligence, Expert Systems, Databases, Automated Reasoning, Finite Satisfiability.

1 Introduction

For many applications of automated reasoning, the tableaux methods [32,16, 34,18] have the following advantages: They not only detect unsatisfiability but also generate models; they are close to common sense reasoning, hence easy to enhance with an explanation tool; and they are quite easy to adapt to the special syntax used in some applications. However, for most applications these methods suffer from the following drawbacks: They are often significantly less efficient than resolution based methods and they sometimes initiate the construction of infinite models, even if finite ones exist.

In this paper, a novel approach is formally introduced, which aims at overcoming these drawbacks. Like the approach [25,11] it refines and extends, this method relies on resolution and "range restriction" for avoiding the "blind instantiation" performed by the γ rule [32,16]. Thanks to range restriction, the proposed method can represent interpretations as sets of ground positive literals. This is beneficial for two reasons. First, it often considerably reduces the search space. Second, it is well suited in application areas such as Artificial Intelligence, Databases, and Logic Programming, where this representation of interpretations and models is usual. Instead of relying on Skolemization, as most refutation methods do, the proposed method uses the extended δ rule, also called δ^* rule, proposed in [10,20,23]. This rule makes the method complete not only for refutation, but also for finite satisfiability. A prototype written in Prolog implements the proposed method and a refinement of it is used in a database application [8, 9]. There, the method described in the present paper is informally recalled, but neither formally stated, nor proved to be sound and complete. A main contribution of the present paper is the proof of these results.

2 Applications

In several application areas, specific techniques have been developed that can be expressed as a systematic search for models of first-order logic specifications.
Diagnosis. The approach to diagnosis described in [28] relies on rules of the form $P_1 \wedge ... \wedge P_n \rightarrow C_1 \vee ... \vee C_m$ interpreted as follows: the premisses $P_1, ..., P_n$ are causes for symptoms C_1, ..., or C_m. Generating a diagnostic thus consists in building up models of both the set of rules and in selecting those models that satisfy the observed symptoms. Tableaux methods are very convenient for this purpose like described e.g. in [4]. Diagnosis in fact requires to seek for models that are as small as possible [5], for simpler explanations are to be preferred to redundant ones: This principle is known as "Occam's razor".
Database View Updates. A database view can be defined as the universal closure of a rule of the form $P_1 \wedge ... \wedge P_n \rightarrow C$. Such a view gives rise to compute instances of C from instances of the P_i, thus making it possible not to blow up the database with "C data". If the view, i.e. the set of derived "C data", is to be updated, changes to the P_i corresponding to the desired view update have to be determined. This is conveniently expressed as a model generation problem [7, 2]. Meaningful solutions to a view update problem obviously have to be finite. Thus, view updates can only be computed by model generators that are complete for finite satisfiability.
Database Schema Design. In general, a database is "populated" from an initial database consisting of empty relations and views, and integrity constraints [19]. It is, however, possible that ill-defined integrity constraints prevent the insertion of any data. A model generator can be applied to detect such cases: populating the database will be possible if and only if its schema has a nonempty and finite model. The system described in [8, 9] gives rise to verify this.
Planning and Design. Solving planning and design problems can as well be seen as model generation. The specifications might describe an environment, the possible movements of a robot, a starting position, and a goal to reach. They can also describe how a complex object can be built from atomic components. In both cases, each finite model describes a solution while infinite models are meaningless.
Natural Language Understanding. Interpreting natural language sentences is often performed by generating the possible models of a logic representation of the considered sentences. Consider e.g. the sentences "*Anna sees a dog. So does Barbara.*" that can be expressed by $\exists x\ sees(Anna, x) \wedge dog(x)$ and $\exists y\ sees(Barbara, y) \wedge dog(y)$. Skolemization as well as the δ rule of standard tableaux methods would only generate one model with two distinct dogs. In contrast, thanks to the extended δ or δ^* rule, the method described in the present paper would more

conveniently generate both, a model with a single dog and a model with two distinct dogs [14].

In the above mentioned applications, the models seeked for must be finite.

Program Verification. In program verification, one tries to prove properties from programs, e.g. loop invariants. Often enough, program drafts do not fulfill their specifications. Model generators can be applied to (a logic representation of) the programs to generate "samples", or "cases" in which a requirement is violated. These samples can then be used for correcting the programs under development. Clearly Occam's razor applies: The simplest samples are preferable over larger ones, that would be interpreted as "redundant" by programmers.

Theorem Proving. Refutation theorem proving can benefit from model generation in a similar manner. If a conjecture C is not a consequence of a set $\mathcal{S}$ of formulas, then, applying a model generator to $\mathcal{S} \cup \{\neg C\}$ will construct counterexamples to the conjectured theorem, i.e. models of $\mathcal{S} \cup \{\neg C\}$. These models can be used for correcting the conjecture C. Here again Occam's razor applies: if counterexamples can be found, the "smallest" ones will better help in understanding the flaw in the conjecture than redundant counterexamples.

Counterexamples to program specifications and conjectures do not have to be finite. For these applications, counterexamples that can be found in finite time, i.e. that are finitely representable, are sufficient. However, in case finite counterexamples exist, it is desirable to detect them. For this purpose, a model generator complete for "finite satisfiability" is needed. This is possible, since finite satisfiability is semidecidable [36]. For as well detecting unsatisfiability, one can rely on a refutation prover coupled with a finite-model finder such as FINDER [31] and SEM [37], as described e.g. in [30]. The present paper proposes instead a *single* deduction method complete for both, unsatisfiability and finite satisfiability. Arguably, this method is more convenient for many applications. In particular, a single method is better amenable to user interaction and explanation as provided by the system [8, 9]. For some applications - e.g. natural language understanding - a single method is necessary [14] and the coupling approach of [30] is not applicable.

Note that the applications mentioned here in general do not give hints for the size of the finite models seeked for. Note also that most applications require that the model generator constructs only "minimal models" in the sense of e.g. [11]. This related issue is beyond the scope of the present paper.

3 Preliminaries

Throughout this paper, a language with a denumerable number of constants, but without function symbols other than constants, is assumed. The interpretations (and models) considered are *term interpretations* (term models, resp.) that, except for their domains, are defined like Herbrand interpretations (models, resp.) [16]. The domain of a term interpretation $\mathcal{I}$ consists in all ground terms, here constants, occurring in the ground atoms satisfied by $\mathcal{I}$ augmented with an additional, arbitrary constant c_0 occurring in no ground atoms satisfied by

$\mathcal{I}$. In contrast, the domain of an Herbrand interpretation consists in all possible ground terms constructable by constants and function symbols in the considered language. The mapping of an interpretation that assigns to every constant (resp. n-ary relation symbol) an element of the domain (resp. a n-ary relation over the domain) will be called *assignment*.

A term interpretation is uniquely characterized by the set $\mathcal{G}$ of ground atoms it satisfies, and will therefore be denoted by $\mathcal{T}(\mathcal{G})$. If $\mathcal{S}$ is a set (finite set, resp.) of formulas and $\mathcal{T}(\mathcal{G})$ a term model of $\mathcal{S}$, there might be constants (finitely many constants, resp.) in $\mathcal{S}$ which do not occur in $\mathcal{G}$. These constants are assumed to be interpreted over the special constant c_0 which without loss of generality can further be assumed not to occur in $\mathcal{S}$. The subset relation $\subseteq$ induces an order $\leq$ on term interpretations: $\mathcal{T}(\mathcal{G}_1) \leq \mathcal{T}(\mathcal{G}_2)$ iff $\mathcal{G}_1 \subseteq \mathcal{G}_2$. A term model of a set of formulas is said to be *minimal*, if it is minimal for $\leq$.

The first-order language considered is assumed to include two atoms $\bot$ and $\top$ that respectively evaluate to false and true in all interpretations. A negated formula $\neg F$ will always be treated as the implication $F \rightarrow \bot$. The multiple quantification $\forall x_1 x_2 \dots x_n F$, also noted $\forall \bar{x} F$ if $\bar{x}$ is the tuple of variables $x_1 x_2 \dots x_n$, is a shorthand notation for $\forall x_1 \forall x_2 \dots \forall x_n F$. The notation $\forall \epsilon F$, where ϵ denotes the empty tuple, is allowed and stands for the formula F. Except when otherwise stated, "formula" is used in lieu of "closed formula". If $\bar{x}$ is a tuple of variables $x_1 \dots x_n$ and if $\bar{c}$ is a tuple of constants $c_1 \dots c_n$, then $[\bar{c}/\bar{x}]$ will denote the substitution $\{c_1/x_1, \dots, c_n/x_n\}$.

In the following familiarity with tableaux methods as introduced in e.g. [32, 16, 18] is assumed.

4 Positive Formulas with Restricted Quantifications

In this section, a fragment of first-order logic, that of "positive formulas with restricted quantifications" (short PRQ formulas), is introduced. Arguably, this fragment is convenient for applications. It is shown to have the same expressive power as full first-order logic. The intuition of PRQ formulas is that of so-called "restricted quantification" in natural language. The first time an object is referred to in a formula, i.e., when a variable is quantified, a positive expression called "range" specifies which kind of object is meant, like e.g. in $\forall x\ (\underline{employee(x)} \rightarrow \exists y\ (\underline{boss(y)} \wedge works\_for(x, y)))$. The underlined expressions are ranges for x and y, respectively. Note also the use of an implication (resp. conjunction) for introducing the range of a universally (resp. existentially) quantified variable. Ranges and PRQ formulas are defined relying on auxiliary notions that are first introduced.

Definition 1.

- Positive conditions *are inductively defined as follows:*

1. *Atoms except* $\bot$ *are positive conditions.*
2. *Conjunctions and disjunctions of positive conditions are positive conditions.*
3. $\exists y\ F$ *is a positive condition if* F *is a positive condition.*

• Ranges for variables $x_1, \ldots, x_n$ *are inductively defined as follows:*

1. *An atom in which all of* $x_1, \ldots,$ *and* x_n *occur is a range for* $x_1, \ldots, x_n$.
2. $A_1 \vee A_2$ *is a range for* $x_1, \ldots, x_n$ *if both* A_1 *and* A_2 *are ranges for* $x_1, \ldots, x_n$.
3. $A_1 \wedge A_2$ *is a range for* $x_1, \ldots, x_n$ *if* A_1 *is a range for* $x_1, \ldots, x_n$ *and* A_2 *is a positive condition.*
4. $\exists y\ R$ *is a range for* $x_1, ..., x_n$ *if* R *is a range for* $y, x_1, ..., x_n$ *and if* $x_i \neq y$ *for all* $i = 1, \ldots, n$.

• Positive Formulas with Restricted Quantifications *(short PRQ formulas) are inductively defined as follows:*

1. *Atoms (in particular* $\bot$ *and* $\top$*) are PRQ formulas.*
2. *Conjunctions and disjunctions of PRQ formulas are PRQ formulas.*
3. *A formula of the form* $P \rightarrow F$ *is a PRQ formula if* P *is a positive condition and* F *a PRQ formula.*
4. *A formula of the form* $\forall x_1 \ldots x_n\ (R \rightarrow F)$ $(n \geq 1)$ *is a PRQ formula if* R *is a range for* $x_1, \ldots, x_n$ *and if* F *is a PRQ formula.*
5. *A formula of the form* $\exists x\ (R \wedge F)$ *is a PRQ formula if* R *is a range for* x *and if* F *is a PRQ formula.*

Example 1. The formula $F = \forall xy\ (p(x) \wedge q(y) \vee r(x, y) \rightarrow \exists z\ (s(z) \wedge t(z)))$ is a PRQ formula, because $p(x) \wedge q(y) \vee r(x, y)$ is a range for x and y and $s(z)$ is a range for z. The formula $G = \forall xy\ (p(x) \vee q(y) \rightarrow s(y))$ in contrast is not a PRQ formula, since the premise of the implication is not a range for x and y.

Note, that ranges are positive conditions. The following Lemma will be used in proving Theorem 4.

Lemma 1. *Let* $\mathcal{M}$ *and* $\mathcal{N}$ *be sets of ground atoms such that* $\mathcal{M} \subseteq \mathcal{N}$ *and* R *a positive condition. If* $\mathcal{T}(\mathcal{M}) \models R$*, then* $\mathcal{T}(\mathcal{N}) \models R$.

Proof. (sketched) By induction on the structure of R. ∎

For most applications the restriction to PRQ formulas is not a severe restriction since in general quantifications in natural languages are "restricted" through – implicit or explicit – sorts. Furthermore for every finite set $\mathcal{F}$ of first-order formulas there exists a finite set PRQ($\mathcal{F}$) of PRQ formulas with the "same" models as $\mathcal{F}$ in the following sense:

Theorem 1. (Expressive Power of PRQ Formulas) *Let* Σ *be the signature of the first-order language under consideration,* D *a unary predicate such that* $D \notin \Sigma$, $\Sigma' = \Sigma \cup \{D\}$*. Then for every finite set* $\mathcal{F}$ *of first-order formulas over* Σ *there exists a finite set PRQ(*$\mathcal{F}$*) of PRQ formulas over* Σ' *such that:*

1. *If* $(\mathcal{D}, m)$ *is a model of* $\mathcal{F}$ *with domain* $\mathcal{D}$ *and assignment function* m *and if* m' *is the mapping over* Σ' *defined as follows:*

$$m'(s) := \begin{cases} m(s) & \text{if } s \neq D \\ \mathcal{D} & \text{if } s = D \end{cases}$$

then $(\mathcal{D}, m')$ *is a model of PRQ(*$\mathcal{F}$*).*

2. If $(\mathcal{D}', m')$ is a model of PRQ($\mathcal{F}$), then there exists $\mathcal{D} \subseteq \mathcal{D}'$ such that $(\mathcal{D}, m' \mid_{\Sigma})$ is a model of $\mathcal{F}$, where $m' \mid_{\Sigma}$ is the restriction of m' to Σ.

Proof. (sketched) Let $\mathcal{F}$ be a finite set of formulas. Recall that there exists a finite set $\mathcal{G}$ of formulas in prenex conjunctive normal form such that $\mathcal{F}$ and $\mathcal{G}$ are logically equivalent. Recall also that a disjunction $D = D_1 \vee \cdots \vee D_n$ of literals is equivalent to the implication $P \rightarrow C$ with **1** $P = P_1 \wedge \ldots \wedge P_k$ if the set $\{\neg P_i \mid i = 1, \ldots, k\}$ of negative literals in D is nonempty, $P = \top$ otherwise, and **2** $C = C_1 \vee \ldots \vee C_m$ if the set $\{C_i \mid i = 1, \ldots, m\}$ of positive literals in D is nonempty, $C = \perp$ otherwise. Call "in implication form" the formula obtained from a formula in prenex conjunctive normal form by transforming each of its conjuncts into the above-mentioned, logically equivalent implication form. Hence, there exists a finite set $\mathcal{F}''$ of formulas in implication form which is logically equivalent to $\mathcal{F}$. Let $\mathcal{F}'$ be the finite set of PRQ formulas obtained by applying the following transformation $\mathcal{R}$ to the formulas in $\mathcal{F}''$: $\mathcal{R}(\forall x F) := \forall x(D(x) \rightarrow \mathcal{R}(F))$, $\mathcal{R}(\exists x F) := \exists x(D(x) \wedge \mathcal{R}(F))$, and $\mathcal{R}(F) := F$ if F is not a quantified formula. One easily verifies that PRQ($\mathcal{F}$) $:= \mathcal{F}' \cup \mathcal{C}(\mathcal{F})$ fulfills the condition of Theorem 1, where $\mathcal{C}(\mathcal{F}) := \{D(c) \mid c$ constant occurring in $\mathcal{F}\}$ if some constants occur in $\mathcal{F}$, $\mathcal{C}(\mathcal{F}) := \{D(c_0)\}$ for some arbitrary constant c_0, otherwise. ■

The predicate D of Theorem 1 generalizes the domain predicate *dom* of [25]. Note that other, more sophisticated transformations of first-order formulas into PRQ formulas than that used in the previous proof are possible which, for efficiency reasons, are more convenient in practice. For space reasons, they are not discussed here.

Corollary 1. *For every finite set $\mathcal{F}$ of first-order formulas there exists a finite set PRQ($\mathcal{F}$) of PRQ formulas such that $\mathcal{F}$ is finitely satisfiable if and only if PRQ($\mathcal{F}$) has a finite term model.*

Proof. From Theorem 1 follows that for every finite set $\mathcal{F}$ of first-order formulas there exists a finite set PRQ($\mathcal{F}$) of PRQ formulas such that $\mathcal{F}$ is finitely satisfiable if and only if PRQ($\mathcal{F}$) is finitely satisfiable. If PRQ($\mathcal{F}$) has a finite model $\mathcal{M}$, then a finite term model of PRQ($\mathcal{F}$) is obtained by a renaming of the elements of the universe of $\mathcal{M}$. ■

5 Extended Positive Tableaux

Extended Positive tableaux, short EP tableaux, are a refinement of the PUHR tableaux defined in [11] as a formalization of the SATCHMO theorem prover [25]. Other related formalizations are given in [13,3]. The refinement of EP tableaux consists of the processing of PRQ formulas instead of (Skolemized) clauses, and in a tableaux expansion rule for existentially quantified subformulas which, as opposed to the standard δ rule [32,16], performs no "run time Skolemization".

Example 2. Consider $\mathcal{S} = \{p(a), \forall x(p(x) \rightarrow \exists y p(y))\}$[1] and a Skolemized version Sk($\mathcal{S}$) of $\mathcal{S}$. Applied to Sk($\mathcal{S}$), the PUHR tableaux method initiates the construction of the infinite model $\{p(a), p(f(a)), p(f(f(a))), \dots\}$. A similar problem arises if the standard δ rule is applied to $\mathcal{S}$: The finite model $\{p(a)\}$ of $\mathcal{S}$ is not detected by the PUHR tableaux method.

Definition 2. (EP Tableaux expansion rules)

$\exists$ (or δ^*) rule:

$$\frac{\exists x E(x)}{E[c_1/x] \mid \dots \mid E[c_k/x] \mid E[c_{new}/x]}$$

where $\{c_1 \dots c_k\}$ is the set of all constants occurring in the expanded node, and where c_{new} is a constant distinct from all c_i for $i = 1, \dots, k$.

PUHR rule:

$$\frac{\forall \bar{x}(R(\bar{x}) \rightarrow F)}{F[\bar{c}/\bar{x}]}$$

$\vee$ rule:

$$\frac{E_1 \vee E_2}{E_1 \mid E_2}$$

$\wedge$ rule:

$$\frac{E_1 \wedge E_2}{\begin{array}{c} E_1 \\ E_2 \end{array}}$$

where $R[\bar{c}/\bar{x}]$ is satisfied by the interpretation specified by the expanded node, i.e., by the set of ground atoms occurring in that node.

In the PUHR rule, $\bar{c}$ is a tuple of constants occurring in the expanded node. These constants are determined by evaluating R against the already constructed interpretation, i.e., the term interpretation determined by the set of ground atoms occurring in the node. This evaluation corresponds to an extension of positive unit hyperresolution. It coincides with (standard) positive unit hyperresolution if $R \rightarrow F$ has the form $P_1 \wedge \dots \wedge P_n \rightarrow C_1 \vee \dots C_m$ where the P_i $(i = 1, \dots, n)$ and C_j $(j = 1, \dots, m)$ are atoms. Recall that the notation $\forall \epsilon (R \rightarrow F)$, where ϵ denotes the empty tuple, is allowed and stands for the formula $R \rightarrow F$. Thus, the PUHR rule handles both, universally quantified and implicative formulas.

Definition 3. (EP Tableaux) *If $\mathcal{S}$ is a set of formulas, Atoms($\mathcal{S}$) will denote the set of ground atoms in $\mathcal{S}$. EP Tableaux for a set $\mathcal{S}$ of PRQ formulas are trees whose nodes are sets of closed formulas. They are inductively defined as follows:*

1. *The tree consisting of the single node $\mathcal{S}$ is an EP Tableau for $\mathcal{S}$.*
2. *Let T be an EP tableau for $\mathcal{S}$, L a leaf of T, and φ a formula in L that is not satisfied in the term interpretation $\mathcal{T}(Atoms(L))$. Then the tree obtained from T by appending one or more children to L according to the expansion rule applicable to φ is an EP tableau for $\mathcal{S}$. Each child of L consists of L and one more formula or two more formulas in the case of φ being a conjunction.*

[1] Strictly, the syntax of Definition 1 would require $\forall x\ (p(x) \rightarrow \exists y\ (p(y) \underline{\wedge \top}))$.

A branch of an EP Tableau is closed *if it contains* $\bot$. *Otherwise, it is* open. *An EP tableau is* open *if at least one of its branches is open; otherwise, it is* closed. *If* $\mathcal{B}$ *is a branch in an EP tableau, then* $\bigcup\mathcal{B}$ *denotes the union of the nodes in* $\mathcal{B}$. *An EP tableau is* satisfiable *if it has a branch* $\mathcal{B}$ *such that* $\bigcup\mathcal{B}$ *is satisfiable.*

Note that the PUHR rule is the only expansion rule which can be applied more than once to the same formula along a branch of an EP tableau. Indeed the condition "that is not satisfied in the term interpretation $\mathcal{T}(Atoms(L))$" prevents repeated applications of rules other than the PUHR rule. Note also that, although only *finite* EP tableaux can be constructed in *finite* time, Definition 3 does not preclude infinite EP tableaux.

Example 3. Let $\mathcal{S}_1 = \{p(a), \forall x\ (p(x) \rightarrow r(x) \vee \exists y\ q(x,y))\}$.[2] The following table denotes a EP tableau for $\mathcal{S}_1$ in the manner of [6] (to which the denomination "tableaux method" goes back): Successor nodes are right from their parent nodes, branching is represented vertically, and the nodes are not labelled with sets of formulas but with the single formula added at the corresponding node.

$\mathcal{S}_1$	$r(a) \vee \exists y q(a,y)$	$r(a)$	
		$\exists y q(a,y)$	$q(a,a)$
			$q(a, c_{\text{new}})$

Example 4. Let $\mathcal{S}_2 = \{empl(c_0), \forall x\ (empl(x) \rightarrow \exists y\ works\text{-}for(x,y)), \forall xy\ (works\text{-}for(x,y) \rightarrow empl(x) \wedge empl(y))\}$[3] The following denotes an infinite EP tableau for $\mathcal{S}_2$ (the predicates are abbreviated to their first letters):

$\mathcal{S}_2$	$\exists y w(c_0, y)$	$w(c_0, c_0)$				
		$w(c_0, c_1)$	$e(c_1)$	$\exists y w(c_1, y)$	$w(c_1, c_0)$	
					$w(c_1, c_1)$	
					$w(c_1, c_2)$	$e(c_2)$ ...

6 Refutation Soundness and Completeness

The results of this section are established using standard techniques (cf. e.g. [16, 11, 18]). In the following, $\mathcal{S}$ denotes a set of PRQ formulas.

Lemma 2. *The application of an expansion rule to a satisfiable EP tableau results in a satisfiable EP tableau.*

Proof. (sketched) For every expansion rule, one easily shows that if a node N of an EP tableau is satisfiable, then there is at least one successor of N which is satisfiable. ∎

Theorem 2. (Refutation Soundness) *If there exists a closed EP tableau for* $\mathcal{S}$, *then* $\mathcal{S}$ *is unsatisfiable.*

[2] Strictly, the syntax of Definition 1 would require $\forall x (p(x) \rightarrow r(x) \vee \exists y (q(x,y) \underline{\wedge \top}))$.
[3] Definition 1 would require $\forall x\ (empl(x) \rightarrow \exists y\ (works\text{-}for(x,y) \underline{\wedge \top}))$.

Proof. Assume $\mathcal{S}$ is satisfiable. By Lemma 2 there exists no closed EP tableaux for $\mathcal{S}$. ■

The following definition formalises the standard concept of fairness cf. e.g. [16]. Recall that the nodes of an EP tableau are sets of PRQ formulas.

Definition 4.
- *Let T be an EP tableau for $\mathcal{S}$, and $\mathcal{B}$ a branch in T. Then $\cup\mathcal{B}$ is said to be* saturated *if the following holds:*

1. *If $E_1 \vee E_2 \in \cup\mathcal{B}$ then $E_1 \in \cup\mathcal{B}$ or $E_2 \in \cup\mathcal{B}$.*
2. *If $E_1 \wedge E_2 \in \cup\mathcal{B}$ then $E_1 \in \cup\mathcal{B}$ and $E_2 \in \cup\mathcal{B}$.*
3. *If $\exists x E(x) \in \cup\mathcal{B}$ then there is $E[x/c_1] \in \cup\mathcal{B}$, or ..., or $E[x/c_n] \in \cup\mathcal{B}$, or $E[x/c_{new}] \in \cup\mathcal{B}$. $c_1, \ldots, c_n$ are all constants occurring in the node that is expanded by the $\exists$-rule, and c_{new} is a constant, not occurring in this node.*
4. ***If $\forall\bar{x}\ (R(\bar{x}) \rightarrow F) \in \cup\mathcal{B}$, then for all substitutions σ, such that $\mathcal{T}(Atoms(\cup\mathcal{B})) \models R\sigma$, $F\sigma \in \cup\mathcal{B}$.***

- *An EP tableau T is called* fair *if $\cup\mathcal{B}$ is saturated for each open branch $\mathcal{B}$ of T.*

Lemma 3. (Model Soundness) *Let T be an EP tableau for $\mathcal{S}$ and $\mathcal{B}$ an open branch of T. If T is fair, then $\mathcal{T}(Atoms(\cup\mathcal{B})) \models \mathcal{S}$.*

Proof. (sketched) By induction on the structure of PRQ formulas. ■

Theorem 3. (Refutation Completeness) *If $\mathcal{S}$ is unsatisfiable, then every fair EP tableau for $\mathcal{S}$ is closed.*

Proof. Immediate consequence of Lemma 3. ■

Corollary 2. *If $\mathcal{S}$ is not finitely satisfiable, then every open branch of a fair EP tableau for $\mathcal{S}$ is infinite.*

Proof. Assume that $\mathcal{S}$ is not finitely satisfiable. Assume there is a fair EP tableau T with a finite open branch $\mathcal{B}$. By Lemma 3 $\mathcal{T}(Atoms(\cup\mathcal{B}))$ is a finite model of $\mathcal{S}$, a contradiction. ■

7 Finite Satisfiability Completeness

The proof of the completeness for finite satisfiability of EP tableaux is more sophisticated than that of other theorems given in the previous sections. It makes use of nonstandard notions, that are first introduced.

Definition 5. (Simple Expansion) *Let $\mathcal{S}$ be a satisfiable set of PRQ formulas, φ an element of $\mathcal{S}$, and $\mathcal{T}(\mathcal{G})$ a term model of $\mathcal{S}$.* Simple expansions $\mathcal{S}'$ of $\mathcal{S}$ with respect to φ and $\mathcal{T}(\mathcal{G})$ *are defined as follows:*

1. *If φ is a ground atom, then $\mathcal{S}' := \mathcal{S}$.*

2. *If* $\varphi = \varphi_1 \wedge \varphi_2$, *then* $\mathcal{S}' := (\mathcal{S} \setminus \{\varphi\}) \cup \{\varphi_1, \varphi_2\}$.
3. *If* $\varphi = \varphi_1 \vee \varphi_2$, *then* $\mathcal{S}' := (\mathcal{S} \setminus \{\varphi\}) \cup \{\varphi_i\}$ *for one* $i \in \{1,2\}$ *such that* $\mathcal{T}(\mathcal{G}) \models \varphi_i$.[4]
4. *If* $\varphi = \exists x\, \varphi_1$, *then* $\mathcal{S}' := (\mathcal{S} \setminus \{\varphi\}) \cup \{\varphi_1[c/x]\}$, *where* c *is a constant, such that* $\mathcal{T}(\mathcal{G}) \models \varphi_1[c/x]$.[5]
5. *If* $\varphi = \forall \bar{x}\, (R(\bar{x}) \rightarrow F)$, *then* $\mathcal{S}' := (\mathcal{S} \setminus \{\varphi\}) \cup \{F[\bar{x}/\bar{c}] \mid \bar{c}$ *a tuple of constants s.t.* $\mathcal{T}(\mathcal{G}) \models R[\bar{x}/\bar{c}]\}$.[6]

Note that for every $\mathcal{S}$, every element φ of $\mathcal{S}$, and every model $\mathcal{T}(\mathcal{G})$ of $\mathcal{S}$, there exists at least one simple expansion of $\mathcal{S}$ w.r.t. φ and $\mathcal{T}(\mathcal{G})$. A simple expansion $\mathcal{S}'$ of $\mathcal{S}$ w.r.t. φ and $\mathcal{T}(\mathcal{G})$ differs from $\mathcal{S}$ whenever φ is nonatomic. The existence of a simple expansion $\mathcal{S}$ w.r.t. φ and $\mathcal{T}(\mathcal{G})$ such that $\mathcal{S} \neq \mathcal{S}'$ does not necessarily mean that some EP tableaux expansion rule can be applied to φ. Indeed, according to Definition 3 an expansion rule can only be applied if $\mathcal{T}(Atoms(\mathcal{S})) \not\models \varphi$. Every simple expansion of a finite set $\mathcal{S}$ of PRQ formulas w.r.t a formula and a *finite* model $\mathcal{T}(\mathcal{G})$ of $\mathcal{S}$ is finite. Because of 5. in Definition 5 this is not necessarily the case if $\mathcal{T}(\mathcal{G})$ is infinite.

Lemma 4. *Let* $\mathcal{S}$ *be a set of PRQ formulas,* $\varphi \in \mathcal{S}$, $\mathcal{T}(\mathcal{E})$ *a finite, minimal term model of* $\mathcal{S}$, *and* $\mathcal{S}'$ *a simple expansion of* $\mathcal{S}$ *w.r.t.* φ *and* $\mathcal{T}(\mathcal{E})$. $\mathcal{T}(\mathcal{E})$ *is a minimal model of* $\mathcal{S}'$.

Proof. (sketched) By a case analysis based on the structure of φ. ∎

Definition 6. (Rank)
• *Let* φ *be a (nonnecessarily closed) PRQ formula and* d *a positive integer. The* d-rank $rk(\varphi, d)$ *of a PRQ formula is inductively defined as follows:*

1. *If* φ *is an atom, then* $rk(\varphi, d) := 0$.
2. *If* $\varphi = \varphi_1 \wedge \varphi_2$, *or* $\varphi = \varphi_1 \vee \varphi_2$, *or* $\varphi = \varphi_1 \rightarrow \varphi_2$, *then* $rk(\varphi, d) := rk(\varphi_1, d) + rk(\varphi_2, d) + 1$.
3. *If* $\varphi = \exists x \psi$, *then* $rk(\varphi, d) := rk(\psi, d) + 1$.
4. *If* $\varphi = \forall \bar{x} \psi$, *then* $rk(\varphi, d) := rk(\psi, d) \times d^n$, *where* n *is the size of the tuple* $\bar{x}$.

• *Let* $\mathcal{S}$ *be a set of PRQ formulas,* $\mathcal{T}(\mathcal{E})$ *a finite minimal model of* $\mathcal{S}$, *and* d *the cardinality of the domain of* $\mathcal{T}(\mathcal{E})$. *The* rank $rk(\mathcal{S}, \mathcal{T}(\mathcal{E}))$ *of* $\mathcal{S}$ *with respect to* $\mathcal{T}(\mathcal{E})$ *is defined by* $rk(\mathcal{S}, \mathcal{T}(\mathcal{E})) := \sum_{\psi \in \mathcal{S}} rk(\psi, d)$ *if* Atoms$(\mathcal{S}) \subset \mathcal{E}$ *and* $rk(\mathcal{S}, \mathcal{T}(\mathcal{E})) := 0$ *if* $Atoms(\mathcal{S}) = \mathcal{E}$.

Note that $rk(\mathcal{S}, \mathcal{T}(\mathcal{E})) = 0$ if and only if $\mathcal{T}(Atoms(\mathcal{S}))$ is a model of $\mathcal{S}$. In other words $rk(\mathcal{S}, \mathcal{T}(\mathcal{E})) > 0$ if and only if some EP tableau expansion rule can be applied to some formula in $\mathcal{S}$.

---

[4] Since $\mathcal{T}(\mathcal{G}) \models \mathcal{S}$ and $\varphi \in \mathcal{S}$, $\mathcal{T}(\mathcal{G}) \models \varphi_i$ for at least one of $i = 1, 2$.
[5] Such a constant necessarily exists since $\mathcal{T}(\mathcal{G}) \models \mathcal{S}$ and $\varphi \in \mathcal{S}$.
[6] If there are no constants $\bar{c}$ such that $\mathcal{T}(\mathcal{G}) \models R[\bar{x}/\bar{c}]$, then $\mathcal{S}' := \mathcal{S} \setminus \{\varphi\}$.

Lemma 5. *Let $\mathcal{S}$ be a finitely satisfiable set of PRQ formulas, $\varphi \in \mathcal{S}$, φ nonatomic, $\mathcal{T}(\mathcal{E})$ a finite minimal model of $\mathcal{S}$, and $\mathcal{S}'$ a simple expansion of $\mathcal{S}$ wrt φ and $\mathcal{T}(\mathcal{E})$. If $rk(\mathcal{S}, \mathcal{T}(\mathcal{E})) \neq 0$, then $rk(\mathcal{S}', \mathcal{T}(\mathcal{E})) < rk(\mathcal{S}, \mathcal{T}(\mathcal{E}))$.*

Proof. (sketched) By a case analysis based on the structure of φ. ∎

Theorem 4. (Completeness for Finite Satisfiability) *Let $\mathcal{T}(\mathcal{E})$ be a finite term model of $\mathcal{S}$. If $\mathcal{T}(\mathcal{E})$ is a minimal model of $\mathcal{S}$, then every fair EP tableau for $\mathcal{S}$ has a finite, open branch $\mathcal{B}$ such that, up to a renaming of constants, $Atoms(\cup\mathcal{B}) = \mathcal{E}$.*

The proof is based on a double induction. This is needed since the PUHR rule can repeatedly be applied to the same formula along the same branch. As a consequence, a measure of the syntactical complexity of the set of formulas, which would be a natural induction parameter, does not decrease after an application of the PUHR rule. This is overcome by a second induction on the number of applications of the PUHR rule to the same formula.

Proof. Let $\mathcal{T}(\mathcal{E})$ be a finite, minimal term model of $\mathcal{S}$. The proof is by induction on $rk(\mathcal{S}, \mathcal{T}(\mathcal{E}))$. Induction hypothesis:

> ($\star$) If $\mathcal{M}$ is a set of PRQ formulas, if $\mathcal{T}(\mathcal{F})$ is a finite, minimal term model of $\mathcal{M}$, and if $rk(\mathcal{M}, \mathcal{T}(\mathcal{F})) < n$, then every fair EP tableau for $\mathcal{M}$ has a finite, open branch $\mathcal{B}$ s. t. up to a renaming of constants $Atoms(\cup\mathcal{B}) = \mathcal{F}$.

Assume that $rk(\mathcal{S}, \mathcal{T}(\mathcal{E})) = 0$. $\mathcal{S}$ has therefore a single minimal term model, namely $\mathcal{T}(Atoms(\mathcal{S}))$, and every fair EP tableau for $\mathcal{S}$ consists of one single node equal to $\mathcal{S}$. Clearly, the result holds.
Assume that $rk(\mathcal{S}, \mathcal{T}(\mathcal{E})) = n > 0$. Let T be a fair EP tableau for $\mathcal{S}$. Since $\mathcal{S}$ is satisfiable, by Theorem 2 T is open. Since $rk(\mathcal{S}, \mathcal{T}(\mathcal{E})) > 0$ there exists at least one formula $\varphi \in \mathcal{S}$ on which an expansion rule can be applied. Since T is fair, its root necessarily has successor(s). Let $\varphi \in \mathcal{S}$ be the formula on which the application of an expansion rule yields the successor(s) of the root of T.
Case 1: $\varphi = \varphi_1 \wedge \varphi_2$, or $\varphi = \varphi_1 \vee \varphi_2$, or $\varphi = \exists x\psi$. By Definition 5 there is at least one successor N of the root such that $N = \{\varphi\} \cup \mathcal{S}'$ where, up to constant renaming in case $\varphi = \exists x\psi$, $\mathcal{S}'$ is a simple expansion of $\mathcal{S}$ w.r.t. φ and $\mathcal{T}(\mathcal{E})$. Since an EP tableau expansion rule cannot be applied more than once to a formula like φ, the tableau rooted at N is an EP tableau T' for the simple expansion $\mathcal{S}'$. T' is fair because so is T. For every simple expansion $\mathcal{S}'$ of $\mathcal{S}$ w.r.t. φ and $\mathcal{T}(\mathcal{E})$, by Lemma 4, $\mathcal{T}(\mathcal{E})$ is a minimal model of $\mathcal{S}'$. Since $rk(\mathcal{S}, \mathcal{T}(\mathcal{E})) > 0$ and φ is nonatomic, by Lemma 5 $rk(\mathcal{S}', \mathcal{T}(\mathcal{E})) < rk(\mathcal{S}, \mathcal{T}(\mathcal{E}))$. Therefore, by induction hypothesis ($\star$) , the tableau rooted at N has a finite open branch $\mathcal{B}'$ such that, up to a renaming of constants, $Atoms(\cup\mathcal{B}') = \mathcal{E}$. Hence, the same holds of T.
Case 2: $\varphi = \forall\bar{x}(R(\bar{x}) \to F)$. Let $\mathcal{S}'$ be the (unique) simple expansion of $\mathcal{S}$ w.r.t. φ and $\mathcal{T}(\mathcal{E})$. Along a branch of the fair EP tableau T for $\mathcal{S}$, the PUHR rule is possibly applied more than once to φ. Therefore, the tree rooted at the successor N of the root of T is not necessarily an EP tableau for $\mathcal{S}'$. In the following it

is shown how parts of T can be regarded as parts of an EP tableau for $\mathcal{S}'$. For $n \in \mathbb{N}$ and a branch $\mathcal{B}$ of T, let $\mathcal{B}^n$ denote the prefix of $\mathcal{B}$ up till (and without) the $(n+1)$-th application of the PUHR rule on φ, if the PUHR rule is applied more than n times to φ in $\mathcal{B}$; otherwise, let $\mathcal{B}^n := \mathcal{B}$. The following is first established by induction on n: For all $n \in \mathbb{N} \setminus \{0\}$,

$(\star\star)$ T has a branch $\mathcal{B}$ s.t. up to a renaming of constants $Atoms(\mathcal{B}^n) \subseteq \mathcal{E}$.

Case 2.1: $n = 1$: The successor N of the root of T results from an application of the PUHR rule to $\varphi = \forall\bar{x}(R(\bar{x}) \rightarrow F)$, i.e., by Definition 2 and 3, there is a set $\mathcal{G}$ of ground atoms and a substitution σ such that $\mathcal{G} \subset \mathcal{S}$, $\mathcal{T}(\mathcal{G}) \models R\sigma$, and $\mathcal{T}(\mathcal{G}) \not\models F\sigma$, and $N = \mathcal{S} \cup \{F\sigma\}$. Since by hypothesis $\mathcal{T}(\mathcal{E})$ is a model of $\mathcal{S}$, $\mathcal{G} \subset \mathcal{E}$ and by Lemma 1 $\mathcal{T}(\mathcal{E}) \models R\sigma$. Furthermore, since $\mathcal{T}(\mathcal{E}) \models \varphi$, $\mathcal{T}(\mathcal{E}) \models F\sigma$. Since $\mathcal{S}'$ is by hypothesis the (unique) simple expansion of $\mathcal{S}$ w.r.t. φ and $\mathcal{T}(\mathcal{E})$, $F\sigma \in \mathcal{S}'$. So, there is an EP tableau T' for $\mathcal{S}'$, which coincides with T from N until the second application of the PUHR rule on φ in all branches. Since by Lemma 4 $\mathcal{T}(\mathcal{E})$ is a minimal model of $\mathcal{S}'$ and since by Lemma 5 $rk(\mathcal{S}', \mathcal{T}(\mathcal{E})) < rk(\mathcal{S}, \mathcal{T}(\mathcal{E}))$, the induction hypothesis $(\star)$ is applicable: There is a branch $\mathcal{B}'$ in T' with, up to constant renaming, $Atoms(\cup\mathcal{B}') = \mathcal{E}$. So, for the corresponding branch $\mathcal{B}$ in T $Atoms(\mathcal{B}^1) \subseteq \mathcal{E}$.
Case 2.2: $n > 1$: Assume that $(\star\star)$ holds for all $m \leq n$. Let $\mathcal{B}_1, \dots, \mathcal{B}_k$ be all such branches of T.
If for some $i = 1, \dots, k$ $\mathcal{B}_i^n = \mathcal{B}_i^{n+1} = \mathcal{B}_i$, i.e. the PUHR rule is applied at most n times to φ along $\mathcal{B}_i$, then by induction hypothesis $(\star\star)$ $Atoms(\mathcal{B}_i^{n+1}) \subseteq \mathcal{E}$. Otherwise, since by induction hypothesis $(\star\star)$ $Atoms(\mathcal{B}_i^n) \subseteq \mathcal{E}$, by Lemma 1 and by definition of $\mathcal{S}'$ each formula $F\sigma_i$ resulting from an $(n+1)$-th application of the PUHR rule to φ in the branch $\mathcal{B}_i$ is in $\mathcal{S}'$. Therefore, an EP tableau T' for $\mathcal{S}'$ can be constructed from the subtree of T rooted at N as follows: First, replace N by $\mathcal{S}'$. Second, keep from each branch $\mathcal{B}_i$ only the prefix $\mathcal{B}_i^{n+1}$. Third, remove from each $\mathcal{B}_i^{n+1}$ those nodes resulting from applications of the PUHR rule to φ. Fourth, cut all other branches immediately before the first application of the PUHR rule to φ. T' is a finite EP tableau for $\mathcal{S}'$, which is not necessarily fair. Since T' is finite, a fair EP tableau T'' for $\mathcal{S}'$ can be obtained by further expanding T'. By Lemma 5, $rk(\mathcal{S}', \mathcal{T}(\mathcal{E})) < rk(\mathcal{S}, \mathcal{T}(\mathcal{E}))$. By induction hypothesis $(\star)$, T'' has a branch $\mathcal{B}'$ with, up to constant renaming, $Atoms(\cup\mathcal{B}') = \mathcal{E}$. By definition of $\mathcal{B}_1, \dots, \mathcal{B}_k$ and T'' there is a branch $\mathcal{B}_i$ in T such that $Atoms(\mathcal{B}_i^{n+1}) = Atoms({\mathcal{B}'}_i^{n+1})$. Hence, $Atoms(\mathcal{B}_i^{n+1}) \subseteq \mathcal{E}$.

Since by hypothesis $\mathcal{T}(\mathcal{E})$ is finite, T has a finite branch $\mathcal{B}$ for which $(\star\star)$ holds. Hence, this branch is open. Since T is fair, by Lemma 3 $\mathcal{T}(Atoms(\cup\mathcal{B})) \models \mathcal{S}$ and since $\mathcal{T}(\mathcal{E})$ is minimal, up to a renaming of constants, $Atoms(\cup\mathcal{B}) = \mathcal{E}$. ∎

It follows from Theorem 4 that a breadth-first expansion of EP tableaux is complete for finite satisfiability. A depth-first expansion of EP tableaux is not complete for finite satisfiability, as Example 5 shows. However, by theorem 3, a depth-first expansion of EP tableaux is complete for unsatisfiability.

Example 5. Consider the following PRQ formulas: $F_1 = s(a,b)$, $F_2 = \forall xy\ (s(x,y) \rightarrow \exists z\ s(y,z))$, $F_3 = \forall xyz\ (s(x,y) \wedge s(y,z) \rightarrow s(x,z))$, $F_4 = \forall xy\ (s(x,y) \wedge s(y,x) \rightarrow \bot)$. Let $\mathcal{S}_3 = \{F_1, F_2, F_3, F_4\}$. The models of $\mathcal{S}_3$ are infinite, as the following EP tableau shows:

$$
\begin{array}{llllll}
\mathcal{S}_3 & \exists z\ s(b,z)) & s(b,a) & \bot & & \\
 & & s(b,b) & \bot & & \\
 & & s(b,c_1) & s(a,c_1) & \exists z\ s(c_1,z)) & s(c_1,a) \quad \bot \\
 & & & & & s(c_1,b) \quad \bot \\
 & & & & & s(c_1,c_1) \quad \bot \\
 & & & & & s(c_1,c_2) \quad \exists z\ s(c_2,z)) \ \ldots
\end{array}
$$

Consider now $G = (F_1 \wedge F_2 \wedge F_3 \wedge F_4) \vee p$ and $\mathcal{S}_4 = \{G\}$. A depth-first, leftmost expansion of an EP tableau for $\mathcal{S}_4$ first starts the construction of an EP tableau for the subformula $(F_1 \wedge F_2 \wedge F_3 \wedge F_4)$, i.e. of an infinite EP tableau similar to an EP tableau for $\mathcal{S}_3$, and thus never expands the finite EP tableau for the subformula p.

8 Implementation

A concise Prolog program, called FINFIMO (FINd all FInite MOdels), in the style of [25] implements a depth-first expansion of EP tableaux. For space reasons, this implementation is not commented here. It is given in [9] and available at: http://www.pms.informatik.uni-muenchen.de/software/finfimo/ . First experiments as well as the database application presented in [8, 9], which is based on an implementation of a breadth-first expansion of EP tableaux [8, 9], point to an efficiency that is fully acceptable for the applications mentioned in Section 2.

9 Related Work

EP tableaux are related to approaches of different kinds: **1** Generators of models of (or up to) a given cardinality, **2** coupling of generators of models of (or up to) a given cardinality with a refutation prover, **3** tableaux methods that rely on the extended δ or δ^* rule first introduced in [10], **4** tableaux methods that make use of the γ rule instead of a PUHR rule, and **5** generators of finitely representable models.

1 Possibly, one of the first generator of models of a given cardinality has been described in [21]. Nowadays, among the best known generator of finite models of (or up to) a given cardinality are FINDER [31] and SEM [37]. Their strength lies in a sophisticated, very efficient implementation of the exhaustive search for models up to a given cardinality. Most generators of finite models up to a given cardinality can continue the search with a higher cardinality, if no models of the formerly given cardinality can be found. However they require an upper bound for the cardinality of the models seeked for. For the applications mentioned in Section 2, e.g. the application described in [8], this might be too strong a requirement. Moreover, model generators for given cardinalities such as FINDER and SEM cannot detect unsatisfiability.

2 Model generators for given cardinalities such as FINDER [31] and SEM [37] can be coupled with a refutation prover as described e.g. in [30], resulting in a system complete for both finite satisfiability and unsatisfiability. Arguably, this approach is less convenient for many applications than the approach described here. In particular, for natural language understanding [14] and for an interactive system as described in [8, 9] a single deduction system as proposed in the present paper seems preferable.
3 The extended δ (or δ^*) rule used in EP tableaux has been proposed in former studies. It has been mentioned several times, to the best of our knowledge first in [10, 20, 23]. It is by no means always the case that this rule results in a loss of efficiency for refutation reasoning, for it sometimes gives rise to sooner detection of (finite) satisfiability. This is e.g. the case with the formula $p(a) \land \forall x\ (p(x) \rightarrow \exists y p(y))$. Although it has a finite model with only one satisfied p fact, most refutation provers as well as tableaux methods using the standard δ rule expand an infinite model. Such an example is by no means unlikely in applications. An "unprovability proof system" is proposed in [35] which is complete for finite falsifiability.[7] Although it relies on two rules that remind of the extended δ or δ^* rule, it is not complete for both, unsatisfiability and finite falsifiability. It is suggested in [35] to couple it with a refutation method to achieve completeness for both properties.
4 The approach described in the present paper differs from most tableaux methods in the use of the PUHR (positive unit hyperresolution) rule, whose introduction in a tableau method has been first proposed in [25]. Formalizations of this approach [25] have been given in e.g. [11, 3]. The PUHR rule avoids the "blind instantiation" of the γ rule in those - frequent - cases, where the D predicate of theorem 1 is not needed. In such cases, the gain in efficiency compared with tableaux methods relying on the γ rule can be considerable [25]. In the implementation described in [23] the blind instantiation of the γ rule is controlled by giving a limit on the number of γ expansions for each γ formula. In practice, conveniently setting such upper bounds might be difficult. A further interest of the approach presented here is its short and easily adaptable implementation given in [9].
5 Other extensions and refinements of tableau methods generate finite representation for (possibly infinite) models [12, 33, 15, 27]. In [27] a method for extracting models of (possibly infinite) branches by means of equational constraints is described. The approaches [33, 15] are based on resolution and therefore are much more efficient than approaches based on the δ rule of classical tableaux methods. In contrast to the method described in the present paper, the method described in [33] only applies to the monadic and Ackermann class. The method of [15] which, like the PUHR [25, 11] and EP tableaux, is based on positive hyperresolution, avoids splitting. In some cases, this results in gains in efficiency. This also makes the method capable of building (finite representations of) *infinite* models for formulas that are not finitely satisfiable. The goal of EP tableaux being completeness for both, refutation and finite satisfiability, the capability of the

[7] Rather unappropriately called "finite unprovability" in [35].

method of [15] to build finite representation of infinite models cannot be taken as comparison criterium: Adding it to the EP tableaux method would make it incorrect with respect to finite satisfiability, indeed. Recall that for the applications outlined in Section 2, that motivated the present paper, finite models are needed and *finitely representable* models are not convenient.

It is difficult to compare the EP tableau method with refutation methods for *clausal* axiom systems, for it does not make much sense to check Skolemized axiom systems for finite satisfiability. Except in trivial cases, Skolemization expands finite Herbrand as well as term models into infinite ones, indeed.

10 Conclusion and Perspectives

Some applications of theorem proving have been discussed that can benefit from a model generator complete for both, refutation and finite satisfiability and furthermore not imposing an upper bound on the size of the models searched for. The approach *Extended Positive (EP) Tableaux*, developed for such applications, has been presented. Like the PUHR Tableaux [25, 11] they extend, EP Tableaux rely on positive unit hyperresolution and "range restriction" for avoiding the "blind instantiation" performed by the γ rule of standard tableaux [16, 34, 18]. Instead of relying on Skolemization, as most refutation methods do, Extended Positive Tableaux use the extended δ (or δ^*) rule of [10, 20, 23]. It was shown that this rule makes the Extended Positive Tableaux method complete not only for refutation, like standard tableaux methods, but also for finite satisfiability. A prototype written in Prolog given in [9] in the style of SATCHMO [25] implements the EP Tableaux method with a depth-first strategy. The system described in [8, 9] is based on a breadth-first expansion of EP tableaux. In these papers, neither are EP tableaux formally introduced, nor are soundness and completeness properties established.

The following issues deserve further investigations.

1 An analysis of EP tableaux and FINFIMO's efficiency is needed. This issue is however a difficult one because there are not much systems that are complete for both, finite satisfiability and unsatisfiability, and no benchmarks are available that are fully relevant for finite satisfiability verification. Note that the extended δ (or $\delta*$) rule sometimes cuts down infinite search spaces and thus sometimes results in a gain in efficiency for refutation reasoning. In other cases, however, it expands a larger search space than the standard δ rule [32, 16]. In some cases, the EP tableaux method expands isomorphic interpretations. It would be interesting to investigate, whether techniques for avoiding this, such as the least number heuristic mentioned in [37] can be integrated in the EP tableaux method.

2 For most applications it would be desirable to have typed variables. An extension based on a simple type system and many-sorted logic seem sufficient for applications such as described in [8].

3 The method could be extended to languages with function symbols. Due to the existential quantifiers it is possible with EP formulas to express functions by relations. No theoretical extensions would be necessary for such an extension and

constraint reasoning techniques as in [1] seem applicable. Nevertheless, explicit function symbols might be more convenient for some applications. On the other hand, there are applications like e.g., that described in [8] that do not need function symbols at all.
4 For applications such as diagnosis [4], natural language understanding [14], and the database issues mentioned in Section 2 [7, 2, 8, 9], it would be preferable to have a method not only complete for finite satisfiability, but also which generates only minimal models, as investigated e.g. in [11] or in [26].

Acknowledgement

The authors are thanksfull to Norbert Eisinger and the referees for usefull remarks and for pointing out some relevant references.

References

1. S. Abdennadher and H. Schütz. Model Generation with Existentially Quantified Variables and Constraints. *Proc. 6th Int. Conf. on Algebraic and Logic Programming*, Springer LNCS 1298, 1997.
2. C. Aravindan and P. Baumgartner. A Rational and Efficient Algorithm for View Deletion in Databases. *Proc. Int. Logic Programming Symposium*, MIT Press, 1997.
3. P. Baumgartner, U. Furbach, and I. Niemelä. Hyper tableaux. *Proc. 5th Europ. Workshop on Logics in AI (JELIA)*, Springer LNCS 1126, 1996.
4. P. Baumgartner, P. Fröhlich, U. Furbach, and W. Nejdl. Tableaux for Diagnosis Applications. *Proc. 6th Workshop on Theorem Proving with Tableaux and Related Methods*, Springer LNAI , 76-90, 1997.
5. P. Baumgartner, P. Fröhlich, U. Furbach, and W. Nejdl. Semantically Guided Theorem Proving for Diagnosis Applications. *Proc. 15th Int. Joint Conf. on Artificial Intelligence (IJCAI)*, 460-465, 1997.
6. E. W. Beth. The Foundations of Mathematics. North Holland, 1959.
7. F. Bry. Intensional Updates: Abduction via Deduction. *Proc. 7th Int. Conf. on Logic Programming*, MIT Press, 561-575, 1990.
8. F. Bry, N. Eisinger, H. Schütz, and S. Torge. SIC: An Interactive Tool for the Design of Integrity Constraints (System Description). *EDBT'98 Demo Session Proceedings*, 45–46, 1998.
9. F. Bry, N. Eisinger, H. Schütz, and S. Torge. SIC: Satisfiability Checking for Integrity Constraints. *Proc. Deductive Databases and Logic Programming (DDLP), Workshop at JICSLP*, 1998.
10. F. Bry and R. Manthey. Proving Finite Satisfiability of Deductive Databases. *Proc. 1st Workshop on Computer Science Logic*, Springer LNCS 329, 44-55, 1987.
11. F. Bry and A. Yahya. Minimal Model Generation with Positive Unit Hyperresolution Tableaux. *Proc. 5th Workshop on Theorem Proving with Tableaux and Related Methods*, Springer LNAI 1071, 1996.
12. R. Caferra and N. Zabel. Building Models by Using Tableaux Extended by Equational Problems. *J. of Logic and Computation*, 3, 3-25, 1993.
13. M. Denecker and D. de Schreye. On the Duality of Abduction and Model Generation in a Framework of Model Generation with Equality. *Theoretical Computer Science*, 122, 1994.

14. N. Eisinger and T. Geisler. Problem Solving with Model-Generation Approaches based on PUHR Tableaux. *Proc. Workshop on Problem-solving Methodologies with Automated Deduction, Workshop at CADE-15*, 1998.
15. C. Fermüller and A. Leitsch. Hyperresolution and Automated Model Building. *J. of Logic and Computation*, 6:2,173-203, 1996.
16. M. Fitting. *First-Order Logic and Automated Theorem Proving.* Springer, 1990.
17. H. Fujita and R. Hasegawa. A Model Generation Theorem Prover in KL1 Using a Ramified -Stack Algorithm. *Proc. 8th Int. Conf. on Logic Programming*, MIT Press, 1991.
18. U. Furbach, ed. *Tableaux and Connection Calculi. Part I.* In *Automated Deduction – A Basis for Applications.* Kluwer Academic Publishers, 1998. To appear.
19. H. Gallaire, J. Minker, and J.-M. Nicolas. Logic and Databases: A Deductive Approach. *ACM Computing Surveys*, 16:2, 1984.
20. J. Hintikka. Model Minimization - An Alternative to Circumscription. *J. of Automated Reasoning*, 4, 1988.
21. K. M. Hörnig. Generating Small Models of First Order Axioms. *Proc. GWAI-81*, Informatik-Fachberichte 47, Springer-Verlag, 1981.
22. M. Kettner and N. Eisinger. The Tableau Browser SNARKS (System Description). *Proc. 14th Int. Conf. on Automated Deduction*, Springer LNAI 1249, 1997.
23. S. Lorenz. A Tableau Prover for Domain Minimization. *J. of Automated Reasoning*, 13, 1994.
24. D. W. Loveland, D. W. Reed, and D. S. Wilson. SATCHMORE: SATCHMO with RElevancy. *J. of Automated Reasoning*, 14, 325–351, 1995.
25. R. Manthey and F. Bry. SATCHMO: A Theorem Prover Implemented in Prolog. *Proc. 9th Int. Conf. on Automated Deduction*, Springer LNAI 310, 1988.
26. I. Niemelä. A Tableau Calculus for Minimal Model Reasoning. *Proc. 5th Workshop on Theorem Proving with Analytic Tableaux and Related Methods*, Springer LNAI 1071, 1996.
27. N. Peltier. Simplifying and Generalizing Formulae in Tableaux. Pruning the Search Space and Building Models. *Proc. 6th Workshop on Theorem Proving with Tableaux and Related Methods*, Springer LNAI 1227, 1997.
28. D. Poole. Normality and Faults in Logic-Based Diagnosis. *Proc. 11th Int. Joint Conf. on Artificial Intelligence*, 1304–1310, 1985.
29. R. Reiter. A Theory of Diagnosis from First Principles. *Artificial Intelligence*, 32, 57–95, 1987.
30. M. Paramasivam and D. Plaisted. Automated Deduction Techniques for Classification in Description Logic Systems. *J. of Automated Reasoning*, 20(3), 1998.
31. J. Slaney. Finder (finite domain enumerator): Notes and Guides. Tech. rep., Australian National University Automated Reasoning Project, Canberra, 1992.
32. R. Smullyan. *First-Order Logic.* Springer, 1968.
33. T. Tammet. Using Resolution for Deciding Solvable Classes and Building Finite Models. *Baltic Computer Science*, Springer LNCS 502, 1991.
34. G. Wrightson. ed. Special Issue on Automated Reasoning with Analytic Tableaux, Parts I and II. *J. of Automated Reasoning*, 13:2 and 3, 173–421, 1994.
35. M. Tiomkin. Proving Unprovability. *Proc. 3rd Symp. Logic in Computer Science*, 22–26, 1988.
36. B. A. Trakhtenbrot. Impossibility of an Algorithm for the Decision Problem in Finite Classes. *Proc. Dokl. Acad. Nauk.*, SSSR 70, 1950.
37. Jian Zhang and Hantao Zhang. SEM: A System for Enumerating Models. *Proc. International Joint Conference on Artificial Intelligence*, 1995.

Requirement-Based Cooperative Theorem Proving

Dirk Fuchs

Fachbereich Informatik, Universität Kaiserslautern, Germany
dfuchs@informatik.uni-kl.de

Abstract. We examine an approach for demand-driven cooperative theorem proving that is well-suited for saturation-based theorem provers. We briefly point out some problems arising from the use of common success-driven cooperation methods, and we propose the application of our approach of requirement-based cooperative theorem proving. This approach aims to allowing more orientation on current needs of provers in comparison with conventional cooperation concepts. We introduce an abstract framework for requirement-based cooperation and describe two instantiations of it: Requirement-based exchange of facts and sub-problem division and transfer via requests. Finally, we report on an experimental study conducted in the areas of superposition and unfailing completion.

1 Introduction

Automated deduction is a search problem that spans huge search spaces. In the past, many different calculi have hence been developed in order to cope with problems from automated theorem proving, e.g. the superposition calculus ([BG94]) or certain kinds of tableau calculi. Furthermore, the general undecidability of problems connected with (automated) deduction entails an indeterminism in the calculi that has to and can only be tackled with heuristics. Hence, usually a large number of calculi, each of them controllable via various heuristics, can be employed when tackling certain problems of theorem proving.

When studying results of certain theorem proving competitions (e.g., [SS97]) it is recognizable that each calculus or heuristic has its specific strengths and weaknesses. As a matter of fact, for the most domains there is not only one strategy capable of proving all problems of the domain in an acceptable amount of time. Therefore, a topic that has recently come into the focus of research is the use of different strategies in parallel (see, e.g., [Ert92]).

A better approach, however, is to employ *cooperative* theorem provers. The aim of cooperative theorem proving is to let several provers work in parallel and to exchange information between them. Thus, probably occurring synergetic effects should entail a further gain of efficiency. Some architectures are proposed for cooperative proving, e.g. in [Sut92,Den95,BH95,Bon96,FD97,Fuc98a,DF98].

Existing cooperation approaches are in main parts *success-driven*: One prover detects a certain information, e.g. a derived fact, that has been useful for it. Then, this information is transferred to the receiver and integrated into its search state. One main problem regarding this cooperation technique is the *lack of orientation on concrete needs or wishes* of receiving provers. Hence, often useless information is exchanged.

Therefore, the aim of this paper is to introduce a cooperation model *for saturation-based provers* that *orients itself on concrete needs* of theorem provers. The main idea of our approach of *requirement-based* theorem proving is only to send information as a *respond* to a *request* of the receiving prover that asks for certain kinds of information. Thus, we want to focus on some kind of *demand-driven* cooperation. We utilize requests so as to concentrate on needs of provers in two ways: Firstly, we point out a method for a requirement-based exchange of facts. Secondly, we will deal with methods to realize problem division and transfer via requests. As we will see, we introduce with the latter an analytic component into provers that do not necessarily work analytically by themselves.

In the following, we introduce basics of automated deduction—in particular saturation-based theorem proving—in section 2. In section 3, we introduce a framework for requirement-based cooperative theorem proving and describe the behavior of our cooperative system. Sections 4 and 5 address concrete aspects of requirements, namely sub-problem transfer via requirements and requirement-based exchange of facts, respectively. After that, we underline the strength of our approach by first empirical studies in section 6. A discussion concludes the paper.

2 Basics of Automated Deduction

In general, automated theorem proving deals with following problem: Given a set of facts Ax (*axioms*), is a further fact λ_G (*goal*) a logical consequence of the axioms? A fact may be a clause, equation, or a general first or higher-order formula.

Provers based on *saturation-based calculi* go the way to continuously producing logic consequences from Ax until a fact covering the goal appears (but also some saturation-based calculi use the goal in inferences). Typically a saturation-based calculus contains several inference rules of an inference system $\mathcal{I}$ which can be applied to a set of facts (which represents a certain search state). *Expansion* inference rules are able to generate new facts from known ones and add these facts to the search state. *Contraction* inference rules delete facts or replace facts by others.

Usually, a theorem prover using a saturation-based calculus maintains a set $\mathcal{F}^P$ of so-called *potential* or *passive facts* from which it selects and removes one fact λ at a time. After the application of some contraction inference rules on λ, it is put into the set $\mathcal{F}^A$ of *activated facts*, or discarded if it was deleted by a contraction rule. Activated facts are, unlike potential facts, allowed to produce new facts via the application of expanding inference rules. The inferred new facts are put into $\mathcal{F}^P$. Initially, $\mathcal{F}^A = \emptyset$ and $\mathcal{F}^P = Ax$. The indeterministic selection or *activation step* is realized by heuristic means. To this end, a heuristic $\mathcal{H}$ associates a natural number $\mathcal{H}(\lambda) \in \mathbb{N}$ with each $\lambda \in \mathcal{F}^P$ and that $\lambda \in \mathcal{F}^P$ with the smallest $\mathcal{H}(\lambda)$ is selected.

We conducted experimental studies with the prover SPASS ([WGR96]) in the area of first-order logic with equality. SPASS is based on the superposition calculus (see [BG94]). The expansion rules of SPASS contain the common rules of the superposition calculus, i.e. superposition left and right, equality resolution, and equality factoring. The contraction rules contain well-known rules like subsumption and rewriting. Moreover, we conducted experiments with the equational prover DISCOUNT ([ADF95]) which is based on *unfailing completion* (see [HR87,BDP89]). In this context the axioms are always universally quantified equations, the proof goal is an arbitrarily quantified equation. The inference system underlying unfailing completion is in main parts a restricted version of the superposition calculus. It contains one expansion inference rule that corresponds to the superposition rule. The contraction rules of unfailing completion are rewriting, subsumption, and tautology deletion.

3 A Framework for Requirement-Based Cooperation

In the following, we will discuss which kinds of requirements may be well-suited for cooperative theorem proving. After that, we describe how requirement-based cooperation can be organized.

3.1 Architecture and abstract process model

The basic idea of requirement-based theorem proving is to establish cooperation between several different saturation-based theorem provers by exchanging requests and responses to requests. Requests describe certain needs of theorem provers, responses to requests contain information of receivers of requests that may be well-suited in order to fulfill some needs formulated in the requests. We consider two different types of requests.

Firstly, if it is possible to *divide a proof problem* into various *(sub-) problems*, a prover can require that some of the sub-problems should be solved by other provers. Hence, requests are used for *sub-problem division and transfer*. Secondly, it is possible to *demand information* of other provers that may be helpful *for solving the actual proof task*. The most profitable information a prover can obtain from others is a set of facts.

The architecture of our system can be described as follows: On each processor in a network of cooperating computers a saturation-based theorem prover conducts a search for a proof goal. We assume that all provers have correct inference rules regarding a common logical consequence relation $\models$. All provers start with a common original proof goal. Since it is possible, however, that provers divide problems into sub-problems it might be that in later steps of the proof run different provers have different (sub-)goals. Each prover is assigned a unique number. We either let only different incarnations of the same prover cooperate—differing from each other only in the search-guiding heuristics they employ—and hence have a network of *homogeneous* provers, or we employ different provers (*heterogeneous* network). We assume that each prover is able to communicate requests and responses directly to each other prover.

The working scheme of the provers is characterized by certain phases. While the provers tackle their problem independently during *working phases* P_w, cooperation takes place during *cooperation phases* P_c. Working phases and cooperation phases alternate with each other. Thus, the sequence of phases is $P_w^0, P_c^0, P_w^1, P_c^1, \ldots$

The activities during a cooperation phase can be divided into four activities of the following *process model:*

1. Determination and transmission of requests to other provers.
2. Transmission of responses to earlier requests of other provers.
3. Receiving and processing foreign requests.
4. Receiving and processing responses to own earlier requests.

This process model does not allow for an immediate processing of incoming requests, i.e. it is not possible to receive a request, to process it, and to transmit a response in just one cooperation phase. Instead, the response must be transmitted in a later cooperation phase. Thus, the cooperation scheme is somewhat inflexible but minimizes the amount of communication.

3.2 Fact-represented requests and responses

In order to make the process model more concrete we make our notions of request and response precise. In the following, $\mathcal{F}_{\mathcal{A}}^{A}$ and $\mathcal{F}_{\mathcal{A}}^{P}$ denote the sets of active and passive facts of a prover $\mathcal{A}$.

Definition 1 (request).
A request from a prover $\mathcal{A}$ to a prover $\mathcal{B}$ is a tuple $req = (id_{req}, \lambda_{req}, S_{req}, t_{req})$. $id_{req} \in I\!N$ is the number of the request, λ_{req} is a fact, S_{req} is a predicate defined on $(\lambda_{req}, \mathcal{F}_{\mathcal{B}}^{A}, \mathcal{F}_{\mathcal{B}}^{P})$, and $t_{req} \in I\!N$ a time index.

The component id_{req} of a request should be—from the point of view of the sender of the request—a unique number which is needed in order to identify requests and responses (see below). The fact λ_{req} *represents* the request, the predicate S_{req} is a *satisfiability condition* of the request *req*: If the predicate S_{req} is true the request is completely processed and can hence be answered by the receiver. Now, we show for both kinds of requests, sub-problem transfer and requests that ask for facts, how they can be represented by λ_{req} and S_{req}.

Firstly, if we want to transfer a sub-goal g by a request *req* we set $\lambda_{req} = g$. The satisfiability condition S_{req} is defined by $S_{req}(\lambda_{req}, \mathcal{F}_{\mathcal{B}}^{A}, \mathcal{F}_{\mathcal{B}}^{P})$ iff λ_{req} is proved by the receiver $\mathcal{B}$ to be a logic consequence of its initial axiomatization.

Secondly, if we ask for certain facts via a request, λ_{req} is a so-called *schema fact*. This schema fact describes in a certain way how facts look like that may be useful for the sender of a request. In our methods, schema facts λ_{req} are valid facts, i.e. logic consequences of Ax. The satisfiability condition S_{req} holds if the receiver $\mathcal{B}$ has at least one fact $\bar{\lambda} \in \mathcal{F}_{\mathcal{B}}^{A} \cup \mathcal{F}_{\mathcal{B}}^{P}$ that corresponds to the schema given through λ_{req}. This is tested via a *correspondence predicate* C, i.e. in this case $S_{req}(\lambda_{req}, \mathcal{F}_{\mathcal{B}}^{A}, \mathcal{F}_{\mathcal{B}}^{P})$ iff $\exists \bar{\lambda} \in \mathcal{F}_{\mathcal{B}}^{A} \cup \mathcal{F}_{\mathcal{B}}^{P} : C(\lambda_{req}, \bar{\lambda})$. Different methods for requesting facts can be developed by using different methods for identifying schema facts and constructing correspondence predicates (see section 5). Note that this is the crucial point of our approach since the schema facts make the wishes of certain provers concrete and the correspondence predicates determine whether a prover is able to fulfill wishes and hence support another.

The time index t_{req} is the maximal number of working phases which are allowed to take place between the receipt of the request and the transmission of its response. The idea behind the use of such a time index is that the receiver of the request should not work mainly on the request but on its own proof attempt. Requests of other provers should be tackled besides the provers own activities. Since we do not want to put too much load on each prover through requests of others we restrict the processing of requests to a fixed duration. Note that we also use the time index t_{req} for defining the predicate S_{req}. We add the condition "time limit t_{req} is exceeded" as a conjunctive condition to S_{req} when dealing with requests for facts. Thus, we achieve that all facts are inserted into the response set that fulfill the correspondence predicate S_{req} after the expiration of the time limit (see below).

Responses are represented by facts, too:

Definition 2 (response).
A response of a prover $\mathcal{A}$ to a prover $\mathcal{B}$ is a triple $rsp = (id_{rsp}, B_{rsp}, \Lambda_{rsp})$. $id_{rsp} \in \mathbb{N}$ is the number of the response, B_{rsp} is a Boolean value, and Λ_{rsp} a set of facts.

The component id_{rsp} of a response equals the number of the respective request that is being answered. The Boolean value B_{rsp} indicates whether or not the responder could process the request successfully (regarding the satisfiability condition S_{req}) within the time limit given by t_{req}. The *response set* Λ_{rsp} is a set of facts which represents the answer to a request. If we respond to a request that transferred a sub-problem usually $\Lambda_{rsp} = \emptyset$. If a prover responds to a request $(id_{req}, \lambda_{req}, S_{req}, t_{req})$ for facts, S_{req} is based on the correspondence predicate C, and I is the set of axioms, Λ_{rsp} contains max_{rsp} facts λ with $I \models \lambda$ and $C(\lambda_{req}, \lambda)$.

By employing these definitions we are able to outline the activities of the process model in more detail:

The determination and transmission of requests is performed as follows: At first it is necessary to identify on the one hand sub-problems that should be tackled by other provers, on the other hand schemata of such facts that appear to be useful. After that, a unique id_{req} as well as a suitable time index t_{req}

is assigned to each sub-problem or schema fact that should be sent to another prover via a request *req*. The next step is to insert each request into a queue *Req* of open requests of the sender and to transmit the request to other provers which are part of the network. More exactly, we transmit all requests that ask for facts to every prover which is part of the network, but transmit each sub-problem only to one other prover.

In order to receive and respond to requests of other provers it is necessary to have also queues Req^i of open requests of other provers i. In order to respond to such requests, requests $req \in Req^i$ must be checked. If *req* is fulfilled a suitable response can be transmitted and *req* can be deleted from the queue Req^i. Otherwise, it is necessary to check whether the time limit is exceeded. If this is true, the response $(id_{req}, false, \emptyset)$ must be communicated to the sender of the request and *req* must be deleted from the queue.

If a prover receives a request *req* from another prover i, firstly *req* is inserted into Req^i. Secondly, the processing of the request is initiated (see sections 4 and 5 for details).

When receiving a response *rsp* to a request it is at first necessary to determine the original request $req \in Req$ with the help of the id_{rsp} component of the response. If the request has been processed successfully one can—if a sub-problem was transmitted by the request—consider the respective sub-problem to be solved or—if facts have been asked for—integrate the response set Λ_{rsp} into the search state. If the request has not been processed successfully the sender can use this information in future (see [Fuc98b]). Finally, *req* has to be deleted from *Req*.

4 Sub-problem Transfer by Requirements

In this section we present a method for transferring sub-problems via requests. We restrict ourselves to the area of first-order theorem proving and henceforth facts are first-order clauses. For literals we define $\sim l$ by $\sim l = l'$, if $l \equiv \neg l'$, and $\sim l = \neg l$, otherwise. For a clause $C = \{l_1, \ldots, l_n\}$, $\sim C$ is the set of clauses $\sim C = \{\{\sim l_1\}, \ldots, \{\sim l_n\}\}$. If C is a clause $V(C)$ denotes the set of different variables in C.

In order to realize requirement-based cooperation by transferring sub-problems we employ our abstract model. It is only needed to make three aspects more precise. Firstly, we have to answer the question how we can identify certain sub-problems. Secondly, we must introduce techniques for managing our different sub-problems. Finally, we have to develop a method well-suited for processing sub-problems of other provers.

4.1 Identifying sub-problems

We start with the identification of sub-problems. In the following, we assume that our proof problem is given as a set $\mathcal{M}$ of clauses whose inconsistency is to be proved by deriving the empty clause $\Box$ from $\mathcal{M}$. This is not a restriction in comparison with our original notion of a proof problem because $Ax \models C$ iff $Ax \cup \sim C$ is inconsistent. Henceforth, we will call the clauses from $\sim C$ *goal clauses*. Now, consider this situation:

Definition 3 (i-AND-partition).
Let $\mathcal{M} = \mathcal{M}' \cup \{C\}$, $\mathcal{M}'$ be a set of clauses, C be a clause. We call $(P_i)_{1 \le i \le n}$ an i-AND-partition of C regarding $\mathcal{M}$ iff

- *$(P_i)_{1 \le i \le n}$ is a partition of C, i.e. $\cup_{1 \le i \le n} P_i = C$, $P_i \cap P_j = \emptyset$ if $i \neq j$*
- *$\mathcal{M}$ is inconsistent iff $\forall i, 1 \le i \le n : \mathcal{M}' \cup \{P_i\}$ is inconsistent.*

If the inconsistency of a set $\mathcal{M}$ should be proved and we have identified an i-AND-partition of a clause $C \in \mathcal{M}$, we have also divided our original proof problem into n sub-problems $\wp_i =$ "$\mathcal{M}' \cup \{P_i\}$ is inconsistent".

Such an approach for dividing a problem into sub-problems is viable because there is an easy method for identifying i-AND partitions of a clause:

Theorem 1. *([Fuc98b]) Let $\mathcal{M}$ be a set of clauses, $\mathcal{M} = \mathcal{M}' \cup \{C\}$ for a set of clauses $\mathcal{M}'$ and a clause C. Let $(P_i)_{1 \le i \le n}$ be a partition of C and $V(P_i) \cap V(P_j) = \emptyset$ for $i \neq j$. Furthermore, let the sets of clauses $\mathcal{N}_i$ $(1 \le i \le n)$ be defined by $\mathcal{N}_i = \mathcal{M}' \cup \{\{\sim l\} : l \in P_j, j < i, V(P_j) = \emptyset\}$. Then it holds:*

1. *$(P_i)_{1 \le i \le n}$ is an i-AND-partition of C regarding $\mathcal{M}$.*
2. *$\mathcal{M}$ is inconsistent iff $\forall i, 1 \le i \le n : \mathcal{N}_i \cup \{P_i\}$ is inconsistent.*

The theorem points out a method for creating new sub-problems: On each processing node—*if we have not already divided the problem into sub-problems*—we check for each activated clause C which is a descendant of a goal clause whether it can be partitioned into $(P_i)_{1 \le i \le n}$, $n \ge 2$, $V(P_i) \cap V(P_j) = \emptyset$ for $i \neq j$. If m is the number of cooperating provers we limit the value n by $1 < n \le m$. Then, each prover that is able to find such a partition of a clause C distributes $n-1$ tasks $\wp_2, \ldots, \wp_n$ to other provers via requests (a prover obtains exactly one of these sub-problems) and tackles the remaining task $\wp_1$. That is, it replaces the clause C with its sub-clause P_1. The tasks which are sent to other provers are stored in a list $R = (\wp_2, \ldots, \wp_n)$. Note that the prover must possibly tackle also these sub-problems. This is because there is no guarantee that the other provers can give positive answers to the requests within their time limits.

A theorem prover that has divided the problem can work more efficiently. This is because it works with smaller clauses. Further, it is possible to obtain interesting lemmas from other provers if they give positive responses to requests.

4.2 Managing and processing sub-problems

The main problem caused by this kind of problem division and transfer is that sender and receiver of a request have to work with clauses that are in general not logic consequences of the initial set of clauses. This is because a sub-clause of a clause need not logically follow from the clause. Thus, we must develop mechanisms so as to work with such clauses.

We start with the *sender* of a request and assume that it tackles the sub-problem $\wp_i =$ "$\mathcal{M}' \cup \{P_i\}$ is inconsistent" by adding P_i to its search state.

In order to work with a semantically invalid clause P_i during the inference process the sender introduces a *tag* for P_i and all descendants of P_i. It indicates that these clauses are no logic consequences of the initial clause set but descendants of semantically invalid clauses. Hence, we do not work any longer with clauses C but with clauses with tag (C, τ). Either $\tau = \epsilon$ or $\tau = n \in \mathbb{N}$. $\tau = \epsilon$ denotes that the clause C is untagged, i.e. it is a logic consequence of the initial clauses. If $\tau = n$, n being the number of the sender, then the clause C is a descendant of P_i. In order to perform inferences we must replace the inference system $\mathcal{I}$ of each prover by an inference system $\mathcal{I}^\tau$.

Definition 4 (Inference system $\mathcal{I}^\tau$).
Let $\mathcal{I}$ be an inference system which works on sets of clauses. Then we construe the inference system $\mathcal{I}^\tau$ working on sets of tagged clauses as follows:

1. *For each expanding inference rule*
   $\{C_1, \ldots, C_n\} \vdash \{C_1, \ldots, C_n, C\}; \mathrm{P}(C_1, \ldots, C_n)$
   in $\mathcal{I}$, $\mathcal{I}^\tau$ contains the rules
   $\{(C_1, \epsilon), \ldots, (C_n, \epsilon)\} \vdash \{(C_1, \epsilon), \ldots, (C_n, \epsilon), (C, \epsilon)\}; \mathrm{P}(C_1, \ldots, C_n)$
   and
   $\{(C_1, \tau_1), \ldots, (C_n, \tau_n)\} \vdash \{(C_1, \tau_1), \ldots, (C_n, \tau_n), (C, \tau)\}; \mathrm{P}(C_1, \ldots, C_n) \wedge$
   $\exists k \in \mathbb{N}: (\exists i : \tau_i = k \wedge \forall i : \tau_i \in \{k, \epsilon\} \wedge \tau = k)$
2. *For each contracting inference rule*
   $\{C_1, \ldots, C_n, C\} \vdash \{C_1, \ldots, C_n, C'\}; \mathrm{P}(C_1, \ldots, C_n, C)$
   in $\mathcal{I}$ (rules that delete clauses are transformed analogously), $\mathcal{I}^\tau$ contains
   $\{(C_1, \epsilon), \ldots, (C_n, \epsilon), (C, \tau\} \vdash \{(C_1, \epsilon), \ldots, (C_n, \epsilon), (C', \tau)\}; \mathrm{P}(C_1, \ldots, C_n, C)$
   and
   $\{(C_1, \tau_1), \ldots, (C_n, \tau_n), (C, \tau)\} \vdash \{(C_1, \tau_1), \ldots, (C_n, \tau_n), (C', \tau')\} \cup \mathcal{D};$
   $\mathrm{P}(C_1, \ldots, C_n, C) \wedge \exists k \in \mathbb{N}: (\exists i : \tau_i = k \wedge \forall i : \tau_i \in \{k, \epsilon\} \wedge \tau \in \{k, \epsilon\} \wedge \tau' = k),$
   $\mathcal{D} = \emptyset$ *if* $\tau \neq \epsilon, \mathcal{D} = (C, \epsilon)$ *if* $\tau = \epsilon$

Hence, expansion inferences are performed in such a way that a clause C which is a result of an expanding inference with premises $C_1, \ldots, C_n$ is tagged, if some clauses C_i are tagged. Untagged clauses can contract every other clause and in that case the tag remains unchanged. If tagged clauses are able to contract an untagged clause it is necessary to store a copy of the (un-contracted) untagged clause. Otherwise, completeness may be lost if the processing of the request is finished and its offspring has been eliminated. Such copies are stored in a list $\mathcal{D}_j$, j being the number of the prover.

Now, if a prover is able to derive the empty clause $\Box$ the tag of the clause is checked. If the clause is untagged a proof of the original goal has been found. If it is tagged with the number of the prover, the current sub-problem has been solved. In the latter case the following activities take place. All clauses which are tagged with the prover's number j are deleted and the clauses $D \in \mathcal{D}_j$ are integrated untagged into the search state. If the list of open sub-problems $R = ()$ all sub-problems have been solved, i.e. also the original problem. Otherwise, a new sub-problem is chosen from R and processed as described.

This modified inference scheme allows us also to process incoming responses easily. If a sub-problem $\wp_j$ has been solved by another prover it must only be eliminated from R. Moreover, if the request clause P_j was ground the clauses being elements of $\{\{\sim l\} : l \in P_j\}$ can be utilized as new lemmas in future.

When working as a *receiver* of open sub-problems we proceed in a similar way. Each sub-goal received from another prover is tagged with the number of this prover and is added to the search state. Inferences between tagged and untagged clauses are performed with inference system $\mathcal{I}^{\tau}$. Hence, we forbid inferences between clauses having different tags in order to avoid inconsistency (see also [Fuc98b]).

If an empty clause is derived by a prover and it is untagged or tagged with the prover's number the activities are as described above. If it is tagged with the number of another prover i all clauses with this tag are deleted and the clauses from $\mathcal{D}_i$ are added to the search state. Furthermore, the sub-problem is considered to be solved, i.e. a positive response can be sent to the request sender in the next cooperation phase.

5 Requirement-Based Exchange of Facts

In this section, we present two different methods for exchanging facts via requests and responses. Note that we restrict ourselves again to the area of first-order theorem proving, i.e. facts correspond in the following to first-order clauses. We assume that all provers employ the superposition calculus and additional contraction rules like subsumption and rewriting. Note that the introduced techniques can easily be transferred to other similar saturation-based calculi.

The principle scheme of a requirement-based exchange of clauses is already known through our abstract model. However, it is necessary to describe two remaining aspects. Firstly, we must introduce methods for detecting request clauses (schema clauses) in each cooperation phase P_c^i. Secondly, we have to deal with the issue of how such requests can be processed by the receivers, i.e. we have to make precise how to compute a response set C_{rsp} of a response rsp to a request req.

5.1 Expansion-based requests

The basic idea of requests for clauses is that a theorem prover tries to get those clauses from other provers that appear to be part of a proof, but are not already derived and seem to be difficult to derive. The main problem in this context is that—because of the general undecidability of first-order theorem proving—it is impossible to predict whether or not a clause is part of a proof. However, often a prover is able to estimate whether some of its own already activated clauses possibly contribute to a proof (see below). Then, if we assume that a prover has identified a set $\mathcal{M}$ of "interesting" activated clauses, clauses other provers can be asked for are such clauses that allow for producing descendants with clauses from $\mathcal{M}$. Perhaps some of this offspring can contribute to a proof. We call such requests *expansion-based requests.*

In detail, in each cooperation phase P_c^i a prover determines a set of request clauses $\mathcal{R}_i = \{C_1^i, \ldots, C_{max_{req}}^i\} \subset \mathcal{F}^A$. Each request clause should be untagged, i.e. a valid clause. As already mentioned, the clauses C_j^i $(1 \leq j \leq max_{req})$ should be the clauses of the prover that appear to be most likely to contribute to a proof, i.e. they should be optimal regarding a judgment function φ which rates the probability that a clause is part of a proof.

We want to deal with the realization of φ in some more detail. Since it is the aim of a prover to derive the empty clause a clause is considered to be the better the less literals it has. Thus, we could use the formula $\varphi(C) = \frac{1}{|C|}$. We adopted and refined the method as follows. In addition to the length of a clause we take into account that clauses having literals with a rather "flat" syntactic structure can often be used for expansion inference steps like resolution. Thus, considering the number of literals and the syntactic structure of the literals we obtain the following weighting function.

Definition 5 (weighting function φ for expansion-based requests). *The weighting function φ for expansion-based requests is defined on clauses by*

$$\varphi(C) = \sum_{i=1}^{n} \varphi_{Lit}(l_i); C = \{l_1, \ldots, l_n\}$$

The function φ_{Lit} is defined as follows: $\varphi_{Lit}(l) = -\varphi_{Lit}^H(l, 0)$, if l is a positive literal, and $\varphi_{Lit}(l) = -\varphi_{Lit}^H(l', 0)$, if $l \equiv \neg l'$. The function φ_{Lit}^H can be computed by

$$\varphi_{Lit}^H(l, d) = \begin{cases} 1 + d & ; l \text{ is a variable} \\ 2 + d + \sum_{i=1}^{n} \varphi_{Lit}^H(t_i, d+1) & ; l \equiv \vartheta(t_1, \ldots, t_n), \vartheta \text{ is a} \\ & \text{function or predicate symbol} \end{cases}$$

φ judges a clause the better the less literals it has, and the less symbols and deep sub-terms each literal has. Thus, the function complies with our demands formulated above.

The second main aspect—besides the determination of request clauses—is the processing of requests by their receivers. Essentially, we have to deal with the problem of computing a response set C_{rsp} regarding a request req. The easiest method for determining a response set C_{rsp} is to insert such valid clauses into C_{rsp} that allow for expanding inferences with C_{req}. A disadvantage of this approach is that certain inferences must be performed twice. On the one hand it is necessary to perform expanding inferences with C_{req} at the receiver site in order to determine clauses $\bar{C}$ being elements of $\mathcal{C}_{rsp}$. On the other hand, the receiver of the response set $\mathcal{C}_{rsp}$ must perform exactly the same inferences with C_{req} when it integrates the clauses of $\mathcal{C}_{rsp}$ into its search state. Thus, our refinement of this simple method is as follows. The main idea is to already perform inferences with C_{req} at the responder site and to transmit only such valid clauses to the sender of the request that are already descendants of C_{req} and some of the clauses of the responder.

These descendants can be computed either independently of the "normal" inferences after the receipt of a request, or simultaneously to the inferences necessary to tackle the proof problem. We chose the latter approach by integrating C_{req} into the search state of the receiver and tagging it with both the number of the sender of the request and the *id* of the request. Then, by extending the tagging mechanism of the preceding section descendants of C_{req} can be computed (see [Fuc98b]).

5.2 Contraction-based requests

There is also another concept for a requirement-based exchange of clauses. Indeed it is difficult to predict whether or not a clause *contributes to a proof* but nevertheless it is possible to recognize whether a clause is *useful for the search for the proof*. If a clause is able to contract many other clauses it is definitely useful for the search process because it helps to save both memory and computation effort. Thus, it is also interesting to require that other provers should send clauses that allow for a lot of contracting inferences. We call these requests *contraction-based requests*.

Especially well-suited for reducing the amount of data and computation are clauses that subsume or rewrite clauses that tend to produce much offspring. Thus, the set of clauses $\mathcal{M} = \{C : C$ is a valid, active, and positive unit, C is among the max_{req} largest generators of clauses$\}$ is determined as a set of request clauses in each cooperation phase. This set offers each receiver of the request clauses the possibility of determining clauses which are able to subsume or rewrite them. These clauses are then especially useful for the search for the proof because they can contract clauses from $\mathcal{M}$ which cause much overhead. Note that we restrict ourselves to positive units mainly due to efficiency reasons. In order to determine a response set $\mathcal{C}_{rsp}$ regarding a request with request clause C_{req} we insert on the one hand clauses into $\mathcal{C}_{rsp}$ which are able to subsume C_{req}, on the other hand clauses which are able to rewrite C_{req}. Hence, we have $\mathcal{C}_{rsp} = \mathcal{C}_{rsp,sub} \cup \mathcal{C}_{rsp,rew}$.

If $\mathcal{F}^{A,v}$ contains all active clauses of the responder that are logic consequences of the initial set of clauses, the set $\mathcal{C}_{rsp,sub}$ regarding a request clause C_{req} is simply given by

$$\mathcal{C}_{rsp,sub} = \{\bar{C} : (\bar{C} \in \mathcal{F}^{A,v}, \exists\sigma : \sigma(\bar{C}) \equiv C_{req})\}$$

Determining a set $\mathcal{C}_{rsp,rew}$ of clauses which are able to rewrite C_{req} is more complicated as before because we must consider the ordering $\succ$ each prover uses for performing inferences. We restrict ourselves in the following again to response sets containing only positive equations. Since we cannot rewrite with the minimal side of an equation (regarding $\succ$), we must at first identify the sides relevant for rewriting and transform clauses $\bar{C} \in \mathcal{F}^{A,v}$ with following function θ to sets $\theta(\bar{C})$: If the sender of the request and the responder have an identical ordering $\succ$, we utilize

$$\theta(\bar{C}) = \begin{cases} \{s\} & ; \bar{C} \equiv s \doteq t, s \succ t \\ \{s,t\} & ; \bar{C} \equiv s = t, s \not\succ t, t \not\succ s \end{cases}$$

Hence, we consider only the left hand side of a rewrite rule but both sides of an equation. Otherwise, if sender and receiver employ different orderings, we employ

$$\theta(\bar{C}) = \{s,t\}; \bar{C} \equiv s = t$$

Then, it is necessary to check whether terms from $\theta(\bar{C})$ match to a sub-term of C_{req}. Such clauses can be inserted into $C_{rsp,rew}$ and send via respond messages.

In the following, $O(C_{req})$ denotes the set of positions in the request clause C_{req} and $C_{req}|p$ the sub-term of C_{req} at position p. Then, we obtain:

$$C_{rsp,rew} = \{\bar{C} : (\bar{C} \in \mathcal{F}^{A,v}, \exists(\sigma, \bar{C}' \in \theta(\bar{C}), p \in O(C_{req})) : \sigma(\bar{C}') \equiv C_{req}|p)\}$$

Note that these response sets can efficiently be computed since the sets of active clauses are usually rather small.

6 Experimental Results

In order to examine the potential of our cooperation concepts we conducted our experimental studies in the light of different domains (ROB, HEN, LCL, LDA) of the problem library TPTP (see [SSY94]). We restricted ourselves to the area of superposition-based theorem proving and coupled the provers SPASS and DISCOUNT. Our test set consisted of pure unit equality problems and of problems specified in full first-order logic. Thus, we can reveal that our cooperation concept achieves cooperation in an area where both provers are complete as well as in an area where one prover is only able to support the other but not to solve the original problem. Hence, we show that our concept is well-suited for provers having equal rights as well as for provers being in a master-slave relation.

Since both calculi—superposition and unfailing completion—are complete for pure equational logic (EQ), SPASS and DISCOUNT can work as partners having equal rights for problems of EQ. Thus, we let each prover send requests and responses to requests to its counterpart. Because of the fact that $|C| = 1$ for all clauses C it is not possible to divide a problem into sub-problems. Thus, we must omit requests dealing with sub-problem transfer. Nevertheless, expansion- and contraction-based requests for exchanging clauses can be utilized. We exchanged expansion-based requests and responses in the following manner. In each cooperation phase each prover determines $max_{req} = 10$ request clauses to be distributed to the other prover. In order to respond to an expansion-based request we inserted $max_{rsp} = 3$ clauses into the respective response set $\mathcal{C}_{rsp}$. It is sensible to give the responder enough time for processing the request. Therefore, we set the time limit $t_{req} = 3$. In order to exchange contraction-based requests we restricted the size of the set $\mathcal{M}$ of largest generators of clauses to $max_{req} = 10$. The time limit t_{req} for contraction-based requests was given by $t_{req} = 1$.

In the area of full first-order logic with equality (PL1EQ) DISCOUNT is not able to prove every valid goal because it can only deal with equations. Nevertheless, SPASS and DISCOUNT can work in some kind of master-slave relation. Because of the fact that DISCOUNT cannot prove every valid goal we decided to let only SPASS send expansion-based requests for clauses. We extended DISCOUNT so as to allow it to perform superposition with its equations and clauses received from SPASS. Contraction-based requests were exchanged by both provers. We have chosen the same parameter setting as in the area of unit equality. Because of the fact that in first-order logic with equality a clause can have a length greater than 1 we can transfer sub-problems from SPASS to DISCOUNT.

In order to allow for a better comparison of our different concepts for sending requests for clauses, we either exchanged only expansion-based requests and responses to the requests or contraction-based requests and responses. Requests that transferred sub-problems were—considering the above restrictions—always exchanged.

In all of our test domains we only considered problems that none of the provers could solve within 10 seconds (medium and hard problems). Moreover, we restricted ourselves to problems with enough unit equations such that the completion of DISCOUNT did not stop. For all examined problems we could observe that one of the two variants of our cooperative system was either better than each of the coupled provers—the runtime was less or the cooperative provers could solve a problem none of the coupled provers could solve when working alone—or we achieved the same result, that is, neither the cooperative system nor one of the coupled provers could cope with the problem. For illustration purposes we present a small representative excerpt of these experiments in table 1 (enriched with some problems taken from other domains). In table 1, the entry "–" denotes that the problem could not be solved within 1000 seconds (all runtimes were achieved on SPARCstations 20/712). Column 4 shows whether the problem is specified in pure equational logic (EQ) or in first-order logic with equality (PL1EQ). Column 5 displays the run time when employing requests for sub-problem transfer and expansion-based requests for clauses, column 6 the respective time when exchanging requests for sub-problem transfer and contraction-based requests for clauses. The last column 7 presents which prover could solve the problem in the cooperating runs.

For all problems we can find at most one cooperation method that allows for a gain of efficiency. Furthermore, sometimes it is even possible to solve prob-

problem	SPASS	DISCOUNT	EQ/PL1EQ	expans.	contr.	proved by
BOO007-4	403.4	–	EQ	330.7	144.4	DISCOUNT
GRP177-2	–	–	EQ	–	123.8	DISCOUNT
GRP179-1	–	–	EQ	447.0	63.5	DISCOUNT
LCL163-1	10.0	12.0	EQ	8.2	6.2	DISCOUNT
ROB005-1	–	109.6	EQ	36.6	60.6	SPASS
ROB008-1	–	98.8	EQ	13.5	85.9	SPASS
ROB022-1	15.1	–	EQ	2.3	3.9	SPASS
ROB023-1	204.6	–	EQ	47.3	44.6	SPASS
CIV001-1	24.9	–	PL1EQ	13.0	25.3	SPASS
LDA011-2	35.1	–	PL1EQ	40.2	30.4	SPASS
ROB011-1	105.3	–	PL1EQ	110.7	54.9	SPASS
ROB016-1	9.8	–	PL1EQ	4.3	5.8	SPASS
HEN009-5	309.9	–	PL1EQ	370.8	233.9	SPASS
HEN010-5	68.7	–	PL1EQ	62.9	70.3	SPASS
HEN011-5	41.2	–	PL1EQ	29.3	20.1	SPASS
LCL143-1	16.1	–	PL1EQ	12.4	11.3	SPASS

Table 1. Coupling SPASS and DISCOUNT by exchanging requests and responses

lems through cooperation that are out of reach for both of the coupled provers. If we compare the results achieved by expansion-based requests with those of contraction-based requests we can see that contraction-based requests are mostly the better alternative. For a deeper analysis we refer to [Fuc98b].

7 Discussion and Future Work

We have presented the approach of requirement-based cooperative theorem proving. This approach realizes some kind of demand-driven cooperation of saturation-based provers. Thus, it is possible to incorporate an orientation on the concrete needs of theorem provers into the cooperation scheme. We described an abstract framework for requirements and particularly two certain aspects of requirement-based cooperation. On the one hand requirement-based exchange of facts, on the other hand sub-problem division and transfer via requests.

Related approaches for an exchange of information between theorem provers are mainly success-driven (e.g., [Sut92], [Den95], [BH95], [Fuc98a]). In contrast to our approach, in these methods information is sent to other provers without considering specific needs of the receivers.

In future, it would be interesting to integrate also analytic provers, e.g. tableau-style provers, into our cooperative system. Since these provers are based on a division of the original problem into sub-problems especially sub-problem transfer via requests might be promising. Then, analytic provers can be used for identifying and transferring sub-problems, saturation-based provers for solving or simplifying them. Thus, requirement-based theorem proving offers the possibility to integrate both top-down and bottom-up theorem proving approaches.

References

[ADF95] J. Avenhaus, J. Denzinger, and M. Fuchs. DISCOUNT: A System For Distributed Equational Deduction. In *Proc. 6th RTA*, pages 397–402, Kaiserslautern, 1995. LNCS 914.

[BDP89] L. Bachmair, N. Dershowitz, and D.A. Plaisted. Completion without Failure. In *Coll. on the Resolution of Equations in Algebraic Structures*. Academic Press, Austin, 1989.

[BG94] L. Bachmair and H. Ganzinger. Rewrite-based equational theorem proving with selection and simplification. *Journal of Logic and Computation*, 4(3):217–247, 1994.

[BH95] M.P. Bonacina and J. Hsiang. The Clause-Diffusion methodology for distributed deduction. *Fundamenta Informaticae*, 24:177–207, 1995.

[Bon96] M.P. Bonacina. On the reconstruction of proofs in distributed theorem proving: a modified Clause-Diffusion method. *Journal of Symbolic Computation*, 21(4):507–522, 1996.

[Den95] J. Denzinger. Knowledge-based distributed search using teamwork. In *Proc. ICMAS-95*, pages 81–88, San Francisco, 1995. AAAI-Press.

[DF98] J. Denzinger and D. Fuchs. Enhancing conventional search systems with multi-agent techniques: a case study. In *Proc. ICMAS-98*, Paris, France, 1998.

[Ert92] W. Ertel. OR-Parallel Theorem Proving with Random Competition. In *Proceedings of LPAR'92*, pages 226–237, St. Petersburg, Russia, 1992. Springer LNAI 624.

[FD97] D. Fuchs and J. Denzinger. Knowledge-based cooperation between theorem provers by TECHS. Technical Report SR-97-11, University of Kaiserslautern, Kaiserslautern, 1997.

[Fuc98a] D. Fuchs. Coupling saturation-based provers by exchanging positive/negative information. In *Proc. 9th RTA*, pages 317–331, Tsukuba, Japan, 1998. LNCS 1379.

[Fuc98b] D. Fuchs. Requirement-based cooperative theorem proving. Technical Report SR-98-02 (ftp://ftp.uni-kl.de/reports_uni-kl/computer_science/SEKI/1998/Fuchs.SR-98-02.ps.gz), University of Kaiserslautern, Kaiserslautern, 1998.

[HR87] J. Hsiang and M. Rusinowitch. On word problems in equational theories. In *Proc. ICALP87*, pages 54–71. LNCS 267, 1987.

[SS97] G. Sutcliffe and C.B. Suttner. The results of the cade-13 ATP system competition. *Journal of Automated Reasoning*, 18(2):271–286, 1997.

[SSY94] G. Sutcliffe, C.B. Suttner, and T. Yemenis. The TPTP Problem Library. In *CADE-12*, pages 252–266, Nancy, 1994. LNAI 814.

[Sut92] G. Sutcliffe. A heterogeneous parallel deduction system. In *Proc. FGCS'92 Workshop W3*, 1992.

[WGR96] C. Weidenbach, B. Gaede, and G. Rock. Spass & Flotter Version 0.42. In *Proc. CADE-13*, pages 141–145, New Brunswick, 1996. LNAI 1104.

$\mho$-Resolution: An Inference Rule for Regular Multiple-Valued Logics*

Sonia M. Leach[1], James J. Lu[2], Neil V. Murray[3], and Erik Rosenthal[4]

[1] Department of Computer Science, Brown University, U.S.A.
sml@cs.brown.edu
[2] Department of Computer Science, Bucknell University, U.S.A.
jameslu@bucknell.edu
[3] Department of Computer Science, State University of New York, U.S.A.
nvm@cs.albany.edu
[4] Department of Mathematics, University of New Haven, U.S.A.
brodsky@charger.newhaven.edu

Abstract. The inference rule $\mho$-*resolution* for regular multiple-valued logics is developed. One advantage of $\mho$-resolution is that linear, regular proofs are possible. That is, unlike existing deduction techniques, $\mho$-resolution admits input deductions (for Horn sets) while maintaining regular signs. More importantly, $\mho$-resolution proofs are at least as short as proofs for definite clauses generated by the standard inference techniques—annotated resolution and reduction—and pruning of the search space occurs automatically.

1 Introduction

Signed logics [18, 10] provide a general[1] framework for reasoning about multiple-valued logics (MVL's). They evolved from a variety of work on non-standard computational logics, including [2, 3, 5, 6, 8, 15, 14, 16, 20, 22]. The key is the attachment of *signs*—subsets of the set of truth values—to formulas in the MVL. This approach is appealing because it facilitates the utilization of classical techniques for the analysis of non-standard logics, which reflects the essentially classical nature of human reasoning. That is, regardless of the domain of truth values associated with a logic, at the meta-level, humans interpret statements about the logic to be either *true* or *false*.

This paper focuses on the class of *regular* signed logics. Regular signed logics are of interest in the knowledge representation and logic programming communities because they correspond to the class of paraconsistent logics known as *annotated logics*, introduced by Subrahmanian [21], Blair and Subrahmanian [2], and Kifer et al. [12, 13, 23]. In [18], regular signed logics were also shown to

* This research was supported in part by the National Science Foundation under grants CCR-9731893, CCR-9404338 and CCR-9504349.

[1] Hähnle, R. and Escalada-Imaz, G. [11] have an excellent survey encompassing deductive techniques for a wide class of MVL's, including (properly) signed logics.

capture fuzzy logics, but in this paper, *regular signed logics* will refer to annotated logics. In particular, the focus is on the definite Horn subset of annotated logic, widely applied within logic programming. The inference rule *$\mho$-resolution* is developed for regular signed logics, and its relative advantages with respect to the standard inference rules for annotated logic programs (ALP's)—annotated resolution and reduction—are described. These include the fact that linear, regular proofs are possible; furthermore, $\mho$-resolution proofs are at least as short as annotated resolution proofs, and pruning of the search space occurs automatically.

The next section is a summary of the basic ideas of signed formulas; greater detail can be found in [20] and in [18].

2 Signed Logics

We assume a language Λ consisting of (finite) logical formulas built in the usual way from a set $\mathcal{A}$ of atoms (predicates and terms at the first order level), a set of connectives, and a set of logical constants. For the sake of completeness, we define a formula in Λ as follows: Atoms are formulas; if Θ is an n-ary connective and if $\mathcal{F}_1, \mathcal{F}_2, \ldots, \mathcal{F}_n$ are formulas, then so is $\Theta(\mathcal{F}_1, \mathcal{F}_2, \ldots, \mathcal{F}_n)$.

Associated with Λ is a set Δ of truth values, and an interpretation for Λ is a function from $\mathcal{A}$ to Δ; i.e., an assignment of truth values to every atom in Λ. A connective Θ of arity n denotes a function $\Theta : \Delta^n \rightarrow \Delta$. Interpretations are extended in the usual way to mappings from the set of formulas to Δ. Alternatively, a formula $\mathcal{F}$ of Λ can be regarded as denoting a mapping from interpretations to Δ.

A *sign* is a subset of Δ, and a *signed formula* is an expression of the form $S\!:\!\mathcal{F}$, where S is a sign and $\mathcal{F}$ is a formula in Λ. If $\mathcal{F}$ is an atom in Λ, we call $S\!:\!\mathcal{F}$ a *signed literal.*

Signed formulas may be thought of as a formalization of meta-reasoning over MVL's [20]. A natural interpretation of the signed formula $S : \mathcal{F}$ is the query, "Is the truth value of $\mathcal{F}$ in S?" The answer to such a query is yes or no; that is, either the formula evaluates to some element in S or does not. Observe that both the query and the answer are at the meta-level; observe also that the question cannot even be formulated at the object level. On the other hand, the question, "What is the truth value of $\mathcal{F}$?" may be interpreted at the object level since the answer is an element of Δ.

For example, let Δ be the interval [0, 1], where elements of Δ represent the degree of belief of some fixed reasoning agent X. Thus, $\{1\}\!:\!P$ can be interpreted as, "Is X certain of the proposition P?" and [0,.1]:P asks, "Is X quite doubtful of P?" These are yes or no questions.

To answer arbitrary queries, we represent queries about formulas in Λ by formulas in a classical logic Λ_S, *the language of signed formulas*; it is defined as follows: The literals are signed formulas and the connectives are (classical) conjunction and disjunction. It should be emphasized that a signed formula

$S{:}\mathcal{F}$ is a literal in Λ_S regardless of the size or complexity of $\mathcal{F}$ and thus has no component parts in the language Λ_S. The set of truth values is $\{true, false\}$.

An arbitrary interpretation for Λ_S may make an assignment of *true* or *false* to any signed formula (i.e., to any literal) in the usual way. Our goal is to focus attention only on those interpretations that relate to the sign in a signed formula. To accomplish this we restrict attention to *Λ-consistent interpretations*. An interpretation I over Λ assigns to each literal, and therefore to each formula $\mathcal{F}$, a truth value in Δ, and the corresponding Λ-consistent interpretation I_c is defined by $I_c(S{:}\mathcal{F}) = true$ if $I(\mathcal{F}) \in S$; $I_c(S{:}\mathcal{F}) = false$ if $I(\mathcal{F}) \notin S$. Note that this correspondence between the set of all interpretations over Λ and the set of Λ-consistent interpretations over Λ_S is 1-to-1. Intuitively, Λ-consistent means an assignment of *true* to all signed formulas whose signs are simultaneously achievable via some interpretation over the original language. Restricting attention to Λ-consistent interpretations yields a new consequence relation: If $\mathcal{F}_1$ and $\mathcal{F}_2$ are formulas in Λ_S, we write $\mathcal{F}_1 \models_\Lambda \mathcal{F}_2$ if whenever I_c is a Λ-consistent interpretation and $I_c(\mathcal{F}_1) = true$, then $I_c(\mathcal{F}_2) = true$. Two formulas $\mathcal{F}_1$ and $\mathcal{F}_2$ in Λ_S are *Λ-equivalent* if $I_c(\mathcal{F}_1) = I_c(\mathcal{F}_2)$ for any Λ-consistent interpretation I_c; we write $\mathcal{F}_1 \equiv_\Lambda \mathcal{F}_2$. The following lemma is immediate.

Lemma 1. Let I_c be a Λ-consistent interpretation, let A be an atom and $\mathcal{F}$ a formula in Λ, and let S_1 and S_2 be signs. Then:
1. $I_c(\emptyset{:}\mathcal{F}) = false$;
2. $I_c(\Delta{:}\mathcal{F}) = true$;
3. $S_1 \subseteq S_2$ if and only if $S_1{:}\mathcal{F} \models_\Lambda S_2{:}\mathcal{F}$ for all formulas $\mathcal{F}$;
4. There is exactly one $\delta \in \Delta$ such that $I_c(\{\delta\}{:}A) = true$. □

The usual results involving Robinson's Unification Theorem are unaffected by this development, and techniques at the ground level can generally be lifted to the general level. As a result, attention is mostly restricted to the ground case for the remainder of the paper.

2.1 Signed Inference

In this section, we describe a method for adapting resolution to produce an inference rule for Λ_S. Similar notions have been developed by Baaz and Fermüller [1] and by Hähnle [9]. Many classical inference rules begin with links (complementary pairs of literals). Such rules typically deal only with formulas in which all negations are at the atomic level. Similarly, the inference techniques described below require that signs be at the "atomic level." To that end, a formula in Λ_S is defined to be *Λ-atomic* if whenever $S : A$ is a literal in the formula, then A is an atom in Λ. Often—with annotated logic formulas, for example—all formulas are assumed to be Λ-atomic.

The inference rule $\mho$-resolution is based on the notion of complementary literals, which is generalized in the next lemma. The lemma is immediate from Part 4 of Lemma 1.

Lemma 2. (The Reduction Lemma) Let $S_1{:}A$ and $S_2{:}A$ be Λ-atomic atoms in Λ_S; then $S_1 : A \wedge S_2 : A \equiv_\Lambda (S_1 \cap S_2) : A$ and $S_1 : A \vee S_2 : A \equiv_\Lambda (S_1 \cup S_2) : A$. □

Consider a Λ-atomic formula $\mathcal{F}$ in Λ_S in conjunctive normal form (CNF). Let $C_j, 1 \leq j \leq r$, be clauses in $\mathcal{F}$ that contain, respectively, Λ-atomic literals $\{S_j{:}A\}$. Thus we may write $C_j = K_j \vee \{S_j{:}A\}$. Then the resolvent R of the C_j's is defined to be the clause

$$\left(\bigvee_{j=1}^{r} K_j\right) \bigvee \left(\left(\bigcap_{j=1}^{r} S_j\right) : A\right).$$

The rightmost disjunct is called the *residue* of the resolution; observe that it is unsatisfiable if its sign is empty and satisfiable if it is not.

In this generality, this definition must be augmented with the following obvious simplification rules that stem from the Reduction Lemma. First, if the residue is unsatisfiable it may simply be deleted from R. Secondly, whenever R contains literals $S_i{:}B$, $1 \leq i \leq k$, we *merge* them into the literal $\bigcup_{i=1}^{k} S_i{:}B$; if $\bigcup_{i=1}^{k} S_i = \Delta$, then R is a tautology and may be deleted from $\mathcal{F}$. Such merging is essentially a generalization to MVL's of ordinary classical ground merging. In this paper, we are concerned with *regular* signs, and we henceforward restrict merging to indentical literals (with identical signs).

The classical notion of subsumption also generalizes to Λ_S: Clause C subsumes clause D if, for every literal $S : A \in C$, there is a literal $S' : A \in D$ such that $S \subseteq S'$. Observe that if $S \subseteq S'$, and if two clauses are resolved on the literals $S : A$ and $S' : A$, then the residue will be $S : A$ (after all, $S \cap S' = S$), so the clause containing $S : A$ must subsume the resolvent. This proves

Lemma 3. The resolvent produced by resolving on two literals in which the sign of one contains the sign of the other is superfluous in the sense that the resolvent is necessarily subsumed by one of its parents. □

2.2 Regular Signed Formulas and Annotated Logics

Assume now that the set of truth values Δ is not simply an unordered set of objects but instead forms a complete lattice under some ordering $\preceq$. The greatest and least elements of Δ are denoted $\top$ and $\bot$, respectively, and Sup and Inf denote, respectively, the supremum (least upper bound) and infimum (greatest lower bound) of a subset of Δ.

Let $(P; \preceq)$ be any partially ordered set, and let $Q \subseteq P$. Then $\uparrow Q = \{y \in P | (\exists x \in Q)\ x \preceq y\}$. Note that $\uparrow Q$ is the smallest *upset* containing Q (see [4]). If Q is a singleton set $\{x\}$, then we simply write $\uparrow x$. We say that a subset Q of P is *regular* if for some $x \in P$, $Q = \uparrow x$ or $Q = (\uparrow x)'$ (the set complement of $\uparrow x$). We call x the *defining element* of the set. In the former case, we call Q *positive*, and in the latter *negative*. Observe that both Δ and $\emptyset$ are regular since $\Delta = \uparrow \bot$ and $\emptyset = \Delta'$. Observe also that if $z = \mathrm{Sup}\{x, y\}$, then $\uparrow x \cap \uparrow y = \uparrow z$. A signed formula is regular if every sign that occurs in it is regular. By Part 1 of Lemma 1, we may assume that no regular signed formulas have any signs of the form $(\uparrow x)'$, where $x = \bot$, since in that case $(\uparrow x)' = \emptyset$.

An *annotated logic* is a signed logic in which only regular signs are allowed.

2.3 Some Remarks on Notation

A regular sign is completely characterized by its defining element, say x, and its *polarity* (whether it is positive or negative). A regular signed atom may be written $\uparrow x : A$, while the complement is the set $(\uparrow x)' : A$. Observe that $(\uparrow x)' : A = \sim (\uparrow x : A)$; that is, the signed atoms are complementary with respect to Λ-consistent interpretations. With annotated logics, the most common notation is $\mathcal{F} : x$ and $\sim \mathcal{F} : x$. There is no particular advantage of one or the other, and it is perhaps unfortunate that both have arisen. We will follow the $x : \mathcal{F}$ convention when dealing with signed logics and use $\mathcal{F} : x$ for annotated logics.[2]

Let us also remark here that, historically, annotated logics have been restricted to Λ-atomic formulas. Though this restriction is unnecessary, it will be obeyed for the remainder of the paper to simplify the presentation. Though formuals are restricted to CNF, they are not restricted to be Horn.

2.4 Signed Resolution for Annotated Logics

A sound and complete resolution proof procedure was defined for clausal annotated logics in [15]. The procedure contains two inference rules that we will refer to as *annotated resolution* and *reduction.*[3] These two inference rules correspond to disjoint instances of signed resolution. Two annotated literals L_1 and L_2 are said to be *complementary* if they have the respective forms $A : \mu$ and $\sim (A : \rho)$, where $\mu \geq \rho$, and annotated resolution is defined as follows: Given the annotated clauses $(L_1 \vee D_1)$ and $(L_2 \vee D_2)$, where L_1 and L_2 are complementary, then the *annotated resolvent* of the two clauses on the annotated literals L_1 and L_2 is $D_1 \vee D_2$.

Two clauses can be so resolved only if the annotation of the positive annotated literal that is resolved upon is greater than or equal to the annotation of the negative literal resolved upon. In that case the two clauses are said to be *resolvable* on the annotated literals L_1 and L_2.

The reduction rule is defined when two occurrences of an atom have positive signs. Suppose $(A : \mu_1 \vee E_1)$ and $(A : \mu_2 \vee E_2)$ are two annotated clauses in which μ_1 and μ_2 are incomparable. Then the annotated clause $(A : Sup\{\mu_1, \mu_2\}) \vee E_1 \vee E_2$ is called a *reductant* of the two clauses, and we say that the two clauses are *reducible* on the annotated literals $A : \mu_1$ and $A : \mu_2$.

It is straightforward to see that the two inference rules are both captured by signed resolution. In particular, annotated resolution corresponds to an application of signed resolution (to regular signed clauses) in which the signs of

[2] The reader can decide whether this will make both communities happy or unhappy.

[3] Kifer and Lozinskii refer to their first inference rule simply as resolution. However, since we are working with several resolution rules in this paper, appropriate adjectives will be used to avoid ambiguity.

the selected literals are disjoint. Reduction on the other hand, corresponds to an application of signed resolution in which the signs of the selected literals are both positive. The next theorem is now immediate. Implicit in the term *signed deduction* in the theorem and in the remainder of the paper is deduction with signed resolution.

Theorem 1. Suppose that $\mathcal{F}$ is a set of annotated clauses and that $\mathcal{D}$ is a deduction of $\mathcal{F}$ using annotated resolution and reduction. Then $\mathcal{D}$ is a signed deduction of $\mathcal{F}$. In particular, if $\mathcal{F}$ is an unsatisfiable set of first order annotated clauses, then there is a signed refutation of $\mathcal{F}$. □

Signed resolution thus provides a general method for implementing annotated logics, but, as is often the case in theorem proving, there is a trade-off between generality and efficiency. Since annotated resolution and reduction are instances of signed resolution, and since there exist signed resolutions for which no corresponding annotation resolutions or reductions exist, the search space induced by signed resolution will typically be larger than the search space induced by annotated resolution and reduction. The following theorem, proved in [18], characterizes that class of signed deductions that corresponds to the deductions obtainable via annotated resolution and reduction.

Theorem 2. Suppose $S_1, \ldots, S_n$ are regular signs whose intersection is empty, and suppose that no proper subset of $\{S_1, \ldots, S_n\}$ has an empty intersection. Then exactly one sign is negative; i.e., for some $j, 1 \leq j \leq n, S_j = (\uparrow x_j)'$, and for $i \neq j, S_i = \uparrow x_i$, where $x_1, \ldots, x_n \in \Delta$. □

The intersection of a positive regular sign and a negative regular sign is regular if and only if it is empty, and two negative signs can have a regular intersection if and only if one is a subset of the other. In view of Lemma 3, the latter situation need never be considered, so we define a signed deduction to be *regular* if every sign that appears in the deduction is regular and if no residue sign is produced by the intersection of two negative signs. Note that this implies that merging of literals is allowed only when regular signs are produced. The next two theorems are immediate. Theorem 4 states that the class of regular signed deductions is precisely the class of deductions using annotated resolution and reduction. As a result, restricting signed resolution to regular clauses captures annotated resolution and reduction without increasing the search space.

Theorem 3. A signed deduction of a regular formula is regular if and only if the sign of every consistent residue is produced by the intersection of two positive regular signs. □

Theorem 4. Let $\mathcal{D}$ be a sequence of annotated clauses. Then $\mathcal{D}$ is an annotated deduction if and only if $\mathcal{D}$ is a regular signed deduction. □

It follows from the theorem that regular signed resolution is complete.

Corollary. Suppose $\mathcal{F}$ is an unsatisfiable set of regular signed clauses. Then there is a regular signed deduction of the empty clause from $\mathcal{F}$. □

2.5 Linear Signed Resolution and Horn Sets

Logic programming[4] may be thought of as providing a procedural interpretation of Horn clauses through linear resolution. Similarly, annotated Horn clauses may be interpreted procedurally through linear signed resolution. In light of Theorem 3 and the requirement that each step in a linear resolution involves the most recent resolvent, linear signed resolution may not in general be regular. Annotated logics, on the other hand, require regularity, which is incompatible with the linear restriction. This creates a difficulty if the goal is to employ annotated logics in a logic programming setting. This is illustrated by the next example.

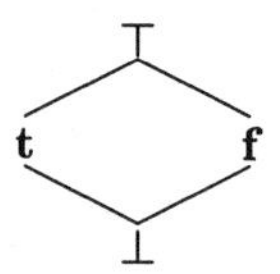

Fig. 1. The Complete Lattice FOUR.

Consider the annotated logic program (ALP) $\mathcal{P} = \{p : \mathbf{t} \leftarrow; p : \mathbf{f} \leftarrow\}$,[5] written over the lattice FOUR pictured in Figure 1 (see [2]). Consider now the query, $\leftarrow p{:}\top$. It is easy to see $\mathcal{P} \models p{:}\top$. However, linear annotated resolution alone does not admit a refutation since the reductant $p{:}\top \leftarrow$ must be computed from the two program clauses. This clause would resolve with the query to yield a refutation, but this clause cannot be inferred with the linear restriction because the goal clause must be used in every step.

Kifer and Subrahmanian circumvent this difficulty by specifying that a deduction consist only of applications of annotated resolution. However, any inference may involve a resolution with an annotated clause obtained by implicit applications of reduction. In the last example, a proof could be obtained by a single deduction between the original query $\leftarrow p{:}\top$ and the reductant $p{:}\top \leftarrow$.

Unfortunately, implicit use of reduction creates new problems. For example, a proof that consists of only annotated resolution steps may contain steps that include clauses not in the original program. This makes the proof difficult to read. Moreover, application of the reduction rule can be expensive since it can occur at any time during a deduction, significantly expanding the search space.

Signed resolution, on the other hand, consists of a single rule of inference, and is therefore amenable to a linear restriction, thus making an SLD-like procedure implementable. However, the proofs may not be regular. Equivalently, any irregular resolvent that arises cannot be annotated with a simple annotation. Hence directly applying a linear restriction would require an extension of the syntax of annotated atoms.

[4] Several researchers have explored annotated logics with an interest in logic programming. A more detailed account can be found in [19].

[5] We use the left arrow to represent definite clauses in the standard way.

2.6 A Linear Procedure with Annotations

There is a means of implementing a linear strategy for signed resolution using annotations since, in view of Theorem 2, it is never necessary to resolve on literals with two negative signs. One idea is to associate with each active goal two annotations, a negative one, α, and a positive one, β. The negative one is always the goal's initial annotation. The positive one, which changes during the deduction, is the supremum of the annotations of the positive literals against which the goal has been resolved. The two annotations α and β represent the sign $(\uparrow\alpha)' \cap \uparrow\beta$. More precisely, if $\leftarrow G : \alpha$ is the goal, then the initial positive annotation is $\perp$. At each stage, if the positive annotation is β, and if the goal is resolved against a rule whose head is $B : \rho$, there are two cases to consider. First, if $\alpha \leq \mathrm{Sup}\{\beta, \rho\}$, then the goal may be deleted. Otherwise, the positive annotation β is replaced by $\mathrm{Sup}\{\beta, \rho\}$, and the goal remains active. This procedure is sound by the Reduction Lemma and is complete (See [18]).

Although this technique does admit linearity, reductions are nevertheless implicitly performed; they are represented by the second annotation. Of course, to be linear, each of the methods we have discussed must either allow irregular signs or must capture reductions (implicitly or explicitly). Unfortunately, many redundant reduction steps are possible with these methods. As we shall see, $\mho$-resolution can avoid many such redundancies and provide a cleaner input-style deduction in which no extra annotations are necessary; i.e., only strictly regular input deductions are required for ALP's using $\mho$-resolution.

3 Decomposition and $\mho$-Resolution

The issues discussed may be summarized as follows:

- For regular signed logics—equivalently, annotated logics—signed resolution provides generality, while annotated resolution with reduction provides efficiency.
- For definite Horn regular signed logics—annotated logic programs—signed resolution is amenable to the linear (and thus input) restriction, while annotated resolution and reduction is not. Signed resolution with the linear restriction does not guarantee regularity.

The problem is to reconcile the shortcomings of these two approaches. When the goal annotation μ is not less than or equal to the program clause annotation ρ, then the intersection of $(\uparrow \mu)'$ and $\uparrow \rho$ is simply not regular and cannot be represented by an annotation. Signed resolution would produce the goal $((\uparrow \mu)' \cap \uparrow \rho) : p$. However, since this goal can be soundly deduced, so can any goal of the form $S : p$, where $S \supseteq ((\uparrow \mu)' \cap \uparrow \rho)$. Intuitively, $\mho$-resolution works by determining an annotation μ_0 such that $(\uparrow \mu_0)'$ is the smallest regular set that contains $((\uparrow \mu)' \cap \uparrow \rho)$. Thus we may soundly infer the new goal $p : \mu_0$ without losing regularity or linearity; completeness is not lost either.

We begin the development of $\mho$-resolution by introducing *decomposition*, an inference rule for annotated logics. Decomposition by itself does not solve the problems alluded to above, but it leads naturally to $\mho$-resolution Consider first the next example.

Example. Let Δ be the lattice FOUR of Figure 1 and let Q be the query $\leftarrow p\!:\!\top$. Further, let P be the program

P_0 $p\!:\!\mathbf{t} \leftarrow q\!:\!\mathbf{t}$
P_1 $p\!:\!\mathbf{f} \leftarrow q\!:\!\mathbf{f}$
P_2 $q\!:\!\mathbf{t} \leftarrow$
P_3 $q\!:\!\mathbf{f} \leftarrow$.

It is easy to see that $P \models p\!:\!\top$. Thus the query $\leftarrow p\!:\!\top$ should be provable. Using the technique of carrying two annotations, we may arrive at a proof using signed resolution. The initial query is provable by first resolving it against $p\!:\!\mathbf{t} \leftarrow q\!:\!\mathbf{t}$ to produce the query $\leftarrow p : \top^{\mathbf{t}}, q : \mathbf{t}$. Resolving next against $p : \mathbf{f} \leftarrow q : \mathbf{f}$ yields $\leftarrow p\!:\!\top^{\top}, q\!:\!\mathbf{t}, q\!:\!\mathbf{f}$, which simplies to $\leftarrow q\!:\!\mathbf{t}, q\!:\!\mathbf{f}$. This query may now be proved through two simple resolution steps against P_2 and P_3.

3.1 Decomposition

Suppose that, instead of maintaining two annotations, after determining that the initial query cannot resolve (using standard annotated resolution) with any program clause, the initial query is decomposed into the two goal query: $\leftarrow p : \mathbf{t}, p\!:\!\mathbf{f}$. Then the two goals can be resolved through annotated resolution against the program clauses in a straightforward way. More generally, in view of the fact that proving $p : \top$ can be accomplished by proving both $p : \mathbf{t}$ and $p : \mathbf{f}$, which are easier to prove, our goal is to set up a rule of inference that performs such a decomposition whenever a suitable clause cannot be found for annotated resolution.

Definition. Let Q be the query $\leftarrow A_1\!:\!\mu_1, ..., A_m\!:\!\mu_m$, and suppose that $A_i\!:\!\rho_1$ and $A_i : \rho_2$ are literals such that $\mu_i \preceq \mathrm{Sup}\{\rho_1, \rho_2\}$. Then $A_i : \mu_i$ is said to *decompose* to $(A_i\!:\!\rho_1, A_i\!:\!\rho_2)$, and the decomposition of Q with respect to A_i is

$$\leftarrow A_1 : \mu_1, ..., A_{i-1} : \mu_{i-1}, A_i : \rho_1, A_i : \rho_2, A_{i+1} : \mu_{i+1}, ..., A_m : \mu_m.$$

Theorem 5. Suppose Q is a ground annotated query. Let Q^d be a decomposition of Q and I_c be a Λ_S-interpretation. If $I_c(Q) = \mathit{true}$, then $I_c(Q^d) = \mathit{true}$. □

Decomposition, together with annotated resolution, constitutes an SLD-style proof procedure, and it does so using only regular signed literals. By decomposing an annotation into two "lower annotations," decomposition is a sort of divide-and-conquer strategy. In the previous example, it was straightforward to choose two annotations into which the annotation $\top$ associated with the query can be decomposed. Not surprisingly, it is not always this easy.

Example. Let Δ be the lattice of Figure 2, and let P be the program

P_1 $p\!:\!V \leftarrow q\!:\!V$
P_2 $q\!:\!1 \leftarrow$
P_3 $q\!:\!2 \leftarrow$,

where V is an arbitrary annotation.

Suppose that Q is the query, $\leftarrow p : \top$. Clearly $P \models p : \top$. If we chose to decompose $p : \top$ into the two literals $p : 1$ and $p : 3$, then the corresponding query $\leftarrow p : 1, p : 3$ does not have a refutation using only annotated resolution. On the other hand, if $p : \top$ is decomposed into the two literals, $p : 1$ and $p : 2$, there is a refutation using annotated resolution.

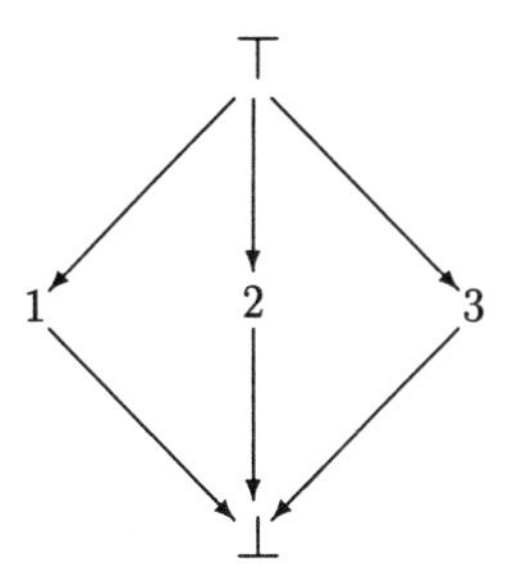

Fig. 2. 1-2-3 Lattice

The example demonstrates how difficulties in choosing a decomposition may arise because the two annotations into which the original should be decomposed is dependent on the structure of Δ and on the program clauses. In the case of FOUR, the choice was immediate, regardless of the annotations of the program clauses. On the other hand, for the 1-2-3-lattice, the usefulness of the choice depended on what information was contained within the program. In general, decomposition may be expensive because the number of guesses may be large and because many of them may not lead to a proof. However, there may be properties embedded in certain classes of lattices that can be used to reduce the number of guesses. In the next section, we define one class of lattices — *ordinary* lattices — and introduce a simple and elegant SLD-style proof procedure based on $\mho$-resolution, which is an enhancement to decomposition in which no guessing is required.

3.2 Ordinary Lattices and $\mho$-Resolution

Decomposition can be employed with any lattice, but its application may be expensive. One way to improve decomposition is to modify it to be a binary inference rule in which the selection of one component is based on the existence of an appropriate head literal with annotation ρ in a program clause. Based on μ and ρ, an annotation γ may be guessed such that $\mu \preceq \mathrm{Sup}\{\gamma, \rho\}$. The inference rule $\mho$-resolution, described below, exploits this observation by further restricting the class of lattices to those with the property that for any two values μ, ρ of the lattice, there is a best choice for γ. We call such a collection of lattices *ordinary lattices*. First, given $\mu, \rho \in \Delta$, we use $\mathcal{M}(\mu, \rho)$ to denote the set of annotations for which the least upper bound of ρ with an element of the set is greater than or equal to μ. Formally, $\mathcal{M}(\mu,\rho) = \{\gamma \in \Delta, | \operatorname{Sup}\{\gamma,\rho\} \succeq \mu\}$.

For the 1-2-3-lattice in Figure 2, if $\mu = \top$ and $\rho = 1$, $\mathcal{M}(\mu,\rho) = \{2,3,\top\}$. For the lattice FOUR in Figure 1, if $\mu = \top$, and $\rho = \mathbf{t}$, $\mathcal{M}(\mu,\rho) = \{\mathbf{f},\top\}$.

Definition. Given $\mu,\rho \in \Delta$, the $\mho$ *operator*[6] is defined as follows: $\mho(\mu,\rho) = \operatorname{Inf}(\mathcal{M}(\mu,\rho)\ )$. A lattice Δ is said to be *ordinary* if $\forall \mu,\rho \in \Delta$, $\mho(\mu,\rho) \in \mathcal{M}(\mu,\rho)$. To simplify the notation, given $\mu,\rho_1,...,\rho_m$, we write $\mho(\mu,\rho_1,...,\rho_m)$ for $\mho(\mho(...\mho(\mho(\mu,\rho_1),\rho_2)...),\rho_m)$.

Notice that the lattice 1-2-3 is not an ordinary lattice since the set $\mathcal{M}(\mu,\rho) = \{2,3,\top\}$ has no least element in the set. The lattice FOUR, however, is an ordinary lattice.

The next property of ordinary lattices is important for proving the completeness of $\mho$-resolution.

Lemma 4. Let Δ be an ordinary lattice. Then $\mho(\mu,\rho_1,...,\rho_m) = \bot$ iff $\mu \preceq \rho$, where $\rho = \operatorname{Sup}\{\rho_1,...,\rho_m\}$. □

If Δ is an ordinary lattice, it is easy to see that, given a query literal $p\!:\!\mu$ and a program clause head annotation $p\!:\!\rho$, the natural choice for the decomposition of $p\!:\!\mu$ based on $p\!:\!\rho$ is $p\!:\!\rho$ and $p\!:\!\mho(\mu,\rho)$. The first part of the decomposition is obvious, and the second part represents the "simplest" remaining condition that must be solved in order to solve the original query. This leads to the desired inference rule.

Definition. Given an ALP over an ordinary lattice, suppose Q is the query

$$\leftarrow A_1\!:\!\mu_1,...,A_m\!:\!\mu_m$$

and C is the program clause $A\!:\!\rho \leftarrow Body$, where A_i and A can be unified with mgu θ, and where $\mho(\mu_i,\rho) \prec \mu_i$. Then the $\mho$-resolvent of Q and C with respect to A_i is the query

$$\leftarrow (A_1\!:\!\mu_1,...,A_{i-1}\!:\!\mu_{i-1},\ A\!:\!\mho(\mu_i,\rho), Body,\ A_{i+1}\!:\!\mu_{i+1},...,A_m\!:\!\mu_m)\theta.$$

A $\mho$-*deduction* of a query from a given ALP and initial query is defined in the usual way. A $\mho$-deduction of a query from an ALP is a $\mho$-*proof* if

$$\leftarrow A_1 : \bot, A_2 : \bot, ..., A_n : \bot$$

is the last clause in the $\mho$-deduction.

Lemma 5. The inference rule $\mho$-resolution is sound for ordinary lattices. □

The next example demonstrates the use of $\mho$-resolution over the ordinary lattice in Figure 3. Notice that the lattice consists of the values $d\mathbf{t}$, $d\mathbf{f}$, and $\top$, in addition to the four elements of FOUR. Ginsberg introduced this lattice

[6] The usual pronunciation of this symbol, which is an upside-down Ω, is "mo"—a long O.

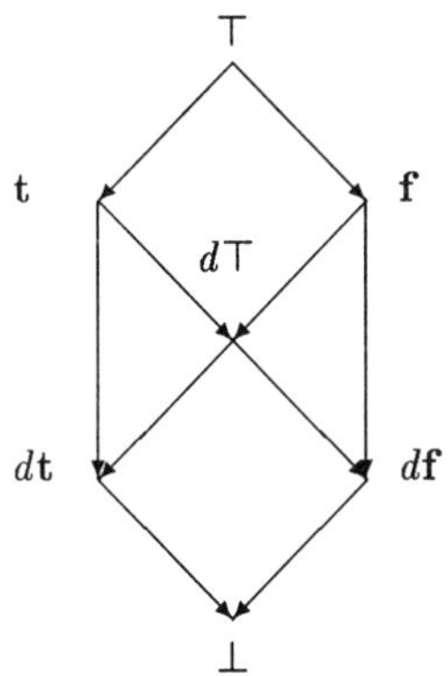

Fig. 3. Lattice with Defaults

to distinguish between truth and default facts [7]. The symbols $d\mathbf{t}$ and $d\mathbf{f}$ may be regarded as "concluded **true**, respectively, **false**, by default." Following this intuition, $d\top$ stands for "**inconsistent** default conclusion" and is distinguished from the stronger inconsistency represented by $\top$.

Example. Let Δ be the lattice of Figure 3, let P be the program

$P_1 \quad p : d\mathbf{t} \leftarrow$

$P_2 \quad p : d\mathbf{f} \leftarrow,$

and let Q be the query $\leftarrow p : d\top$. Then a $\mho$-proof proceeds as follows.

$Q^* = Q_0$:	$\leftarrow p : d\top$	initial query;
Q_1 :	$\leftarrow p : d\mathbf{f}$	$\mho$-resolvent of P_1 and Q_0; note: $\mathcal{M}(d\top, d\mathbf{t}) = \{d\mathbf{f}, d\top\}$
Q_2 :	$\leftarrow p : \bot$	$\mho$-resolvent of P_2 and Q_1; note: $\mathcal{M}(d\mathbf{f}, d\mathbf{f}) = \Delta$

The inference rule $\mho$-resolution is a variant of the more general decomposition; it exploits properties intrinsic to the structure of the ordinary lattices. As such, $\mho$-resolution may be regarded as a semantically-based inference rule. One nice feature of $\mho$-resolution is that it allows simple SLD-style proof procedures for annotated logic programs over ordinary lattices. This procedure eliminates the expensive reduction rule, yet it does not require irregular deductions. Pruning of the search occurs naturally in $\mho$-resolution. For instance, consider the following program P over the lattice FOUR.

Example.

$$P = \{\underbrace{p:\mathbf{f} \leftarrow q:\mathbf{f}}_{C_1}, \underbrace{p:\mathbf{t} \leftarrow q:\mathbf{t}}_{C_2}, \underbrace{p:\mathbf{t}}_{C_3}, \underbrace{q:\mathbf{f}}_{C_4}\}.$$

Observe that $P \models p:\mathbf{t}$. However, the query $\leftarrow p:\mathbf{t}$ cannot $\mho$-resolve with the head of C_1, $p:\mathbf{f}$, since $\mho(\mathbf{t}, \mathbf{f}) = \mathbf{t}$. Thus, C_2 and C_3 are acceptable candidates for $\mho$-resolution with the query, but C_1 is not.

In essence, C_1 is not required for solving the query, and detection of such irrelevancies is built into $\mho$-resolution. In the same example, other inference techniques will not prevent resolution of the query with C_1, or the use of C_1 for reduction. This may represent considerable savings, since the body of C_1 could be arbitrarily large and/or could lead to unnecessary deductions. Indeed, proof space reduction is a primary motivation for the introduction of $\mho$-resolution. In the next section, we show that $\mho$-resolution guarantees not only a "smaller" proof space, but one in which only the "best" proofs are admitted. The basic result is that, for each proof of a query using annotated resolution and reduction, there is a $\mho$-proof of the same query using at most the same number of inferences. An example showing how search space pruning is built into $\mho$-resolution is also described.

3.3 Proof Space and Search Space Considerations

In this section, deductions are viewed as sequences of inferences rather than as sequences of clauses. This is merely a matter of convenience. Deductions in standard annotated logic programming contain both annotated resolutions and reductions, and we wish to refer easily to the particular inference rule applied at a given step.

It is impractical (and unnecessary) to deal with the space of *all* annotated resolution/reduction deductions. For one thing, given any deduction, a different deduction can be obtained merely by adding redundant steps. Thus we will consider deductions that are minimal in the sense that removal of any one step results in a sequence that does not formally comprise a deduction. Note that this in no way implies that such deductions are *shortest* in any sense.

For another, the steps of a given deduction may be reordered in many ways. For example, the literals in the goal clause of an ALP computation may be solved in different orders. This amounts to reordering the annotated resolution steps of a deduction, and such a reordering produces a distinct deduction. However, a given annotated resolution step may require a reductant that is produced by a series of reduction steps. These reductions (treated as implicit in annotated logic programming) can be performed in any order and at any point prior to the resolution. Although different reduction orders will be treated as different deductions, we will assume that, in aggregate, all reductions necessary to enable a resolution occur immediately prior to that resolution step. It is obvious that any deduction, say $\mathcal{D}'$, not allowed under these assumptions is in fact an essentially redundant variation of one, say $\mathcal{D}$, that is allowed, and $\mathcal{D}'$ is no shorter than $\mathcal{D}$.

In sum, we assume that an annotated logic programming deduction $\mathcal{D}$ consists of a sequence of annotated resolution steps: $\mathcal{S}_1, \mathcal{S}_2, \ldots, \mathcal{S}_n$. Furthermore, if $\mathcal{S}$ is any one of those steps, we assume that $\mathcal{S}$ has the form $\mathcal{R}_1, \mathcal{R}_2, \ldots, \mathcal{R}_k, \mathcal{R}$, $k \geq 0$, where the $\mathcal{R}_i$'s are reductions and $\mathcal{R}$ is annotated resolution. Let $A\!:\!\mu$ be the goal removed by step $\mathcal{R}$, and let $A\!:\!\rho_0,\ A\!:\!\rho_1,\ \ldots,\ A\!:\!\rho_k$ be the heads of the $k+1$ program clauses used in $\mathcal{S}$. We will show that corresponding to these k reductions followed by a resolution, there is a sequence of $k+1$ $\mho$-resolutions, the last of which produces the goal $A:\bot$. (Starting with the goal clause, both deductions involve exactly the $k+1$ program clauses used for the k reductions. In both cases, the resulting goal list will be expanded to include all goals in the bodies of those program clauses. So the goal resolved away by annotated resolution is the only one of concern.)

The series of $k+1$ $\mho$-resolutions will, by definition yield the goal $A : \mho(\mu, \rho_0, ..., \rho_k)$. By the definition of annotated resolution,

$\mu \preceq \text{Sup}(\rho_0, \rho_1, \ \ldots, \ \rho_k)$. But then by Lemma 4, $\mho(\mu, \rho_0, ..., \rho_k) = \bot$, and we have proved:

Theorem 6. Let $\mathcal{P}$ be an annotated logic program, let $\mathcal{Q}$ be a query, and let $\mathcal{D}_{\mathcal{AR}}$ be a proof of $\mathcal{Q}$ from $\mathcal{P}$ by annotated resolution and reduction. Then there is a proof $\mathcal{D}_{\mho}$ of $\mathcal{Q}$ from $\mathcal{P}$ by $\mho$-resolution that is no longer than $\mathcal{D}_{\mathcal{AR}}$. □

Corollary. Suppose $\mathcal{F}$ is an unsatisfiable set of regular signed Horn clauses. Then there is a $\mho$-resolution input deduction of the empty clause from $\mathcal{F}$. □

3.4 Characterizing Ordinary Lattices

Theorem 6 says that $\mho$-resolution works well for annotated logic programs *when it applies*. If there are few applications which use ordinary lattices, $\mho$-resolution loses its significance. One large class of ordinary lattices is the collection of linear lattices. They are easily shown to be ordinary, and this class of lattices is important for applications of annotated logic to quantitative reasoning. However, when the truth domain is linear, no reductions are necessary, and $\mho$-resolution reduces to annotated resolution.

A richer class of truth domains is the collection of distributive lattices (defined below). A number of applications of annotated logics involve distributive lattices [17]. Finite distributive lattices are ordinary; whether this result generalizes to infinite distributive lattices remains unanswered.

Definition. A lattice Δ is said to be *distributive* if it satisfies the distributive laws:

$$(\forall \alpha, \beta, \gamma \in \Delta) \ \ \text{Inf}\{\alpha, \ (\text{Sup}\{\beta, \gamma\})\} = \text{Sup}\{(\text{Inf}\{\alpha, \beta\}), \ (\text{Inf}\{\alpha, \gamma\})\}$$

and

$$(\forall \alpha, \beta, \gamma \in \Delta) \ \ \text{Sup}\{\alpha, \ (\text{Inf}\{\beta, \gamma\})\} = \text{Inf}\{(\text{Sup}\{\alpha, \beta\}), \ (\text{Sup}\{\alpha, \gamma\})\}$$

The lattice in Figure 3 is ordinary, and it can be shown that it is also distributive.

Theorem 7. Finite distributive lattices are ordinary. □

References

1. Baaz, M., and Fermüller, C.G., Resolution-based theorem proving for many-valued logics, *J. of Symbolic Computation*, 19(4):353–391, 1995.
2. Blair, H.A. and Subrahmanian, V.S., Paraconsistent Logic Programming, *Theoretical Computer Science*, 68:135–154, 1989.
3. Carnielli, W.A., Systematization of finite many-valued logics through the method of tableaux. *J. of Symbolic Logic* 52(2):473–493, 1987.
4. Davey, B.A., and Priestley, H.A., *Introduction to Lattices and Order*, Cambridge Mathematical Textbooks, (1990)

5. Doherty, P., *NML3 – A Non-Monotonic Formalism with Explicit Defaults.* Linköping Studies in Science and Technology. Dissertations No. 258. Department of Computer and Information Science, Linköping University, 1991.
6. Fitting, M., First-order modal tableaux, *J. Automated Reasoning* 4:191-213, 1988.
7. Ginsberg, M.L., Multivalued Logics, *Readings in Non-Monotonic Reasoning*, 251–255, 1987.
8. Hähnle, R., *Automated Deduction in Multiple-Valued Logics*, International Series of Monographs on Computer Science, vol. 10. Oxford University Press, 1994.
9. Hähnle, R., Short conjunctive normal forms in finitely-valued logics, *J. of Logic and Computation*, 4(6): 905–927, 1994.
10. Hähnle, R., Tableaux methods for many-valued logics, *Handbook of Tableau Methods* (M d'Agostino, D. Gabbay, R. Hähnle and J. Posegga eds.), Kluwer, Dordrecht. To appear.
11. Hähnle, R. and Escalada-Imaz, G., Deduction in many-valued logics: a survey, *Mathware & Soft Computing*, IV(2), 69–97, 1997.
12. Kifer, M., and Li, A., On the Semantics of Rule-based Expert Systems with Uncertainty, *Proceedings of the 2nd International Conference on Database Theory*, 102–117, 1988.
13. Kifer, M., and Lozinskii, E.L., RI: A Logic for Reasoning in Inconsistency, *Proceedings of the Fourth Symposium of Logic in Computer Science*, Asilomar, 253-262, 1989.
14. Kifer, M., and Subrahmanian, V.S., Theory of generalized annotated logic programming and its applications, the *J. of Logic Programming* 12, 335–367, 1992.
15. Kifer, M., and Lozinskii, E., A logic for reasoning with inconsistency, *J. of Automated Reasoning* 9, 179–215, 1992.
16. Lu, J.J., Henschen, L.J., Subrahmanian, V.S., and da Costa, N.C.A., Reasoning in paraconsistent logics, *Automated Reasoning: Essays in Honor of Woody Bledsoe* (R. Boyer ed.), Kluwer Academic, 181–210, 1991.
17. Lu, J.J., Nerode, A. and Subrahmanian, V.S. Hybrid Knowledge Bases, *IEEE Transactions on Knowledge and Data Engineering*, 8(5):773–785, 1996.
18. Lu, J.J., Murray, N.V., and Rosenthal, E., A Framework for Automated Reasoning in Multiple-Valued Logics, *J. of Automated Reasoning* 21:39–67, 1998. To appear.
19. Lu, J.J., Logic programming with signs and annotations, *J. of Logic and Computation*, 6(6):755 – 778, 1996.
20. Murray, N.V., and Rosenthal, E., Adapting classical inference techniques to multiple-valued logics using signed formulas, *Fundamenta Informaticae* 21:237–253, 1994.
21. Subrahmanian, V.S., On the Semantics of Quantitative Logic Programs, in: *Proceedings of the 4th IEEE Symposium on Logic Programming*, Computer Society Press, 1987.
22. Subrahmanian, V.S., Paraconsistent Disjunctive Databases, *Theoretical Computer Science*, 93, 115–141, 1992.
23. Thirunarayan, K., and Kifer, M., A Theory of Nomonotonic Inheritance Based on Annotated Logic, *Artificial Intelligence*, 60(1):23–50, 1993.

A Matrix Characterization for $\mathcal{MELL}$

Heiko Mantel[1] and Christoph Kreitz[2]

[1] German Research Center for Artificial Intelligence (DFKI),Saarbrücken, Germany
mantel@dfki.de
[2] Department of Computer Science,Cornell-University, Ithaca, USA
kreitz@cs.cornell.edu

Abstract. We present a matrix characterization of logical validity in the multiplicative fragment of linear logic with exponentials. In the process we elaborate a methodology for proving matrix characterizations correct and complete. Our characterization provides a foundation for matrix-based proof search procedures for $\mathcal{MELL}$ as well as for procedures which translate machine-found proofs back into the usual sequent calculus.

1 Introduction

Linear logic [12] has become known as a very expressive formalism for reasoning about action and change. During its rather rapid development linear logic has found applications in logic programming [14,19], modeling concurrent computation [11], planning [18], and other areas. Its expressiveness, however, results in a high complexity. Propositional linear logic is undecidable. The multiplicative fragment ($\mathcal{MLL}$) is already $\mathcal{NP}$-complete [16]. The complexity of the multiplicative exponential fragment ($\mathcal{MELL}$) is still unknown. Consequently, proof search in linear logic is difficult to automate. Girard's sequent calculus [12], although covering all of linear logic, contains too many redundancies to be useful for efficient proof search. Attempts to remove permutabilities from sequent proofs [1,10] and to add proof strategies [23] have provided significant improvements. But because of the use of sequent calculi some redundancies remain. Proof nets [7], on the other hand, can handle only a fragment of the logic.

Matrix characterizations of logical validity, originally developed as foundation of the *connection method* for classical logic [2,3,5], avoid many kinds of redundancies contained in sequent calculi and yield a compact representation of the search space. They have been extended successfully to intuitionistic and modal logics [24] and serve as a basis for a uniform proof search method [20] and a method for translating matrix proofs back into sequent proofs [21,22]. Resource management similar to multiplicative linear logic is addressed by the *linear connection method* [4]. Fronhöfer [8] gives a matrix characterization of $\mathcal{MLL}$ that captures some aspects of weakening and contraction but does not appear to generalize any further. In [15] we have developed a matrix characterization for $\mathcal{MLL}$ and extended the uniform proof search and translation procedures accordingly.

In this paper we present a matrix characterization for the full multiplicative exponential fragment including the constants **1** and $\perp$. This characterization uses Andreoli's focusing principle [1] as one of its major design steps and does not appear to share the limitations of the previous approaches. Our approach also includes a methodology for developing such a characterization. By introducing a series of intermediate calculi the development of a matrix characterization for $\mathcal{MELL}$ becomes manageable. Each newly introduced calculus adds one compactification principle and is proven correct and complete with respect to the previous one. We expect that this methodology will generalize to further fragments of linear logic as well as to a wide spectrum of other non-classical logics.

We first create a compact representation Σ_1' of Girard's sequent calculus [12] by adopting Smullyan's tableaux notation to $\mathcal{MELL}$ (Section 2). By introducing the notion of multiplicities, i.e. an eager handling of contraction and a lazy handling of weakening, we arrive at a *dyadic* calculus Σ_2' which we then refine to a *triadic* calculus Σ_3' by removing redundancies which are due to permutabilities (Section 3). In Section 4 we develop a calculus Σ_{pos} which operates on *positions* in a formula tree instead of on the subformulas themselves. In order to express the peculiarities of some connectives we insert *special positions* into the formula tree. Finally, in Section 5, we arrive at the matrix characterization, technically the most demanding but also the most compact of all calculi. Proofs are only sketched briefly. Details can be found in the first author's technical report [17].

2 Multiplicative Exponential Linear Logic

Linear logic [12] treats formulas like resources that disappear after their use unless explicitly marked as reusable. Technically, it can be seen as the outcome of removing the rules for contraction and weakening from the classical sequent calculus and re-introducing them in a controlled manner. Linear negation $^\perp$ is involutive like classical negation. The two traditions for writing the sequent rule for conjunction result in two different conjunctions $\otimes$ and & and two different disjunctions $⅋$ and $\oplus$. The constant `true` splits up into **1** and $\top$ and `false` into $\perp$ and **0**. The unary connectives ? and ! mark formulas for a controlled application of weakening and contraction. Quantifiers $\forall$ and $\exists$ are added as usual.

Linear logic can be divided into the multiplicative, additive, and exponential fragment. While in the multiplicative fragment resources are used exactly once, resource sharing is enforced in the additive fragment. Exponentials mark formulas as reusable. All fragments exist on their own right and can be combined freely. The full power of linear logic comes from combining all of them.

Throughout this article we will focus on multiplicative exponential linear logic ($\mathcal{MELL}$), the combination of the multiplicative and exponential fragments, leaving the additive fragment and the quantifiers out of consideration. $^\perp$, $\otimes$, $⅋$, $\multimap$, **1**, $\perp$, !, and ? are the connectives of $\mathcal{MELL}$. Linear negation $^\perp$ expresses the difference between resources that are to be used up and resources to be produced. $F^\perp$ means that the resource F must be produced. Having a resource $F_1 \otimes F_2$ means having F_1 as well as F_2. $F_1 \multimap F_2$ allows the construction of F_2 from F_1. $F_1 ⅋ F_2$ is equivalent to $F_1^\perp \multimap F_2$ and to $F_2^\perp \multimap F_1$. Having a resource **1** has no impact while nothing can be constructed when $\perp$ is used up. A resource $!F$ acts like a machine which produces any number of copies of F. During the construction of $!F$ only such machines can be used. ? is the dual to !.

$$\frac{}{\langle A,+\rangle,\langle A,-\rangle}\ axiom \qquad \frac{}{\tau}\ \tau \qquad \frac{\Upsilon}{\Upsilon,\omega}\ \omega \qquad \frac{\Upsilon, succ_1(o)}{\Upsilon, o}\ o$$

$$\frac{\Upsilon, succ_1(\alpha), succ_2(\alpha)}{\Upsilon,\alpha}\ \alpha \qquad \frac{\Upsilon_1, succ_1(\beta) \quad \Upsilon_2, succ_2(\beta)}{\Upsilon_1,\Upsilon_2,\beta}\ \beta$$

$$\frac{\Upsilon, succ_1(\nu)}{\Upsilon,\nu}\ \nu \qquad \frac{\nu, succ_1(\pi)}{\nu,\pi}\ \pi \qquad \frac{\Upsilon}{\Upsilon,\nu}\ w \qquad \frac{\Upsilon,\nu,\nu}{\Upsilon,\nu}\ c$$

Table 1. Sequent calculus Σ_1' for $\mathcal{MELL}$ in uniform notation

The validity of a linear logic formula can be proven syntactically by using a sequent calculus. For multi-sets Γ and Δ of formulas $\Gamma \longrightarrow \Delta$ is called a *sequent*. It can be understood as the specification of a transformation which constructs Δ from Γ. The formulas in Γ are connected implicitly by $\otimes$ while the formulas in Δ are connected implicitly by $\invamp$.

By adopting Smullyan's uniform notation to $\mathcal{MELL}$ we receive a compact representation of sequent calculi, which simplifies proofs about their properties. A *signed formula* $\varphi = \langle F,k\rangle$ denotes an occurrence of F in Δ or Γ. Depending on the *label* F and its *polarity* $k \in \{+,-\}$, a signed formula will receive a *type* α, β, ν, π, o, τ, ω, or *lit* according to the tables below. The functions $succ_1$ and $succ_2$ return the major signed subformulas of a signed formula. Note that during the decomposition of a formula the polarity switches only for $^\perp$ and $\multimap$. We use type symbols as meta-variables for signed formulas of the respective type, e.g. α stands for a signed formula of type α.

lit	$\langle A,-\rangle$	$\langle A,+\rangle$

τ	$\langle\perp,-\rangle$	$\langle\mathbf{1},+\rangle$

ω	$\langle\mathbf{1},-\rangle$	$\langle\perp,+\rangle$

α	$\langle F_1\otimes F_2,-\rangle$	$\langle F_1 \invamp F_2,+\rangle$	$\langle F_1 \multimap F_2,+\rangle$
$succ_1(\alpha)$	$\langle F_1,-\rangle$	$\langle F_1,+\rangle$	$\langle F_1,-\rangle$
$succ_2(\alpha)$	$\langle F_2,-\rangle$	$\langle F_2,+\rangle$	$\langle F_2,+\rangle$

β	$\langle F_1\otimes F_2,+\rangle$	$\langle F_1 \invamp F_2,-\rangle$	$\langle F_1 \multimap F_2,-\rangle$
$succ_1(\beta)$	$\langle F_1,+\rangle$	$\langle F_1,-\rangle$	$\langle F_1,+\rangle$
$succ_2(\beta)$	$\langle F_2,+\rangle$	$\langle F_2,-\rangle$	$\langle F_2,-\rangle$

o	$\langle F^\perp,-\rangle$	$\langle F^\perp,+\rangle$
$succ_1(o)$	$\langle F,+\rangle$	$\langle F,-\rangle$

ν	$\langle !F,-\rangle$	$\langle ?F,+\rangle$
$succ_1(\nu)$	$\langle F,-\rangle$	$\langle F,+\rangle$

π	$\langle ?F,-\rangle$	$\langle !F,+\rangle$
$succ_1(\pi)$	$\langle F,-\rangle$	$\langle F,+\rangle$

A sequent calculus Σ_1' based on this uniform notation is depicted in table 1. In a rule the sequents above the line are the *premises* and the one below the *conclusion*. A *principal formula* is a formula that occurs in the conclusion but not in any premise. Formulas that occur in a premise but not in the conclusion are called *active*. All other formulas compose the *context*. Σ_1' is shown correct and complete wrt. Girard's original sequent calculus [12] by a straightforward induction over the structure of proofs.

In *analytic proof search* one starts with the sequent to be proven and reduces it by application of rules until the *axiom*-rule or the τ-rule can be applied. There are several choice points within this process. First, a principal formula must be chosen. Unless the principal formula has type ν, this choice determines which rule must be applied. If a β-rule is applied the context of the sequent must be partitioned onto the premises (*context splitting*). Several solutions have been proposed in order to optimize these choices [1,10,23,6,13]. Additional difficulties arise from the rules *axiom*, τ, and π. The rules *axiom* and τ require an empty context which expresses that all formulas must be used up in a proof. The π rule requires that all formulas in the context are of type ν. Though the connectives of linear logic make proof search more difficult they also give rise to new possibilities. Some applications for linear logic programming are illustrated in [19].

$$\dfrac{\dfrac{\dfrac{\dfrac{\dfrac{\dfrac{}{a_{000}, a_{0100}}\,axiom}{a_{000}, o_{010}}\,o}{a_{000}, \nu_{01}}\,\nu \qquad \dfrac{\dfrac{\dfrac{\dfrac{}{a_{0010}, a'_{0100}}\,axiom}{a_{0010}, o'_{010}}\,o}{a_{0010}, \nu'_{01}}\,\nu}{\pi_{001}, \nu'_{01}}\,\pi}{\beta_{00}, \nu_{01}, \nu'_{01}}\,\beta}{\beta_{00}, \nu_{01}}\,c}{\alpha_0}\,\alpha$$

$lab(\varphi')$	φ'
$(A \otimes !A) ⅋\ ?(A^\perp)$	α_0
$A \otimes !A$	β_{00}
A	a_{000}
$!A$	π_{001}
A	a_{0010}
$?(A^\perp)$	ν_{01}, ν'_{01}
$A^\perp$	o_{010}, o'_{010}
A	a_{0100}, a'_{0100}

Fig. 1. Example Σ_1'-proof.

Example 1. Figure 1 presents a Σ_1'-proof of $\varphi = \langle (A \otimes !A) ⅋\ ?(A^\perp), + \rangle$. We abbreviate subformulas φ' of φ by position markers as shown in the table on the right. The proof requires that the contraction rule c is applied before the β-rule.

3 N-adic Sequent Calculi for $\mathcal{MELL}$

In this section we define two intermediate sequent calculi which are closely related to Andreoli's dyadic calculus Σ_2 and triadic calculus Σ_3 [1] but differ in the way structural rules are handled. While Andreoli uses a lazy strategy for both contraction and weakening, our calculi Σ_2' and Σ_3', which are *not* intended for proof search, are based on an eager strategy for contraction. Eager contraction corresponds to the concept of multiplicities in matrix characterizations [24].

3.1 Dyadic Calculus

$$\dfrac{}{\Theta : \langle A, + \rangle, \langle A, - \rangle}\,axiom \qquad \dfrac{}{\Theta : \tau}\,\tau \qquad \dfrac{\Theta : \Upsilon, succ_1(o)}{\Theta : \Upsilon, o}\,o \qquad \dfrac{\Theta : \Upsilon}{\Theta : \Upsilon, \omega}\,\omega$$

$$\dfrac{\Theta : \Upsilon, succ_1(\alpha), succ_2(\alpha)}{\Theta : \Upsilon, \alpha}\,\alpha \qquad \dfrac{\Theta_1 : \Upsilon_1, succ_1(\beta) \quad \Theta_2 : \Upsilon_2, succ_2(\beta)}{\Theta_1, \Theta_2 : \Upsilon_1, \Upsilon_2, \beta}\,\beta$$

$$\dfrac{\Theta, succ_1(\nu)^{\mu(\nu)} : \Upsilon}{\Theta : \Upsilon, \nu}\,\nu \qquad \dfrac{\Theta : succ_1(\pi)}{\Theta : \pi}\,\pi \qquad \dfrac{\Theta : \Upsilon, \varphi}{\Theta, \varphi : \Upsilon}\,focus$$

Table 2. Dyadic sequent calculus Σ_2' for $\mathcal{MELL}$ in uniform notation

In sequent proofs there are two possible notions of occurrence of a formula φ: an occurrence of φ as subformula in some formula tree or its occurrences within a derivation. The difference between these two becomes only apparent when contraction is applied. In $\mathcal{MELL}$ only formulas of type ν are *generic*, i.e. may be contracted. Since we are aiming at a matrix characterization, we apply contraction in an eager way. For this purpose we introduce a function μ which determines the *multiplicity* of an occurrence of a formula, i.e. the number of copies of that occurrence in a proof.[1] Let Θ and Υ be multi-sets of signed formulas. A *dyadic sequent* $S = \Theta : \Upsilon$ has two zones which are separated by a colon. Θ is called the *unbounded zone* and Υ the *bounded zone* of S.

The sequent calculus Σ_2' for dyadic sequents depicted in table 2 employs eager contraction. Derivations of a dyadic sequent S are defined with respect to

[1] Wallen's multiplicities for modal and intuitionistic logics are based on occurrences within a *formula tree* [24]. Our notion respects the resource sensitivity of linear logic.

a fixed multiplicity function μ. This function is important whenever the ν-rule is applied. Additionally, Σ_2' uses lazy weakening. There is no explicit weakening rule but weakening is done implicitly in the rules *axiom* and τ. The π rule requires that all context formulas are in the unbounded zone instead of requiring them to have a specific type as in Σ_1'. A special rule *focus* moves a formula from the unbounded into the bounded zone. A signed formula φ is *derivable* in Σ_2' if $\cdot : \varphi$ is derivable for some multiplicity, where $\cdot$ denotes the empty multi-set.

Theorem 2 (Completeness). *Let $\mathcal{P}_1$ be a Σ_1'-proof for $S_1 = \Upsilon, \boldsymbol{\nu}_{\overline{c}}$ where $\boldsymbol{\nu}_{\overline{c}}$ consists of all signed formulas of type ν in S_1 to which the contraction rule is not applied in $\mathcal{P}_1$. Then there is a multiplicity μ_2 such that the dyadic sequent $S_2 = \mathit{succ}_1(\boldsymbol{\nu}_{\overline{c}}) : \Upsilon$ can be derived in Σ_2'.*[2]

Theorem 3 (Correctness). *Let $\mathcal{P}_2$ be a Σ_2'-proof for $S_2 = \Theta_1^+, \Theta_2^- : \Upsilon$ with multiplicity μ_2 where Θ_1^+ and Θ_2^- contain only positive and negative signed formulas, respectively. Then there exists a Σ_1'-proof $\mathcal{P}_1$ for the unary sequent $S_1 = ?\Theta_1^+, !\Theta_2^-, \Upsilon$.*

3.2 Triadic Calculus

$$\frac{}{\Theta : \langle A,+\rangle, \langle A,-\rangle \Uparrow \cdot}\ \mathit{axiom} \qquad \frac{}{\Theta : \tau \Uparrow \cdot}\ \tau \qquad \frac{\Theta : \Upsilon \Uparrow \Xi}{\Theta : \Upsilon \Uparrow \Xi, \omega}\ \omega$$

$$\frac{\Theta : \Upsilon \Downarrow \mathit{succ}_1(o)}{\Theta : \Upsilon \Downarrow o}\ o\Downarrow \qquad \frac{\Theta : \Upsilon \Uparrow \Xi, \mathit{succ}_1(o)}{\Theta : \Upsilon \Uparrow \Xi, o}\ o\Uparrow$$

$$\frac{\Theta : \Upsilon \Uparrow \Xi, \mathit{succ}_1(\alpha), \mathit{succ}_2(\alpha)}{\Theta : \Upsilon \Uparrow \Xi, \alpha}\ \alpha \qquad \frac{\Theta_1 : \Upsilon_1 \Downarrow \mathit{succ}_1(\beta) \quad \Theta_2 : \Upsilon_2 \Downarrow \mathit{succ}_2(\beta)}{\Theta_1, \Theta_2 : \Upsilon_1, \Upsilon_2 \Downarrow \beta}\ \beta$$

$$\frac{\Theta, \mathit{succ}_1(\nu)^{\mu(\nu)} : \Upsilon \Uparrow \Xi}{\Theta : \Upsilon \Uparrow \Xi, \nu}\ \nu \qquad \frac{\Theta : \cdot \Uparrow \mathit{succ}_1(\pi)}{\Theta : \cdot \Downarrow \pi}\ \pi$$

$$\frac{\Theta : \Upsilon \Downarrow \varphi}{\Theta, \varphi : \Upsilon \Uparrow \cdot}\ \mathit{focus}_1 \qquad \frac{\Theta : \Upsilon \Downarrow \varphi}{\Theta : \Upsilon, \varphi \Uparrow \cdot}\ \mathit{focus}_2 \qquad \frac{\Theta : \Upsilon, \varphi \Uparrow \Xi}{\Theta : \Upsilon \Uparrow \Xi, \varphi}\ \mathit{defocus} \qquad \frac{\Theta : \Upsilon \Uparrow \varphi}{\Theta : \Upsilon \Downarrow \varphi}\ \mathit{switch}$$

In focus_2 φ must not be of type *lit* or τ. In *defocus* φ must be of type *lit*, τ, β, or π. In *switch* φ must be of type *lit*, τ, ω, α, or ν.

Table 3. Triadic sequent calculus Σ_3' for $\mathcal{MELL}$ in uniform notation

During proof search in sequent calculi the order of some rule applications may be permuted. For linear logic, the permutabilities and non-permutabilities of sequent rules have been investigated in [1,10]. Andreoli's *focusing principle* [1] allows to fix the order of permutable rules without losing completeness. A distinctive feature of this principle is that the reduction ordering is determined for *layers of formulas* rather than for individual formulas. Let φ be a signed formula, Ξ be a sequence and Θ and Υ be multi-sets of signed formulas. A *triadic sequent* $S = \Theta : \Upsilon \Downarrow \varphi$ or $S = \Theta : \Upsilon \Uparrow \Xi$ has three zones. Θ is called the *unbounded zone*, Υ the *bounded zone*, and φ or Ξ the *focused* zone. The sequent is either in *synchronous* ($\Downarrow$) or in *asynchronous* mode ($\Uparrow$).

[2] For convenience we extend functions and connectives to multi-sets of signed formulas. $\mathit{succ}_1(\boldsymbol{\nu}_{\overline{c}})$ abbreviates $\{\mathit{succ}_1(\nu) \mid \nu \in \boldsymbol{\nu}_{\overline{c}}\}$, $?\Theta$ denotes $\{\langle ?F, k\rangle \mid \langle F, k\rangle \in \Theta\}$, etc.

The sequent calculus Σ_3' for triadic sequents depicted in table 3 employs the focusing principle. Derivations are defined with respect to a fixed multiplicity function μ. The multiplicity is important whenever the rule ν is applied. A signed formula φ is derivable in Σ_3' if the sequent $\cdot : \varphi \Uparrow \cdot$ is derivable for some μ.

In Σ_3' there are two focusing rules, *focus*$_1$ and *focus*$_2$, which move a signed formula into the focus. Both rules switch the sequent into synchronous mode. Depending on the structure of the formula this enforces a sequence of rules to be applied next. Since selection of these rules is deterministic, the permutabilities of rule applications are removed from the search space. The matrix characterization developed in this article exploits this focusing principle. However, it yields a representation with even less redundancies than a calculus like Σ_3' can.

Theorem 4 (Completeness). *Let $\mathcal{P}_2$ be a Σ_2'-proof for $S_2 = \Theta : \Upsilon$ with multiplicity μ_2 and let Υ' be a linearization of Υ. Then there is a multiplicity μ_3 such that the triadic sequent $S_3 = \Theta : \cdot \Uparrow \Upsilon'$ is derivable in Σ_3'.*

Theorem 5 (Correctness). *Let $\mathcal{P}_3$ be a Σ_3'-proof for $S_3 = \Theta : \Upsilon \Uparrow \Xi$ (or $\Theta : \Upsilon \Downarrow \varphi$) with multiplicity μ_3. Then there is a Σ_2'-proof $\mathcal{P}_2$ for the dyadic sequent $S_2 = \Theta : \Upsilon, \Xi'$ (or $\Theta : \Upsilon, \varphi$) for some multiplicity μ_2, where Ξ' is the multi-set that contains the same signed formulas as the sequence Ξ.*

4 A Position Calculus for $\mathcal{MELL}$

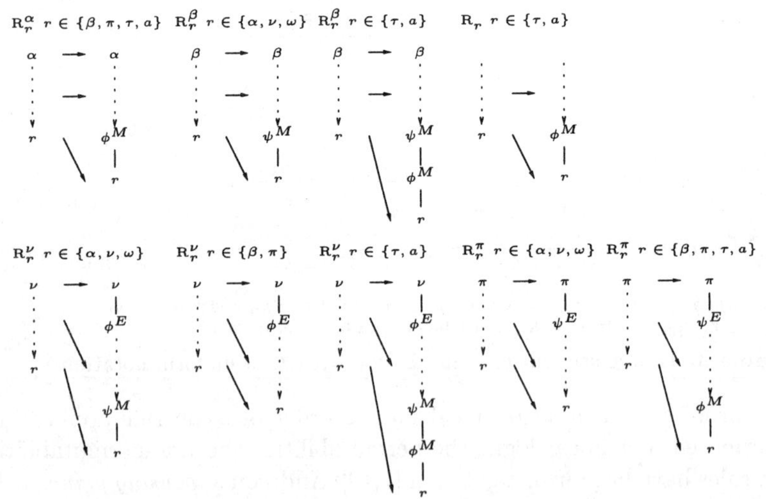

Fig. 2. Rules for inserting special positions into basic position trees

In the previous section we have used established techniques for removing redundancies in the search space of a specific sequent proof. In order to reason about logical validity as such, the non-classical aspects of sequent proofs must be expressed in a more compact way. Wallen [24] uses prefixes of special positions for modal logics and intuitionistic logic. We adapt this approach to $\mathcal{MELL}$ and introduce position calculi as intermediate step in the development of matrix characterizations. We capture the difference between an occurrence of a formula within a proof or as subformula by basic positions and positions.

Basic Position Trees Let V_b be an arbitrary set of *basic positions*. The *basic position tree* for a signed formula φ is a directed tree $\mathcal{T}_b = (V_b, E_b)$ which originates from a formula tree by inserting additional nodes.[3] These nodes are inserted by applying rewrite rules from figure 2 until none of them is applicable. The functions $succ_1$ and $succ_2$ are redefined for basic position trees such that $succ_1(bp)$ is the left and $succ_2(bp)$ the right successor of the node bp. For a basic position $bp \in V_b$, the corresponding formula, main connective, signed formula, polarity, and type can be retrieved by the functions *lab*, *con*, *sform*, *pol*, and *Ptype*, respectively. For the inserted nodes we introduce new *special types* ϕ^M, ψ^M, ϕ^E, and ψ^E which are assigned according to the rewrite rules. For an inserted basic position of type ϕ^M, ψ^M, or ϕ^E the value of the functions *lab*, *con*, and *pol* equals the value of the successor node and for type ψ^E it is the value of the predecessor node. We use type symbols as meta-variables for basic positions of the respective type and a as a meta-variable for basic positions of type *lit*.

A rewrite rule R can be applied to a tree $\mathcal{T}$ if its left hand side matches a subtree $\mathcal{T}'$ of $\mathcal{T}$. In this case $\mathcal{T}'$ is rewritten according to the pattern on the right hand side of R. The dotted lines in the patterns match arbitrary subtrees that contain only nodes of type o. A special case are the rules R_a and R_τ. They can only be applied if there are just positions of type o between the root and the leaf. The other rewrite rules separate layers of subformulas within a formula tree: $\mathrm{R}_{t_2}^{t_1}$ inserts special positions wherever a subformula of type t_1 has a subformula of type t_2. The rewrite system is confluent and noetherian.

Example 6. We illustrate the application of the rule R_β^α. Below, we have depicted the formula tree $\mathcal{T}_1$ for $\langle (A \otimes !A) ⅋ ?(A^\perp), + \rangle$. β_{00} is a successor of α_0 with no nodes in between. The subtree consisting of α_0, β_{00}, and the edge between them matches the left hand side of R_β^α. The tree is rewritten to $\mathcal{T}_2$ and can further be rewritten by applying R_a^β, R_a^π, and R_a^ν. The resulting basic position tree is $\mathcal{T}_3$.

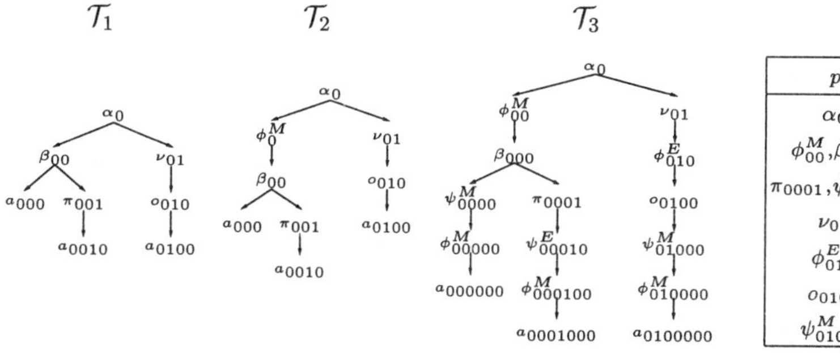

p	$con(p)$
α_0	⅋
ϕ^M_{00}, β_{000}	$\otimes$
$\pi_{0001}, \psi^E_{00010}$	!
ν_{01}	?
ϕ^E_{010}	$\perp$
o_{0100}	$\perp$
ψ^M_{01000}	$\perp$

Special positions represent the possible behavior of a layer within a derivation. Positions of type ψ^M and ψ^E are called *constants* while those of type ϕ^M and ϕ^E are *variables*. Inserted variables express that the corresponding formula may be part of the bounded (ϕ^M) or unbounded (ϕ^E) zone. During a sequence of

[3] We denote basic positions by strings over $\{0, 1\}$. 0 is the root position of a tree. Extending a tree by 0 or 1 yields the basic position for the left or right successor node.

proof rule applications that ends with a constant position, a part of the context may have to be of a specific type. The requirements of the π–rule, for instance, are expressed by a ψ^E position.

Position Trees, Position Forests, and Position Sequents A *position tree* for a signed formula wrt. a multiplicity μ is constructed from a basic position tree by subsequently replacing subtrees of positions ν with $\mu(\nu) = n$ with n copies of that tree starting from the root. New positions are assigned to the copies.[4]

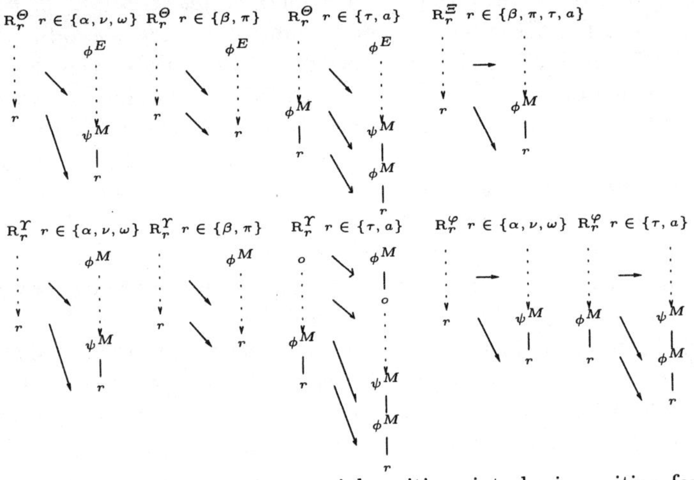

R^{Θ}_r $r \in \{\alpha, \nu, \omega\}$ R^{Θ}_r $r \in \{\beta, \pi\}$ R^{Θ}_r $r \in \{\tau, a\}$ R^{Ξ}_r $r \in \{\beta, \pi, \tau, a\}$

R^{Υ}_r $r \in \{\alpha, \nu, \omega\}$ R^{Υ}_r $r \in \{\beta, \pi\}$ R^{Υ}_r $r \in \{\tau, a\}$ R^{φ}_r $r \in \{\alpha, \nu, \omega\}$ R^{φ}_r $r \in \{\tau, a\}$

Fig. 3. Rules for inserting special positions into basic position forests

As formulas can be represented as formula trees, sequents can be represented as *sequent forests*, i.e. collections of formula trees that are divided into different zones. A *basic position forest* for a triadic sequent is a collection of basic position trees which consists of three zones (unbounded, bounded, and focused) and has a mode ($\Downarrow$ or $\Uparrow$). The trees are modified by the rewrite rules in figure 3, which modify trees only at their roots. The exponent of a rewrite rule defines the zone in which it can be applied. $R^{\Theta}_?, R^{\Upsilon}_?, R^{\Xi}_?$, and $R^{\varphi}_?$ can be applied to trees in the unbounded zone, bounded zone, focused zone (mode $\Uparrow$) and focused zone (mode $\Downarrow$), respectively. A *position forest* is a collection of position trees which is divided into three zones, has a mode, and which is constructed from a basic position forest together with a multiplicity. We use *position sequent* as well as *position matrix* as a synonym for position tree. We denote trees by their roots if the rest of the tree is obvious from the context. Note, that by the definition of the rewrite rules a root in the unbounded zone is always of type ϕ^E. A root in the bounded zone is of type ϕ^M. A root in the focused zone in mode $\Uparrow$ has a type in $\{o, \omega, \alpha, \nu, \phi^M\}$ and a root in mode $\Downarrow$ has a type from $\{o, \beta, \pi, \psi^M, \psi^E\}$.

Example 7. Position sequents can be represented graphically. Let ϕ^M_0 be the root of a position tree which corresponds to $\langle (A \otimes !A) ⅋ ?(A^\perp), + \rangle$ with multiplicity $\mu(\nu_{0001})$=2. The position sequent for $\cdot : \phi^M_0 \Uparrow \cdot$ is depicted in figure 4.

[4] We denote positions by strings over $\{0, 1, 0^m\}_{m \in \mathbb{N}}$. Positions in different copies of a subtree are distinguished by their exponents.

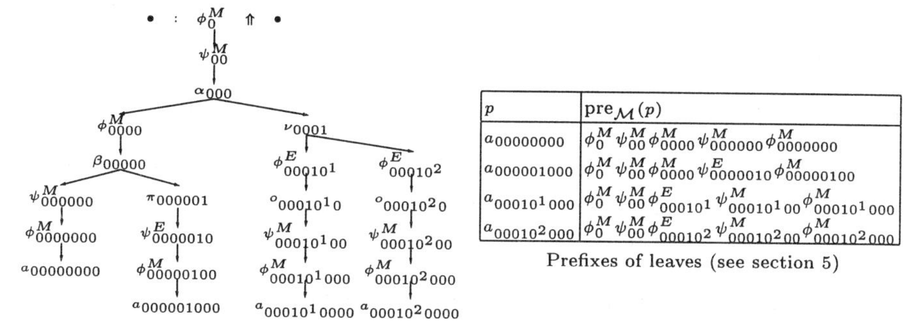

p	$\text{pre}_{\mathcal{M}}(p)$
$a_{00000000}$	$\phi_0^M \psi_{00}^M \phi_{0000}^M \psi_{000000}^M \phi_{0000000}^M$
$a_{000001000}$	$\phi_0^M \psi_{00}^M \phi_{0000}^M \psi_{0000010}^E \phi_{00000100}^M$
$a_{00010^1 000}$	$\phi_0^M \psi_{00}^M \phi_{00010^1}^E \psi_{00010^1 00}^M \phi_{00010^1 000}^M$
$a_{00010^2 000}$	$\phi_0^M \psi_{00}^M \phi_{00010^2}^E \psi_{00010^2 00}^M \phi_{00010^2 000}^M$

Prefixes of leaves (see section 5)

Fig. 4. Position sequent for the sequent $\cdot : \phi_0^M \Uparrow \cdot$

Position Calculus A sequent calculus for position sequents is depicted in table 4. Derivations of a position sequent S are defined with respect to a fixed multiplicity function μ for S. The position calculus makes apparent that, as pointed out earlier, inserted positions express how the context is affected in certain rules. The rewrite rules in figure 3 guarantee that the reduction of a position sequent by a rule results again in a position sequent. In a proof, the mode of a position sequent can be switched either by ψ^M or ψ^E. There is no *defocus* rule as in Σ_3' nor does the π rule cause a switch. Defocusing is done by the rules ϕ^M and ν. For all other rules there is a corresponding rule in Σ_3'.

$$\frac{}{\Phi^E : \phi_1^M, \phi_2^M \Uparrow \cdot}\ axiom \qquad \text{where } Ptype(succ_1(\phi_1^M)) = Ptype(succ_1(\phi_2^M)) = lit, \ lab(\phi_1^M) = lab(\phi_2^M), \text{ and } pol(\phi_1^M) \neq pol(\phi_2^M)$$

$$\frac{\Phi^E : \Phi^M \Downarrow succ_1(o)}{\Phi^E : \Phi^M \Downarrow o}\ o\Downarrow \qquad \frac{\Phi^E : \Phi^M \Uparrow \Xi, succ_1(o)}{\Phi^E : \Phi^M \Uparrow \Xi, o}\ o\Uparrow$$

$$\frac{\Phi^E : \Phi^M \Uparrow \Xi}{\Phi^E : \Phi^M \Uparrow \Xi, \omega}\ \omega \qquad \frac{}{\Phi^E : \phi_1^M \Uparrow \cdot}\ \tau \quad \text{where } Ptype(succ_1(\phi_1^M)) = \tau$$

$$\frac{\Phi^E : \Phi^M \Uparrow \Xi, succ_1(\alpha), succ_2(\alpha)}{\Phi^E : \Phi^M \Uparrow \Xi, \alpha}\ \alpha \qquad \frac{\Phi_1^E : \Phi_1^M \Downarrow succ_1(\beta) \quad \Phi_2^E : \Phi_2^M \Downarrow succ_2(\beta)}{\Phi_1^E, \Phi_2^E : \Phi_1^M, \Phi_2^M \Downarrow \beta}\ \beta$$

$$\frac{\Phi^E : \Phi^M \Downarrow succ_1(\pi)}{\Phi^E : \Phi^M \Downarrow \pi}\ \pi \qquad \frac{\Phi^E, succ_1^1(\nu), \dots, succ_1^{\mu(\nu)}(\nu) : \Phi^M \Uparrow \Xi}{\Phi^E : \Phi^M \Uparrow \Xi, \nu}\ \nu$$

$$\frac{\Phi^E : \Phi^M \Downarrow succ_1(\phi_1^E)}{\Phi^E, \phi_1^E : \Phi^M \Uparrow \cdot}\ focus_1 \qquad \frac{\Phi^E : \Phi^M \Downarrow succ_1(\phi_1^M)}{\Phi^E : \Phi^M, \phi_1^M \Uparrow \cdot}\ focus_2 \quad \text{where } Ptype(succ_1(\phi_1^M)) \notin \{lit, \tau\}$$

$$\frac{\Phi^E : \Phi^M, \phi_1^M \Uparrow \Xi}{\Phi^E : \Phi^M \Uparrow \Xi, \phi_1^M}\ \phi^M \qquad \frac{\Phi^E : \Phi^M \Uparrow succ_1(\psi_1^M)}{\Phi^E : \Phi^M \Downarrow \psi_1^M}\ \psi^M \qquad \frac{\Phi^E : \cdot \Uparrow succ_1(\psi_1^E)}{\Phi^E : \cdot \Downarrow \psi_1^E}\ \psi^E$$

Table 4. Position calculus Σ_{pos} for $\mathcal{MELL}$ in uniform notation

Theorem 8 (Completeness). *Let $\mathcal{P}_3$ be a Σ_3'-proof for $S_3 = \Theta : \Upsilon \Uparrow \Xi$ (or $S_3 = \Theta : \Upsilon \Downarrow \varphi$) with multiplicity μ_3. Then there is a multiplicity $\widetilde{\mu_3}$ such that the corresponding position sequent $\widetilde{S_3}$ is derivable in Σ_{pos}.*

Theorem 9 (Correctness). *Let $\mathcal{P}$ be a Σ_{pos}-proof for $S = \Phi^E : \Phi^M \Uparrow \Xi$ (or $S = \Phi^E : \Phi^M \Downarrow \varphi$) with multiplicity μ. Then there is a multiplicity $\widetilde{\mu}$ such that the corresponding triadic sequent $\widetilde{S} = \text{sform}(\Phi^E) : \text{sform}(\Phi^M) \Uparrow \text{sform}(\Xi)$ (or $\widetilde{S} = \text{sform}(\Phi^E) : \text{sform}(\Phi^M) \Downarrow \text{sform}(\varphi)$) is derivable in Σ_3'.*

5 The Matrix Characterization for $\mathcal{MELL}$

On the basis of the position calculus we can now develop a compact characterization of logical validity. For this purpose we extend the concept of position forests by important technical notions, state the requirements for validity in $\mathcal{MELL}$, and prove them to be sufficient and complete. We then summarize all the insights gained in this paper in a single characterization theorem.

Fundamental Concepts. A matrix characterization of logical validity of a formula φ is expressed in terms of properties of certain sets of subformulas of φ. We use positions to achieve a compact representation of φ and its subformulas.

A *matrix* $\mathcal{M}$ is a position forest, constructed from a basic position forest and a multiplicity μ. The set of positions in a matrix $\mathcal{M}$ is denoted by $Pos(\mathcal{M})$. The set of *axiom positions* $AxPos(\mathcal{M})$ contains all positions with principal type τ or *lit*. The set of *weakening positions* $WeakPos(\mathcal{M})$ contains all positions with principal type ω and all positions of type ν with $\mu(\nu) = 0$. The set of *leaf positions* is defined as $LeafPos(\mathcal{M}) = AxPos(\mathcal{M}) \cup WeakPos(\mathcal{M})$. $\beta(\mathcal{M})$ is the set of all positions in $Pos(\mathcal{M})$ of type β. The sets $\Phi^M(\mathcal{M})$, $\Psi^M(\mathcal{M})$, $\Phi^E(\mathcal{M})$, and $\Psi^E(\mathcal{M})$ are defined accordingly. The set of *special positions* is defined as $SpecPos(\mathcal{M}) = \Phi^M(\mathcal{M}) \cup \Psi^M(\mathcal{M}) \cup \Phi^E(\mathcal{M}) \cup \Psi^E(\mathcal{M})$.

A *weakening map* for $\mathcal{M}$ is a subset of $\Phi^E(\mathcal{M}) \cup WeakPos(\mathcal{M})$. This novel concept is required because of the restricted application of weakening in $\mathcal{MELL}$.

A *path* is a set of positions. The set $Paths(\mathcal{T})$ of *paths through a position tree* $\mathcal{T}$ is defined recursively by

- $P = \{0\}$, the set containing the root of $\mathcal{T}$, is a path through $\mathcal{T}$.
- If $P \cup \{p\}$ is a path through $\mathcal{T}$ then the following are paths

$$
\begin{array}{ll}
P \cup \{succ_1(p), succ_2(p)\} & \text{if } Ptype(p) = \alpha \\
P \cup \{succ_1(p)\} \text{ and } P \cup \{succ_2(p)\} & \text{if } Ptype(p) = \beta \\
P \cup \{succ_1(p)\} & \text{if } Ptype(p) \in \{o, \pi, \phi^M, \psi^M, \phi^E, \psi^E\} \\
P \cup \bigcup_{i \leq \mu(p)} \{succ_1^i(p)\} & \text{if } Ptype(p) = \nu \text{, and } \mu(\nu) > 0
\end{array}
$$

The *set of paths through a set of position trees* $\mathcal{F}_s$ is defined recursively by

$$
\begin{array}{ll}
Paths(\emptyset) & = \emptyset, \\
Paths(\{\mathcal{T}\}) & = Paths(\mathcal{T}), \text{ and} \\
Paths(\{\mathcal{T}\} \cup \mathcal{F}'_s) & = \{P_1 \cup P_2 \mid P_1 \in Paths(\mathcal{T}), P_2 \in Paths(\mathcal{F}'_s)\}.
\end{array}
$$

The *set of paths through a matrix* is defined by

$$
\begin{array}{l}
Paths(\Phi^E : \Phi^M \Downarrow \varphi) = Paths(\Phi^E \cup \Phi^M \cup \{\varphi\}) \text{ and} \\
Paths(\Phi^E : \Phi^M \Uparrow \Xi) = Paths(\Phi^E \cup \Phi^M \cup \Xi) \quad .
\end{array}
$$

A *path of leaves* through $\mathcal{M}$ is a subset of $LeafPos(\mathcal{M})$. $LPaths(\mathcal{M})$ denotes the *set of all paths of leaves through* $\mathcal{M}$. Since leaf positions are not decomposed in the definition of paths, a path of leaves contains only irreducible positions.

A *connection* in a matrix $\mathcal{M}$ is a subset of $AxPos(\mathcal{M})$. It is either a two-element set $\{p_1, p_2\}$ where p_1 and p_2 are positions with different polarity, $lab(p_1) = lab(p_2)$, and $Ptype(p_1) = lit = Ptype(p_2)$, or a one-element set $\{p_1\}$ with $Ptype(p_1) = \tau$. *A connection C is on a path P* if $C \subseteq P$.

The *prefix* $pre_{\mathcal{M}}(p)$ of a position $p \in Pos(\mathcal{M})$ is defined by

– $pre_{\mathcal{M}}(0) = \begin{cases} r_0 & \text{if } r = Ptype(0) \in \{\phi^M, \psi^M, \phi^E, \psi^E\} \\ \varepsilon & \text{otherwise} \end{cases}$

– $pre_{\mathcal{M}}(p'i) = \begin{cases} pre_{\mathcal{M}}(p')r_{p'i} & \text{if } r = Ptype(p'i) \in \{\phi^M, \psi^M, \phi^E, \psi^E\} \\ pre_{\mathcal{M}}(p') & \text{otherwise} \end{cases}$

If $p_1 \ll p_2$, i.e. p_1 is a predecessor of p_2 in the position tree, then $pre_{\mathcal{M}}(p_1)$ is an initial substring of $pre_{\mathcal{M}}(p_2)$. We denote this by $pre_{\mathcal{M}}(p_1) \angle pre_{\mathcal{M}}(p_2)$. The prefixes of leaves of the example position sequent are displayed in figure 4.

A *multiplicative prefix substitution* is a mapping $\sigma_M : \Phi^M \to (\Phi^M \cup \Psi^M)^*$. An *exponential prefix substitution* is a mapping $\sigma_E : \Phi^E \to (\Phi^M \cup \Psi^M \cup \Phi^E \cup \Psi^E)^*$. A *multiplicative exponential prefix substitution* is a mapping $\sigma : (\Phi^M \cup \Phi^E) \to (\Phi^M \cup \Psi^M \cup \Phi^E \cup \Psi^E)^*$ which maps elements from Φ^M to strings from $(\Phi^M \cup \Psi^M)^*$ only. Substitutions are extended homomorphically to strings and are assumed to be computed by unification.

Complementarity. Matrix characterizations for classical [5] and non-classical [24] logics are based on a notion of *complementarity*. Essentially this means that every path through a matrix must contain a unifiable connection. These requirements also hold for linear logic but have to be extended by a few additional properties. We shall specify all these requirements now.

In the following we always assume $\mathcal{M}$ to be a matrix, $\mathcal{C}$ and $\mathcal{W}$ to be a set of connections and a weakening map for $\mathcal{M}$, and σ to be a prefix substitution.

- The *spanning property* is the most fundamental requirement. Each path of leaves must contain a connection. A set of connections $\mathcal{C}$ *spans* a matrix $\mathcal{M}$ iff for every path $P \in LPaths(\mathcal{M})$ there is a connection $C \in \mathcal{C}$ with $C \subseteq P$.
- The *unifiability property* states that connected leaves must be made identical wrt. their prefixes. Furthermore, because of the restricted application of weakening in $\mathcal{MELL}$, that each position in a weakening map must be related to a connection. $\langle \mathcal{C}, \mathcal{W} \rangle$ is *unifiable* if there exists a prefix substitution σ such that (1) $\sigma(pre_{\mathcal{M}}(p_1)) = \sigma(pre_{\mathcal{M}}(p_2))$ for all $C \in \mathcal{C}$ and $p_1, p_2 \in C$ and (2) for all $wp \in \mathcal{W}$ there is some $C = \{p, \dots\} \in \mathcal{C}$ with $\sigma(pre_{\mathcal{M}}(wp)) \angle \sigma(pre_{\mathcal{M}}(p))$. In this case σ is called a *unifier* for $\langle \mathcal{C}, \mathcal{W} \rangle$.
  A unifier σ of $\langle \mathcal{C}, \mathcal{W} \rangle$ can always be modified to a unifier σ' such that $\sigma'(pre_{\mathcal{M}}(\phi^E)) = \sigma'(pre_{\mathcal{M}}(p))$ holds for all $\phi^E \in \mathcal{W}$ and some connection $C = \{p, \dots\} \in \mathcal{C}$. A *grounded substitution* σ' is constructed from a substitution σ by removing all variable positions from values. If σ is a unifier for $\langle \mathcal{C}, \mathcal{W} \rangle$ then the corresponding grounded substitution σ' is a unifier as well.
- The *linearity property* expresses that no resource is to be used twice, i.e. contraction is restricted. A resource cannot be connected more than once and cannot be connected at all if that part of the formula is weakened. $\langle \mathcal{C}, \mathcal{W} \rangle$ is *linear* for $\mathcal{M}$ iff (1) any two connections $C_1 \neq C_2 \in \mathcal{C}$ are disjoint and (2) no predecessor ϕ^E of a position p in a connection $C \in \mathcal{C}$ belongs to the set $\mathcal{W}$.
- The *relevance property* requires that each resource must be used at least once. A resource is used if it is connected or weakened. $\langle \mathcal{C}, \mathcal{W} \rangle$ has the *relevance* property for $\mathcal{M}$ if for $p \in LeafPos(\mathcal{M})$ either (1) $p \in C$ for some $C \in \mathcal{C}$, or (2) $p \in \mathcal{W}$, or (3) some predecessor ϕ^E of p belongs to $\mathcal{W}$.

– The *cardinality property* expresses that the number of branches in a sequent proof is adequate. It substitutes the *minimality property* in [8], which would require a complicated test for $\mathcal{MELL}$. A pair $\langle \mathcal{C}, \mathcal{W} \rangle$ has the *cardinality property* for $\mathcal{M}$ if $|\mathcal{C}| + \sum_{\phi^E \in \mathcal{W}} \beta(\phi^E) = \beta(\mathcal{M}) + 1$.

Definition 10. A matrix $\mathcal{M}$ is *complementary* iff there are a set of connections $\mathcal{C}$, a weakening map $\mathcal{W}$ and a prefix substitution σ such that (1) $\mathcal{C}$ spans $\mathcal{M}$, (2) σ is a unifier for $\langle \mathcal{C}, \mathcal{W} \rangle$, and (3) $\langle \mathcal{C}, \mathcal{W} \rangle$ is linear for $\mathcal{M}$ and has the relevance and cardinality properties. We also say that $\mathcal{M}$ *is complementary for* $\mathcal{C}$, $\mathcal{W}$, and σ.

The complementarity of a matrix ensures the existence of a corresponding Σ_{pos}-proof. Each requirement captures an essential aspect of such a proof. Thus, there are relations between basic concepts in matrix proofs and Σ_{pos}-proofs. Paths are related to sequents. A connection on a path expresses the potential to close a Σ_{pos}-branch by an application of *axiom* or τ which involves the connected positions. A weakening map $\mathcal{W}$ contains all positions which are explicitly weakened by the rules ω and ν (for $\mu(\nu) = 0$) or implicitly weakened in *axiom* and τ. The unifiability of prefixes guarantees that connected positions can move into the same Σ_{pos} branch and that positions in $\mathcal{W}$ can be weakened in some branch. Linearity and relevance resemble the lack of contraction and weakening for arbitrary formulas, while cardinality expresses the absence of the rule of mingle, i.e. a proof can only branch at the reduction of β-type positions.

Since complementarity captures the essential aspects of Σ_{pos}-proofs but no unimportant details the search space is once more compactified. Problems like e.g. context splitting at the reduction of β-type positions simply do not occur.

Soundness. The unifiability property requires that each position in a weakening map $\mathcal{W}$ is related to some connection $C \in \mathcal{C}$. Let $AssSet(C)$ be the union of C and the set of all positions in $\mathcal{W}$ which are related to C.

A matrix $\mathcal{M}$ can be seen as a collection of trees, i.e. a forest. Let $\mathcal{T}_1$ and $\mathcal{T}_2$ be position trees in $\mathcal{M}$. We add for each connection C edges which link all positions in $AssSet(C)$. In order to identify the connected components of the resulting graph we define a relation $AssRel$ by

$$AssRel(\mathcal{T}_1, \mathcal{T}_2) \;\; \textit{iff} \;\; \exists C \in \mathcal{C}. \exists p_1 \in Pos(\mathcal{T}_1), p_2 \in Pos(\mathcal{T}_2). p_1, p_2 \in AssSet(C)$$

Let $\sim$ be the reflexive transitive closure of $AssRel$. A *connected component* in $\mathcal{M}$ is a set of position trees which is an equivalence class of $\sim$.

We define a function $FCons$ which reduces a set of connections $\mathcal{C}$ to those connections C whose elements are contained within a certain position forest $\mathcal{F}$. We define $FCons(\mathcal{F}, \mathcal{C}) = \{C \in \mathcal{C} \mid \forall p \in C.p \in Pos(\mathcal{F})\}$.

Theorem 11. *Let $\langle \mathcal{C}, \mathcal{W} \rangle$ be linear and relevant for the matrix $\mathcal{M}$ and σ be a unifier for $\langle \mathcal{C}, \mathcal{W} \rangle$. Then $|\beta(\mathcal{F})| < |FCons(\mathcal{F}, \mathcal{C})| + \sum_{\phi^E \in \mathcal{W}} |\beta(\phi^E)|$ holds for any connected component $\mathcal{F}$ in $\mathcal{M}$.*

Theorem 11 is proven by induction over the size of $\beta(\mathcal{F})$. In the induction step one first shows that there is always a node of type β, the separator of a β–chain, which can be removed such that the preconditions of the theorem hold. The induction hypothesis can be applied after removing that node.

Corollary 12. *Let $\langle \mathcal{C}, \mathcal{W} \rangle$ be linear for the matrix $\mathcal{M}$ and have the relevance and cardinality properties for $\mathcal{M}$. Let σ be a unifier for $\langle \mathcal{C}, \mathcal{W} \rangle$. Then there is exactly one connected component in $\mathcal{M}$.*

Lemma 13. *Let $\langle \mathcal{C}, \mathcal{W} \rangle$ be linear for the matrix $\mathcal{M}$ and have the relevance and cardinality properties for $\mathcal{M} = \Phi^E : \Phi^M \Downarrow \psi_1^E$. Let σ be a grounded prefix substitution which unifies $\langle \mathcal{C}, \mathcal{W} \rangle$. Then for any $p \in LeafPos(\mathcal{M})$ there is a string s such that $\sigma(pre_{\mathcal{M}}(p)) = \psi_1^E.s$ holds.*

Corollary 12 and lemma 13 show how the prefix substitution guarantees a proper context management. Due to the definition of prefix substitutions the multi-set Φ^M in lemma 13 must be empty. This ensures that the rule ψ^E is applicable in the constructed position calculus proof.

Theorem 14 (Correctness). *If a matrix $\mathcal{M}$ is complementary for a set $\mathcal{C}$ of connections, a weakening map $\mathcal{W}$, and a substitution σ then there is a position calculus proof $\mathcal{P}$ for $\mathcal{M}$.*

Proof. Define the *weight* of a matrix $\mathcal{M}$ by weighing the number of positions in $\mathcal{M}$ and use well-founded induction wrt. the weight of matrices. In the induction step we perform a complete case analysis of the structure of $\mathcal{M}$. The difficult cases where β or ψ^E must be applied is shown with the help of corollary 12 and lemma 13. The remaining proof is tedious but straightforward. Details can be found in [17].

Completeness. For a Σ_{pos}-proof $\mathcal{P}$ of a matrix $\mathcal{M}$ we construct a set *ConSet($\mathcal{P}$)* of connections, a weakening map *WeakMap($\mathcal{P}$)*, and a relation $\sqsubset_{\mathcal{P}} \subseteq SpecPos(\mathcal{M})^2$. The connections in *ConSet($\mathcal{P}$)* are constructed from applications of *axiom* and τ in $\mathcal{P}$. If *axiom* is applied on $\Phi^E : \phi_1^M, \phi_2^M \Uparrow \cdot$ then $\{succ_1(\phi_1^M), succ_1(\phi_2^M)\}$ is in *ConSet($\mathcal{P}$)*. If τ is applied on a sequent $\Phi^E : \phi_1^M \Uparrow \cdot$ then $\{succ_1(\phi_1^M)\}$ is in *ConSet($\mathcal{P}$)*. *WeakMap($\mathcal{P}$)* contains those elements of *WeakPos($\mathcal{M}$)* which are explicitly weakened by ω or ν and those elements from $\Phi^E(\mathcal{M})$ which are implicitly weakened in *axiom* or τ. $\sqsubset_{\mathcal{P}}$ resembles the order in which special positions are reduced. We write $p \sqsubset_{\mathcal{P}} p'$ if p is reduced before p', i.e. the reduction occurs closer to the root of the proof tree $\mathcal{P}$. For any proof $\mathcal{P}$, $\sqsubset_{\mathcal{P}}$ is irreflexive, antisymmetric, and transitive, thus an ordering and $\ll$ (for $\mathcal{M}$) is a subordering of $\sqsubset_{\mathcal{P}}$.

Lemma 15. *Let $\mathcal{M}$ be a matrix, $\mathcal{P}$ be a position calculus proof for $\mathcal{M}$, and $p_1, p_2 \in (\Psi^M(\mathcal{M}) \cup \Psi^E(\mathcal{M}))$ be positions in $\mathcal{M}$ with $p_1 \neq p_2$. If there is a position $p \in SpecPos(\mathcal{M})$ with $p_1 \sqsubset_{\mathcal{P}} p$ and $p_2 \sqsubset_{\mathcal{P}} p$ then either $p_1 \sqsubset_{\mathcal{P}} p_2$ or $p_2 \sqsubset_{\mathcal{P}} p_1$ holds.*

We finally construct the substitution $\sigma_{\mathcal{P}}$ from $\sqsubset_{\mathcal{P}}$. Let $\phi \in (\Phi^M(\mathcal{M}) \cup \Phi^E(\mathcal{M}))$ be a variable special position in $\mathcal{M}$. We define $\sigma_{\mathcal{P}}(\phi) = Z = \psi_1 \dots \psi_n$ where the string $Z \in (\Psi^M(\mathcal{M}) \cup \Psi^E(\mathcal{M}))^*$ has the following properties.

- *sortedness:* $\psi_i \sqsubset_{\mathcal{P}} \psi_{i+1}$ holds for all $i \in \{1, \dots, n-1\}$.
- *prior reduction:* $\psi_i \sqsubset_{\mathcal{P}} \phi$ holds for all $i \in \{1, \dots, n\}$.
- *exclusivity:* $p \ll \phi \Rightarrow p \sqsubset_{\mathcal{P}} \psi_1$ holds for all $p \in SpecPos(\mathcal{M})$.
- *maximality:* For any $\psi \in \Psi^M(\mathcal{M}) \cup \Psi^E(\mathcal{M})$ not in Z with $\psi \sqsubset_{\mathcal{P}} \phi$ exists a $p \ll \phi$ with $\psi \sqsubset_{\mathcal{P}} p$.

According to the structure of Σ_{pos}-proofs, $\sigma_{\mathcal{P}}$ really is a prefix substitution. The construction guarantees that for any $\phi^M \in \Phi^M$ holds $\sigma_{\mathcal{P}} \in (\Phi^M(\mathcal{M}) \cup \Psi^M(\mathcal{M}))^*$.

Theorem 16 (Completeness). *Let $\mathcal{P}$ be a position calculus proof for a matrix $\mathcal{M}$. Then $\mathcal{M}$ is complementary for ConSet($\mathcal{P}$), WeakMap($\mathcal{P}$), and $\sigma_{\mathcal{P}}$.*
Proof. Each of the following properties is proven by induction over the structure of $\mathcal{P}$. (1) *ConSet($\mathcal{P}$)* spans $\mathcal{M}$. (2) $\sigma_{\mathcal{P}}$ is a *unifier* for $\langle$*ConSet($\mathcal{P}$)*, *WeakMap($\mathcal{P}$)*$\rangle$. (3) $\langle$*ConSet($\mathcal{P}$)*, *WeakMap($\mathcal{P}$)*$\rangle$ is linear, has the relevance property for $\mathcal{M}$, and has the cardinality property for $\mathcal{M}$. The individual proofs are lengthy because induction over $\mathcal{P}$ and case analysis is necessary. Details can be found in [17].

The Characterization. The characterization theorem proven in this section is the foundation for matrix based proof search methods. It yields a compactified representation of the search space which can be exploited by proof search methods in the same way as for other logics [20]. The method has been extended uniformly to multiplicative linear logic, as shown in [15]. Along the same lines an extension to $\mathcal{MELL}$ is possible.

Theorem 17 (Characterization Theorem). *A formula φ is valid in $\mathcal{MELL}$ if and only if the corresponding matrix is complementary for some multiplicity.*
Proof. Correctness follows from theorems 14, 9, 5, 3, and the correctness of Σ_1'. Completeness follows from theorems 16, 8, 4, 2, and the completeness of Σ_1'.

Example 18. Let $\mathcal{M}$ be the matrix for $\varphi = \langle (A \otimes !A) \parr ?(A^\perp), + \rangle$ from figure 4. We choose $\mathcal{C} = \{\{a_{00000000}, a_{00010^20000}\}, \{a_{000001000}, a_{00010^10000}\}\}$, $\mathcal{W} = \emptyset$, and

$$\sigma = \{\phi^M_{0000} \backslash \varepsilon,\ \phi^M_{0000000} \backslash \psi^M_{00010^200},\ \phi^M_{00000100} \backslash \psi^M_{00010^100},\ \phi^E_{00010^1} \backslash \psi^E_{0000010},\ \phi^M_{00010^1000} \backslash \varepsilon,\ \phi^E_{00010^2} \backslash \psi^M_{000000},\ \phi^M_{00010^2000} \backslash \varepsilon\}\ .$$

Then $\mathcal{M}$ is complementary for $\mathcal{C}$, $\mathcal{W}$, and σ. Consequently φ is valid in $\mathcal{MELL}$.

6 Conclusion

We have presented a matrix characterization of logical validity for the full multiplicative exponential fragment of linear logic ($\mathcal{MELL}$). It extends our characterization for $\mathcal{MLL}$ [15] by the exponentials ? and ! and the multiplicative constants **1** and $\perp$. Our extension, as pointed out in [8], is by no means trivial and goes beyond all existing matrix characterizations for fragments of linear logic.

In the process we have also outlined a methodology for developing matrix characterizations from sequent calculi and for proving them correct and complete. It introduces a series of intermediate calculi, which step-wisely remove redundancies from sequent proofs while capturing their essential parts, and arrives at a matrix characterization as the most compact representation for proof search.

If applied to modal or intuitionistic logics, this methodology would essentially lead to Wallen's matrix characterization [24]. In order to capture the resource sensitivity of linear logic, however, we have introduced several refinements. The notion of multiplicities is based on positions instead of basic positions. Different types of special positions are used. The novel concept of weakening maps makes us able to deal with the aspects of resource management.

Fronhöfer has developed matrix characterizations for various variations of the multiplicative fragment [8]. Compared to his work for linear logic our characterization captures additionally the multiplicative constants and the controlled application of weakening and contraction. In fact, we are confident that our methodology will extend to further fragments of linear logic as well as to other resource sensitive logics, such as affine or relevant logics.

In the future we plan to extend our characterization to quantifiers, which again is a non-trivial problem although much is known about them in other logics. Furthermore, the development of *matrix systems* as a general theory of matrix characterizations has become possible. These systems would include a uniform framework for defining notions of complementarity and a methodology for supporting the proof of characterization theorems. Matrix systems might also enable us to integrate induction into connection-based theorem proving.

Matrix characterizations are known to be a foundation for efficient proof search procedures for classical, modal and intuitionistic logics [20] and $\mathcal{MLL}$ [15]. We expect that these proof procedures can now be extended to $\mathcal{MELL}$ and a wide spectrum of other logics, as soon as our methodology has led us to a matrix characterization for them.

Acknowledgements. We would like to thank Serge Autexier for his patience and his valuable comments while we were discussing the details of this work.

References

1. J.-M. Andreoli. Logic programming with focusing proofs in linear logic. *Journal of Logic and Computation*, 2(3):297–347, 1992.
2. P. ANDREWS. Theorem-Proving via General Matings. *Jour. of the ACM* 28 (2), 193–214, 1981.
3. W. Bibel. On matrices with connections. *Jour. of the ACM* 28, 633–645, 1981.
4. W. Bibel. A deductive solution for plan generation. *New Generation Computing* 4:115–132, 1986.
5. W. Bibel. *Automated Theorem Proving.* Vieweg, 1987.
6. I. Cervesato, J.S. Hodas, F. Pfenning. Efficient resource management for linear logic proof search. In *Extensions of Logic Programming*, LNAI 1050, pages 67–81. Springer, 1996.
7. V. Danos & L. Regnier. The structure of the multiplicatives. *Arch. Math. Logic* 28:181–203, 1989.
8. B. Fronhöfer. *The action-as-implication paradigm.* CS Press, 1996.
9. D. Galmiche. Connection methods in linear logic fragments and proof nets. Technical report, CADE–13 workshop on proof search in type-theoretic languages, 1996.
10. D. Galmiche & G. Perrier. On proof normalization in linear logic. *TCS*, 135:67–110, 1994.
11. V. Gehlot and C. Gunter. Normal process representatives. *Sixth Annual Symposium on Logic in Computer Science*, pages 200–207, 1991.
12. J.-Y. Girard. Linear logic. *TCS*, 50:1–102, 1987.
13. J. Harland and D. Pym. Resource-Distribution via Boolean Constraints. *14th Conference on Automated Deduction*, LNCS 1249, pp. 222–236. Springer, 1997.
14. J.S. Hodas & D. Miller. Logic programming in a fragment of linear logic. *Journal of Information and Computation*, 110(2):327–365, 1994.
15. C. Kreitz, H. Mantel, J. Otten, S. Schmitt. Connection-Based Proof Construction in Linear Logic. *14th Conference on Automated Deduction*, LNCS 1249, pp. 207–221. Springer, 1997.
16. P. Lincoln and T. Winkler. Constant-only multiplicative linear logic is NP-complete. *TCS*, 135:155-169, 1994.
17. H. Mantel. Developing a Matrix Characterization for $\mathcal{MELL}$. Technical Report, DFKI Saarbrücken, 1998. `http://www.dfki.de/vse/staff/mantel/Papers/98tr-dev-mc-mell.ps.gz`
18. M. Masseron, C. Tollu, J. Vauzeilles. Generating plans in linear logic. In *Foundations of Software Technology and Theoretical Computer Science*, LNCS, Springer,1991.
19. D. Miller. FORUM: A Multiple-Conclusion Specification Logic. *TCS*, 165(1):201-232, 1996.
20. J. Otten & C. Kreitz. A Uniform Proof Procedure for Classical and Non-classical Logics. *KI-96: Advances in Artificial Intelligence*, LNAI 1137, pp. 307–319. Springer, 1996.
21. S. SCHMITT, C. KREITZ. Converting non-classical matrix proofs into sequent-style systems. *CADE–13*, LNAI 1104, pp. 418–432, Springer, 1996.
22. S. SCHMITT, C. KREITZ. A uniform procedure for converting non-classical matrix proofs into sequent-style systems. *Journal of Information and Computation*, submitted.
23. T. Tammet. Proof strategies in linear logic. *Jour. of Automated Reasoning*, 12:273–304, 1994.
24. L. Wallen. *Automated deduction in nonclassical logic.* MIT Press, 1990.

A Resolution Calculus for Dynamic Semantics

Christof Monz and Maarten de Rijke

ILLC, University of Amsterdam, The Netherlands
{christof, mdr}@wins.uva.nl

Abstract. This paper applies resolution theorem proving to natural language semantics. The aim is to circumvent the computational complexity triggered by natural language ambiguities like pronoun binding, by interleaving pronoun binding with resolution deduction. To this end, disambiguation is only applied to expressions that actually occur during derivations. Given a set of premises and a conclusion, our resolution method only delivers pronoun bindings that are needed to derive the conclusion.

1 Introduction

Natural language processing (NLP), has a long tradition in Artificial Intelligence, but it still remains to be one of the hardest problems in the area. Research areas such as semantic representation and theorem proving with natural language have to deal with a problem that is characteristic of natural languages, namely ambiguity. There are several kinds of ambiguity, see for instance [RN95] for an overview. In the present paper, we focus on pronoun binding,[1] a certain instance of ambiguity, as exemplified by (1) below.

(1) A man sees a boy. He whistles.

Often, there are lots of possibilities to bind a pronoun and it is not clear which one to choose. The pronoun *he* in the short discourse in (1) can be bound in two ways as given in (2), where co-indexation indicates referential identity.

(2) a. A man_i sees a boy. He_i whistles.
 b. A man sees a boy_i. He_i whistles.

For some cases heuristics are applicable which prefer certain bindings to others, but at present there is no approach making use of heuristics which is general enough to cover all problems.

Dynamic semantics [Kam81,GS91] allows to give a perspicuous solution to some problems involving pronoun binding. Since we are interested in binding occurrences of pronouns to expressions mentioned earlier in a discourse, we take

[1] Throughout this paper we use the term *binding* to express the referential identification of a pronoun and another referential expression occurring in the discourse. Common terms are also *co-indexation* or *pronoun resolution*. We especially did not use *pronoun resolution* to avoid confusion with resolution as a deduction principle.

a slight modification of Dynamic Predicate Logic (DPL) [GS91], where it is not presupposed that pronouns are already co-indexed. Actually, pronoun binding falls into the realm of constructing semantic representations of natural language discourses, and one of the main purposes of constructing these representations is to reason with them. Now, the question arises which form the input of the theorem prover should have. Should a theorem prover work only on totally disambiguated expressions? Total disambiguation results in an explosion of readings, because of the multiplicative behavior of ambiguity. On the other hand, to prove a conclusion φ from a set of premises Γ it may be enough to use only premises from a small subset Δ of Γ, and it may be sufficient, and much more efficient, to disambiguate only Δ instead of the whole set of premises Γ. In general, we do not know in advance which subset of premises might be enough to derive a certain conclusion, but during a derivation often certain (safe) strategies may be applied that prevent some premises from being used since they cannot lead to the conclusion, anyway. Common strategies to constrain the search space in resolution deduction are e.g., the *set-of-support strategy* and *ordered resolution.* Our goal is to constrain the set of premises that have to be disambiguated by interleaving deduction and disambiguation. Roughly speaking, premises are only disambiguated if they are used by a deduction rule.

The rest of the paper is structured as follows. Section 2 provides some rudimentary background in dynamic semantics and explains what kind of structural information is necessary to restrict pronoun binding. In addition, the basics of resolution deduction are introduced. Section 3 discusses some of the problems of the (standard) resolution method when applied to natural language. The method of labeled unification and resolution is presented to overcome these problems. Section 4 briefly relates our work to some other approaches to pronoun binding. Section 5 provides some conclusions and prospects for further work.

2 Background

Before we turn to our method of labeled resolution deduction and its applications to discourse semantics, we briefly present the idea of dynamic semantics. The second subsection shortly explains the classic resolution method for (static) first-order logic.

2.1 Dynamic Reasoning

Dynamic reasoning differs from classical reasoning to the extent that sequences of formulas are considered instead of sets of formulas. To model discourse relations like pronoun binding it is important to take the order of sentences into account because two sequences which have the same members, but differ in order, may have a different meaning. (Compare 'A man walks in the park. He whistles.' and 'He whistles. A man walks in the park.')

DPL is a semantic framework which works on sequences of formulas and it allows to represent pronoun binding, where the antecedent of the pronoun and the pronoun itself may occur in different formulas. This is accomplished by assigning the existential quantifier flexible binding. In (3.b) a DPL representation of the short discourse in (3.a) is given.

(3) a. A man$_i$ sees a boy. He$_i$ whistles.
b. $\exists x\,(man(x) \wedge \exists y\,(boy(y) \wedge see(x,y)))\ \wedge\ whistle(x)$

The pronoun *he* is represented by the variable x which is the same as the one bound by the existential quantifier, but it occurs outside of its scope. To bind x in $whistle(x)$ it is necessary to give the existential quantifier flexible scope.

One of the advantages of dynamic approaches like DPL is that they allow for a formal definition of possible antecedents for a pronoun. Without giving too many details, we just note that negations function as barriers for flexible binding. Therefore, an existential quantifier occurring in the scope of a negation cannot bind a pronoun that occurs outside of the negation, as shown by (4).

(4) *John doesn't own a car$_i$. It$_i$ is in front of his house.

The three properties (a) existential quantifiers can bind variables occurring to the right-hand side of their traditional scope, (b) conjunctions preserve the flexible scope, and (c) negations are barriers for dynamic binding, allow us to define the properties of the other logical connectives $\vee$, $\rightarrow$ and $\forall$. $[\![\cdot]\!]$ is a function that assigns to each formula its semantic value.

(5)
$$\begin{aligned}
[\![\varphi \vee \psi]\!] &= [\![\neg(\neg\varphi \wedge \neg\psi)]\!]\\
[\![\varphi \rightarrow \psi]\!] &= [\![\neg(\varphi \wedge \neg\psi)]\!]\\
[\![\forall x\,\varphi]\!] &= [\![\neg\exists x\,\neg\varphi]\!]
\end{aligned}$$

Given these definitions, we see that disjunction is a barrier both internally and externally, implication is a barrier externally but internally it allows for flexible binding, and universal quantification does not allow for external binding.

We differ in two respects from DPL. First, we do not allow two or more occurrences of $\exists x$ within a single text. The problem is that the second occurrence of $\exists x$ resets the value of x, and thereby previous restrictions on x are lost. We assume for simplicity that all bound variables are disjoint. This is not a severe restriction and an algorithm for constructing semantic representations for natural language sentences can easily accomplish this. The second difference with DPL is that we do not assume co-indexation of quantifiers and the pronouns which they bind. In (3) the variable for *he* is already assumed to be x and in DPL the question of pronoun binding is pushed to some kind of preprocessing. But finding the right binding is far from being an easy task and it is very complex from a computational point of view. The pronoun in (3) could also be represented by y, indicating that that *he* refers to *a boy*. E.g., a discourse containing twenty indefinites followed by a sentence with two pronouns, has $20 \cdot 20 = 400$ possible bindings, disregarding any linguistic constraints which rule out some of the bindings.

To this end, we postpone pronoun binding and represent pronouns in the semantic representation by free variables. Variables for pronouns are displayed in boldface and are of a different kind than regular variables. Pronoun variables are bound by the ?-operator. It differs from $\exists$ and $\forall$, because it only binds its argument, but does not quantify over it. Actually, it is not necessary to have a special operator for pronouns, and we only introduced it here for the sake of convenience to identify the position where the pronoun is introduced. Our representation of (1), repeated as (6.a) below, is given in (6.b). As mentioned before, co-indexation of pronouns and antecedents is not carried out.

(6) a. A man sees a boy. He whistles.
  b. $\exists x\,(man(x) \wedge \exists y\,(boy(y) \wedge see(x,y))) \wedge ?\mathbf{u}\, whistle(\mathbf{u})$

The task whether $\mathbf{u}$ has to be substituted by x or by y is postponed to the deduction component, as motivated in Section 1.

Unlike the existential quantifier, the ?-operator does not have the property of flexible binding. We get the following equivalence:

$$[\![\neg ?\mathbf{u}\varphi]\!] \;=\; [\![?\mathbf{u}\neg\varphi]\!]$$

To define *accessibility* we can now say that a variable x is accessible from a pronoun $\mathbf{u}$ if no barrier occurs between the quantifier introducing x and $?\mathbf{u}$. A formal definition of accessibility is given in the next section. The equations in (5) show that $\vee$, $\rightarrow$ and $\forall$ introduce barriers because of the way they are defined in terms of negation. This is exemplified by (7) below.

(7) *Every farmer owns a donkey$_i$. It$_i$ is grey.

Dispensing with the presupposition that pronouns and antecedents are already co-indexed re-introduces the concept of ambiguity to our framework. This makes it necessary to give a definition of the semantics of ambiguous formulas. It is common to define their semantics in terms of their possible disambiguations, see [Rey93], and here we follow the same approach. A *total disambiguation* is a mapping δ from ambiguous dynamic formulas to classical first-order formulas. Disambiguation encompasses two steps. First, we have to find a proper antecedent for a pronoun. To define proper antecedents, we use the notion of accessibility. Second, we have to map unambiguous dynamic formulas to classical formulas. This means that we have to turn flexible quantification into static quantification, and this involves re-bracketing and quantifier movement. [GS91] give an algorithm that computes for each DPL-formula φ a formula φ' which is in normal binding form, i.e., all pronouns are quantified over in the classical sense, and which is valid in first-order logic iff φ is valid in DPL. For instance, the normal binding form of (8.b) is (9).

(8) a. If a farmer$_i$ owns a donkey$_j$, then he$_i$ beats it$_j$.
  b. $\exists x\,(f(x) \wedge \exists y\,(d(y) \wedge o(x,y))) \rightarrow b(x,y)$
(9) $\forall x\,\forall y\,(f(x) \wedge d(y) \wedge o(x,y) \rightarrow b(x,y))$

To define the validity of ambiguous formulas, we say that an ambiguous formula φ is *valid*, i.e., for all models M it holds that $M \models_a \varphi$, if there is a disambiguation δ, such that $M \models \delta(\varphi)$, for all models M. In words: φ is valid iff there exists a disambiguation which is valid in first-order logic.

Unfortunately we do not have enough space to give a more detailed account of dynamic semantics, but we refer the reader to [Kam81,GS91].

2.2 The Resolution Method

The *resolution method* [Rob65] has become quite popular in automated theorem proving, because it is very efficient and it is easily augmentable by lots of strategies which restrict the search space, see e.g., [Lov78]. On the other hand, the resolution method has the disadvantage of presupposing that its input has to be in *clause form*, which is a set of clauses, interpreted as a conjunction. A *clause* is a set of literals, interpreted as a disjunction. Probably the most attractive feature of resolution is that it has only one inference rule, the resolution rule:

$$\frac{C \cup \{\neg \mathrm{P}_1, \ldots, \neg \mathrm{P}_n\} \quad D \cup \{\mathrm{Q}_1, \ldots, \mathrm{Q}_m\}}{(C \cup D\pi)\sigma} \ (res)$$

where
- $\mathrm{Q}_1, \ldots, \mathrm{Q}_m$ are atomic
- π is a substitution such that $C \cup \{\neg \mathrm{P}_1, \ldots, \neg \mathrm{P}_n\}$ and $D\pi \cup \{\mathrm{Q}_1\pi, \ldots, \mathrm{Q}_m\pi\}$ are variable disjoint
- σ is the most general unifier of $\{\mathrm{P}_1, \ldots, \mathrm{P}_n, \mathrm{Q}_1\pi, \ldots, \mathrm{Q}_m\pi\}$

To prove that $\Gamma \models \varphi$ holds we transform $(\bigwedge \Gamma) \wedge \neg\varphi$ in clause form and try to derive a contradiction (the empty clause) from it by using the resolution rule.

For a comprehensive introduction to resolution see for instance [Lov78].

3 Dynamic Resolution

Applying the classical resolution method to a dynamic semantics causes problems. Below we will first discuss some of them and then see how we have to design our dynamic resolution method to overcome these problems.

3.1 Adapting the Resolution Method

There are two problems that we have to find a solution for. First, transforming formulas to clause form causes a loss of structural information. Therefore, it is sometimes impossible to distinguish between variables that can serve as antecedents for a pronoun and variables than can not. The second problem concerns the duplication of literals which may occur during clause from transformation and the assumption of the resolution method that clauses are variable disjoint. Although the same pronoun may have two occurrences in different clauses, we do not want them to be bound by different antecedents.

Turning to the first problem, in (10) the pronoun **u** cannot be bound by the existential quantifier, whereas the pronoun **z** can be bound by it.

(10)a. Every farmer who owns a donkey beats it. It suffers.

b. $\forall x\,(f(x) \wedge \exists y\,(d(y) \wedge o(x,y)) \rightarrow ?\mathbf{z}\,b(x,\mathbf{z})))\wedge ?\mathbf{u}\,s(\mathbf{u})$

(11) $\{\ \{\neg f(x), \neg d(y), \neg o(x,y), b(x,\mathbf{z})\},\ \{s(\mathbf{u})\}\ \}$

How can we tell which identifications are allowed by looking at the corresponding clause form in (11)? How do we know whether a term is accessible?

We use *labels* to carry the information about accessible variables. Each pronoun variable is annotated with a label that indicates the set of accessible variables. Besides the set of first-order or proper variables (*VAR*), first-order formulas (*FORM*), and pronoun variables (*PVAR*), we are going to introduce the sets of labeled pronoun variables (*LPVAR*) and labeled formulas (*LFORM*). Labeled pronoun variables are of the form $V:\mathbf{u}$, where $V \subseteq VAR$ and $\mathbf{u}$ is a pronoun variable. *LFORM* is the set of first-order formulas plus formulas containing labeled pronoun variables. To be able to recognize the antecedents later on, each variable is annotated with its name, $(x^x, y^y, \ldots)$, and during skolemization only the variable is changed, but the label remains unchanged.

To see which variables inside of a formula φ can serve as antecedents for pronouns, [GS91] introduce the function AQV which returns the set of *actively quantifying variables* when applied to φ.

Definition 1. *Let FORM be the set of classical first-order formulas and VAR the set of first-order variables. The function* AQV : *FORM* $\rightarrow$ *POW*(*VAR*) *is defined recursively:*

$$\begin{aligned}
\mathsf{AQV}(R(x_1 \ldots x_n)) &= \emptyset \\
\mathsf{AQV}(\neg\varphi) &= \emptyset \\
\mathsf{AQV}(\varphi \wedge \psi) &= \mathsf{AQV}(\varphi) \cup \mathsf{AQV}(\psi) \\
\mathsf{AQV}(\varphi \rightarrow \psi) &= \emptyset \\
\mathsf{AQV}(\varphi \vee \psi) &= \emptyset \\
\mathsf{AQV}(\forall x\,\varphi) &= \emptyset \\
\mathsf{AQV}(\exists x\,\varphi) &= \mathsf{AQV}(\varphi) \cup \{x\} \\
\mathsf{AQV}(?\mathbf{u}\,\varphi) &= \mathsf{AQV}(\varphi)
\end{aligned}$$

Using the above definition we define the notion of accessible variables.

Definition 2 (Annotation with Accessible Variables). *To annotate* $\mathbf{u}$ *in* $?\mathbf{u}\,\psi$, *we drop the binding operator* $?\mathbf{u}$ *and substitute all occurrences of the pronoun variable in* ψ *by its annotated counterpart. The annotation function* annot : *VAR* $\times$ *FORM* $\rightarrow$ *LFORM is defined recursively, where* $V \subseteq VAR$*:*

$$\begin{aligned}
\mathsf{annot}(V, R(x_1 \ldots x_n)) &= R(x_1 \ldots x_n) \\
\mathsf{annot}(V, \neg\varphi) &= \neg\mathsf{annot}(V, \varphi) \\
\mathsf{annot}(V, \varphi \wedge \psi) &= \mathsf{annot}(V, \varphi) \wedge \mathsf{annot}(V \cup \mathsf{AQV}(\varphi), \psi) \\
\mathsf{annot}(V, \varphi \rightarrow \psi) &= \mathsf{annot}(V, \varphi) \rightarrow \mathsf{annot}(V \cup \mathsf{AQV}(\varphi), \psi) \\
\mathsf{annot}(V, \varphi \vee \psi) &= \mathsf{annot}(V, \varphi) \vee \mathsf{annot}(V, \psi) \\
\mathsf{annot}(V, \forall x\varphi) &= \forall x\ \mathsf{annot}(V \cup \{x\}, \varphi) \\
\mathsf{annot}(V, \exists x\varphi) &= \exists x\ \mathsf{annot}(V \cup \{x\}, \varphi) \\
\mathsf{annot}(V, ?\mathbf{u}\varphi) &= \mathsf{annot}(V, \varphi[\mathbf{u}/V:\mathbf{u}])
\end{aligned}$$

The actual annotation takes place in the last case, where the pronoun is substituted. The other cases thread the actively quantifying variables through the formula. To annotate a whole discourse $\varphi_1 \wedge \cdots \wedge \varphi_n$, the variable parameter of annot is initialized with $\emptyset$, $\mathsf{annot}(\emptyset, \varphi_1 \wedge \cdots \wedge \varphi_n)$. A term t^x is *accessible from* a pronoun $\mathbf{u}$ iff x is element of the set of the accessible variables of $\mathbf{u}$.

Reconsider the last example, *every farmer who owns a donkey beats it. It suffers.* Applying annotation yields:[2]

$$\begin{aligned}&\mathsf{annot}(\emptyset, \forall x\,(f(x) \wedge \exists y\,(d(y) \wedge o(x,y)) \rightarrow ?\mathbf{z}\,b(x,\mathbf{z})) \wedge ?\mathbf{u}\,s(\mathbf{u}))\\ &= \forall x\,(f(x) \wedge \exists y\,(d(y) \wedge o(x,y)) \rightarrow b(x, \{x,y\}{:}\mathbf{z}))) \wedge s(\emptyset{:}\mathbf{u})\end{aligned}$$

Applying clause form transformation to the annotated formulas yields:

(12) $\{\,\{\neg f(x), \neg d(y), \neg o(x,y), b(x, \{x,y\}{:}\mathbf{z})\}, \{s(\emptyset{:}\mathbf{u})\}\,\}$

We can also see that (10.a) is not well-formed because there are no accessible pronouns for the second pronoun *it*, i.e., the label of $\mathbf{u}$ is the empty set.

Now we turn to the second problem: how do we make sure that the same pronoun, occurring in different clauses, is bound to the same antecedent? As we said earlier, we do not want to assume pronouns to be bound in a set of premises when we apply resolution. The reason is that pronoun binding is highly ambiguous and often it is not necessary to bind all pronouns in a set of premises to derive a certain conclusion from it. Another issue, which we briefly hinted at in Section 2, is that pronouns should be treated as free variables of a special kind, not to be dealt with in the same manner as universally quantified variables (which also happen to be represented by free variables). This is illustrated by the following example, which shows an invalid entailment.

(13)a. $\exists x\,\exists y\,((A(x) \vee A(y)) \wedge (?\mathbf{z}A(\mathbf{z}) \rightarrow (B \wedge C))) \not\models_a B \vee C$
b. $\{\,\{A(f^x), A(g^y)\}, \{\neg A(\mathbf{z}), B\}, \{\neg A(\mathbf{z}), C\}, \{\neg B\}, \{\neg C\}\,\}$

The transformation in (13) causes a duplication of the literal $\neg A(\mathbf{z})$, and we have to make sure that the pronoun is instantiated the same way in both cases.

(14)
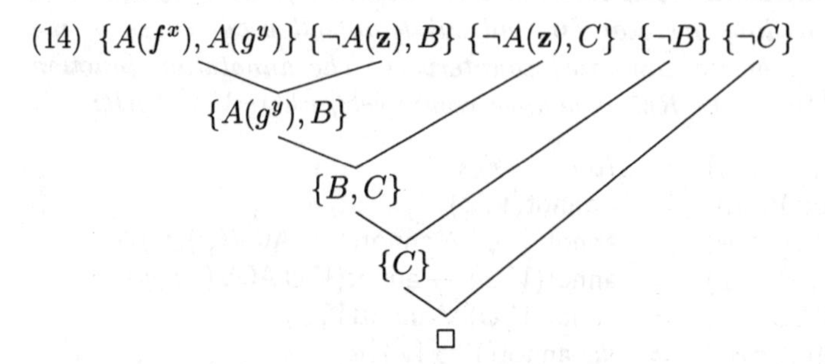

[2] For simplicity, we neglect the fact that pronouns and their antecedents have to agree in gender, number, etc.

In (14) $\mathbf{z}$ is instantiated with f^x in the first resolution step and then with g^y in the second. The resolution rule as it was stated in the preceding section assumes that clauses to be resolved are variable disjoint. We have to modify the resolution rule such that the same pronoun variable is allowed to occur in both clauses. Additionally, the instantiation of a pronoun variable for constructing the most general unifier in a resolution step is applied globally, i.e., to all clauses.

(15) $\{A(f^x), A(g^y)\}\ \{\neg A(\mathbf{z}), B\}\ \{\neg A(\mathbf{z}), C\}\ \ \{\neg B\}\ \{\neg C\}$

$\{A(g^y), B\}\quad \{A(f^x), C\}$

$\{A(g^y)\}\quad \{A(f^x)\}$

Global instantiation correctly prevents us from deriving a contradiction in (15).

3.2 Labeled Resolution

Unification is a fundamental technique in the resolution method. Since we are also dealing with labeled variables, we have to think how the unification mechanism has to be adapted. In the course of this subsection, it will turn out that pronoun binding can be reduced to unification.

Labeled Unification. We use the unification algorithm of Martelli and Montanari [MM82] as a basis and adapt it in such a way that it can deal with labeled pronoun variables.

What does it mean to unify a set of equations $E = \{s_1 \approx t_1, \ldots, s_n \approx t_n\}$, where s_i or t_i can also be a labeled pronoun variable? We have to distinguish three possible cases: (i) neither s_i nor t_i is a labeled pronoun variable, then labeled unification and normal unification are the same thing, (ii) one of them is a pronoun and the other is not, and (iii) both are pronouns. Case (ii) is the normal pronoun binding, where one tries to identify a pronoun with a proper variable. Case (iii) is not an instance of pronoun binding, but an identification of two pronouns, i.e., whatever is the antecedent of the first pronoun, it is also the antecedent of the other one.

Definition 3 (Labeled Unifier). *We call a substitution σ a* labeled unifier *or* unifier* *of a set of equations $E = \{s_1 \approx t_1, \ldots, s_n \approx t_n\}$ iff*

1. $s_1\sigma = t_1\sigma, \ldots, s_n\sigma = t_n\sigma$
2. *if* $(V\!:\!\mathbf{u})\sigma = t^x$, *then* $x \in V$
3. *if* $(V\!:\!\mathbf{u})\sigma = V'\!:\!\mathbf{v}$ *then* $V' \subseteq V$

We use $\approx$ to express equality in our object language, whereas $=$ denotes equality in the meta language.

Condition 1 is the normal condition of unifiability, namely that the terms of an equation have to be identical after substitution. The second condition says that unifiers have to obey accessibility, for instance $\sigma := [\{x, y\}\!:\!\mathbf{u}/g^z]$ is not a unifier of $\{\{x,y\}:\mathbf{u} \approx g^z\}$, because g^z is not accessible from $\mathbf{u}$, as $z \notin \{x,y\}$. To ensure that identification of pronouns always restricts the set of accessible antecedents, we need condition 3.

Definition 4 (Most General Labeled Unifier). *A labeled unifier σ of a set of equations $E = \{s_1 \approx t_1, \ldots, s_n \approx t_n\}$ is the* most general labeled unifier *or* mgu* *of E if*

1. *if θ is a unifier* of E then there is substitution τ such that $\theta = \sigma\tau$*
2. *if $(V:\mathbf{u})\sigma = V_1:\mathbf{v}$, $(V:\mathbf{u})\theta = V_2:\mathbf{v}$, $V_1, V_2 \subseteq V$, and $V_1, V_2 \neq \emptyset$ then $V_2 \subseteq V_1$*

Again, the first condition is standard in regular unification. Condition 2 says that the most general unifier* has to restrict the set of accessible antecedents as little as possible when identifying pronouns. To unify $V_1:\mathbf{u}$ and $V_2:\mathbf{v}$ it suffices to take any non-empty subset of the intersection of V_1 and V_2, but this fact may prohibit some antecedents from being accessible, although they are in fact accessible for both pronouns.

Definition 5 (The Labeled Unification Algorithm). *First, the unification function* unify* *is applied to a pair of atoms, and then it tries to unify the set of corresponding argument pairs. The algorithm terminates successfully if it did not terminate with failure and no further equations are applicable.*

1. $\mathsf{unify}^*(R(s_1...s_n), R(t_1..t_n))$
   $= \mathsf{unify}^*(\{s_1 \approx t_1...s_n \approx t_n\})$
2. $\mathsf{unify}^*(\{f(s_1...s_n) \approx f(t_1...t_n)\} \cup E)$
   $= \mathsf{unify}^*(\{s_1 \approx t_1...s_n \approx t_n\} \cup E)$
3. $\mathsf{unify}^*(\{f(s_1...s_n) \approx g(t_1...t_m)\} \cup E)$, $f \neq g$ *or* $n \neq m$
   $=$ *terminate with failure*
4. $\mathsf{unify}^*(\{x \approx x\} \cup E$
   $= \mathsf{unify}^*(E)$
5. $\mathsf{unify}^*(\{t \approx x\} \cup E)$, $t \notin VAR$
   $= \mathsf{unify}^*(\{x \approx t\} \cup E)$
6. $\mathsf{unify}^*(\{x \approx t\} \cup E)$, $x \neq t$, $t \notin LPVAR$, x *in* t
   $=$ *terminate with failure*
7. $\mathsf{unify}^*(\{x \approx t\} \cup E)$, $x \neq t$, $t \notin LPVAR$, x *not in* t, x *in* E
   $= \mathsf{unify}^*(\{x \approx t\} \cup E[x/t])$
8. $\mathsf{unify}^*(\{V:\mathbf{u} \approx t^x\} \cup E)$, $x \in V$, $V:\mathbf{u}$ *in* E
   $= \mathsf{unify}^*(\{V:\mathbf{u} \approx t^x\} \cup E[V:\mathbf{u}/t^x])$
9. $\mathsf{unify}^*(\{V_1:\mathbf{u} \approx V_2:\mathbf{v}\} \cup E)$, $V_1 \cap V_2 \neq \emptyset$, $V_1 \cap V_2 \subset V_2$
   $= \mathsf{unify}^*(\{V_1:\mathbf{u} \approx V_1 \cap V_2:\mathbf{v}, V_2:\mathbf{v} \approx V_1 \cap V_2:\mathbf{v}\}$
   $\cup E[V_1:\mathbf{u}/V_1 \cap V_2:\mathbf{v}, V_2:\mathbf{v}/V_1 \cap V_2:\mathbf{v}])$

The first six equations of the algorithm are the same as in [MM82], except for additional side conditions which make sure that t is not a labeled variable. The interesting cases are 8 and 9. In 8 a pronoun is bound to an antecedent and in 9 two pronouns are identified, i.e., they have the same possible antecedents, namely those which are accessible for both of them. This is accomplished by identifying the pronoun variables and substituting the set of possible antecedents by the intersection of the possible antecedents of each pronoun.

Identification of pronouns underlies different constraints than binding a pronoun to a proper antecedent. To identify two pronouns **u** and **v**, it is not required that **u** is accessible from **v**, or the other way around. But they can only be identified if they have at least one proper accessible antecedent in common.

(16) Buk is a poet. For every man there is a woman who hates him.
$\models_a$ There is a woman who hates him.
(17) $p(b) \wedge \forall x(w(x) \rightarrow \exists y(w(y) \wedge ?\mathbf{u}\, h(y, \mathbf{u})))$
$\models_a \exists z(w(z) \wedge ?\mathbf{v}\, h(z, \mathbf{v}))$

For instance, in (16) the conclusion is only valid if the first and the second occurrence of *him* are identified. In Section 2 it was said that universal quantification is a barrier for flexible binding, and therefore the second occurrence of *him* cannot be bound to the first one. On the other hand, both of them have a proper antecedent in common, namely the constant b representing the proper name *Buk*. In addition, the first occurrence of *him* has the variable x as an accessible antecedent, introduced by the universal quantification *every man*. If one wants to identify them, one has to take the intersection of both sets of accessible antecedents and hence drop x as a possible antecedent. Observe that identification of pronouns still leaves some space for underspecification, because the intersection of two pronouns does not have to be a singleton. Of course, identifying two pronouns, where more than one antecedent is accessible for both, forces them to be bound to the same element of the intersection. Both can be bound to any element of the intersection, but it has to be the same one for both pronouns.

If the unification algorithm terminates successfully for a pair of literals P,Q, the solved set determines a substitution σ that is the mgu* of P,Q:

$$\sigma := \{s/t \mid s \approx t \in \mathsf{unify}^*(\mathrm{P}, \mathrm{Q})\}.$$

A set of equations $\{s_1 \approx t_1, \ldots, s_n \approx t_n\}$ is called *solved* if

1. $s_i \in \mathit{VAR} \cup \mathit{LPVAR}$ and the s_i are pairwise disjoint
2. no s_i occurs in a term t_j $(1 \leq i, j \leq n)$.

Lemma 1 (Correctness of the Unification* Algorithm). *Let E be a set of equations and* $\mathsf{unify}^*(E) = E'$, *then*

(i) E is unifiable iff E' is unifiable**
(ii) σ is the mgu of E iff σ is the mgu* of E'*

Proof. (i) We have to show that actions 2, 4, 5, 7, 8, and 9 preserve unifiability*, when unify* is applied to a unifiable* set E. For 2, 4, and 5, this is obvious. To show it for 7, note that $\tau := [x/t]$ is a unifier* of x and t. If σ is a unifier* of $\{x \approx t\} \cup E$ then σ is of the form $\tau\rho$. Because $\tau\tau = \tau$, it holds that $\sigma = \tau\rho = \tau\tau\rho = \tau\sigma$. Therefore σ unifies* $\{x \approx t\} \cup E$ iff σ unifies* $\{x \approx t\} \cup E[x/t]$. 8 is analogous to 7, plus the additional side condition that $x \in V$. The last case is

9. If $\{V_1 : \mathbf{u} \approx V_2 : \mathbf{v}\} \cup E$ is unifiable*, then it is with a unifier* σ of the form $\tau\rho$ with $\tau := [V_1 : \mathbf{u}/V_1 \cap V_2 : \mathbf{v}, V_2 : \mathbf{u}/V_1 \cap V_2 : \mathbf{v}]$.
Again, $\sigma = \tau\rho = \tau\tau\rho = \tau\sigma$ and then σ also unifies*
$\{V_1 : \mathbf{u} \approx V_1 \cap V_2 : \mathbf{v}, V_2 : \mathbf{v} \approx V_1 \cap V_2 : \mathbf{v}\} \cup E[V_1 : \mathbf{u}/V_1 \cap V_2 : \mathbf{v}, V_2 : \mathbf{v}/V_1 \cap V_2 : \mathbf{v}]$.

(ii) The actions 2, 4, 5, 7, and 8 turn a set of equations into an equivalent one. For σ to be the mgu* of $\{V_1 : \mathbf{u} \approx V_2 : \mathbf{v}\} \cup E$ means according to our definition that σ has to be of the form $\tau\rho$, where

$$\tau := [V_1 : \mathbf{u}/V_1 \cap V_2 : \mathbf{v}, V_2 : \mathbf{u}/V_1 \cap V_2 : \mathbf{v}].$$

But then σ is also the mgu* of

$$\{V_1 : \mathbf{u} \approx V_1 \cap V_2 : \mathbf{v}, V_2 : \mathbf{v} \approx V_1 \cap V_2 : \mathbf{v}\} \cup E[V_1 : \mathbf{u}/V_1 \cap V_2 : \mathbf{v}, V_2 : \mathbf{v}/V_1 \cap V_2 : \mathbf{v}].\square$$

Lemma 2 (Termination of the Unification* Algorithm). *The unification* algorithm terminates for each finite set of equations.*

Proof. If rules 3 and 6 are applied, we are done. Otherwise, rule 7 can be applied only once, because after application the side condition is no longer fulfilled. In 9 it is presupposed that $V_1 \cap V_2$ is a proper subset of V_2; this ensures that an application of 9 really reduces the set of possible antecedents. Because 9 can be applied only a finite number of times, it can reintroduce a term $V : \mathbf{u}$ only finitely often, therefore rule 8 can also be applied only finitely many times. Rules 1, 5, and 6 are only applied once, and the number of possible applications of rule 2 is finite as well, because terms contain only finitely many symbols. Therefore all rules can be applied only finitely many times, and termination follows. □

Proposition 1 (Total Correctness of the Unification* Algorithm). *The unification* algorithm computes for each finite set of equations E a solved set, that has the same mgu* as E in finitely many steps iff E is unifiable*.*

Proof. The fact that the unification* algorithm preserves unifiability* and that it terminates has been proven in Lemma 1 and 2, respectively. It remains to be shown that the set of equations computed by the algorithm is a solved set. In 7, 8, and 9, the left side of the equation is always substituted in E by the right side of the equation. If the left side is identical to the right side, the equation is erased by rule 4. Therefore, no left side of an equation occurs somewhere else. □

The Resolution Method. Having defined labeled unification, it is straightforward to adapt the resolution principle. The only thing we have to change is to make sure that variable disjointness applies only to proper variables (elements of *VAR*). The function *VAR* returns the set of proper variables, when applied to a set of clauses Δ : $VAR(\Delta) = \{x \in VAR \mid x \text{ occurs in } \Delta\}$. The resolution rule accomplishing pronoun binding (res_p) is defined as follows:

$$\frac{C \cup \{\neg \mathrm{P}_1, \ldots, \neg \mathrm{P}_n\} \quad D \cup \{\mathrm{Q}_1, \ldots, \mathrm{Q}_m\}}{(C \cup D\pi)\sigma} \; (res_p)$$

where
- $Q_1, \ldots, Q_m$ are atomic
- π is a substitution such that $VAR(C \cup \{\neg P_1, \ldots, \neg P_n\}) \cap (VAR(D \cup \{Q_1, \ldots, Q_m\}))\pi = \emptyset$
- σ is the mgu* of $\{P_1, \ldots, P_n, Q_1\pi, \ldots, Q_m\pi\}$

Definition 6 (The Proof Algorithm). *Our proof algorithm* prf *consists of three steps:*

1. *annotate the conjunction of the premises and the negation of the conclusion;*
2. *apply clause form transformation; and*
3. *apply the resolution rule until a contradiction can be derived, or no new resolvents can be generated.*

An Example. We will only give a very short, and therefore very simple example of a labeled resolution derivation. We hope that it illustrates some of the aspects of labeled resolution mentioned before.

Consider example (16) again, here repeated as (18), where (19) is the corresponding semantic representation.

(18) Buk is a poet. For every man there is a woman who hates him.
$\models_a$ There is a woman who hates him.
(19) $p(b) \wedge \forall x(w(x) \rightarrow \exists y(w(y) \wedge ?\mathbf{u}\, h(y, \mathbf{u})))$
$\models_a \exists z(w(z) \wedge ?\mathbf{v}\, h(z, \mathbf{v}))$

Annotating (19):

$$\mathsf{annot}(\emptyset, p(b) \wedge \forall x(w(x) \rightarrow \exists y(w(y) \wedge ?\mathbf{u}\, h(y, \mathbf{u}))) \wedge \neg\exists z(w(z) \wedge ?\mathbf{v}\, h(z, \mathbf{v}))) =$$
$$p(b) \wedge \forall x(w(x) \rightarrow \exists y(w(y) \wedge h(y, \{b, x\} : \mathbf{u}))) \wedge \neg\exists z(w(z) \wedge h(z, \{b\} : \mathbf{v})))$$

Clause form transformation:

$\{p(b^b)\}, \{m(h^h)\}, \{\neg m(x^x), w(f^y)\}, \{\neg m(x^x), h(f^y, \{b, x\} : \mathbf{u})\}, \{\neg w(z^z), \neg h(z^z, \{b\} : \mathbf{v})\}$, where the additional clause $\{m(h^h)\}$ stems from the assumption that the domain of men is nonempty.

Resolution:

$\{p(b^b)\}$ $\{m(h^h)\}$ $\{\neg m(x^x), w(f^y)\}$ $\{\neg m(x^x), h(f^y, \{b, x\} : \mathbf{u})\}$ $\{\neg w(z^z), \neg h(z^z, \{b\} : \mathbf{v})\}$

$\{w(f^y)\}$ $\{\neg m(x^x), \neg w(f^x)\}$

$\{\neg w(f^y)\}$

$\square$

Actually, the only remarkable step in the derivation is resolving

$$\{\neg m(x), h(f, \{b, x\} : \mathbf{u})\} \text{ and } \{\neg w(z), \neg h(z, \{b\} : \mathbf{v})\}$$

with $\{\neg m(x), \neg w(f)\}$ as the resolvent. Here, the two labeled pronoun variables can be identified, because the intersection of their accessible antecedents is nonempty. The corresponding mgu* of

$$\{\neg m(x^x), h(f^y, \{b,x\}:\mathbf{u}), \neg w(z^z), \neg h(z^z, \{b\}:\mathbf{v})\}$$

is $\sigma := [x^x/z^z, z^z/f^y, \{b,x\}:\mathbf{u}/\{b\}:\mathbf{v}]$.

Note also, that although $p(b)$ introduced the antecedent b, it is not used in the derivation because all information that is necessary to derive the contradiction is captured by the labels. This is the advantage of using labels; it allows us to express non-local dependency relations in our framework, which is essential for dealing with pronoun binding in dynamic semantics where a pronoun and its antecedent can occur in different formulas.

Evaluation from a Linguistic Point of View. In general, it is not enough if one gives just the information that there is a binding that allows to derive a conclusion, but one also wants to know *which* binding. It is easy to augment our method in a way such that it accomplishes this simply by memorizing the substitutions of pronoun variables that occur during a derivation.

From a linguistic point of view, one is also interested in comparing different bindings. If we force the proof procedure to backtrack every time it has found a binding which allows to derive a contradiction, we can generate all possible bindings. Probably some of the bindings are preferable to others by taking linguistic heuristics for pronoun resolution into account, see for instance [GJW95], but this is beyond the scope of the present paper.

3.3 Results

Before we prove completeness and soundness of our method, we have to explain what these notions mean in our setting.

To show that the resolution principle is correct we have to find the right loop invariant. We will show that if the parent clauses of a resolution step are strongly satisfiable, then so is the resolvent.

Definition 7 (Strong Satisfiability). *We say that a clause C is* strongly satisfiable *if there is a model M and for all substitutions θ from PVAR to VAR), then there is a literal $L \in C\theta$, such that $M \models L$.*

Lemma 3. *Let $C \cup \{\neg \mathrm{P}_1, \ldots, \neg \mathrm{P}_n\}$ and $D \cup \{\mathrm{Q}_1, \ldots, \mathrm{Q}_m\}$ be variable disjoint and strongly satisfiable. If σ is the mgu* of $\{\mathrm{P}_1, \ldots, \mathrm{P}_n, \mathrm{Q}_1, \ldots, \mathrm{Q}_m\}$, then $C\sigma \backslash \{\neg \mathrm{P}_1\sigma\} \cup D\sigma \backslash \{\mathrm{P}_1\sigma\}$ is strongly satisfiable.*

Proof. The set of possible disambiguations of the resolvent is a subset of the possible disambiguations of the parent clauses, because possible antecedents are unified, and in case of pronoun unification only the intersection of possible antecedents has to be considered. Now, two cases have to be distinguished.

(i) $M \not\models \mathrm{P}_i\sigma$. Because $\mathrm{P}_i\sigma$ is an instance of P_i and $D \cup \{\mathrm{Q}_1, \ldots, \mathrm{Q}_m\}$ is strongly satisfiable, it holds that $M \models D\sigma \backslash \{\mathrm{P}_1\sigma\}$. But $D\sigma \backslash \{\mathrm{P}_1\sigma\}$ is a subset of the resolvent and therefore $M \models C\sigma \backslash \{\neg \mathrm{P}_1\sigma\} \cup D\sigma \backslash \{\mathrm{P}_1\sigma\}$.

(ii) $M \models \mathrm{P}_i\sigma$. Again, $\mathrm{P}_i\sigma$ is an instance of P_i and $C \cup \{\neg\mathrm{P}_1, \ldots, \neg\mathrm{P}_n\}$ is strongly satisfiable. Hence, it holds that $M \models C\sigma \backslash \{\neg\mathrm{P}_1\sigma\}$ and thereby $M \models C\sigma \backslash \{\neg\mathrm{P}_1\sigma\} \cup D\sigma \backslash \{\mathrm{P}_1\sigma\}$. □

Corollary 1 (Soundness). *If* prf *(see Definition 6) produces the empty clause on input* $\neg\varphi$*, then* φ *is valid.*

Proof. If we can derive □ from a set of clauses C, where C is the clause form of $\neg\varphi$, then we can show by induction that C is not strongly satisfiable, i.e., there is no model M such that $M \models C\theta$ for all possible substitutions. Hence, for all models M, there is a disambiguation δ, such that $M \not\models \delta(\neg\varphi)$, which is equivalent to $M \models_a \varphi$, the definition of φ being valid. □

Lemma 4. *Let* θ *be a total disambiguation of* φ*, and assume that* $\varphi\theta$ *is unsatisfiable. Then there is a (classical) resolution deduction of* □ *from* $\varphi\theta$*.*

Lemma 5. *Let* θ *be a total disambiguation of* φ*. If there is a resolution deduction of* □ *from* $\varphi\theta$*, then* prf *generates the empty clause on input* φ*.*

Proof. The idea of the proof is to turn the classical resolution proof of □ from $\delta(\varphi)$ into a labeled resolution proof of □ from the original formula φ by repeating the resolution steps and inserting the required substitutions (i.e., partial disambiguations) just before any steps where they were used in the original proof.

Although the idea of this proof is simple, the details are too numerous to be included here. □

Corollary 2 (Completeness). *If* φ *is valid, then* prf *generates the empty clause on input* $\neg\varphi$*.*

4 Related Work

Most work in the area of ambiguity and discourse semantics focuses on representational issues, but see [vEJ96,MdR98] for calculi for quantificational ambiguities. Approaches that deal with pronoun binding are mostly trying to bind pronouns by applying some heuristics. The work that is closest to ours is the approach of Kohlhase and Konrad [KK98] who deal with pronoun binding in the setting of natural language corrections by using higher-order unification, and a higher-order tableaux method [Koh95] to reason about possible bindings. Van Eijck [vE98] presents a sequent calculus for *DPL* which deals with some of the complications we avoided in this paper; for instance multiple quantification of the same variable. Some of the ways in which dynamic updating can restrict possible pronoun bindings are considered in [Mon98].

5 Conclusion

In this paper we have presented a resolution calculus for reasoning with ambiguities triggered by pronouns and the different ways to bind them. Deduction steps and pronoun bindings are interleaved with the effect that only pronouns that are used during a derivation are bound to a possible antecedent. Labels allow us to capture relevant structural information of the original formula on a very local level, namely by annotating variables. Therefore structural manipulation, a prerequisite of any efficient proof method, does no harm.

Our ongoing work focuses on two aspects. First, we have to see how our resolution method behaves when other strategies restricting the search space are added; e.g., set-of-support strategy, ordered unification, or subsumption checking. Second, we are in the process of implementing the annotation and unification* algorithms and are trying to integrate them into a resolution theorem prover.

Acknowledgment. The research in this paper was supported by the Spinoza project 'Logic in Action' at ILLC, University of Amsterdam.

References

[GJW95] B. Grosz, A. Joshi, and S. Weinstein. Centering: A framework for modelling the local coherence of discourse. *Computational Linguistics*, 21(2), 1995.

[GS91] J. Groenendijk and M. Stokhof. Dynamic Predicate Logic. *Linguistics and Philosophy*, 14:39–100, 1991.

[Kam81] H. Kamp. A theory of truth and semantic representation. In J. Groenendijk et al., editor, *Formal Methods in the Study of Language*. Mathematical Centre, Amsterdam, 1981.

[KK98] M. Kohlhase and K. Konrad. Higher-order automated theorem proving for natural language semantics. SEKI-Report SR-98-04, Universität des Saarlandes, 1998.

[Koh95] M. Kohlhase. Higher-order tableaux. In P. Baumgartner et al., editor, *Theorem Proving with Analytic Tableaux and Related Methods,*TABLEAUX'95, LNAI, pages 294–309. Springer, 1995.

[Lov78] D. W. Loveland. *Automated Theorem Proving: A Logical Bases.* North-Holland, Amsterdam, 1978.

[MdR98] C. Monz and M. de Rijke. A tableaux calculus for ambiguous quantification. In H. de Swart, editor, *Automated Reasoning with Analytic Tableaux and Related Methods,*TABLEAUX'98, LNAI 1397, pages 232–246. Springer, 1998.

[MM82] A. Martelli and U. Montanari. An efficient unification algorithm. *ACM Transactions on Programming Languages and Systems*, 4:258–282, 1982.

[Mon98] C. Monz. Dynamic semantics and underspecification. In H. Prade, editor, *Proceedings of the 13th European Conference on Artificial Intelligence (ECAI'98)*, pages 201–202. John Wiley & Sons, 1998.

[Rey93] U. Reyle. Dealing with ambiguities by underspecification: Construction, representation, and deduction. *Journal of Semantics*, 10(2):123–179, 1993.

[RN95] S.J. Russell and P. Norvig. *Artificial Intelligence: A Modern Approach.* Prentice Hall, 1995.

[Rob65] J. A. Robinson. A machine oriented logic based on the resolution principle. *Journal of the ACM*, 12(1):23–41, 1965.

[vE98] J. van Eijck. A calculus for dynamic predicate logic. Unpublished manuscript, 1998.

[vEJ96] J. van Eijck and J. Jaspars. Ambiguity and reasoning. Technical Report CS-R9616, Centrum voor Wiskunde en Informatica, Amsterdam, 1996.

Algorithms on Atomic Representations of Herbrand Models

Reinhard Pichler

Institut für Computersprachen, Technische Universität Wien, Austria
reini@logic.at

Abstract. The importance of models within automated deduction is generally acknowledged both in constructing countermodels (rather than just giving the answer "NO", if a given formula is found to be not a theorem) and in speeding up the deduction process itself (e.g. by semantic resolution refinement).
However, little attention has been paid so far to the efficiency of algorithms to actually work with models. There are two fundamental decision problems as far as models are concerned, namely: the equivalence of 2 models and the truth evaluation of an arbitrary clause within a given model. This paper focuses on the efficiency of algorithms for these problems in case of Herbrand models given through atomic representations. Both problems have been shown to be coNP-hard in [Got 97], so there is a certain limit to the efficiency that we can possibly expect. Nevertheless, what we can do is find out the real "source" of complexity and make use of this theoretical result for devising an algorithm which, in general, has a considerably smaller upper bound on the complexity than previously known algorithms, e.g.: the partial saturation method in [FL 96] and the transformation to equational problems in [CZ 91].
The main result of this paper are algorithms for these two decision problems, where the complexity depends non-polynomially on the number of atoms (rather than on the total size) of the input model equivalence problem or clause evaluation problem, respectively. Hence, in contrast to the above mentioned algorithms, the complexity of the expressions involved (e.g.: the arity of the predicate symbols and, in particular, the term depth of the arguments) only has polynomial influence on the overall complexity of the algorithms.

1 Introduction

Models play an increasingly important role in automated theorem proving. Their applicability is basically twofold: Firstly, rather than just proving *that* some input formula is not a theorem, it would be desirable for a theorem prover to provide some insight as to *why* a given formula is not a theorem. To this end, the theorem prover tries to construct a countermodel rather than just giving the answer "NO". Consequently, over the past few years, automated model building has evolved as an important discipline within the field of automated deduction. The second application of models arises from the idea of guiding the proof search by providing some additional knowledge on the domain from which the input formula is taken. This knowledge can be represented in the form of a model, which may then be used e.g. in semantic resolution.

In any case, an appropriate representation of models is called for. The following properties are indispensable prerequisites for any practically relevant model representation:

1. "reasonable" expressive power: A minimum requirement on the expressive power is the possibility to finitely represent infinite models.
2. algorithms for two decision problems: There are two decision problems involved in the actual work with models, namely: deciding the equivalence of models and evaluating an arbitrary clause in such a model.

In [FL 96], *atomic representations of Herbrand models* (= AR's) are shown to be a very useful formalism: On the one hand, it is proven that the above mentioned requirements are met and, on the other hand, an algorithm is presented, which allows the automatic model construction for satisfiable clause sets of a certain syntax class.

However, the *efficiency* of algorithms concerning model equivalence and clause evaluation, which is an important issue for the practical applicability of a model representation, has received little attention so far. What we are interested here is the (time) complexity of such algorithms for AR's. Actually, both decision problems (i.e.: the model equivalence and the evaluation of a clause to true) have been shown to be coNP-hard even for the special case of linear atomic representations and even if the Herbrand universe contains no function symbols (cf. [Got 97]). Therefore, we cannot expect to find a polynomial algorithm without giving a positive answer to the $P = NP$-problem. However, what we can do is find out the real "source" of complexity and make use of this theoretical result for devising an algorithm which is, in general, considerably more efficient than previously known ones, e.g.: the partial saturation method in [FL 96] and the transformation to equational problems in [CZ 91]. The main result of this work are algorithms for solving the model equivalence problem and the clause evaluation problem for AR's, where the time complexity is non-polynomial only in the number of atoms. In contrast to the methods of [FL 96] and [CZ 91], the complexity of the expressions involved (e.g.: the arity of the predicate symbols and, in particular, the term depth of the arguments) only has polynomial influence on the complexity of our algorithms.

This paper is organized as follows: After briefly revising some basic terminology in chapter 2, we shall provide algorithms for the two decision problems mentioned above, namely the model equivalence problem (in the chapters 3 and 4) and the clause evaluation problem (in chapter 5), respectively: In chapter 3, we present a transformation of the original model equivalence problem into another type of problem, which we shall refer to as the *term tuple cover problem*, i.e.: Given a set $M = \{(t_{11}, \ldots, t_{1k}), \ldots, (t_{n1}, \ldots, t_{nk})\}$ of k-tuples of terms over some Herbrand universe H. Is every ground term tuple $(s_1, \ldots, s_k) \in H^k$ an instance of some tuple $(t_{i1}, \ldots, t_{ik}) \in M$? An algorithm for the solution of the term tuple cover problem will be presented in chapter 4. The clause evaluation problem for AR's will be tackled in chapter 5. Again we first transform the orignal problem to another kind of problem which we shall call the *term tuple inclusion problem*, i.e.: Given term tuple sets $E^{(1)} = \{\boldsymbol{u}_1^{(1)}, \dots, \boldsymbol{u}_{n_1}^{(1)}\}, \dots, E^{(m)} = \{\boldsymbol{u}_1^{(m)}, \dots, \boldsymbol{u}_{n_m}^{(m)}\}$ and $M = \{\boldsymbol{t}_1, \dots, \boldsymbol{t}_n\}$. Is every common H-ground instance $\boldsymbol{s}$ of $E^{(1)}, \dots, E^{(m)}$ also an instance of M? (Note that we call a ground term tuple $\boldsymbol{v}$ an instance of a set of tuples $W = \{\boldsymbol{w}_1, \dots, \boldsymbol{w}_n\}$, iff there exists some $w_i \in W$, s.t. $\boldsymbol{v}$ is an instance of $\boldsymbol{w}_i$.) A solution for the term tuple inclusion problem is then provided by transforming it to a set of term tuple cover problems. Finally, in chapter 6 we shall summarize the main results of this paper and identify directions for future work. For reference, some related methods are briefly sketched and their complexity is compared with our algorithm in the appendix.

2 Preliminary Definitions

The following basic definitions from [FL 96] are also central to our considerations: An *atomic representation of a Herbrand model* (= AR) over some Herbrand universe H is a set $\mathcal{A} = \{A_1, \dots, A_n\}$ of atoms over H with the following intended meaning: a ground atom over H evaluates to true, iff it is an instance of some atom $A_i \in \mathcal{A}$. In a *linear atomic representation* (= LAR), all atoms are *linear*, i.e.: they have no multiple variable occurrences. Two AR's $\mathcal{A}$ and $\mathcal{B}$ are equivalent, iff they represent the same (Herbrand) model, i.e.: the same ground atoms evaluate to true in both models. We say that a set $\mathcal{C} = \{C_1, \dots, C_n\}$ of clauses *H-subsumes* a clause D, i.e.: $\{C_1, \dots, C_n\} \leq_{ss}^H D$, iff all H-ground instances of D are subsumed by some clause $C_i \in \mathcal{C}$. For a term t over H, we denote the set of H-ground instances of t by $G_H(t)$.

The generalization of these concepts from simple terms to term tuples is straightforward, e.g.: By $G_H(t_1, \dots, t_k)$ we denote the set of *H-ground instances generated by the term tuple* $(t_1, \dots, t_k)$.

3 Transformation of the Model Equivalence Problem

In [FL 96], the following criterion for the equivalence of AR's is stated:

Lemma 3.1. (H-subsumption criterion) *Let $\mathcal{A} = \{A_1, \dots, A_n\}$ and $\mathcal{B} = \{B_1, \dots, B_m\}$ be AR's w.r.t. some Herbrand universe H. Then $\mathcal{A}$ and $\mathcal{B}$ are equivalent, iff $\{A_1, \dots, A_n\} \leq_{ss}^H B_j$ for every $j \in \{1, \dots, m\}$ and $\{B_1, \dots, B_m\}$ $\leq_{ss}^H A_i$ for every $i \in \{1, \dots, n\}$.*

This characterization of model equivalence provides the starting point for our considerations. The following theorem shows how the H-subsumption criterion can be further transformed:

Theorem 3.1. (transformation of the H-subsumption problem) *Let B, $A_1, \ldots, A_n$ be atoms over some Herbrand universe H. Furthermore, let $V(B) = \{x_1, \ldots, x_k\}$ denote the variables occurring in B and suppose that $V(B) \cap V(A_i) = \emptyset$ for all $i \in \{1, \ldots, n\}$ (i.e.: B and the A_i's have no variables in common). Finally, let Θ denote the set of unifiers of pairs (A_i, B), i.e.:*

$$\Theta = \{\vartheta \mid (\exists i) \text{ s.t. } A_i \text{ and } B \text{ are unifiable with } \vartheta' = mgu(A_i, B) \text{ and } \vartheta = \vartheta'|_{V(B)}\}$$

Then $\{A_1, \ldots, A_n\} \leq_{ss}^{H} B$ holds, iff $\bigcup_{\vartheta \in \Theta} G_H(x_1\vartheta, \ldots, x_k\vartheta) = H^k$ does.

Proof (sketch): B is H-subsumed, iff all ground instances are subsumed by some A_i. Obviously, only those instances of the A_i's play a role, which are unifiable with B, i.e.: For every ground substitution σ, $B\sigma$ must be subsumed by some $A_i\vartheta'_i = B\vartheta'_i = B\vartheta_i$, where $\vartheta'_i = mgu(A_i, B)$ and $\vartheta_i = \vartheta'_i|_{V(B)}$. But this is the case, iff for every H-ground substitution σ, $(x_1\sigma, \ldots, x_k\sigma)$ is a ground instance of some $(x_1\vartheta_i, \ldots, x_k\vartheta_i)$ with $\vartheta_i \in \Theta$. ◇

Remark: As far as complexity is concerned, the original model equivalence problem and the resulting collection of term tuple cover problems are basically the same: Both problems are coNP-hard, the number of (term tuple cover-) subproblems and the number of term tuples within each subproblem are restricted by the number of atoms, the total length of each term tuple cover problem is restricted by the length of the original model equivalence problem (provided that an efficient unification algorithm is used, which represents terms as directed acyclic graphs rather than as strings of symbols), etc.

4 The Term Tuple Cover Problem

In this chapter we shall construct an algorithm which solves the term tuple cover problem. The target of such an algorithm is to transform a given term tuple set into a "particularly simple form", for which it is easy to decide whether it represents a positive or negative instance of the term tuple cover problem. The following definition of "solved term tuple sets" makes this idea precise.

Definition 4.1. (solved term tuple set) *We call a term tuple set M* solved, *iff either $M = \emptyset$ or M contains a tuple $(x_1, \ldots, x_k)$ of pairwise distinct variables.*

Note that if $M = \emptyset$, then no ground instance at all is covered by M and therefore, M trivially represents a negative instance of the term tuple cover problem. Likewise, if M contains a tuple $(x_1, \ldots, x_k)$ of pairwise distinct variables, then M trivially represents a positive instance of the term tuple cover problem.

The central idea of the transformation to solved term tuple sets is the following: *Divide the original problem (with n term tuples) into subproblems s.t. the number of term tuples in each subproblem is strictly smaller than n and the number of subproblems is bounded by n rather than by the total input length.* Our algorithm will comprise two principal components, namely: A **division into subproblems**, which is based on an appropriate partition of the Herbrand universe H, and **redundancy criteria** which control both the number and the size of the resulting subproblems. The basic partition of H is already well-known from previous algorithms, while the latter one is new. Moreover, the redundancy criteria are the main reason why our algorithm is, in general, considerably more efficient than the other algorithms.

The basic form of our division of a term tuple cover problem into subproblems will be based on the following partition of the Herbrand universe H, which is also central to the explosion rule for solving equational problems in [CL 89] and to the orthogonalization method in [FL 96]: Let $FS(H)$ denote the set of function symbols of H (constants are considered as function symbols of arity 0). Then every ground term of H has exactly one $f \in FS(H)$ as leading symbol. Hence, H can be partitioned as $H = \bigcup_{f \in FS(H)} G_H\Big(f(x_1, \ldots, x_{\alpha(f)})\Big)$. In the following theorem we use this partition of H to split a term tuple cover problem into subproblems:

Theorem 4.1. (basic division into subproblems) *Let H be some Herbrand universe whose set of function symbols is denoted by $FS(H)$ (constants are considered as function symbols of arity 0). Furthermore, let $M = \{\boldsymbol{t}_1, \ldots, \boldsymbol{t}_n\}$ be a set of term k-tuples over H and let $p \in \{1, \ldots, k\}$ denote a component of the tuples. For every $f \in FS(H)$ with arity $\alpha(f)$, we define the "subproblem" $M_f^{(p)}$ as follows:*

$$\begin{aligned} M_f^{(p)} &= \{(t_{i1}, \ldots, t_{i(p-1)}, s_{i1}, \ldots, s_{i\alpha(f)}, t_{i(p+1)}, \ldots, t_{ik}) \mid t_{ip} = f(s_{i1}, \ldots, s_{i\alpha(f)})\} \\ &\cup \{(t_{i1}, \ldots, t_{i(p-1)}, x_1, \ldots, x_{\alpha(f)}, t_{i(p+1)}, \ldots, t_{ik})\sigma \mid t_{ip} \text{ is a variable, the } x_j \\ &\qquad \text{are new pairwise distinct variables and } \sigma = \{t_{ip} \leftarrow f(x_1, \ldots, x_{\alpha(f)})\}\} \end{aligned}$$

Then M covers H^k, iff $M_f^{(p)}$ covers $H^{k-1+\alpha(f)}$ for every $f \in FS(H)$.

Proof: (sketch) Let $p \in \{1, \ldots, k\}$ be an arbitrary component of the k-tuples. Then the above mentioned partition of H via the possible leading symbols of the terms in H can be generalized to the following partition of H^k:

$$H^k = \bigcup_{f \in FS(H)} H^{p-1} \times G_H\Big(f(x_1, \ldots, x_{\alpha(f)})\Big) \times H^{k-p}$$

Note that the tuples from M whose p-th component is a functional term with leading symbol $g \neq f$ play no role in covering $H^{p-1} \times G_H\Big(f(x_1, \ldots, x_{\alpha(f)})\Big) \times H^{k-p}$. Hence, in order to test whether some term tuple set M covers $H^{p-1} \times G_H\Big(f(x_1, \ldots, x_{\alpha(f)})\Big) \times H^{k-p}$, only the term tuples with a variable in the p-th component or with a functional term with leading symbol f have to be considered. Moreover, from the tuples with a variable z in the p-th component, only the instance where "z" is instantiated to the term $f(x_1, \ldots, x_{\alpha(f)})$ for some new, pairwise distinct variables x_i is needed.

By restricting the term tuple set M in this way, we get a set M', where all tuples have a functional term with leading symbol f in the p-th component, i.e.:

$$M' = \{\boldsymbol{t}_1, \ldots, \boldsymbol{t}_n\} \text{ with } \boldsymbol{t}_i = (t_{i1}, \ldots, t_{i(p-1)}, f(s_{i1}, \ldots, s_{i\alpha(f)}), t_{i(p+1)}, \ldots, t_{ik}).$$

Then M' covers $H^{p-1} \times G_H\Big(f(x_1, \ldots, x_{\alpha(f)})\Big) \times H^{k-p}$, iff $M'' = \{(t_{i1}, \ldots, t_{i(p-1)}, s_{i1}, \ldots, s_{i\alpha(f)}, t_{i(p+1)}, \ldots, t_{ik}) \mid t_i \in M'\}$ covers $H^{k-1+\alpha}$. But the latter condition corresponds to the term tuple subproblem $M_f^{(p)}$. ◇

In the following example, we put this theorem to work:

Example 4.1. (basic division into subproblems) Let H be the Herbrand universe with signature $FS(H) = \{f, g, a\}$. Furthermore, let an instance of the term tuple cover problem be given through the set $M = \{(f(x), y), (g(x), y), (x, x), (x, f(y)), (x, g(y))\}$. Then the subproblems $M_a^{(1)} = \{(a), (f(y)), (g(y))\}$, $M_f^{(1)} = \{(x, y), (x, f(x)), (x, f(y)), (x, g(y))\}$ and $M_g^{(1)} = \{(x, y), (x, g(x)), (x, f(y)), (x, g(y))\}$ correspond to the requirement that M covers the subsets $G_H(a) \times H$, $G_H(f(x)) \times H$ and $G_H(g(x)) \times H$, respectively, of H^2.

In the above example, the division into subproblems was not problematical at all: Note that all of the 3 subproblems $M_a^{(1)}$, $M_f^{(1)}$ and $M_g^{(1)}$ have strictly less term tuples than the original problem M. The reason why things ran so smoothly in example 4.1 is that two different function symbols (namely f and g) occurred as leading symbols of the terms in the first component. However, there is no guarantee, that two different function symbols actually do occur in some component. It is, therefore, the purpose of this chapter to provide appropriate solutions for the cases where only one function symbol or none occurs.

Both for the division into subproblems and for the redundancy criteria, we have to distinguish two cases, namely Herbrand universes with 2 or more function symbols of non-zero arity and Herbrand universes with only 1 such function symbol. Surprisingly enough, the latter case turns out to be much more difficult to handle than the former one. Moreover, we get a much better upper bound on the time complexity in the former case: By theorem 4.5, the term tuple cover problem over some Herbrand universe with 2 or more function symbols can be solved in time exponential in the number of term tuples while the upper bound obtained in theorem 4.6 for the latter case corresponds to the factorial of the number of term tuples.

4.1 Two or More Function Symbols

In this section we shall prove two redundancy criteria, which hold for any infinite Herbrand universe. Nevertheless, only in case of a Herbrand universe with 2 or more function symbols of non-zero arity, they are strong enough to allow the construction of an efficient algorithm. In section 4.2, we shall see that they do not suffice in case of only 1 such function symbol.

It has already been mentioned that the splitting into strictly smaller subproblems according to theorem 4.1 requires that 2 different function symbols occur as leading symbols in the p-th component. Now suppose that all term tuples of M have a variable as p-th component. Then all tuples would have to be considered in each subproblem. Actually, we do not worry about the case where the variables in the p-th component occur nowhere else in their tuple, since then we are on the right way towards the solved form from definition 4.1. However, if some variable x_{ip} occurs more than once in the i-th tuple, then some action has to be taken. The following theorem shows that we can handle this situation by deleting all tuples with a multiply occurring variable in the p-th component:

Theorem 4.2. (redundancy criterion based on variable components) *Let H be an arbitrary, infinite Herbrand universe and let $M = \{\mathbf{t}_1, \ldots, \mathbf{t}_n\}$ be a set of term k-tuples over H. Suppose that all terms occurring in the p-th component of the tuples from M are variables. Then every term tuple $\mathbf{t}_i \in M$ whose variable from the p-th component occurs somewhere else in $\mathbf{t}_i$ is* redundant *and may, therefore, be deleted, i.e.:*
Let $M^{(p)} := \{\mathbf{t}_i = (t_{i1}, \ldots, t_{i(p-1)}, x, t_{i(p+1)}, \ldots, t_{ik}) \in M \mid$ s.t. x is a variable that occurs somewhere else in $\mathbf{t}_i\}$. Then M covers all of H^k, iff $M - M^{(p)}$ does.

Proof: We assume w.l.o.g. that $p = 1$ (since otherwise we would swap the p-th component with the first one). Suppose that M covers all of H^k. Furthermore, let $\mathbf{t} = (t_1, \ldots, t_k)$ be an arbitrary ground term tuple which is covered by some $\mathbf{t}_i \in M^{(p)}$. We have to prove that $\mathbf{t}$ is also covered by some $\mathbf{t}_j \in (M - M^{(p)})$:

Let d denote the term depth of $\mathbf{t}$, i.e.: $d = \max(\{\tau(t_\gamma) \mid 1 \leq \gamma \leq k\})$ and choose an arbitrary term $s \in H$ with $\tau(s) > d$. Since we only consider the case of an infinite Herbrand universe H here, such a ground term $s \in H$ actually does exist. Then the term tuple $\mathbf{s} = (s, t_2, \ldots, t_k)$ is also contained in H^k and, therefore, covered by some $\mathbf{t}_j = (x, t_{j2}, \ldots, t_{jk}) \in M$, i.e.: $\mathbf{s} = \mathbf{t}_j\sigma$ with $\sigma = \{x \leftarrow s\} \circ \eta$ for some H-ground substitution η.

We claim that $\mathbf{t}_j \in (M - M^{(p)})$. For suppose, on the contrary, that $\mathbf{t}_j \in M^{(p)}$. Then x also occurs in some component t_{jq} of $\mathbf{t}_j$. Hence $\tau(t_{jq}\sigma) \geq \tau(x\sigma) = \tau(s) > d$. But this contradicts the assumption that $\tau(t_q) \leq d$ for all $q \geq 2$ and $\mathbf{t}_j\sigma = (s, t_2, \ldots, t_k)$.

Hence $\mathbf{t}_j \in (M - M^{(p)})$, i.e.: x occurs nowhere else in $\mathbf{t}_j$. But then we can substitute another term for the variable x without changing the remaining components $t_{jq}\sigma$ with $q \geq 2$. Hence, $\mathbf{t} = (t_1, \ldots, t_k) = \mathbf{t}_j\sigma'$ with $\sigma' = \{x \leftarrow t_1\} \circ \eta$. Therefore, $\mathbf{t}$ is also covered by $\mathbf{t}_j \in (M - M^{(p)})$. ◇

If only one function symbol $f \in FS(H)$ occurs as leading symbol of the p-th component then the subproblem $M_f^{(p)}$, which corresponds to the condition that M covers $H^{p-1} \times G_H\Big(f(x_1, \ldots, x_{\alpha(f)})\Big) \times H^{k-p}$, has the same number of term tuples as the original problem M. The following redundancy criterion shows, that these difficulties can be resolved by deleting all tuples with a non-variable term in the p-th component:

Theorem 4.3. (redundancy criterion based on non-variable components) *Let H be an arbitrary, infinite Herbrand universe and let $f \in FS(H)$ be a function symbol of non-zero arity. Furthermore, let $M = \{\mathbf{t}_1, \ldots, \mathbf{t}_n\}$ be a set of term k-tuples over H. Suppose that there exist term tuples in M with a non-variable term in the p-th component but the function symbol f does not occur as leading symbol of any of these terms. Then every term tuple $\mathbf{t} \in M$ with* *a non-variable term in the p-th component is* redundant *and may, therefore, be deleted, i.e.: Let* $M^{(p)} := \{\boldsymbol{t}_i = (t_{i1}, \ldots, t_{ik}) \in M \mid t_{ip}$ *is a non-variable term*$\}$. *Then* M *covers all of* H^k, *iff* $M - M^{(p)}$ *does.*

Proof: Again we assume w.l.o.g. that $p = 1$. Suppose that M covers all of H^k. Furthermore, let $\boldsymbol{t} = (t_1, \ldots, t_k)$ be an arbitrary ground term tuple which is covered by some $\boldsymbol{t}_i \in M^{(p)}$. We have to prove that $\boldsymbol{t}$ is also covered by some $\boldsymbol{t}_j \in (M - M^{(p)})$:

$\boldsymbol{t}_i \in M^{(p)}$. Hence, by the definition of $M^{(p)}$, t_{i1} is a non-variable term. Therefore, t_1 is a term with leading symbol g for some $g \in FS(H)$ s.t. $g \neq f$. Let $d = \max(\{\tau(t_\gamma) \mid 1 \leq \gamma \leq k\})$ denote the term depth of $\boldsymbol{t}$ and choose an arbitrary term $s \in H$ with $\tau(s) \geq d$. Then the term tuple $\boldsymbol{s} = (f(s, \ldots, s), t_2, \ldots, t_k)$ is also contained in H^k and, therefore, covered by some $\boldsymbol{t}_j = (t_{j1}, \ldots, t_{jk}) \in M$. But $\boldsymbol{s}$ has a term with leading symbol f in the first component and, therefore, $\boldsymbol{s}$ cannot be covered by a term tuple from $M^{(p)}$. Hence, $\boldsymbol{t}_j \in (M - M^{(p)})$. But then, t_{j1} is a variable x and, consequently, $\boldsymbol{s} = \boldsymbol{t}_j\sigma$ with $\sigma = \{x \leftarrow f(s, \ldots, s)\} \circ \eta$ for some H-ground substitution η.

Analogously to the proof of theorem 4.2, we can show that the variable x occurs nowhere else in $\boldsymbol{t}_j$. For suppose on the contrary, that x also occurs in t_{jq} for some $q \geq 2$. Then $\tau(t_{jq}\sigma) \geq \tau(x\sigma) = \tau(f(s, \ldots, s)) > d$. But this contradicts the assumption that $\tau(t_q) \leq d$ for all $q \geq 2$ and $\boldsymbol{t}_j\sigma = (f(s, \ldots, s), t_2, \ldots, t_k)$.

But then, (again like in the proof of theorem 4.2) we can substitute another term for the variable x without changing the remaining components $t_{jq}\sigma$ with $q \geq 2$. Hence, $\boldsymbol{t} = \boldsymbol{t}_j\sigma'$ with $\sigma' = \{x \leftarrow t_1\} \circ \eta$. Therefore, $\boldsymbol{t}$ is also covered by $\boldsymbol{t}_j \in (M - M^{(p)})$. ◇

Remark: The redundancy criteria from the theorems 4.2 and 4.3 allow the deletion of a subset $M^{(p)}$ of M by inspecting the p-th component of every tuple. Hence, for either criterion, we shall refer to the elements of $M^{(p)}$ as the *tuples redundant on p*. Note that (for a fixed Herbrand universe H) these redundancy criteria can be easily tested in polynomial time (the latter one can be tested in quadratic time while, for the former one, linear time is sufficient).

We are now ready to construct an algorithm for solving the term tuple cover problem for an arbitrary Herbrand universe with at least 2 function symbols of non-zero arity. In analogy with [CL 89], we shall use the notation "$\rightarrow$" and "$\rightharpoondown$", in order to refer to two different kinds of transformation rules, namely: rules which transform a term tuple set into another term tuple set or into a collection of term tuple sets, respectively.

Definition 4.2. (transformation rules) *We define a rule system consisting of the following three rules:*

1. *V (= redundancy based on variable components):* $M \rightarrow_V M - M^{(p)}$, *where p is a component in M which contains only variables and* $M^{(p)} = \{\boldsymbol{t}_i = (t_{i1}, \ldots, t_{i(p-1)}, x, t_{i(p+1)}, \ldots, t_{ik}) \in M \mid$ *s.t. x is a variable that occurs somewhere else in* $\boldsymbol{t}_i\}$ *denotes the term tuples which are redundant on p by the redundancy criterion of theorem 4.2. The rule "V" may only be applied, if* $M^{(p)} \neq \emptyset$.

2. *NV (= redundancy based on non-variable components): $M \to_{NV} M - M^{(p)}$, where p is a non-variable component in M s.t. there exists a function symbol $f \in FS(H)$ of non-zero arity which does not occur as leading symbol of the $p-th$ component of any tuple in M. Furthermore, $M^{(p)} = \{\boldsymbol{t}_i = (t_{i1}, \ldots, t_{ik}) \in M \mid t_{ip}$ is a non-variable term$\}$ denotes the term tuples which are redundant on p by the redundancy criterion of theorem 4.3. The rule "NV" may only be applied, if $M^{(p)} \neq \emptyset$.*
3. *S (= splitting into subproblems): $M \rightharpoonup_S M_f^{(p)}$, provided that there are at least 2 different function symbols occurring as leading symbols in the p-th component. $f \in FS(H)$ is a function symbol in H (constants are considered as function symbols of arity 0) and $M_f^{(p)}$ denotes the subproblem from theorem 4.1, which corresponds to the requirement that M covers $H^{p-1} \times G_H(f(x_1, \ldots, x_\alpha) \times H^{k-p}$.*

In the following theorem we prove that the above rule system has the desired property of transforming an arbitrary term tuple problem into the solved form from definition 4.1. An upper bound on the (time) complexity of this transformation will then be provided in theorem 4.5.

Theorem 4.4. *Let H be a Herbrand universe with at least 2 function symbols of non-zero arity. Then non-deterministic applications of the rule system from definition 4.2 to an arbitrary instance of the term tuple cover problem terminates with an equivalent collection of term tuple sets in solved form.*

Proof: For the correctness of the rule system we have to show, on the one hand, that every single rule is correct and, on the other hand, that the resulting term tuple sets are all in solved form, when no more rule is applicable: The correctness of every single rule has already been proven, i.e.: The rules "V" and "NV" transform a term tuple set into an equivalent set by the theorems 4.2 and 4.3. Likewise, by theorem 4.1, the rule "S" replaces the original term tuple set by an equivalent collection of term tuple sets.

Now suppose that $\mathcal{M}$ is a collection of term tuple sets s.t. no more rule application is possible. If $M \in \mathcal{M}$ is empty then, by definition, M is solved. So suppose that $M \in \mathcal{M}$ is non-empty. We have to show that then M contains a term tuple $\boldsymbol{t}$ which consists of pairwise distinct variables only. To this end we exclude the following two cases: If no function symbol at all occurs in the tuples of M but none of the tuples consists of pairwise distinct variables, then there exists a tuple $\boldsymbol{t} \in M$ with the same variable occurring in 2 different components p and q. Then the rule "V" is applicable to the p-th component, since there are only variables in the p-th component of the remaining tuples and, therefore, $\boldsymbol{t}$ is redundant on p by theorem 4.2. Likewise, if there does a exist a function symbol in M, then M contains a tuple $\boldsymbol{t}$ with a functional term t_p in the p-th component. Hence, either the rule "NV" (if one function symbol of non-zero arity is missing as leading symbol in the p-th component) or the rule "S" is applicable (if at least 2 different function symbols occur as leading symbols in the p-th component). But this again contradicts the assumption that no rule is applicable to M. Note that this is the only place in the proof where we actually make use of the fact that H contains at least 2 distinct function symbols of non-zero arity, for otherwise it may happen that neither the rule "NV" nor the rule "S" is applicable to a component where some function symbol occurs.

The termination of non-deterministic applications of the above rules can be shown by a multiset argument based on the following observation: Whenever a rule "V", "NV" or "S" is applied, then the original term tuple set is replaced by a finite number "c" of term tuple sets with strictly less term tuples. (In case of the rules "V" and "NV", $c = 1$, whereas $c = |FS(H)|$ in case of "S"). ◇

Theorem 4.5. (complexity estimation) *Let H be a Herbrand universe with at least 2 function symbols of non-zero arity. Then the term tuple cover problem can be decided in time $O(c^n * pol(N))$, where $c = |FS(H)|$, n denotes the number of term tuples and $pol(N)$ is some polynomial function in the total length N of an input problem instance.*

Proof: (sketch) In order to analyse the complexity of the rule system from definition 4.2, we have a closer look at the multiset argument in the termination proof of theorem 4.4: Whenever one of the transformation rules is applied to some term tuple set M, then this set M branches into at most c sets each containing strictly less term tuples. Hence, the whole transformation process corresponds to a tree whose depth is restricted by the number n of term tuples in the original set and where the degree of the nodes is restricted by c. But then the number of rule applications (which corresponds to the number of internal nodes of the tree) is restricted by c^n. ◇

4.2 One Function Symbol

As we have seen in the previous section, redundancy criteria are necessary to ensure that the original term tuple cover problem can be split into strictly smaller subproblems. However, the basic idea of partitioning the Herbrand universe could be carried over from previsouly known algorithms to ours without modifications. The following example illustrates that, in case of a Herbrand universe H with only 1 function symbol of non-zero arity, we even have to modify the way in which H is partitioned.

Example 4.2. (unsuitable partition of H) Let $M = \{(f^2(x), y), (x, f^2(y)), (x, x), (f(x), x), (x, f(x))\}$ be an instance of the term tuple cover problem over the Herbrand universe H with signature $FS(H) = \{f, a\}$.

Then M covers H^2, iff $M_a = \{(a, f^2(y)), (a, a), (a, f(a))\}$ covers $G_H(a) \times H$ and $M_f = \{(f^2(x), y), (f(x), f^2(y)), (f(x), f(x)), (f(x), x), (f(x), f^2(x))\}$ covers $G_H(f(x)) \times H$.

The subproblem $M_f^{(1)} = \{(f^2(x), y), (f(x), f^2(y)), (f(x), f(x)), (f(x), x), (f(x), f^2(x))\}$ from example 4.2 contains all tuples of M, where $M_f^{(1)}$ corresponds to the requirement that M covers $G_H(f(x)) \times H$ (cf. theorem 4.1). In section 4.1, it was possible to resolve this kind of situation by means of redundancy criteria. Note, however, that in the above situation we cannot hope to delete one of the tuples of M via a new redundancy criterion, since all of the 5 tuples of M are actually necessary to cover H^2. Hence, in contrast to the previous section, we even need a different partition of H in order to guarantee, that all subproblems have strictly less term tuples.

Due to space limitations, we can only give the complexity result obtained for the case of a Herbrand universe with a single function symbol of non-zero arity. In [Pic 98], an extended version of this paper is available, where the algorithm and the required lemmas are worked out in detail.

Theorem 4.6. (complexity estimation) *Let H be a Herbrand universe with only 1 function symbol of non-zero arity and let n denote the number of term tuples. Then the term tuple cover problem can be decided in time $O(n! * pol(N))$, where $pol(N)$ is some polynomial function in the total length N of an input problem instance.*

5 The Clause Evaluation Problem

5.1 Transformation of the Clause Evaluation Problem

Analogously to the transformation of the model equivalence problem in chapter 3, we shall transform the clause evaluation problem to an equivalent term tuple problem. However, the term tuple cover problem is not appropriate in this case, due to the existence of negative literals. Hence, the clause evaluation problem will be transformed into another type of problem which we shall call the *term tuple inclusion problem*, i.e.: Given term tuple sets $E^{(1)} = \{\boldsymbol{u}_1^{(1)}, \dots, \boldsymbol{u}_{n_1}^{(1)}\}, \dots, E^{(m)} = \{\boldsymbol{u}_1^{(m)}, \dots, \boldsymbol{u}_{n_m}^{(m)}\}$ and $M = \{\boldsymbol{t}_1, \dots, \boldsymbol{t}_n\}$. Is every common H-ground instance $\boldsymbol{s}$ of $E^{(1)}, \dots, E^{(m)}$ also an instance of M? (Note that we call a ground term tuple $\boldsymbol{v}$ an instance of a set of tuples $W = \{\boldsymbol{w}_1, \dots, \boldsymbol{w}_n\}$, iff there exists some $w_i \in W$, s.t. $\boldsymbol{v}$ is an instance of $\boldsymbol{w}_i$.)

The following transformation based on mgu's of atoms in a clause C and the atoms A_i of an atom representation $\mathcal{A}$ is a generalization of theorem 3.1:

Theorem 5.1. (transformation of the clause evaluation problem) *Let $\mathcal{A} = \{A_1, \dots, A_n\}$ be an AR of the model $\mathcal{M}_{\mathcal{A}}$ over some Herbrand universe H and let $C = L_1 \vee \dots \vee L_l \vee \neg M_1 \vee \dots \vee \neg M_m$ be a clause over H. Let $V(C) = \{x_1, \dots, x_k\}$ denote the variables occurring in C and suppose that $V(C) \cap V(A_i) = \emptyset$ for all $i \in \{1, \dots, n\}$ (i.e.: C and the A_i's have no variables in common). Furthermore, let Φ_j be defined as the set of unifiers of M_j with some A_i and let Ψ be the set of unifiers of pairs (A_i, L_j), i.e.:*

$\Phi_j = \{\varphi \mid (\exists i)$ *s.t.* A_i *and* M_j *are unifiable with* $\varphi' = mgu(A_i, M_j)$ *and* $\varphi = \varphi'|_{V(C)}\}$ *(with* $1 \leq j \leq m$*)*

$\Psi = \{\psi \mid (\exists i)(\exists j)$ *s.t.* A_i *and* L_j *are unifiable with* $\psi' = mgu(A_i, L_j)$ *and* $\psi = \psi'|_{V(C)}\}$

Then C evaluates to **T** *in $\mathcal{M}_A$, iff* $\left(\bigcup_{\varphi \in \Phi_1} G_H(x_1\varphi, \ldots, x_k\varphi)\right) \cap \ldots \cap \left(\bigcup_{\varphi \in \Phi_m} G_H(x_1\varphi, \ldots, x_k\varphi)\right) \subseteq \left(\bigcup_{\psi \in \Psi} G_H(x_1\psi, \ldots, x_k\psi)\right)$

Proof (sketch): Analogously to the proof of theorem 3.1, the evaluation of ground atoms in $\mathcal{M}_A$ can be expressed through appropriate conditions on term tuples and their H-ground instances, i.e.: Let σ be an H-ground substitution with domain $V(C)$ and let $(t_1, \ldots, t_k) = (x_1\sigma, \ldots, x_k\sigma)$ be a ground term tuple in H^k. Then the following relations hold:

1. For every j with $1 \leq j \leq m$: $(t_1, \ldots, t_k) \in \bigcup_{\varphi \in \Phi_j} G_H(x_1\varphi, \ldots, x_k\varphi) \Leftrightarrow M_j\sigma$ evaluates to **T** in $\mathcal{M}_A$.
2. Hence, $(t_1, \ldots, t_k) \in \left(\bigcup_{\varphi \in \Phi_1} G_H(x_1\varphi, \ldots, x_k\varphi)\right) \cap \ldots \cap \left(\bigcup_{\varphi \in \Phi_n} G_H(x_1\varphi, \ldots, x_k\varphi)\right) \Leftrightarrow M_1\sigma \wedge \ldots \wedge M_m\sigma$ evaluates to **T** in $\mathcal{M}_A$ $\Leftrightarrow \neg M_1\sigma \vee \ldots \vee \neg M_m\sigma$ evaluates to **F** in $\mathcal{M}_A$
3. $(t_1, \ldots, t_k) \in \bigcup_{\psi \in \Psi} G_H(x_1\psi, \ldots, x_k\psi) \Leftrightarrow$ there is some j with $1 \leq j \leq l$, s.t. $L_j\sigma$ evaluates to **T** in $\mathcal{M}_A$ $\Leftrightarrow L_1\sigma \vee \ldots \vee L_l\sigma$ evaluates to **T** in $\mathcal{M}_A$

The transformation into the term tuple inclusion problem is based on the following idea: C evaluates to **T** $\Leftrightarrow$ all H-ground instances of C evaluate to **T** $\Leftrightarrow$ for every H-ground substitution σ: if all negative literals $M_j\sigma$ of $C\sigma$ evaluate to **F**, then some positive literal $L_j\sigma$ of $C\sigma$ evaluates to **T** $\Leftrightarrow$ for every H-ground term tuple $\boldsymbol{t} = (t_1, \ldots, t_k) \in H^k$: If $\boldsymbol{t} \in \left(\bigcup_{\varphi \in \Phi_1} G_H(x_1\varphi, \ldots, x_k\varphi)\right) \cap \ldots \cap \left(\bigcup_{\varphi \in \Phi_m} G_H(x_1\varphi, \ldots, x_k\varphi)\right)$, then $\boldsymbol{t} \in \bigcup_{\psi \in \Psi} G_H(x_1\psi, \ldots, x_k\psi)$. ◇

In the special case where C contains no negative literals, then the empty intersection H^k is tested for inclusion, i.e.: $H^k \subseteq \bigcup_{\psi \in \Psi} G_H(x_1\psi, \ldots, x_k\psi)$, which is equivalent to the term tuple cover problem $\bigcup_{\psi \in \Psi} G_H(x_1\psi, \ldots, x_k\psi) = H^k$

Remark: As far as complexity is concerned, the original clause evaluation problem and the resulting term tuple inclusion problem are basically the same: Both problems are coNP-hard, the number of term tuples is restricted by the square of the number of atoms, the total length of each term tuple inclusion problem is restricted by the square of the length of the original clause evaluation problem (provided that an efficient unification algorithm is used, which represents terms as directed acyclic graphs rather than as strings of symbols), etc.

5.2 The Term Tuple Inclusion Problem

In chapter 4, we have proven that the complexity of the term tuple cover problem depends primarily on the number of term tuples. Consequently, the number of atoms was identified as the main source of complexity for the model equivalence problem. In order to derive a similar result for the term tuple inclusion problem and the clause evaluation problem, we transform a given term tuple inclusion problem into an equivalent collection of term tuple cover problems in the following way:

1. Distributivity of $\cap$ and $\cup$: In the term tuple inclusion problem for the sets $E^{(1)} = \{\boldsymbol{u}_1^{(1)}, \ldots, \boldsymbol{u}_{n_1}^{(1)}\}$, ..., $E^{(m)} = \{\boldsymbol{u}_1^{(m)}, \ldots, \boldsymbol{u}_{n_m}^{(m)}\}$, $M = \{\boldsymbol{t}_1, \ldots, \boldsymbol{t}_n\}$,

the following kind of set inclusion has to be tested:

$$(A_{11} \cup \ldots \cup A_{1n_1}) \cap \ldots \cap (A_{m1} \cup \ldots \cup A_{mn_m}) \subseteq B,$$

where $A_{ij} = G_H(\boldsymbol{u}_j^{(i)})$ and $B = (\bigcup_{l=1}^{n} G_H(\boldsymbol{t}_l)$. Then this intersection of unions can be transformed into an equivalent union of intersections by the distributivity of $\cap$ and $\cup$, i.e.:

$$(A_{11}\cup\ldots\cup A_{1n_1})\cap\ldots\cap(A_{m1}\cup\ldots\cup A_{mn_m}) = \bigcup_{\alpha_1=1}^{n_1} \cdots \bigcup_{\alpha_m=1}^{n_m} (A_{1\alpha_1}\cap\ldots\cap A_{m\alpha_m})$$

But a union of sets $\bigcup_{i\in I} C_i$ is contained in another set B, iff every set C_i is. Hence, the original inclusion problem is equivalent to the following collection of simple inclusion problems:

$$(\forall \alpha_1 \in \{1, \ldots, n_1\}) \ldots (\forall \alpha_m \in \{1, \ldots, n_m\}) : (A_{1\alpha_1} \cap \ldots \cap A_{m\alpha_m}) \subseteq B$$

2. Intersection of sets $G_H(\boldsymbol{e}_1)$: Let $G_H(\boldsymbol{e}_1)\cap\ldots\cap G_H(\boldsymbol{e}_m)$ with $\boldsymbol{e}_i = \boldsymbol{u}_{\alpha_i}^{(i)}$ denote one of the intersections which result from the previous transformation step. Then this intersection can be represented as the set of ground instances of a single term tuple in the following way: We first rename the variables in the tuples $\boldsymbol{e}_i$, s.t. these tuples are pairwise variable disjoint. Then the set of common ground instances of these tuples can be computed via unification:
   (a) If $\boldsymbol{e}_1, \ldots, \boldsymbol{e}_m$ are not unifiable, then $G_H(\boldsymbol{e}_1) \cap \ldots \cap G_H(\boldsymbol{e}_m) = \emptyset$.
   (b) If $\boldsymbol{e}_1, \ldots, \boldsymbol{e}_m$ are unifiable with mgu η, then $G_H(\boldsymbol{e}_1) \cap \ldots \cap G_H(\boldsymbol{e}_m) = G_H(\boldsymbol{e}_1\eta)$.

   But then every term tuple inclusion problem of the form

   $$G_H(\boldsymbol{e}_1) \cap \ldots \cap G_H(\boldsymbol{e}_m) \subseteq (\bigcup_{l=1}^{n} G_H(\boldsymbol{t}_l)$$

   can be either deleted (if $\boldsymbol{e}_1, \ldots, \boldsymbol{e}_m$ are not unifiable) or transformed into an inclusion problem of the form $G_H(\boldsymbol{e}_1\eta) \subseteq (\bigcup_{l=1}^{n} G_H(\boldsymbol{t}_l)$.
3. H-subsumption Let $\boldsymbol{s} = \boldsymbol{e}_1\eta$ denote one of the term tuples which are obtained by the previous step. Then the condition $G_H(\boldsymbol{s}) \subseteq \bigcup_{i=1}^{n} G_H(\boldsymbol{t}_i)$ corresponds to an H-subsumption criterion for term tuples, namely: $\{\boldsymbol{t}_1, \ldots, \boldsymbol{t}_n\} \leq_{ss}^{H} \boldsymbol{s}$. Now suppose that $\boldsymbol{s}$ and the $\boldsymbol{t}_i$'s have no variables in common. Furthermore, let $V(\boldsymbol{s}) = \{x_1, \ldots, x_l\}$ denote the variables occurring in $\boldsymbol{s}$ and let Θ be defined as the set of unifiers of pairs $(\boldsymbol{t}_i, \boldsymbol{s})$, i.e.:

   $$\Theta = \{\vartheta \mid \exists i \text{ s.t. } \boldsymbol{t}_i \text{ and } \boldsymbol{s} \text{ are unifiable with } \vartheta' = mgu(\boldsymbol{t}_i, \boldsymbol{s}) \text{ and } \vartheta = \vartheta'|_{V(\boldsymbol{s})}\}$$

   Then, analogously to theorem 3.1, this H-subsumption criterion can be transformed into the term tuple cover problem $\bigcup_{\vartheta\in\Theta} G_H(x_1\vartheta, \ldots, x_l\vartheta) = H^l$

By combining the above transformation with the complexity results for the term tuple cover problem from chapter 4, we arrive at an upper bound on the complexity of our term tuple inclusion problem. Due to space limitations, we only consider the case of a Herbrand universe with 2 or more function symbols of non-zero arity below:

Theorem 5.2. (complexity estimation) *Let H be a Herbrand universe with at least 2 function symbols of non-zero arity. Furthermore, let $\mathcal{P}$ be an instance of the term tuple inclusion problem (over H), where $c = |FS(H)|$, n denotes the number of term tuples and $pol(N)$ is some polynomial function in the total length N of $\mathcal{P}$. Then the term tuple inclusion problem for $\mathcal{P}$ can be solved in time $O(c^n * pol(N))$.*

Proof: (sketch) If we solve the term tuple inclusion problem by first transforming it into an equivalent set of term tuple cover problems, then the overall complexity of this algorithm is mainly determined by the number of term tuple cover problems and by the cost of solving each term tuple cover problem. Let $\mathcal{P} = (E^{(1)}, \ldots, E^{(m)}, M)$ be an instance of the term tuple inclusion problem with $|M| = n'$ and $|E^{(1)}| + \ldots + |E^{(m)}| = n''$. Then the number of term tuple cover problems obtained by the above transformation corresponds to $|E^{(1)}| * \ldots * |E^{(m)}|$, which (for fixed n'') becomes maximal, if all sets $|E^{(i)}|$ consist of 3 elements. Hence, the number of term tuple cover problems is restricted by $3^{\frac{n}{3}''}$. On the other hand, the number of term tuples in a single term tuple cover problem is restricted by $|M| = n'$. Furthermore, the total size of every term tuple cover problem is linearly restricted by the total size of the original term tuple inclusion problem, provided that we use an efficient unification algorithm. Hence, together with theorem 4.5, we get the complexity bound $O(3^{\frac{n}{3}''} * c^{n'} * pol(N)) \leq O(c^{n''} * c^{n'} * pol(N)) = O(c^n * pol(N))$ for the term tuple inclusion problem. ◇

6 Concluding Remarks and Future Work

From the theoretical point of view, the number of atoms (rather than the total length of the input problem) has been identified as the real complexity "source" both of the model equivalence problem and the clause evaluation problem of AR's. From the practical point of view, the foundation has been laid for considerably more efficient algorithms than previously known ones. The main reason for this improvement are the various redundancy criteria proven in chapter 4.

In this paper, we have mainly concentrated on the worst case complexity of the algorithms under investigation. However, in practice, also heuristics concerning the *rule appilcation strategy* and *further simplification rules* play a crucial role, even if they do not affect the worst case complexity. One possible simplification would be to delete all term tuples $\boldsymbol{t}_i$ from a term tuple set M, which are an instance of some other tuple $\boldsymbol{t}_j \in M$. The search for further improvements of this sort as well as any *implementational details* have been left for future work.

Remember that in chapter 1, the close relationship between atomic representations and constraint solving was already mentioned. In particular, the problems of model equivalence and clause evaluation can be first transformed into equational problems and then tackled by constraint solving methods. On the other hand, one may try to go the other direction and find out in what way the ideas presented in this paper (in particular, the redundancy criteria) can be applied to *constraint solving*, e.g.: in [LM 87] or [CL 89].

This work only deals with atomic representations of models. However, many more formalisms for representing models can be found in the literature (cf. [Mat 97]). Like in the case of AR's, no particular emphasis is usually put on the efficiency of algorithms to actually work with these formalisms. Hence, a thorough complexity analysis and the search for reasonably efficient algorithms would be also desirable for *other model representations.*

References

[CL 89] H. Comon, P. Lescanne: Equational Problems and Disunification, Journal of Symbolic Computation, Vol 7, pp. 371-425 (1989).

[CZ 91] R.Caferra, N.Zabel: Extending Resolution for Model Construction, in Proceedings of JELIA '90, LNAI 478, pp. 153-169, Springer (1991).

[FL 96] C.Fermüller, A.Leitsch: Hyperresolution and Automated Model Building, Journal of Logic and Computation, Vol 6 No 2, pp. 173-230 (1996).

[Got 97] G.Gottlob: The Equivalence Problem for Herbrand Interpretations, unpublished note (1997).

[Lei 97] A.Leitsch: The Resolution Calculus, Texts in Theoretical Computer Science, Springer (1997).

[LM 87] J.-L.Lassez, K. Marriott: Explicit Representation of Terms defined by Counter Examples, Journal of Automated Reasoning, Vol 3, pp. 301-317 (1987).

[Mat 97] R.Matzinger: Comparing Computational Representations of Herbrand Models, in Proceedings of KGC'97, LNCS 1289, Springer (1997).

[Pic 98] R.Pichler: Algorithms on Atomic Representations of Herbrand Models, technical report TR-CS-RP-98-1 of the Technical University of Vienna, available as `ftp://ftp.logic.at/pub/reini/armod.ps` (1998).

Appendix

A An Overview of Related Methods

It has already been mentioned that, for the decision problems treated in this work, algorithms are also provided in [CZ 91] and [FL 96]. In both papers, an algorithm for deciding the H-subsumption problem forms the basis both for solving the model equivalence problem and the clause evaluation problem. Furthermore, the way the H-subsumption problem is treated is decisive for the overall complexity of these algorithms. Therefore, in this chapter, we shall concentrate on the H-subsumption problem when the main ideas of these algorithms are outlined and their complexity is compared with our algorithm:

The H-subsumption algorithm in [**FL 96**] is based on the following theorem: Let $\mathcal{C}$ and $\mathcal{D}$ be sets of clauses s.t. the minimum depth of variable occurrences in $\mathcal{D}$ is greater than the term depth of $\mathcal{C}$, then the H-subsumption and ordinary subsumption coincide. Hence, the H-subsumption problem $\mathcal{A} \leq_{ss}^{H} B$ for an atom set $\mathcal{A}$ and an atom B can be decided as follows: First, B is transformed into an equivalent atom set $\mathcal{B}$ by partial saturation, s.t. the minimum depth of variable occurrences in $\mathcal{B}$ is greater than the term depth of $\mathcal{A}$. Then, it is tested whether every atom from $\mathcal{B}$ is an instance of some atom from $\mathcal{A}$.

If the condition on the minimum depth of variable occurrences is already fulfilled by B, then the algorithm from [FL 96] is polynomial. However, in the worst case, the partial saturation leads to a set $\mathcal{B}$ with exponentially many atoms. If the signature contains at least one function symbol of arity greater than or equal to 2 then even the size of the atoms in $\mathcal{B}$ may become exponential. In any case, the exponentiality both of the time and space complexity refers to the size of the atoms of the original H-subsumption problem rather than to the number of atoms.

In [**CZ 91**], the H-subsumption problem $\{P(\boldsymbol{t}_1), \ldots, P(\boldsymbol{t}_n)\} \leq^H_{ss} P(\boldsymbol{s})$ is reduced to the equational problem

$$(\forall \boldsymbol{y}_1 : \boldsymbol{t}_1 \neq \boldsymbol{s}) \wedge \ldots \wedge (\forall \boldsymbol{y}_n : \boldsymbol{t}_n \neq \boldsymbol{s}),$$

where $\boldsymbol{y}_i$ denotes the vector of variables in $\boldsymbol{t}_i$. The satisfiability of this equational problem is then tested by the method from [CL 89]. The rules in [CL 89] may be applied non-deterministically. At any rate, in the absence of redundancy criteria similar to ours, the only way to deal with the universally quantified variables y_i is to first transform arbitrary disequations containing universally quantified variables into disequations of the form $y_i \neq t$, s.t. y_i does not occur in t. Then such a variable y_i can be eliminated by the transformation rule U_2 (= "universality of parameters") from [CL 89], i.e.:

$$\forall \boldsymbol{y} : P \wedge (y_i \neq t \wedge R) \rightarrow_{U_2} \forall \boldsymbol{y} : P \wedge R[y_i \leftarrow t].$$

In order to eliminate all occurrences of universally quantified variables at a depth greater than 0, the explosion rule has to be applied repeatedly. The idea behind this transformation is illustrated by the following example:

The problem $\forall y : f(f(y)) \neq x$ over the signature $FS(H) = \{f, a\}$ is equivalent to the disjunction of the problems $\exists u : x = f(u) \wedge (\forall y : f(f(y)) \neq x)$ and $x = a \wedge (\forall y : f(f(y)) \neq x)$. These problems can be simplified to $\exists x : x = f(u) \wedge (\forall y : f(y) \neq u)$ and $x = a$, respectively. Then the depth of occurrence of the universally quantified variable y has been strictly decreased in all subproblems.

In our original equational problem $(\forall \boldsymbol{y}_1 : \boldsymbol{t}_1 \neq \boldsymbol{s}) \wedge \ldots \wedge (\forall \boldsymbol{y}_n : \boldsymbol{t}_n \neq \boldsymbol{s})$, this idea of eliminating all occurrences of universally quantified variables at a depth greater than 0 has to be applied to every disequation $\boldsymbol{t}_i \neq \boldsymbol{s}$. The number of explosion rule applications required depends linearly on the number of non-variable positions in the term tuple $\boldsymbol{t}_i$. We therefore get the upper bound $O((c * m)^n * pol(N))$ for the worst case time complexity of this equational problem solving method, where c is a constant, m is an upper bound on the size of the tuples $\boldsymbol{t}_i$ and n denotes the number of tuples.

In [**LM 87**], a problem strongly related to our term tuple cover problem is tackled, namely: Let an "implicit generalization" over some Herbrand universe be given as $t/\{t\theta_1, \ldots, t\theta_n\}$ with the intended meaning that it represents all ground instances of t which are not an instance of any $t\theta_i$. Apart from some other purposes like the transformation of an "implicit generalization" into an equivalent "explicit" one (for details, cf. [LM 87]), the algorithm from [LM 87] is used to decide whether such a generalization is empty. Note that the original version of this algorithm can be easily extended to term tuples $\boldsymbol{t}$ and $\boldsymbol{t}\theta_i$. Furthermore, the term tuple cover problem $M = \{\boldsymbol{t}_1, \ldots, \boldsymbol{t}_n)\}$ corresponds to the emptiness problem for the "implicit generalization" $\boldsymbol{x}/\{\boldsymbol{t}_1, \ldots, \boldsymbol{t}_n)\}$, where $\boldsymbol{x} = (x_1, \ldots, x_k)$ is an arbitrary k-tuple of pairwise distinct variables. Then the H-subsumption problem can be solved as follows via the "uncover"-algorithm from [LM 87]: Choose some linear term tuple $\boldsymbol{t}_i$ from the right hand side of the implicit generalization and perform a partitioning of H^k s.t. one partition corresponds to the ground instances of $\boldsymbol{t}_i$ (The idea of this partitioning basically comes down to iterated applications of the explosion rule from [CL 89]). Let P denote the set of terms generated by the partitioning algorithm from [LM 87] and let $P' = P - \{\boldsymbol{t}_i\}$. Then the original implicit generalization is equivalent to the following collection of generalizations:

$$\{p/\{\text{mgi}(p, \boldsymbol{t}_1), \ldots, \text{mgi}(p, \boldsymbol{t}_{i-1}), \text{mgi}(p, \boldsymbol{t}_{i+1}), \ldots, \text{mgi}(p, \boldsymbol{t}_n)\} \mid p \in P'\},$$

where "mgi" denotes the most general common instance of two term tuples. The number of tuples on the right hand side is strictly decreased on every recursive call of the procedure "uncover". If eventually an implicit generalization is produced s.t. the right hand side is empty or contains only non-linear term tuples, then this generalization (and, hence, also the original one) is found to be non-empty. If, on the other hand, the whole set of generalizations eventually becomes empty, then the original generalization actually is empty.

Note that the number of terms in the partitioning set P depends linearly on the number of non-variable positions in the term tuple $\boldsymbol{t}_i$. Hence, analogously to the equational problem solving method from [CL 89], we get the the upper bound $O((c * m)^n * pol(N))$ for the worst case time complexity of the "uncover"-algorithm, where c is a constant, m is an upper bound on the size of the tuples $\boldsymbol{t}_i$ and n denotes the number of tuples. Even though the two algorithms from [CL 89] and [LM 87] behave very similarly in the worst case, the termination criterion of the latter algorithm provides a significant improvement, i.e.: rather than eliminating all tuples from the right hand side of an implicit generalization, it suffices to eliminate the linear tuples only.

In contrast to our algorithm, there exists no constant c s.t. the worst case complexity of the above mentioned algorithms can be bounded by $O(c^n * pol(N))$. In particular, when expressions of high term complexity are involved, the advantage of our algorithm w.r.t. the others is obvious. Moreover, several ideas presented in the other algorithms, can be easily incorporated into our algorithm, e.g.: The termination criterion from [LM 87] can be used to extend our notion of solved forms from definition 4.1 in the following way: A term tuple set M will be called *solved*, iff either M contains no linear tuple or M contains a tuple $(x_1, \ldots, x_k)$ of pairwise distinct variables.

On the Intertranslatability of Autoepistemic, Default and Priority Logics, and Parallel Circumscription

Tomi Janhunen

Digital Systems Laboratory, Helsinki University of Technology, Finland
Tomi.Janhunen@hut.fi

Abstract. This paper concentrates on comparing the relative expressive power of five non-monotonic logics that have appeared in the literature. The results on the computational complexity of these logics suggest that these logics have very similar expressive power that exceeds that of classical monotonic logic. A refined classification of non-monotonic logics by their expressive power can be obtained using translation functions that satisfy additional requirements such as faithfulness and modularity used by Gottlob. Basically, we adopt Gottlob's framework for our analysis, but propose a weaker notion of faithfulness. A surprising result is deduced in light of Gottlob's results: Moore's autoepistemic logic is less expressive than Reiter's default logic and Marek and Truszczyński's strong autoepistemic logic. The expressive power of priority logic by Wang et al. is also analyzed and shown to coincide with that of default logic. Finally, we present an exact classification of the non-monotonic logics under consideration in the framework proposed in the paper.

1 Introduction

A variety of *non-monotonic logics* have been proposed as formalizations of *non-monotonic reasoning* (NMR). Among these formalizations are *circumscription* by McCarthy [19], *default logic* by Reiter [24], *autoepistemic logic* by Moore [20], *strong autoepistemic logic* by Marek and Truszczyński [17] as well as *priority logic* by Wang, You and Yuan [28]. The main goal of this paper is to compare these five non-monotonic logics on the basis of their *expressive power*, i.e. their capability of representing various problems from the NMR domain.

A way of measuring the expressive power of a non-monotonic logic is to analyze the computational complexity of its decision problems, and to rank these decision problems in the *polynomial time hierarchy* (PH) [1]. In fact, complexity issues have received much attention in the NMR community recently, and the decision problems of default logic, (strong) autoepistemic logic and circumscription have been systematically analyzed [4, 5, 8, 18, 21, 26]. To summarize these results in the propositional case, the major decision problems of these four non-monotonic logics are complete problems on the second level of PH. These complexity results suggest that (i) the expressive powers of non-monotonic logics exceed that of classical monotonic logic and (ii) the non-monotonic logics mentioned are of equal expressive power – if measured by the levels of PH.

The expressibility issue can also be addressed in terms of *translation functions* between theories of non-monotonic logics. For instance, a variety of translation functions have been proposed to transform a default theory into an autoepistemic one [2, 9, 12, 13, 17, 25, 27] and back [11, 18] such that sets of conclusions are preserved to a reasonable degree in the translation. Also, translation functions between various kinds of default theories have been considered [2, 6, 15]. In fact, complexity results are based on *polynomial transformations* between decision problems of non-monotonic logics and thus give rise to translation functions between non-monotonic theories, too. Unfortunately, the aim of such transformations is to preserve the yes/no-answers of decision problems and nothing more. This leaves room for translations that depend globally on the theory under translation so that a local modification to the theory changes the translation totally. However, it is possible to introduce further constraints for translation functions. A very promising requirement – *modularity* – is introduced by Imielinski [10] and then used by Gottlob [9] and Niemelä [23]. Roughly speaking, a modular translation function is in a sense *systematic*: local changes in a theory cause only local changes in its translation. Most importantly, it has been shown that a modular translation function between certain non-monotonic logics is not possible. This indicates that the non-monotonic logics involved differ in expressive power, although their decision problems are equally complex.

This paper takes the expressive power of five non-monotonic logics into reconsideration. A central concept – the notion of a polynomial, faithful and modular translation function – is adopted from Gottlob's work [9]. However, the notion of faithfulness is revised in an important way: it is assumed that sets of conclusions are preserved up to a fixed propositional language. This allows one to add, e.g., new propositional atoms in a translation if necessary and this is indeed the case with a number of translation functions addressed in this paper. Moreover, it is shown that polynomial, faithful and modular translations do not exist in certain cases in order to establish strict differences in expressive power. The comparisons made in the paper lead to an exact classification of non-monotonic logics. A particular novelty in this respect is that *Moore's autoepistemic logic is less expressive than Reiter's default logic.* Gottlob [9] employs a stronger notion of faithfulness and concludes the opposite. This demonstrates in an interesting way how the requirements imposed on translation functions affect the results on expressibility. Also, new light is shed on the interconnection of priority logic and default logic by showing that these logics are of equal expressive power.

The plan of this paper is as follows. In Section 2, we review the basic notions of non-monotonic logics mentioned. After this, the criteria for translation functions are set up in Section 3. In Section 4, actual translation functions are presented to rank non-monotonic logics by their expressive power. Some comparisons are also made with related work. Finally, the resulting classification of non-monotonic logics by their expressive power is illustrated and discussed in Section 5.

2 Logics of Interest

In this section, we review the basic definitions and notions of non-monotonic logics [14, 17, 20, 24, 28] that appear in the rest of this work. To allow a uniform treatment of these logics in sections to come, we have made the definitions similar as follows. (i) Only the propositional case is considered. Definitions are given relative to a propositional language $\mathcal{L}$ which is based on a finite or at most countable set of propositional atoms $\mathcal{A}$. (ii) A propositional subtheory $T \subseteq \mathcal{L}$ is distinguished for each *non-monotonic theory*, i.e. a theory of a non-monotonic logic. (iii) The sets of conclusions associated with non-monotonic theories are identified. Such sets are often called *extensions* or *expansions* and they determine the *semantics* of non-monotonic theories. Generally speaking, a non-monotonic theory may have a unique extension, several extensions, or sometimes even no extensions. We consider both *brave* and *cautious* reasoning strategies with extensions. In the former strategy, finding a single extension for a theory is of interest, while the intersection of extensions is considered in the latter.

2.1 Default Logic (DL)

A *default theory* [24] is a pair $\langle D, T\rangle$ where $T \subseteq \mathcal{L}$ and D is a set of *default rules* (or *defaults*) of the form $\alpha : \beta_1, \ldots, \beta_n / \gamma$ such that $n \geq 0$ and the *prerequisite* α, the *justifications* $\beta_1, \ldots, \beta_n$ and the *consequent* γ of the rule are sentences of $\mathcal{L}$. Marek and Truszczyński [18] reduce a set of defaults D with respect to a propositional theory $E \subseteq \mathcal{L}$ to a set of *inference rules* D_E which contains an inference rule α/γ whenever there is a default rule $\alpha : \beta_1, \ldots, \beta_n/\gamma \in D$ such that $E \cup \{\beta_i\}$ is consistent for all $0 < i \leq n$. We need also the closure of a theory $T \subseteq \mathcal{L}$ under a set of inference rules R, denoted by $\mathrm{Cn}^R(T)$, which is the least theory $E \subseteq \mathcal{L}$ satisfying (i) $T \subseteq E$, (ii) the set of propositional consequences $\mathrm{Cn}(E) = \{\phi \in \mathcal{L} \,|\, E \models \phi\} \subseteq E$ and (iii) $\{\gamma \,|\, \alpha/\gamma \in R \text{ and } \alpha \in E\} \subseteq E$.[1] The sets of conclusions associated with a default theory $\langle D, T\rangle$ are defined as follows.

Definition 1 (Marek and Truszczyński [18]). *A theory $E \subseteq \mathcal{L}$ is an extension of a default theory $\langle D, T\rangle$ if and only if $E = \mathrm{Cn}^{D_E}(T)$.*

2.2 Autoepistemic Logic (AEL)

An *autoepistemic language* $\mathcal{L}_{\mathbf{B}}$ is the unimodal extension of $\mathcal{L}$ with a modal operator $\mathbf{B}$ for beliefs whereas an *autoepistemic theory* $\Sigma \subseteq \mathcal{L}_{\mathbf{B}}$ [20]. Sentences of the form $\mathbf{B}\phi$ are known as *belief atoms* and the set of logical consequences $\mathrm{Cn}(\Sigma) \subseteq \mathcal{L}_{\mathbf{B}}$ is defined in the standard way by treating belief atoms as additional propositional atoms. In this paper, a pair of theories $\langle \Sigma, T\rangle$ where $T \subseteq \mathcal{L}$ is a propositional theory is also called an autoepistemic theory (then $\Sigma \cup T$ is a theory in Moore's sense). Moore's idea is to capture the sets of beliefs Δ of an *ideal and rational agent* which believes exactly the logical consequences of $\Sigma \subseteq \mathcal{L}_{\mathbf{B}}$ and its

[1] Marek and Truszczyński [18] propose a proof system to capture this closure.

beliefs $\mathbf{B}\Delta = \{\mathbf{B}\phi \mid \phi \in \Delta\}$ and disbeliefs $\neg\mathbf{B}\overline{\Delta} = \{\neg\mathbf{B}\phi \mid \phi \in \mathcal{L}_{\mathbf{B}} - \Delta\}$ obtained by *introspection*. Such sets of beliefs/conclusions are called *stable expansions*.

Definition 2 (Moore [20]). *A theory $\Delta \subseteq \mathcal{L}_{\mathbf{B}}$ is a stable expansion of an autoepistemic theory $\Sigma \subseteq \mathcal{L}_{\mathbf{B}}$ if and only if $\Delta = \mathrm{Cn}(\Sigma \cup \mathbf{B}\Delta \cup \neg\mathbf{B}\overline{\Delta})$.*

2.3 Strong Autoepistemic Logic (SAEL)

Theories of strong autoepistemic logic [17] are similar to those of AEL. Given $\Sigma \subseteq \mathcal{L}_{\mathbf{B}}$, we let $\mathrm{Cn}_{\mathbf{B}}(\Sigma)$ denote the closure of Σ under propositional inference and the standard *necessitation rule:* from ϕ infer $\mathbf{B}\phi$. More formally, $\mathrm{Cn}_{\mathbf{B}}(\Sigma)$ is the least theory $\Delta \subseteq \mathcal{L}_{\mathbf{B}}$ satisfying (i) $\Sigma \subseteq \Delta$, (ii) $\mathrm{Cn}(\Delta) \subseteq \Delta$ and (iii) $\mathbf{B}\Delta \subseteq \Delta$.[2] This leads to the definition of *iterative expansions* below. Iterative expansions of $\Sigma \subseteq \mathcal{L}_{\mathbf{B}}$ are also stable expansions of Σ, but not necessarily vice versa [17], and thus Σ is assigned a different semantics under iterative expansions.

Definition 3 (Marek and Truszczyński [17]). *A theory $\Delta \subseteq \mathcal{L}_{\mathbf{B}}$ is an iterative expansion of $\Sigma \subseteq \mathcal{L}_{\mathbf{B}}$ if and only if $\Delta = \mathrm{Cn}_{\mathbf{B}}(\Sigma \cup \neg\mathbf{B}\overline{\Delta})$.*

2.4 Parallel Circumscription (CIRC)

We present a generalization of McCarthy's approach [19], namely *parallel circumscription* by Lifschitz [14]. A *minimal model theory* is a triple $\langle P, F, T\rangle$ where $P \subseteq \mathcal{A}$ and $F \subseteq \mathcal{A}$ are mutually disjoint sets of atoms and $T \subseteq \mathcal{L}$. The idea behind parallel circumscription [14] is to distinguish propositional models $\mathcal{M}$ that are minimal in the sense of Definition 4: as many atoms of P should be false in $\mathcal{M}$ as possible. Note that the atoms of F remain *fixed* in the minimization process while the atoms in $\mathcal{A} - (P \cup F)$ may *vary* freely.

Definition 4 (Lifschitz [14]). *A propositional model $\mathcal{M} \subseteq \mathcal{A}$ of $T \subseteq \mathcal{L}$ is $\langle P, F\rangle$-minimal if and only if there is no propositional model $\mathcal{M}' \subseteq \mathcal{A}$ of T such that $\mathcal{M}' \cap F = \mathcal{M} \cap F$ and $\mathcal{M}' \cap P \subset \mathcal{M} \cap P$.*

There is no explicit notion of extensions involved in parallel circumscription, but let us propose an implicit one. Given a propositional model $\mathcal{M}$, we let $\mathrm{True}(\mathcal{M})$ denote the theory $\{\phi \in \mathcal{L} \mid \mathcal{M} \models \phi\}$. A $\langle P, F\rangle$-minimal model $\mathcal{M}$ gives rise to a $\langle P, F\rangle$-extension $E \subseteq \mathcal{L}$ of T which is the intersection of the theories $\mathrm{True}(\mathcal{M}')$ for all $\langle P, F\rangle$-minimal models $\mathcal{M}'$ of T such that $\mathcal{M}' \cap (P \cup F) = \mathcal{M} \cap (P \cup F)$. In this setting, the $\langle P, F\rangle$-minimal models of T are divided into (equivalence) classes that give rise to $\langle P, F\rangle$-extensions. Obviously, there may be several models in one class, since the atoms in $\mathcal{A} - (P \cup F)$ may vary freely. What comes to the cautious reasoning strategy, the correspondence of $\langle P, F\rangle$-extensions and $\langle P, F\rangle$-minimal models given in Proposition 1 is straightforward to establish. The notion of $\langle P, F\rangle$-extensions proposed is also appropriate if the brave[3] reasoning strategy is used in conjunction with parallel circumscription.

[2] Gottlob's **B**-proofs [8] capture this closure.

[3] Note, e.g., that Eiter and Gottlob [5] consider the complexity of propositional circumscription according to the cautious strategy only.

Proposition 1. *Given a minimal model theory $\langle P, F, T\rangle$, the intersection of $\langle P, F\rangle$-extensions of the theory $T \subseteq \mathcal{L}$ coincide with the intersection of the theories* $\mathrm{True}(\mathcal{M})$ *for all $\langle P, F\rangle$-minimal models of T.*

2.5 Priority Logic (PL)

A theory of *priority logic* [28] is a triple $\langle R, P, T\rangle$ where R is a set of (monotonic) inference rules[4] in $\mathcal{L}$, $P \subseteq R \times R$ gives a priority relation among the rules of R and $T \subseteq \mathcal{L}$. The idea behind prioritisation of inference rules is that if two rules r_1 and r_2 from R are in the relation P (denoted by $r_1 \prec r_2$ in [28]), then the application of the rule r_2 *blocks* that of r_1 and the rule r_2 has a higher priority than r_1 in this sense. For a set of rules R and a theory $E \subseteq \mathcal{L}$, we write $\mathrm{App}(R, E)$ to denote the set $\{\alpha/\gamma \in R \mid E \models \alpha\}$ which contains the rules of R that are applicable given E. To interpret the priority relation P, we define $\mathrm{Nb}(R, P, R') \subseteq R$ as the set of rules $r \in R$ which are *not blocked* given that the rules of R' are applicable, i.e. there is no $r' \in R'$ such that $\langle r, r'\rangle \in P$. These notions suffice to define extensions for a priority theory $\langle R, P, T\rangle$. The set of rules R' in the definition is called a *stable argument* by Wang et al. [28].

Definition 5. *A theory $E \subseteq \mathcal{L}$ is an extension of a priority theory $\langle R, P, T\rangle$ if and only if $E = \mathrm{Cn}^{R'}(T)$ for $R' \subseteq R$ satisfying $R' = \mathrm{App}(\mathrm{Nb}(R, P, R'), E)$.*

3 Requirements Imposed on Translation Functions

From now on, we restrict ourselves to **finite theories** of non-monotonic logics introduced in Section 2. In this section, we introduce the basic requirements for translation functions that map a theory of one non-monotonic logic to a theory of another. The requirements will be named as *polynomiality*, *faithfulness* and *modularity.* In the forthcoming mathematical formulations of these requirements, we let $\langle X, T\rangle$ stand for a non-monotonic theory where T is its propositional subtheory and X stands for any set(s) of syntactic elements which are specific to the non-monotonic logic in question (such as a set of defaults D in DL). The non-monotonic theories introduced in Section 2 are clearly of this form. Our first requirement involves the *length* of a non-monotonic theory $\langle X, T\rangle$, denoted by $||\langle X, T\rangle||$, which is the number of symbol occurrences needed to represent $\langle X, T\rangle$.

Definition 6 (Polynomiality). *A translation function* Tr *is polynomial, iff for all $\langle X, T\rangle$, the time required to compute* $\mathrm{Tr}(\langle X, T\rangle)$ *is polynomial in* $||\langle X, T\rangle||$.

To give an example of a such a function, we introduce a linear function that transforms a minimal model theory into an autoepistemic one.

[4] Wang et al. [28] use rules of the form $\gamma \leftarrow \alpha_1, \ldots, \alpha_n$ with multiple prerequisites, but such rules can be represented as $\alpha_1 \wedge \ldots \wedge \alpha_n/\gamma$ under propositional closure.

Definition 7 (Niemelä [22]). *For all minimal model theories* $\langle P, F, T\rangle$, *let* $\mathrm{Tr}_{\mathrm{N}}(\langle P, F, T\rangle) = \langle\{\neg\mathbf{B}\mathrm{a} \rightarrow \neg\mathrm{a} \mid \mathrm{a} \in P \cup F\} \cup \{\neg\mathbf{B}\neg\mathrm{a} \rightarrow \mathrm{a} \mid \mathrm{a} \in F\}, T\rangle$.

The next question is whether a translation function Tr preserves the semantics of a non-monotonic theory $\langle X, T\rangle$. We have used the following criteria to formulate our forthcoming definition. (i) Since the semantics of $\langle X, T\rangle$ is determined by its extensions and both brave and cautious reasoning strategies should be supported, a one-to-one correspondence of extensions is a natural solution. (ii) Only propositionally consistent extensions are taken into account in this one-to-one relationship, because we have in mind translation functions (such as Tr_1 in Definition 10) whose faithfulness depends on this restriction. (iii) Moreover, we are assuming that the language $\mathcal{L}$ of T is a fixed propositional language which is used for knowledge representation in a given domain. The propositional languages associated with $\langle X, T\rangle$ and $\mathrm{Tr}(\langle X, T\rangle)$ may extend $\mathcal{L}$, but we project the extensions of these theories with respect to $\mathcal{L}$. In particular, this means that a translation function can add new atoms, but within the bounds of our polynomiality requirement. This seems a crucial option in order to support different kinds of knowledge representation and reasoning techniques. For instance, the translation function Tr_{N} introduces belief atoms for these reasons.

Definition 8 (Faithfulness). *A translation function* Tr *is faithful, iff for all* $\langle X, T\rangle$, *the propositionally consistent extensions of* $\langle X, T\rangle$ *and* $\mathrm{Tr}(\langle X, T\rangle)$ *are in one-to-one correspondence and coincide up to the propositional language* $\mathcal{L}$ *of* T.

This definition ensures that given a faithful translation function Tr, any brave or cautious conclusion $\phi \in \mathcal{L}$ obtained from $\langle X, T\rangle$ can also be obtained from the translation $\mathrm{Tr}(\langle X, T\rangle)$, and vice versa. Note that this presumes that only propositionally consistent extensions are taken into account in the brave strategy. Let us yet point out that our notion is useful only if a notion of extensions is available for the (non-monotonic) logics involved. Fortunately, this is the case with logics addressed in this paper. Our notion of faithfulness is also closely related to the one by Gottlob [9]. The differences are that Gottlob does not allow new atoms to be introduced in a translation and he takes also the propositionally inconsistent extensions into account. Consequently, a translation function that is faithful in Gottlob's sense is also faithful in our sense. The converse does not hold in general – which is to be demonstrated in Theorem 4 and Example 1.

By Niemelä's results [22] and the notion of $\langle P, F\rangle$-extensions proposed in this paper, the translation function introduced in Definition 7 is faithful: the $\langle P, F\rangle$-extensions of T and the propositionally consistent stable expansions of $\mathrm{Tr}_{\mathrm{N}}(\langle P, F, T\rangle)$ are in a one-to-one correspondence and coincide up to the language $\mathcal{L}$ of T. An inconsistent stable expansion $\Delta = \mathcal{L}_{\mathbf{B}}$ appears only if T is inconsistent and there are no $\langle P, F\rangle$-extensions. Our last requirement follows.

Definition 9 (Modularity, Gottlob [9]). *A translation function* Tr *is modular, iff for all* $\langle X, T\rangle$, $\mathrm{Tr}(\langle X, T\rangle) = \langle X', T' \cup T\rangle$ *where* $\langle X', T'\rangle = \mathrm{Tr}(\langle X, \emptyset\rangle)$.

Our modularity requirement is a generalization of the one that Gottlob formulated for translations from DL into AEL [9]. In particular, a modular translation function provides a fixed translation for X (i.e. the non-monotonic theory

$\mathrm{Tr}(\langle X, \emptyset \rangle))$ which is independent of T. Therefore, if T is updated, there is no need to recompute the fixed part $\mathrm{Tr}(\langle X, \emptyset \rangle)$ in order to compute $\mathrm{Tr}(\langle X, T \rangle)$. Note also that the translation function Tr_{N} of Definition 7 is modular in this sense, because P and F are translated into a fixed autoepistemic theory.

For the sake of brevity, we say that a **translation function is PFM** if it satisfies the three requirements set up in Definitions 6, 8 and 9. A fundamental property of PFM translations is pointed out in the following.

Proposition 2. *A composition of PFM translation functions is also PFM.*

PFM translation functions provide us the basis for analyzing the relative expressive power of non-monotonic logics. The motivation for this is that if there is a PFM translation function that maps theories of a non-monotonic logic L_1 to theories of a non-monotonic logic L_2, then we consider L_2 to be *at least as expressive as* L_1. This gives rise to a preorder among (non-monotonic) logics. For instance, AEL is at least as expressive as CIRC, because Tr_{N} is PFM. If – *in addition* – there are **no** PFM translation functions in the opposite direction, then we say that L_1 is *less expressive* than L_2. If there are PFM translation functions in both directions, then L_1 and L_2 are of equal expressive power. As concluded by Gottlob [9], this view identifies the expressive power of non-monotonic logics with their capability of representing different propositional closures in $\mathcal{L}$.

As a final issue in this section, we compare our approach with another by Gogic, Kautz, Papadimitriou and Selman [7]. They propose a framework for analyzing the *succinctness* of knowledge representation (i.e. the space required in knowledge representation) and thus also the expressive power of formalisms involved, but different kinds of translation functions are used. (i) Gogic et al. use a different polynomiality requirement: the length of the translation has to be polynomial in the length of the theory. This allows even exponential computations to obtain a translation as long as only a polynomial blow-up results in the translation. Our requirement restricts the translation time and thus also the translation space to be polynomial. (ii) Gogic et al. formulate their notion of faithfulness as a requirement that the propositional models of the theory under translation are preserved. If a translation function is faithful in our sense, then it is in their sense, too, provided that models of extensions are taken into account up to $\mathcal{L}$. The converse does not hold in general, since our notion of faithfulness presumes that a notion of extensions is available for the non-monotonic logics involved. (iii) Gogic et al. do not employ a modularity requirement.

4 Classifying Non-monotonic Logics

Having set up the notion of a PFM translation function, such translation functions are exhibited in this section in order to classify non-monotonic logics by their expressive power. Moreover, counter-examples are provided to show that such translations are not possible in certain cases. Such non-equivalence proofs have already been devised for non-monotonic logics by Imielinski [10], Gottlob

[9] and Niemelä [23]. In the forthcoming subsections, we perform a pairwise comparison of non-monotonic logics in the following order: CIRC, AEL, SAEL, DL and PL. But as a starter, we relate classical propositional logic (CL) with CIRC. This result is supported by other complexity and intranslatability results [3–5].

Theorem 1. *CL is less expressive than CIRC.*

Proof. A propositional theory $T \subseteq \mathcal{L}$ is translated into a minimal model theory $\mathrm{Tr}_0(T) = \langle \emptyset, \emptyset, T \rangle$ having the same propositional language $\mathcal{L}$. The only $\langle \emptyset, \emptyset \rangle$-extension of T is $\mathrm{Cn}(T)$, i.e. the natural "extension" of T in propositional logic. Thus it is easy to see that Tr_0 is PFM and CIRC is at least as expressive as CL. Let us then assume there is also a PFM translation function Tr in the other direction. Let $\mathcal{A} = \{\mathrm{a}, \mathrm{b}\}$ and $\mathcal{L}$ the corresponding propositional language. Then consider a minimal model theory $\langle P, F, T \rangle$ based on $\mathcal{L}$ where $T = \{\mathrm{a} \rightarrow \mathrm{b}\}$, $P = \{\mathrm{a}, \mathrm{b}\}$ and $F = \emptyset$. This has a unique $\langle P, F \rangle$-minimal model $\mathcal{M} = \emptyset$ so that a unique $\langle P, F \rangle$-extension $E = \mathrm{Cn}(\{\neg \mathrm{a}, \neg \mathrm{b}\})$ results. Thus the propositional translation $\mathrm{Tr}(\langle P, F, T \rangle)$ must entail $\neg$a and $\neg$b. However, if we update T to $T' = T \cup \{a\}$, there is a unique $\langle P, F \rangle$-extension $E' = \mathrm{Cn}(\{\mathrm{a}, \mathrm{b}\})$ of T'. By modularity, the translation $\mathrm{Tr}(\langle P, F, T' \rangle)$ has to be $\mathrm{Tr}(\langle P, F, T \rangle) \cup \{\mathrm{a}\}$ which is necessarily propositionally inconsistent. Thus Tr cannot be faithful, a contradiction.

4.1 Comparison of CIRC and AEL

The translation function Tr_{N} satisfies our requirements by Niemelä's results [22].

Theorem 2 (Niemelä [22]). *The translation function* Tr_{N} *is PFM.*

This indicates that reasoning corresponding to $\langle P, F \rangle$-minimal models is easily captured in terms of stable expansions of the translation and that AEL is at least as expressive as CIRC. In Theorem 3, we adopt a counter-example given by Niemelä [23] to show that there is no translation function meeting our criteria in the opposite direction. Thus CIRC is less expressive than AEL.

Theorem 3. *There is* **no** *PFM translation function from autoepistemic theories under stable expansions into minimal model theories.*

Proof. Let $\mathcal{A} = \{\mathrm{a}, \mathrm{b}\}$ and let $\mathcal{L}$ and $\mathcal{L}_{\mathbf{B}}$ be the respective propositional and autoepistemic languages. Let us make a hypothesis that there is a fixed polynomial translation of $\Sigma = \{\mathbf{B}\mathrm{a} \rightarrow \mathrm{b}\} \subseteq \mathcal{L}_{\mathbf{B}}$ into sets of atoms P and F and a propositional theory $\mathrm{Tr}(\Sigma)$ such that for all $T \subseteq \mathcal{L}$, the propositionally consistent stable expansions of $\langle \Sigma, T \rangle$ and $\langle P, F \rangle$-extensions of $\mathrm{Tr}(\Sigma) \cup T$ are in one-to-one correspondence and coincide up to $\mathcal{L}$. The language $\mathcal{L}'$ of $\mathrm{Tr}(\Sigma) \cup T$ is assumed to be based on $\mathcal{A}' \supseteq \mathcal{A}$ and the sets of atoms P and F are subsets of $\mathcal{A}'$.

For $T = \emptyset$, there is exactly one propositionally consistent stable expansion $\Delta = \{\neg \mathbf{B}\mathrm{a}, \neg \mathbf{B}\mathrm{b}, \ldots\}$ of $\langle \Sigma, T \rangle$ such that $\mathrm{a} \rightarrow \mathrm{b} \notin \Delta$. It follows by our hypothesis that there is a unique $\langle P, F \rangle$-extension E of $\mathrm{Tr}(\Sigma) \cup T$ such that $\Delta \cap \mathcal{L} = E \cap \mathcal{L}$. This implies that $\mathrm{a} \rightarrow \mathrm{b} \notin E$. So there is a $\langle P, F \rangle$-minimal model $\mathcal{M}$ of $\mathrm{Tr}(\Sigma) \cup T$

such that $\mathcal{M} \not\models \mathrm{a} \to \mathrm{b}$, i.e. $\mathcal{M} \models \mathrm{a}$ and $\mathcal{M} \not\models \mathrm{b}$. Then let $T' = \{\mathrm{a}\}$ so that also $\mathcal{M} \models \mathrm{Tr}(\Sigma) \cup T'$. It is easy to see that $\mathcal{M}$ is a $\langle P, F\rangle$-minimal model of $\mathrm{Tr}(\Sigma) \cup T'$, since otherwise $\mathcal{M}$ would not be a $\langle P, F\rangle$-minimal model of $\mathrm{Tr}(\Sigma) \cup T$. It follows that b does not belong to the corresponding (propositionally consistent) $\langle P, F\rangle$-extension E' of $\mathrm{Tr}(\Sigma) \cup T'$. By our hypothesis, there is a propositionally consistent stable expansion Δ' of $\langle \Sigma, T'\rangle$ such that $\mathrm{b} \notin \Delta'$. But this is a contradiction, since the only stable expansion of $\langle \Sigma, T'\rangle$ is $\{\mathrm{a}, \mathbf{B}\mathrm{a}, \mathrm{b}, \mathbf{B}\mathrm{b} \ldots\}$. □

4.2 Comparison of AEL and SAEL

The author [11] presents a translation function that allows one to capture the stable expansions of an autoepistemic theory with the iterative expansions of the translation. We let $\mathrm{RBa}(\phi)$ denote the set of belief atoms that appear in an autoepistemic sentence ϕ recursively and define $\mathrm{RBa}(\Sigma) = \bigcup\{\mathrm{RBa}(\phi) \mid \phi \in \Sigma\}$ for sets of autoepistemic sentences Σ. To give a simple example, we note that $\mathrm{RBa}(\mathbf{B}(\mathbf{B}\mathrm{p} \wedge \mathbf{B}\mathrm{q})) = \{\mathbf{B}(\mathbf{B}\mathrm{p} \wedge \mathbf{B}\mathrm{q}), \mathbf{B}\mathrm{p}, \mathbf{B}\mathrm{q}\}$. The intuition behind the translation is that the positive introspection ($\mathbf{B}\Delta$) in the definition of stable expansions is realized using instances $\neg\mathbf{B}\neg\mathbf{B}\phi \to \mathbf{B}\phi$ of the axiom schema **5**.

Definition 10 (Janhunen [11]). *For all autoepistemic theories $\langle \Sigma, T\rangle$, the translation* $\mathrm{Tr}_1(\langle \Sigma, T\rangle) = \langle \Sigma \cup \{\neg\mathbf{B}\neg\mathbf{B}\phi \to \mathbf{B}\phi \mid \mathbf{B}\phi \in \mathrm{RBa}(\Sigma)\}, T\rangle$.

Theorem 4. *The translation function* Tr_1 *given in Definition 10 is PFM.*

Proof. The polynomiality and modularity of Tr_1 are easily seen from the definition. Faithfulness follows by the results of Marek et al. [16] and the author [11], namely the propositionally consistent stable expansions of $\langle \Sigma, T\rangle$ and the iterative expansions of $\mathrm{Tr}_1(\langle \Sigma, T\rangle)$ coincide. This implies the one-to-one correspondence of propositionally consistent expansions as required by Definition 8 so that the translation function Tr_1 is also faithful. Note that Tr_1 is not faithful in Gottlob's sense [9], since an autoepistemic theory $\langle \Sigma, \emptyset\rangle$ where $\Sigma = \{\mathbf{B}\mathrm{p} \to \mathrm{p} \wedge \mathrm{r},\ \mathbf{B}\mathrm{q} \to \mathrm{q} \wedge \neg\mathrm{r}\}$ [11] has a propositionally inconsistent stable expansion $\Delta = \mathcal{L}_{\mathbf{B}}$ which is not an iterative expansion of $\mathrm{Tr}_1(\langle \Sigma, \emptyset\rangle)$. □

Theorem 5 shows that a PFM translation in the opposite direction is not possible. The proof is obtained by modifying Gottlob's proof [9] which shows that a modular translation from DL into AEL cannot be realized (an analog of this result is considered later as Corollary 1). Theorems 4 and 5 signify together that AEL is *less expressive* than SAEL.

Theorem 5. *There is* **no** *PFM translation function from autoepistemic theories under iterative expansions into such theories under stable expansions.*

Proof. Let $\mathcal{A} = \{\mathrm{a}, \mathrm{b}\}$ be a set of atoms and let $\mathcal{L}$ and $\mathcal{L}_{\mathbf{B}}$ be the respective propositional and autoepistemic languages. Let us then make a hypothesis that there is a fixed polynomial translation of $\Sigma = \{\mathbf{B}\mathrm{a} \to \mathrm{b}, \mathbf{B}(\mathrm{a} \to \mathrm{b}) \to \mathrm{a}\} \subseteq \mathcal{L}_{\mathbf{B}}$ into $\Sigma' \subseteq \mathcal{L}'_{\mathbf{B}}$ and $T' \subseteq \mathcal{L}'$ such that for all $T \subseteq \mathcal{L}$ the propositionally consistent iterative expansions of $\langle \Sigma, T\rangle$ and the propositionally consistent stable expansions of $\langle \Sigma', T' \cup T\rangle$ are in one-to-one correspondence and coincide up to $\mathcal{L}$. The languages $\mathcal{L}'$ and $\mathcal{L}'_{\mathbf{B}}$ are assumed to be based on a set of atoms $\mathcal{A}' \supseteq \mathcal{A}$. Note that we may assume a single translation $\mathrm{Tr}(\Sigma) = \Sigma' \cup T'$ without a loss of generality, since $\langle \Sigma', T' \cup T\rangle$ and $\langle \mathrm{Tr}(\Sigma), T\rangle$ are effectively the same in AEL. Let us then introduce propositional theories $T_0 = \emptyset$, $T_1 = \{\mathrm{a}\}$, $T_2 = \{\mathrm{a} \rightarrow \mathrm{b}\}$ and $T_3 = \{\mathrm{a}, \mathrm{a} \rightarrow \mathrm{b}\}$. For $i \in \{1,2,3\}$, the autoepistemic theory $\langle \Sigma, T_i\rangle$ has a unique, propositionally consistent iterative expansion $\Delta = \{\mathrm{a}, \mathbf{B}\mathrm{a}, \mathrm{b}, \mathbf{B}\mathrm{b}, \mathrm{a} \rightarrow \mathrm{b}, \mathbf{B}(\mathrm{a} \rightarrow \mathrm{b})\}$. As Tr is faithful, there are unique propositionally consistent stable expansions Δ'_i of $\langle \mathrm{Tr}(\Sigma), T_i\rangle$ that coincide with Δ up to $\mathcal{L}$.

Since $\mathrm{a} \rightarrow \mathrm{b} \in \Delta \cap \mathcal{L}$, it follows that $\mathrm{a} \rightarrow \mathrm{b} \in \Delta'_1$. Since Δ'_1 is a stable expansion of $\langle \mathrm{Tr}(\Sigma), T_1\rangle$, it holds that $\Delta'_1 = \mathrm{Cn}(\mathrm{Tr}(\Sigma) \cup T_1 \cup \mathbf{B}\Delta'_1 \cup \neg\mathbf{B}\overline{\Delta'_1})$ and thus $\mathrm{Tr}(\Sigma) \cup T_1 \cup \mathbf{B}\Delta'_1 \cup \neg\mathbf{B}\overline{\Delta'_1} \models \mathrm{a} \rightarrow \mathrm{b}$. It follows that also $\Delta'_1 = \mathrm{Cn}(\mathrm{Tr}(\Sigma) \cup T_3 \cup \mathbf{B}\Delta'_1 \cup \neg\mathbf{B}\overline{\Delta'_1})$, i.e. Δ'_1 is a stable expansion of $\langle \mathrm{Tr}(\Sigma), T_3\rangle$. Because $a \in \Delta \cap \mathcal{L}$, it follows similarly that Δ'_2 is a stable expansion of $\langle \mathrm{Tr}(\Sigma), T_3\rangle$. Then $\Delta'_1 = \Delta'_2 = \Delta'_3$ is necessarily the case, as Δ'_3 is the unique stable expansion of $\langle \mathrm{Tr}(\Sigma), T_3\rangle$. So let Δ' denote any of Δ'_i with $i \in \{1,2,3\}$. Since $\mathrm{b} \in \Delta \cap \mathcal{L}$, we know that $\mathrm{b} \in \Delta'$. Then it follows that $\mathrm{Tr}(\Sigma) \cup \{\mathrm{a}\} \cup \mathbf{B}\Delta' \cup \neg\mathbf{B}\overline{\Delta'} \models \mathrm{b}$ and the deduction theorem of propositional logic implies $\mathrm{Tr}(\Sigma) \cup \mathbf{B}\Delta' \cup \neg\mathbf{B}\overline{\Delta'} \models \mathrm{a} \rightarrow \mathrm{b}$. Thus $\Delta' = \mathrm{Cn}(\mathrm{Tr}(\Sigma) \cup \mathbf{B}\Delta' \cup \neg\mathbf{B}\overline{\Delta'})$, i.e. Δ' is a propositionally consistent stable expansion of $\langle \mathrm{Tr}(\Sigma), T_0\rangle$. Then $\langle \Sigma, T_0\rangle$ has a propositionally consistent iterative expansion which contains both a and b. But this is contradiction, since the only iterative expansion of $\langle \Sigma, T_0\rangle$ is $\Delta'' = \{\neg\mathbf{B}\mathrm{a}, \neg\mathbf{B}\mathrm{b}, \neg\mathbf{B}(\mathrm{a} \rightarrow \mathrm{b}), \ldots\}$. □

4.3 Comparison of SAEL and DL

The author [11] has proposed an idea of representing autoepistemic introspection in terms of default rules. In this approach, default rules of the forms $\frac{\phi:}{\mathbf{B}\phi}$ and $\frac{:\neg\phi}{\neg\mathbf{B}\phi}$ capture the *positive* and the *negative* introspection of an autoepistemic sentence $\phi \in \mathcal{L}_{\mathbf{B}}$, respectively. A translation function is obtained as follows.

Definition 11 (Janhunen [11]). *For all autoepistemic theories* $\langle \Sigma, T\rangle$, *let* $\mathrm{Tr}_2(\langle \Sigma, T\rangle) = \langle \{\frac{\phi:}{\mathbf{B}\phi} \mid \mathbf{B}\phi \in \mathrm{RBa}(\Sigma)\} \cup \{\frac{:\neg\phi}{\neg\mathbf{B}\phi} \mid \mathbf{B}\phi \in \mathrm{RBa}(\Sigma)\}, \Sigma \cup T\rangle$.

The propositional language $\mathcal{L}'$ of the translation $\mathrm{Tr}_2(\langle \Sigma, T\rangle)$ is assumed to contain atoms that correspond to the belief atoms in $\mathrm{RBa}(\Sigma)$ exactly.

Theorem 6. *The translation function* Tr_2 *given in Definition 11 is PFM.*

Proof. The translation function Tr_2 is clearly polynomial and modular. For the faithfulness of the translation, we refer to results shown by the author elsewhere [11, Theorem 13 and Proposition 16]. First of all, the author shows that the iterative expansions of an autoepistemic theory $\Sigma \subseteq \mathcal{L}_{\mathbf{B}}$ and the extensions of a translation $\langle \{\frac{\phi:}{\mathbf{B}\phi} \mid \phi \in \mathcal{L}_{\mathbf{B}}\} \cup \{\frac{:\neg\phi}{\neg\mathbf{B}\phi} \mid \phi \in \mathcal{L}_{\mathbf{B}}\}, \Sigma\rangle$ coincide (note that this is an extended and infinite translation). A translation obtained by Tr_2 is limited to belief atoms in $\mathrm{RBa}(\Sigma)$ and thus it captures essentially *full sets* [8] of Σ. This implies the one-to-one correspondence between the iterative expansions of

$\langle \Sigma, T\rangle$ and the extensions of $\mathrm{Tr}_2(\langle \Sigma, T\rangle)$. In addition, the propositional parts of expansions and extensions in question coincide. □

A number of principles have been proposed to translate default theories into autoepistemic ones. Basically, the problem is to translate a default $\alpha : \beta_1, \ldots, \beta_n/\gamma$ into an autoepistemic sentence. Konolige [13] introduces a translation $\mathbf{B}\alpha \wedge \neg\mathbf{B}\neg\beta_1 \wedge \ldots \wedge \neg\mathbf{B}\neg\beta_n \rightarrow \gamma$ for a default. Unfortunately, the resulting translation function Tr_K for default theories is not faithful in general, as shown by Marek and Truszczyński [17]. As a response to this problem, they handle justifications β_i differently: $\neg\mathbf{B}\neg\beta_i$ is replaced by $\neg\mathbf{B}\mathbf{B}\neg\beta_i$ in their translation. Later, Truszczyński [27] ends up with a translation $\mathbf{B}\neg\mathbf{B}\neg\beta_i$ for justifications. This gives rise to a translation function for default theories as follows.

Definition 12 (Truszczyński [27]). *For all default theories $\langle D, T\rangle$, define* $\mathrm{Tr}_\mathrm{T}(\langle D, T\rangle) = \langle \{\mathbf{B}\alpha \wedge \mathbf{B}\neg\mathbf{B}\neg\beta_1 \wedge \ldots \wedge \mathbf{B}\neg\mathbf{B}\neg\beta_n \rightarrow \gamma \mid \frac{\alpha:\beta_1,\ldots,\beta_n}{\gamma} \in D\}, T\rangle$.

Theorem 7 (Marek and Truszczyński [18]). *The translation function Tr_T given in Definition 12 is PFM.*

It is worth mentioning that the above result holds as long as the notion of faithfulness takes only the propositionally consistent expansions of the translation into account. Niemelä [22] demonstrates that for a set of defaults $D = \{\frac{\neg\mathrm{b}:}{\neg\mathrm{a}}, \frac{:\mathrm{b}}{\neg\mathrm{b}}\}$ and a theory $T = \{\mathrm{a}\}$ the translation $\mathrm{Tr}_\mathrm{T}(\langle D, T\rangle)$ has an inconsistent iterative expansion while $\langle D, T\rangle$ has no extensions. However, Gottlob [9] proposes a variant of Tr_T that avoids such inconsistent iterative expansions.

Since PFM translations exist in both directions, we conclude that SAEL and DL have an equal expressive power according to the measure set up in Section 3. Note also that the theorems presented so far constitute an indirect proof of the following corollary. Therefore, Gottlob's intranslatability result [9] remains valid although a weaker notion of faithfulness is applied.

Corollary 1 (Gottlob [9]). *There is* **no** *PFM translation function from default theories into autoepistemic theories under stable expansions.*

Proof. Assume there is such a function. By Theorem 6 and Proposition 2, there is a PFM translation function that maps autoepistemic theories under iterative expansions to ones under stable expansions. But this contradicts Theorem 5. □

In spite of this result, Gottlob [9] sets up a non-modular translation function Tr_G to capture the extensions of a default theory $\langle D, T\rangle$ with the stable expansions of $\mathrm{Tr}_\mathrm{G}(\langle D, T\rangle)$.[5] Then he provides a counter-example showing that faithful translations in the other direction are not possible and concludes then that DL is less expressive than AEL. This conclusion is in contrast with Theorems 4 and 6 and Proposition 2 which indicate that there is a PFM translation function (the composition of Tr_1 and Tr_2) for this purpose. The difference between the two views is due to the notions of faithfulness considered. Gottlob assumes that the

[5] Schwarz [25] proposes an alternative translation for this purpose.

language $\mathcal{L}$ of the default theory $\langle D, T\rangle$ obtained as a translation of an autoepistemic theory $\Sigma \subseteq \mathcal{L}_{\mathbf{B}}$ (under stable expansions) is the propositional sublanguage of $\mathcal{L}_{\mathbf{B}}$. However, we are ready to introduce new atoms to extend $\mathcal{L}$.

To illustrate the effect of new atoms, we construct a default theory in order to capture the (propositionally consistent) stable expansions of an autoepistemic theory $\Sigma = \{\mathbf{B}\mathrm{p} \rightarrow \mathrm{p}\}$ used in Gottlob's counter-example [9].

Example 1. Let $\mathcal{A} = \{\mathrm{p}\}$ and let $\mathcal{L}$ and $\mathcal{L}_{\mathbf{B}}$ be the respective propositional and autoepistemic languages. Then let $\Sigma = \{\mathbf{B}\mathrm{p} \rightarrow \mathrm{p}\} \subseteq \mathcal{L}_{\mathbf{B}}$. The stable expansions of $\langle \Sigma, \emptyset\rangle$ are $\Delta_1 = \{\neg\mathbf{B}\mathrm{p}, \mathbf{B}\neg\mathbf{B}\mathrm{p}, \ldots\}$ and $\Delta_2 = \{\mathbf{B}\mathrm{p}, \mathrm{p}, \neg\mathbf{B}\neg\mathbf{B}\mathrm{p}, \ldots\}$ so that $\Delta_1 \cap \mathcal{L} = \mathrm{Cn}(\emptyset)$ and $\Delta_2 \cap \mathcal{L} = \mathrm{Cn}(\{\mathrm{p}\})$. Since $\mathrm{Cn}(\emptyset) \subseteq \mathrm{Cn}(\{\mathrm{p}\})$, Gottlob [9] concludes that there is no default theory with extensions $\mathrm{Cn}(\emptyset)$ and $\mathrm{Cn}(\{\mathrm{p}\})$, because the extensions of a default theory form an antichain [24].

As the first step, we apply Tr_1 and add an instance of the schema **5** to Σ and obtain $\Sigma' = \{\mathbf{B}\mathrm{p} \rightarrow \mathrm{p}, \neg\mathbf{B}\neg\mathbf{B}\mathrm{p} \rightarrow \mathbf{B}\mathrm{p}\}$. It is easy to see that Δ_1 and Δ_2 are the iterative expansions of $\langle \Sigma', \emptyset\rangle$. In particular, note that $\neg\mathbf{B}\mathrm{p} \notin \Delta_2$ implies that $\neg\mathbf{B}\neg\mathbf{B}\mathrm{p} \in \neg\mathbf{B}\overline{\Delta_2}$ so that $\mathbf{B}\mathrm{p}$ and p are $\mathbf{B}$-provable from $\Sigma' \cup \neg\mathbf{B}\overline{\Delta_2}$, since Σ' contains the critical instance $\neg\mathbf{B}\neg\mathbf{B}\mathrm{p} \rightarrow \mathbf{B}\mathrm{p}$ of **5**.

The next step is to apply Tr_2. An extended propositional language $\mathcal{L}'$ based on a set of atoms $\mathcal{A}' = \mathcal{A} \cup \{\mathbf{B}\mathrm{p}, \mathbf{B}\neg\mathbf{B}\mathrm{p}\}$ is introduced. The set of defaults introduced by Tr_2 is $D = \{\frac{\mathrm{p}:}{\mathbf{B}\mathrm{p}}, \frac{:\neg\mathrm{p}}{\neg\mathbf{B}\mathrm{p}}, \frac{\neg\mathbf{B}\mathrm{p}:}{\mathbf{B}\neg\mathbf{B}\mathrm{p}}, \frac{:\neg\neg\mathbf{B}\mathrm{p}}{\neg\mathbf{B}\neg\mathbf{B}\mathrm{p}}\}$. Consequently, the extensions of the resulting default theory $\langle D, \Sigma'\rangle$ are $E_1 = \mathrm{Cn}(\Sigma' \cup \{\neg\mathbf{B}\mathrm{p}, \mathbf{B}\neg\mathbf{B}\mathrm{p}\})$ and $E_2 = \mathrm{Cn}(\Sigma' \cup \{\mathbf{B}\mathrm{p}, \neg\mathbf{B}\neg\mathbf{B}\mathrm{p}\})$, because the reductions of D are $D_{E_1} = \{\mathrm{p}/\mathbf{B}\mathrm{p}, \top/\neg\mathbf{B}\mathrm{p}, \neg\mathbf{B}\mathrm{p}/\mathbf{B}\neg\mathbf{B}\mathrm{p}\}$ and $D_{E_2} = \{\mathrm{p}/\mathbf{B}\mathrm{p}, \neg\mathbf{B}\mathrm{p}/\mathbf{B}\neg\mathbf{B}\mathrm{p}, \top/\neg\mathbf{B}\neg\mathbf{B}\mathrm{p}\}$. It follows that $E_1 \cap \mathcal{L} = \mathrm{Cn}(\emptyset)$ and $E_2 \cap \mathcal{L} = \mathrm{Cn}(\{\mathrm{p}\})$. Thus the stable expansions of $\langle \Sigma, \emptyset\rangle$ and the extensions of $\langle D, \Sigma'\rangle$ coincide up to $\mathcal{L}$. In particular, the extended language $\mathcal{L}'$ allows the relationship $E_1 \cap \mathcal{L} \subseteq E_2 \cap \mathcal{L}$, although $E_1 \not\subseteq E_2$.

Bonatti and Eiter [2] analyze the expressive power of non-monotonic logics as query languages for disjunctive databases. A comparison with our results is possible in the propositional case, if a restriction to empty databases is made. Then Theorems 6.3 and 7.3 in [2] speak about the intertranslatability of non-monotonic theories. These theorems involve two translation functions, $\mathrm{Tr}_{\mathrm{BE}}$ and Tr_{K}. The latter function Tr_{K} is due to Konolige [13], and it allows one to capture the extensions of a prerequisite-free[6] default theory $\langle D, T\rangle$ in terms of stable expansions of the translation $\mathrm{Tr}_{\mathrm{K}}(\langle D, T\rangle)$. The results by Marek and Truszczyński [18, Section 12.5] and Gottlob [9] suggest that this translation is PFM in our sense, implying that AEL is at least as expressive as prerequisite-free default logic (PDL). It follows by Corollary 1 and compositionality that there is **no** PFM translation from default theories into prerequisite-free ones. Interestingly, the translation function $\mathrm{Tr}_{\mathrm{BE}}$ is proposed to remove prerequisites from a default theory. However, such a translation cannot be PFM by our remarks above. It seems that $\mathrm{Tr}_{\mathrm{BE}}$ is polynomial and modular so that $\mathrm{Tr}_{\mathrm{BE}}$ cannot be faithful in our sense. Indeed, the idea behind $\mathrm{Tr}_{\mathrm{BE}}$ is to simulate the defaults of the original default theory $\langle D, T\rangle$ without actually applying them and consequently the

[6] A default $\alpha : \beta_1, \ldots, \beta_n/\gamma$ is called prerequisite-free, if $\alpha = \top$.

extensions produced for the translation $\mathrm{Tr_{BE}}(\langle D,T\rangle)$ do not coincide with the extensions of $\langle D,T\rangle$ up to $\mathcal{L}$ (i.e. the language of $\langle D,T\rangle$). Moreover, Theorem 6.3 in [2] does not establish a one-to-one correspondence of extensions.

Marek et al. [15] and Engelfriet et al. [6] propose mappings to translate a default theory $\langle D,T\rangle$ into a prerequisite-free one such that extensions are preserved. However, these translations introduce a new default for each *quasi-proof* which is a sequence of defaults from D. Consequently, these translations are not polynomial in general and no contradiction arises with Corollary 1 as discussed above. Let us also note that the latter approach [6] deals with *infinitary* defaults whereas only finite defaults and default theories are considered here.

4.4 Comparison of DL and PL

Wang, You and Yuan [28] propose a translation of a default theory $\langle D,T\rangle$ into a priority theory $\langle R,P,T\rangle$. The idea is to break a default $\alpha:\beta_1,\ldots,\beta_n/\gamma \in D$ to inference rules α/γ, $\neg\beta_1/\neg\beta_1,\ldots,\neg\beta_n/\neg\beta_n$ to be included in R. The priority relation P is chosen such that the rule α/γ has a lower priority than the rules $\neg\beta_1/\neg\beta_1,\ldots,\neg\beta_n/\neg\beta_n$. As reported by Wang et al., this translation is faithful only if *dissimilar* sets of defaults D are considered, i.e. sets of defaults D which do not contain two defaults with exactly same prerequisite α and consequent γ. Wang et al. argue that this restriction is not significant, since it is possible to differentiate the prerequisites of defaults without changing their semantics. An unrestricted translation function introduces a new atom p_d for each $d \in D$.

Definition 13. *For all default theories $\langle D,T\rangle$, the translation $\mathrm{Tr_W}(\langle D,T\rangle) = \langle R,P,T\rangle$ where R and P are such that for each $d = \frac{\alpha:\beta_1,\ldots,\beta_n}{\gamma} \in D$, (i) the rules $\alpha\wedge(\mathrm{p}_d\vee\neg\mathrm{p}_d)/\gamma$ and $\neg\beta_1/\neg\beta_1,\ldots,\neg\beta_n/\neg\beta_n$ belong to R and (ii) the rule $\alpha\wedge(\mathrm{p}_d\vee\neg\mathrm{p}_d)/\gamma$ is in the relation P with the rules $\neg\beta_1/\neg\beta_1,\ldots,\neg\beta_n/\neg\beta_n$.*

Theorem 8. *The translation function $\mathrm{Tr_W}$ given in Definition 13 is PFM.*

Proof. It is clear that $\mathrm{Tr_W}$ is polynomial and modular. To establish the faithfulness of $\mathrm{Tr_W}$, we note that adding the tautology $\mathrm{p}_d\vee\neg\mathrm{p}_d$ to the prerequisite of a default $d \in D$ does not affect the applicability of the default d by any means. Let D' denote D modified in this way. It is clear that the extensions of $\langle D,T\rangle$ and $\langle D',T\rangle$ coincide up to the language $\mathcal{L}$ of $\langle D,T\rangle$. Since D' is definitely dissimilar, the translation function $\mathrm{Tr_W}$ is faithful by the one-to-one correspondence of extensions established by Wang et al. [28, Theorem 8]. □

A translation function in the other direction can also be obtained and it seems that one cannot do without new atoms in this case. The idea is that an atom a_r is introduced to denote that a rule r of a priority theory $\langle R,P,T\rangle$ is applied. Then the priority relation P of the priority theory is easily representable in terms of the justifications of defaults. Note that finiteness of R is essential in this translation: a single default is sufficient to represent a rule $r \in R$.

Definition 14. *For all priority theories* $\langle R, P, T\rangle$, *the translation* $\mathrm{Tr}_3(\langle R, P, T\rangle)$ *is* $\langle D, T\rangle$ *where* D *contains for each rule* $r = \alpha/\gamma \in R$ *a default* $\frac{\alpha:\neg \mathrm{a}_{r_1},\ldots,\neg \mathrm{a}_{r_n}}{\gamma \wedge \mathrm{a}_r}$ *where* $r_1, \ldots, r_n$ *are all the rules of* R *such that* $\langle r, r_1\rangle \in P, \ldots, \langle r, r_n\rangle \in P$.

Theorem 9. *The translation function* Tr_3 *given in Definition 14 is PFM.*

Proof sketch. It is obvious that Tr_3 is polynomial and modular. Let us then sketch how Tr_3 is proved faithful. Consider a priority theory $\langle R, P, T\rangle$ and the set of defaults D introduced by Tr_3. Let $\mathcal{L}$ be the language of $\langle R, P, T\rangle$ based on a set of atoms $\mathcal{A}$. Thus the language $\mathcal{L}'$ of the resulting default theory $\langle D, T\rangle$ is based on a set of atoms $\mathcal{A}' = \mathcal{A} \cup \{\mathrm{a}_r \mid r \in R\}$.

(i) Consider an extension $E \subseteq \mathcal{L}$ of $\langle R, P, T\rangle$ based on a set of rules $R' \subseteq R$ satisfying the stability condition of Definition 5. Define $E' = \mathrm{Cn}(E \cup A') \subseteq \mathcal{L}'$ where A' is the set of atoms $\{\mathrm{a}_r \mid r \in R'\}$. Then the reduct $D_{E'}$ contains the inference rule $\alpha/\gamma \wedge \mathrm{a}_r$ if and only if $r = \alpha/\gamma$ belongs to $\mathrm{Nb}(R, P, R')$. Consequently, it can be shown that E' is the unique extension of $\langle D, T\rangle$ with the property $\{r \in R \mid \mathrm{a}_r \in E'\} = R'$. (ii) Then assume that there is an extension $E' \subseteq \mathcal{L}'$ of $\langle D, T\rangle$ and let $R' = \{r \in R \mid \mathrm{a}_r \in E'\}$. It follows that R' satisfies the stability condition. It follows that $E = \mathrm{Cn}^{R'}(T) = E' \cap \mathcal{L}$ is an extension of $\langle R, P, T\rangle$. The steps (i) and (ii) above establish a one-to-one correspondence of extensions. Moreover, these extensions coincide up to the language $\mathcal{L}$. □

The results of Theorems 8 and 9 entitle us to conclude that default logic and priority logic are of equal expressive power. It is also worth pointing out that the translations presented lead to straightforward reductions between the decision problems of DL and PL corresponding to brave and cautious strategies. Thus our results and the complexity results on DL [8] have the following corollary.

Corollary 2. *The decision problems of PL corresponding brave and cautious reasoning strategies are* $\mathbf{\Sigma}_2^{\mathrm{p}}$*-complete and* $\mathbf{\Pi}_2^{\mathrm{p}}$*-complete problems, respectively.*

5 Conclusions

A framework of polynomial, faithful and modular (PFM) translation functions is proposed in this paper to classify non-monotonic logics by their expressive power. If there is a PFM translation function that maps theories of one non-monotonic logic L_1 into theories of another L_2, the sets of conclusions induced by a theory of L_1 – which determine the semantics of the theory – are effectively captured by the sets of conclusions induced by a theory of L_2. This is interpreted to indicate that the non-monotonic logic L_2 is at least as expressive as L_1. A number of translation functions are considered, and three novel translation functions Tr_1, Tr_2 and Tr_3 are proposed in the paper for classification purposes. The first two are merely obtained by modifying existing translation functions while the last is completely new. It is established that these translation functions are PFM. Two impossibility proofs are also provided to establish strict relationships in the expressive power of non-monotonic logics under consideration.

To conclude, the comparisons made in the paper give rise to a classification illustrated in Figure 1. Classical propositional logic CL is also included in the figure to complete our view. Solid arrows denote PFM translation functions from one non-monotonic logic to another that are considered in this work. Dotted arrows in the figure denote translation functions obtained as compositions of others. Such compositions are not necessarily optimal for a particular purpose. Note, for instance, that many unnecessary atoms and sentences would be introduced if a minimal model theory $\langle P, F, T\rangle$ were translated into a default theory using $\mathrm{Tr_N}$, $\mathrm{Tr_1}$ and $\mathrm{Tr_2}$ that involve the intermediate representations of $\langle P, F, T\rangle$ as an autoepistemic theory. Nevertheless, the resulting translation function $\mathrm{Tr_N} \circ \mathrm{Tr_1} \circ \mathrm{Tr_2}$ is still PFM by compositionality.

The *non-monotonic* logics under consideration are divided in three equivalence classes by their expressive power. The strongest class contains SAEL, DL and PL. The class below this contains AEL which is *less expressive* than SAEL, DL and PL. The third and the least expressive class contains CIRC which is less expressive than AEL. Below these three classes, there is the fourth class containing CL which is less expressive than any of the non-monotonic logics considered. The relationships depicted in Figure 1 refine the classification of non-monotonic logics based on earlier results [2, 7, 9, 10, 23] on the expressive power of non-monotonic logics. Finally, we want to emphasize that the ranking of non-monotonic logics by their expressive power is very sensitive to the requirements imposed on translation functions. It is demonstrated in this paper how a slight change in the notion of faithfulness changes the relative ordering of AEL and DL to the opposite compared to Gottlob's results [9].

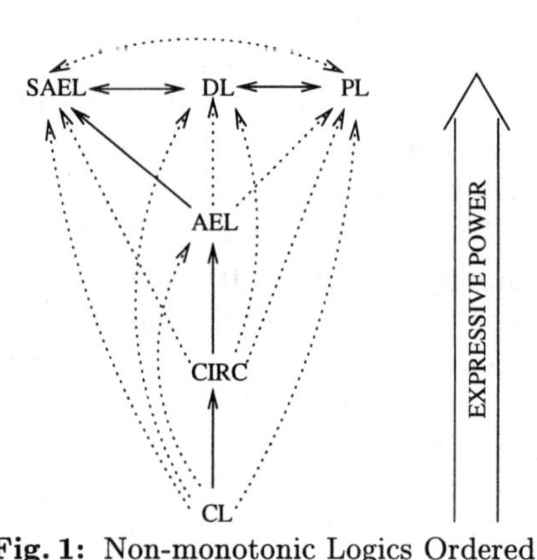

Fig. 1: Non-monotonic Logics Ordered by Their Expressive Power

Future Work. The notion of modularity considered in the paper is rather weak, i.e. only changes in the propositional subtheory are tolerated. We expect that the translations considered are also modular in a stronger sense and thus a stronger notion of modularity can be introduced such that the classification depicted in Figure 1 remains intact. For instance, changes in the defaults of a default theory $\langle D, T\rangle$ cause only local changes in the respective sentences of the translation $\mathrm{Tr_T}(\langle D, T\rangle)$. Moreover, there are also other non-monotonic logics as well as variants of those considered in the paper. These logics should be analysed in terms of PFM translation functions in order to classify them in the hierarchy. For instance, PDL is interesting in this respect on the basis of Section 4.3.

Acknowledgments

The author thanks Ilkka Niemelä for interesting discussions on a draft of this paper and anonymous referees for their comments and suggestions for improvements. The counter-example given in Theorem 1 is due to Niemelä. The author is also grateful to Thomas Eiter for his help on comparing the results of this paper with those in [2]. This research has been supported by Academy of Finland and Helsinki Graduate School in Computer Science and Engineering.

References

1. J.L. Balcázar, I. Díaz, and J. Gabarró. *Structural Complexity I.* Springer-Verlag, Berlin, 1988.
2. P.A. Bonatti and T. Eiter. Querying disjunctive database through nonmonotonic logics. *Theoretical Computer Science*, 160:321–363, 1996.
3. M. Cadoli, F.M. Donini, and M. Schaerf. Is intractability of nonmonotonic reasoning a real drawback. *Artificial Intelligence*, 88:215–251, 1996.
4. M. Cadoli and M. Lenzerini. The complexity of propositional closed world reasoning and circumscription. *Journal of Computer and System Sciences*, 2:255–310, 1994.
5. T. Eiter and G. Gottlob. Propositional circumscription and extended closed world reasoning are Π_2^p-complete. *Theoretical Computer Science*, 114:231–245, 1993.
6. J. Engelfriet, V. Marek, J. Treur, and M. Truszczyński. Infinitary default logic for specification of nonmonotonic reasoning. In J. Alferes, L. Pereira, and E. Orlowska, editors, *Proceedings of the 5th European Workshop on Logics is Artificial Intelligence*, pages 224–236. Springer Verlag, October 1996. LNAI, vol. 1126.
7. Goran Gogic, Henry Kautz, Christos Papadimitriou, and Bart Selman. The comparative linguistics of knowledge representation. In *Proceedings of the 14th International Joint Conference on Artificial Intelligence*, pages 862–869, Montreal, Canada, August 1995. Morgan Kaufmann Publishers.
8. G. Gottlob. Complexity results for nonmonotonic logics. *Journal of Logic and Computation*, 2(3):397–425, June 1992.
9. G. Gottlob. Translating default logic into standard autoepistemic logic. *Journal of the Association for Computing Machinery*, 42(2):711–740, 1995.
10. T. Imielinski. Results on translating defaults to circumscription. *Artificial Intelligence*, 32:131–146, 1987.
11. T. Janhunen. Representing autoepistemic introspection in terms of default rules. In *Proceedings of the European Conference on Artificial Intelligence*, pages 70–74, Budapest, Hungary, August 1996. John Wiley.
12. T. Janhunen. Separating disbeliefs from beliefs in autoepistemic reasoning. In J. Dix, U. Furbach, and A. Nerode, editors, *Proceedings of the 4th International Conference on Logic Programming and Non-monotonic Reasoning*, pages 132–151, Dagstuhl, Germany, July 1997. Springer-Verlag. LNAI 1265.
13. K. Konolige. On the relation between default and autoepistemic logic. *Artificial Intelligence*, 35:343–382, 1988.
14. V. Lifschitz. Computing circumscription. In *Proceedings of the 9th International Joint Conference on Artificial Intelligence*, pages 121–127, Los Angeles, California, USA, August 1985. Morgan Kaufmann Publishers.

15. V. Marek, J. Treur, and M. Truszczyński. Representation theory for default logic. *Annals of Mathematics in Artificial Intelligence*, 21:343–358, 1997.
16. W. Marek, G.F. Shvarts, and M. Truszczyński. Modal nonmonotonic logics: Ranges, characterization, computation. In *Proceedings of the 2nd International Conference on Principles of Knowledge Representation and Reasoning*, pages 395–404, Cambridge, MA, USA, April 1991. Morgan Kaufmann Publishers.
17. W. Marek and M. Truszczyński. Modal logic for default reasoning. *Annals of Mathematics and Artificial Intelligence*, 1:275–302, 1990.
18. W. Marek and M. Truszczyński. *Nonmonotonic Logic: Context-Dependent Reasoning*. Springer-Verlag, Berlin, 1993.
19. J. McCarthy. Circumscription—a form of non-monotonic reasoning. *Artificial Intelligence*, 13:27–39, 1980.
20. R.C. Moore. Semantical considerations on nonmonotonic logic. In *Proceedings of the 8th International Joint Conference on Artificial Intelligence*, pages 272–279, Karlsruhe, FRG, August 1983. Morgan Kaufmann Publishers.
21. I. Niemelä. On the decidability and complexity of autoepistemic reasoning. *Fundamenta Informaticae*, 17(1,2):117–155, 1992.
22. I. Niemelä. A unifying framework for nonmonotonic reasoning. In *Proceedings of the 10th European Conference on Artificial Intelligence*, pages 334–338, Vienna, Austria, August 1992. John Wiley.
23. I. Niemelä. Autoepistemic logic as a unified basis for nonmonotonic reasoning. Doctoral dissertation. Research report A24, Helsinki University of Technology, Digital Systems Laboratory, Espoo, Finland, August 1993.
24. R. Reiter. A logic for default reasoning. *Artificial Intelligence*, 13:81–132, 1980.
25. G. Schwarz. On embedding default logic into Moore's autoepistemic logic. *Artificial Intelligence*, 80:349–359, 1996.
26. J. Stillman. It's not my default: the complexity of membership problems in restricted propositional default logics. In *Proceedings of the 8th National Conference on Artificial Intelligence*, pages 571–578, Boston, Massachusetts, USA, July 1990. The MIT Press.
27. M. Truszczyński. Modal interpretations of default logic. In *Proceedings of the 12th International Joint Conference on Artificial Intelligence*, pages 393–398, Sydney, Australia, August 1991. Morgan Kaufmann Publishers.
28. X. Wang, J.-H. You, and L.Y. Yuan. Nonmonotonic reasoning by monotonic inferences with priority constraints. In *Proceedings of the 2nd International Workshop on Non-monotonic Extensions of Logic Programming*, pages 91–109. Springer, 1996. LNAI 1216.

An Approach to Query-Answering in Reiter's Default Logic and the Underlying Existence of Extensions Problem

Thomas Linke and Torsten Schaub

Institut für Informatik, Universität Potsdam, Germany
{linke,torsten}@cs.uni-potsdam.de

Abstract. We introduce new concepts for default reasoning in the context of query-answering in regular default logic. For this purpose, we develop a proof-oriented approach for deciding whether a default theory has an extension containing a given query. The inherent problem in Reiter's default logic is that it necessitates the inspection of all default rules for answering no matter what query. Also, default theories are known to lack extensions occasionally. We address these two problems by sloting in a compilation phase before the actual query-answering phase. The examination of the entire set of default rules is then done only once in the compilation phase; this allows us to inspect only the ultimately necessary default rules during the actual query answering phase. In fact, the latter inspection must not only account for the derivability of the query, but moreover it must guarantee the existence of an encompassing extension. We address this traditionally important problem by furnishing novel criteria guaranteeing the existence of extensions that are arguably simpler and go well beyond existing approaches.

1 Introduction

In many AI applications default reasoning plays an important role since many subtasks involve reasoning from incomplete information. This is why there is a great need for corresponding implementations that allow us to integrate default reasoning capabilities into complex AI systems. For addressing this problem, we have chosen Reiter's *default logic* [11] as the point of departure: Default logic augments classical logic by *default rules* that differ from standard inference rules in sanctioning inferences that rely upon given as well as absent information. Knowledge is represented in default logic by *default theories* (D, W) consisting of a set of formulas W and a set of default rules D. A default rule $\frac{\alpha:\beta}{\gamma}$ has two types of antecedents: A *prerequisite* α which is established if α is derivable and a *justification* β which is established if β is consistent in a certain way. If both conditions hold, the *consequent* γ is concluded by default. A set of such conclusions (sanctioned by default rules and classical logic) is called an *extension* of an initial set of facts: Given a set of formulas W and a set of default rules D, any such extension E is a deductively closed set of formulas containing W such that, for any $\frac{\alpha:\beta}{\gamma} \in D$, if $\alpha \in E$ and $\neg\beta \notin E$ then $\gamma \in E$. (A formal introduction to default logic is given in Section 2.)

In what follows, we are interested in the basic approach to query-answering in default logic that allows for determining whether a formula is in *some* extension of a given default theory. The inherent problem with this in regular default logic is that it necessitates the inspection of all default rules, no matter which query is posed. This is related to the fact that default theories may not possess extensions at all. Hence, query-answering has so far only been addressed indirectly: On the one hand, we find approaches that are primarily interested in the construction of extensions; queries are then either answerable by simple membership tests or by directing the construction towards extensions containing the query [7, 14, 9, 3]. On the other hand, we find approaches based on variants of default logic that guarantee the existence of extensions and that allow for local proof procedures [12]. Unfortunately, these variants do not offer the same expressiveness as the original approach [1]. In the general case, there is thus no way of avoiding exhaustive computations while preserving the expressiveness of Reiter's default logic. Our key to this problem is given by the discovery that, for query-answering, the inspection of the entire default theory must only be done once, no matter which and how many queries are posed subsequently. This leads us to the following idea: We slot in a compilation phase before the actual query-answering process so that afterwards queries are answerable in a rather local fashion, insofar that the resulting procedures must examine only those default rules necessary for proving the query.

Consider an example where birds fly, birds have wings, penguins are birds, and penguins don't fly along with a formalization through general default theory

$$(D,W) = \left(\left\{\frac{b:\neg ab_b}{f}, \frac{b:w}{w}, \frac{p:b}{b}, \frac{p:\neg ab_p}{\neg f}\right\}, \{\neg f \rightarrow ab_b, f \rightarrow ab_p, p\}\right). \tag{1}$$

(For short we denote the default rules in (1) by $\delta_1, \delta_2, \delta_3, \delta_4$, resp.) An analysis of these rules should provide us with the information that the application of the first rule depends on the blockage of the last one (and vice versa), while the second and third rule can be applied no matter which of the other rules apply. We may thus derive $\neg f$ by ignoring the second and the third rule, while assuring that the first one is blocked. Notably, our initial analysis must be truly global and also extend to putatively unrelated parts of the theory. To see this, simply add the rule $\frac{:x}{\neg x}$, destroying all previous extensions consistent with x. Now, the application of each rule in (1) depends additionally on the blockage of $\frac{:x}{\neg x}$. Thus, given p, we can only apply $\frac{p:\neg ab_p}{\neg f}$ to derive $\neg f$, if $\frac{b:\neg ab_b}{f}$ and $\frac{:x}{\neg x}$ are blocked.

We note that the application of rules depends on the blockage of other rules. The outcome of our analysis must thus provide information on which rules may block other rules (or even themselves). But although we may allot somehow unlimited time to a compilation phase, we cannot provide unlimited space for its result. Our approach complies with this principle and offers as a result a so-called *block graph* whose size is quadratic in the number of defaults, although its computation may need exponential time in the worst case.[1] The block graph represents the essential information about blockage between default rules. In the query-answering phase this information is then used to focus the computation of *default proofs* on ultimately necessary defaults only. For instance, given the block graph of Theory (1), we may derive f from $(D, W \cup \{b\})$

[1] The reader is reminded that query-answering in default logic has even two distinct sources of exponential complexity; it is Σ_2^P-complete [6].

by means of $\frac{b:\neg ab_b}{f}$. While this involves testifying the blockage of $\frac{p:\neg ab_p}{\neg f}$, the two other rules may be completely ignored. This is what makes our approach different from traditional, extension-oriented ones: Once we have compiled the blocking information, subsequent query-answering must only consider default rules *by need*.

In addition to the rules needed for deriving a query, like $\frac{b:\neg ab_b}{f}$ for *f*, however, this involves considering furthermore those rules threatening the rules in the proof, like $\frac{p:\neg ab_p}{\neg f}$, or even those menacing its encompassing extension, like $\frac{:x}{\neg x}$. Both type of rules are identifiable by means of the block graph. The first ones are necessarily among the predecessors (in the block graph) of the rules used for deriving the query. For guaranteeing an encompassing extension, we propose a range of novel criteria that can be directly read off the block graph, too. In fact, these criteria are arguably simpler and go well beyond existing approaches addressing the traditionally important problem of existence of extensions. This is formally proven in the full version of this paper.

The rest of the paper is organized as follows. In Section 2 we provide a formal introduction to default logic. Section 3 introduces blocking (supporting) sets and the block graph of a default theory, which actually is the result of the compilation phase. Section 5 develops a new representation of extensions, which relies on blocking sets and is suitable for query-answering. It also allows for defining local default proofs for different kinds of default theories in Section 6. Section 7 grasps the presented approach in its entirety and discusses its modularity.

2 Background

We start by completing our initial introduction to Reiter's default logic: A *default rule* $\frac{\alpha:\beta}{\gamma}$ is called *normal* if β is equivalent to γ; it is called *semi-normal* if β implies γ. We sometimes denote the *prerequisite* α of a default rule δ by $Pre(\delta)$, its *justification* β by $Just(\delta)$ and its *consequent* γ by $Cons(\delta)$.[2] A set of default rules D and a set of formulas W form a *default theory*[3] $\Delta = (D, W)$, that may induce one, multiple or even no *extensions* [11]:

Definition 1. *Let (D, W) be a default theory. For any set of formulas S, let $\Gamma(S)$ be the smallest set of formulas S' such that*

E1 $W \subseteq S'$,
E2 $Th(S') = S'$,
E3 *For any* $\frac{\alpha:\beta}{\gamma} \in D$, *if* $\alpha \in S'$ *and* $\neg\beta \notin S$ *then* $\gamma \in S'$.

A set of formulas E is an extension of (D, W) iff $\Gamma(E) = E$.

Any such extension represents a possible set of beliefs about the world. For example, Default theory (1) has two extensions: $Th(W \cup \{b, w, \neg f\})$ and $Th(W \cup \{b, w, f\})$, while theory $(D \cup \{\frac{:x}{\neg x}\}, W)$ (where D and W are taken as in (1)) has no extension.

For a set of formulas S and a set of defaults D, we define the *set of generating default rules* as $GD(D, S) = \{\delta \in D \mid S \vdash Pre(\delta) \text{ and } S \nvdash \neg Just(\delta)\}$. The two last extensions are generated by $\{\delta_2, \delta_3, \delta_4\}$ and $\{\delta_1, \delta_2, \delta_3\}$, respectively. We call a

[2] This notation generalizes to sets of default rules in the obvious way.

[3] If clear from the context, we sometimes refer with Δ to D and W (and vice versa) without mention.

set of default rules D *grounded* in a set of formulas W iff there exists an enumeration $\langle\delta_i\rangle_{i\in I}$ of D such that for all $i \in I$, $W \cup \mathit{Cons}(\{\delta_0,\ldots,\delta_{i-1}\}) \vdash \mathit{Pre}(\delta_i)$. Note that $E = \mathit{Th}(W \cup \mathit{Cons}(\mathit{GD}(D,E)))$ forms an extension of (D,W) if $\mathit{GD}(D,E)$ is grounded in W.

For simplicity, we assume for the rest of the paper that default theories (D,W) comprise finite sets only. Additionally, we assume that for each default rule δ in D, we have that $W \cup \mathit{Just}(\delta)$ is consistent. This can be done without loss of generality, because we can clearly eliminate all rules δ' from D for which $W \cup \mathit{Just}(\delta')$ is inconsistent, without altering the set of extensions.

3 Compressing Blocking Relations by Block Graphs

Our approach is founded on the concept of *blocking sets*: Given a default theory (D,W) and a default rule $\delta \in D$, intuitively, a *blocking set* for δ is a minimal set of default rules $B \subseteq D$ such that the joint application of its rules denies the application of δ. Such a blocking set provides a *candidate* for disabling the putatively applicable default rule δ. For this purpose, it is actually sufficient to refute a rule's justification, ignoring its prerequisite. This is because an existing derivation of a prerequisite can only be counterbalanced by refuting the justification of one of its default rules. This motivates Condition **BS1** in the formal definition of blocking sets, given below.

In order to become effective, a blocking set must belong to the generating default rules of an extension. Thus, it must be grounded and the respective justifications must be consistent with the extension. Groundedness is indispensable and also easily verifiable, leading to Condition **BS2**. Global consistency however is context-dependent since it makes reference to (possible) extensions encompassing the blocking set. In fact, in many cases, we are rather interested in showing that a critical blocking set *does not* contribute to a certain extension, either because it threatens a default proof at hand or even because it menaces the extension as such. Therefore, this (context-dependent) consistency condition must be finally addressed in the context of the respective query-answering process (see Section 6), so that we confine ourselves to a rather local notion of consistency reflected by Condition **BS3**.

This leads us to the following definition of (putative) blocking sets:

Definition 2. *Let $\Delta = (D,W)$ be a default theory. For $\delta \in D$, we define the set $\mathcal{B}_\Delta(\delta)$ of all blocking sets for δ in Δ as follows.*

If $B \subseteq D$, then $B \in \mathcal{B}_\Delta(\delta)$ iff B is a set such that

BS1 $W \cup \mathit{Cons}(B) \vdash \neg \mathit{Just}(\delta)$,
BS2 *B is grounded in W,*
BS3 **BS1** *and* **BS2** *for no δ' and no B' where $\delta' \in B \cup \{\delta\}$ and $B' = B \setminus \{\delta''\}$ for $\delta'' \in B$.*

Note that the set of consequences, $\mathit{Cons}(B)$, is not required to be consistent. This is needed, for instance, to detect groups of default rules whose joint application blocks any other default, like $\{\frac{:a}{c}, \frac{:b}{\neg c}\}$.

The problem of deciding whether a set B satisfies **BS1** and **BS2** is co-NP-complete [13]. **BS3** is a NP-complete problem. Notably, **BS3** indicates that B is minimal wrt set inclusion and that B does not contain both a default and some of its blocking sets.

Minimality is confirmed by confuting **BS1** or **BS2** for $B' = B \setminus \{\delta\}$ and $\delta' = \delta$ for all $\delta \in B$; while the denial of the latter is addressed by taking $\delta' \in B$ and $B' = B \setminus \{\delta''\}$. The set of all blocking sets for a(ll) default rule(s) can thus be computed with an algorithm iterating over 2^D with appeal to an NP-oracle.

We show in the full paper that in the worst case, a theory with n rules may comprise $O(2^n)$ blocking sets. However, the number of blocking sets is not related to the number of extensions of a given theory. To see this, observe that Theory $(\{\frac{:a_i}{\neg c_i}, \frac{:c_i}{\neg a_i} \mid i = 1..n\}, \emptyset)$ has 2^n extensions but only $2n$ blocking sets. That is, although we encounter an exponential number of extensions, we have only a linear number of blocking sets.

For illustration, consider Theory (1) along with its blocking sets given in (2):

$$\begin{array}{ll} \mathcal{B}_\Delta(\delta_1) = \{\{\delta_4\}\} & \mathcal{B}_\Delta(\delta_2) = \emptyset \\ \mathcal{B}_\Delta(\delta_3) = \emptyset & \mathcal{B}_\Delta(\delta_4) = \{\{\delta_1, \delta_3\}\} \end{array} \tag{2}$$

For example, $\{\delta_4\}$ is the only blocking set for δ_1, because it comprises a possible refutation of ab_b, the justification of δ_1. In general, a single default rule may have multiple blocking sets. For example, adding $\frac{:u}{v \wedge \neg v}$ to Theory (1) augments each set $\mathcal{B}_\Delta(\delta_i)$ by $\{\frac{:u}{v \wedge \neg v}\}$. The addition of $\frac{:x}{\neg x}$ to (1) leaves blocking sets (2) unaffected and yields $\mathcal{B}_\Delta(\frac{:x}{\neg x}) = \{\{\frac{:x}{\neg x}\}\}$ reflecting self blockage.

Observe that **BS3** allows us to discard blocking sets that block their own constituent rules. For instance, $\{\delta_1, \delta_3, \delta_4\}$ is a putative blocking set of δ_2; it is ruled out by **BS3** since it contains both δ_1 and one of its blocking sets, $\{\delta_4\}$.

Now, given the concept of blocking sets, we are ready to define the outcome of our compilation phase: The *block graph* of a default theory.

Definition 3. *Let $\Delta = (D, W)$ be a default theory. The block graph $\Gamma_\Delta = (V_\Delta, A_\Delta)$ of Δ is a directed graph with vertices $V_\Delta = D$ and arcs*

$$A_\Delta = \{(\delta, \delta') \mid \text{there is some } B \in \mathcal{B}_\Delta(\delta') \text{ with } \delta \in B\}.$$

So, there is an arc (δ', δ) between default rules δ' and δ in the block graph iff δ' belongs to some blocking set for δ. For Default theory (1), we obtain the block graph given in Figure 1; it has arcs (δ_4, δ_1), (δ_1, δ_4) and (δ_3, δ_4).

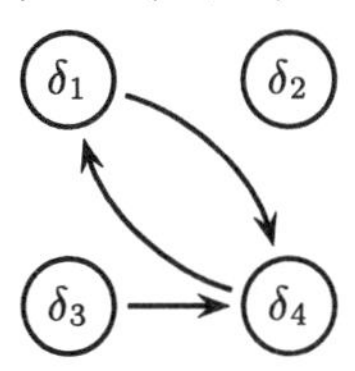

Fig. 1. Block graph of Default theory (1).

We observe that the size of the block graph is quadratic in the number of default rules, although there may be an exponential number of blocking sets. The block graph thus comprises the essential information from the blocking sets; this is accomplished by abstracting from the membership of default rules in specific blocking sets. It is actually unnecessary to keep any (possibly exponential number of) blocking sets beyond the compilation phase, because they can be effectively reconstructed from Γ_Δ during the query-answering phase.

Notably, we do not even have to keep or recompute blocking sets during the compilation phase itself, since the block graph is computable incrementally:

Lemma 1. *Let $\Delta = (D, W)$ and $\Delta' = (D', W)$ be default theories with $D' \subseteq D$. If $\delta \in D'$, then $B \in \mathcal{B}_{\Delta'}(\delta)$ implies $B \in \mathcal{B}_{\Delta}(\delta)$.*

The usage of the block graph is further detailed in the following sections.

A key issue during query-answering is to cover the safe application of default rules needed for proving a query. For this, we must guarantee that all (possible) blocking sets of such rules are blocked themselves. This leads us to the concept of *supporting sets*, which are intuitively simply blocking sets for blocking sets. First, we extend the notion of blocking sets to sets of rules: For a default theory $\Delta = (D, W)$ and sets $B, B' \subseteq D$, we call B' a *blocking set* for B, written $B' \succ_{\Delta} B$, if there is some default rule $\delta \in B$ such that $B' \in \mathcal{B}_{\Delta}(\delta)$.

Definition 4. *Let $\Delta = (D, W)$ be a default theory. For $\delta \in D$, we define the set $\mathcal{S}_{\Delta}(\delta)$ of all supporting sets for δ as*

$$\mathcal{S}_{\Delta}(\delta) = \{B'_1 \cup \ldots \cup B'_n \mid B'_i \subseteq D \text{ such that } B'_i \succ_{\Delta} B_i \text{ and } \mathcal{B}_{\Delta}(\delta) = \{B_1, \ldots, B_n\}\}$$

provided $\mathcal{B}_{\Delta}(\delta) \neq \emptyset$. Otherwise, we define it as $\mathcal{S}_{\Delta}(\delta) = \{\emptyset\}$.

Observe that $\mathcal{S}_{\Delta}(\delta) = \emptyset$ whenever $\mathcal{B}_{\Delta}(\delta') = \emptyset$ for all δ' in some $B_i \in \mathcal{B}_{\Delta}(\delta)$, because then for B_i there is no set of default rules B'_i such that $B'_i \succ_{\Delta} B_i$, that is, $B'_1 \cup \ldots \cup B'_n$ is undefined.

The purpose of supporting sets is to rule out blocking sets as subsets of the generating default rules: Once a supporting set for δ has been applied (ie. it belongs to the generating default rules), δ itself can be applied safely. The supporting sets in Theory (1) are given in (3):

$$\begin{array}{ll} \mathcal{S}_{\Delta}(\delta_1) = \{\{\delta_3, \delta_1\}\} & \mathcal{S}_{\Delta}(\delta_2) = \{\emptyset\} \\ \mathcal{S}_{\Delta}(\delta_3) = \{\emptyset\} & \mathcal{S}_{\Delta}(\delta_4) = \{\{\delta_4\}\} \end{array} \tag{3}$$

For example, consider the supporting set for δ_1: We have to find *one* blocking set for each blocking set in $\mathcal{B}_{\Delta}(\delta_1) = \{\{\delta_4\}\}$. In this easy case, we have to find some blocking set for δ_4, yielding $\{\delta_3, \delta_1\}$. Here, $\{\delta_3, \delta_1\}$ is the only supporting set for δ_1. Similarly, for δ_4, we have to find a blocking set B' for $\{\delta_3, \delta_1\}$ (see (2)). That is, we must have $B' \in \mathcal{B}_{\Delta}(\delta_3)$ or $B' \in \mathcal{B}_{\Delta}(\delta_1)$. Because $\mathcal{B}_{\Delta}(\delta_3)$ is empty, we get $\{\delta_4\} \in \mathcal{B}_{\Delta}(\delta_1)$ as the only supporting set for δ_4. The occurrence of δ_4 in its supporting set is due to the fact that there is a direct conflict between δ_4 and its blocking set. This says that δ_4 is safely applicable on its own. In general this need not be the case. For example, in default theory $(\{\frac{:b}{a}, \frac{:c}{\neg b}, \frac{:d}{\neg c},\}, \emptyset)$ the last rule forms the single supporting set for the first one.

4 Existence of Extensions

For query-answering, it is clearly important to know whether a default theory has an extension, since reasonable conclusions must reside in such an extension. In fact, determining whether a default theory has an extension or not is on itself a major computational problem, pertinent to default logic. In previous works, broad subclasses of default theories always possessing extensions have been identified, among them we find *normal* [11], *ordered* [5], *ps-even* [10],[4], and *strongly stratified* [2] default theories. In this section, we provide a class of default theories guaranteeing the existence of extensions that are more general than the previous ones (see Section 7). The corresponding criteria are thus not only a salient part of our approach to query-answering, but they moreover represent an important contribution on their own.[5]

The basic idea is to further exploit the structure of blocking graphs for providing sufficient conditions on whether the underlying default theory has an extension.

To begin with, we call a default theory (D, W) *non-conflicting*, if it has no blocking sets, that is, if its block graph has no arcs; otherwise we call it *conflicting*. Non-conflicting default theories have unique extensions and trivially allow for query-answering without consistency checks[6].

Lemma 2. *Every non-conflicting default theory has a single extension.*

For instance, default theory $(\{\frac{:a}{a}, \frac{:b}{b}\}, \emptyset)$ is non-conflicting, yielding a block graph with no arcs. The same holds for Theory (1), when eliminating either $\frac{b:\neg ab_b}{f}$ or $\frac{p:\neg ab_p}{\neg f}$.

More interestingly, we call a default theory *well-ordered*, if its block graph is acyclic. The next result shows that well-ordered default theories have single extensions.

Theorem 1. *Every well-ordered default theory has a single extension.*

For instance, default theory $(\{\frac{:a}{a}, \frac{:b\wedge\neg a}{b}\}, \emptyset)$ is well-ordered; its block graph contains a single arc, indicating that the first rule may block the second one (but *not* vice versa).

We call a default theory *even*, if its block graph contains cycles with even length only.[7] Our first major result of this section states that even default theories always have extensions:

Theorem 2. *Every even default theory has an extension.*

For instance, default theory $(\{\frac{:a\wedge\neg b}{a}, \frac{:b\wedge\neg a}{b}\}, \emptyset)$ is even; its block graph contains two arcs, indicating that the first rule may block the second one, *and* vice versa.

Evenness is also enjoyed by our initial default theory in (1), as can be easily verified by regarding its block graph in Figure 1. Unlike this, default theory $(\{\frac{:x}{\neg x}\}, \emptyset)$ is not even, since its block graph contains an odd cycle, namely one of length one.

In fact, the elimination of putative extensions is somehow always due to odd cycles. Not all of them, however, do lead to the destruction of extensions. For example, the theory consisting of $\frac{:a\wedge\neg b}{a}$, $\frac{:b\wedge\neg c}{b}$, $\frac{:c\wedge\neg a}{c}$, and no facts has no extension; its block graph contains a cycle of length three, say $(\delta_c, \delta_b, \delta_a)$ (where the indexes refer to consequents). Adding formula $c \to b$ yields actually a theory, whose only extension contains c and b. Although the block graph of this theory contains the an odd cycle, it does now contain additionally arc (δ_a, δ_c), which counterbalances the self-blocking-behavior of the odd cycle.

---

[4] We use ps-even for referring to the notion of *evenness* due to Papadimitriou and Sideri [10].

[5] For continuity, we delay a detailed comparison with the aforementioned approaches to Section 7.

[6] Observe that although non-conflicting default theories ignore justifications, they are still non-monotonic because extensions may be invalidated after augmenting a non-conflicting theory.

[7] The *length* of a directed cycle in a graph is the total number of arcs occurring in the cycle.

The salient advantage of evenness and well-orderedness is that these properties are simple, that is, they can be tested in polynomial time, and they rely on a simple data structure, namely a block graph. And even more importantly, they apply to general default theories and are syntax-independent, unlike other approaches [5, 10] that apply to semi-normal default theories only[8] and give different results on equivalent yet syntacticly different theories (see Section 7 for more details). Even though we put forward evenness (or well-orderedness) for testing existence of extensions, it is already worth mentioning that this is no characteristic feature of our approach to query-answering, since it is actually modular in allowing for alternative checks, provided that they are more appropriate.

5 Formal foundations of query-answering

This section lays the formal foundations of our approach to query-answering. We have already seen in the introductory section that we need additional efforts for backing up a default proof. This involves two tasks, namely, protecting the constituent default rules of a default proof and assuring an encompassing extension.

The first task is accomplishable by blocking and supporting sets:

Definition 5. *Let $\Delta = (D, W)$ be a default theory. A set of default rules $D' \subseteq D$ is protected in D iff for each $\delta \in D'$ we have that (i) $S \subseteq D'$ for some $S \in \mathcal{S}_\Delta(\delta)$ and (ii) $B \subseteq D'$ for no $B \in \mathcal{B}_\Delta(\delta)$.*

In words, a set of defaults is protected if it contains some supporting set for each constituent default and if it contains no blocking set for any of its defaults.

The second task is actually more crucial. This is because there are default theories without extensions and, in view of local query-answering, the test for an encompassing extension should be accomplishable without actually computing such an extension. For this purpose, we can build on the criteria developed in the last section, which allow us to guarantee the existence of an extension of a default theory by looking at its block graph.

By combining the two previous tasks, we obtain an alternative characterization of extensions. For this, define for $\Delta = (D, W)$,

$$\Delta \ominus D' \quad \text{as} \quad (D \setminus (D' \cup \overline{D'}), W \cup Cons(D')) \tag{4}$$

where $D' \subseteq D$ and $\overline{D'} = \{\delta \in D \mid W \cup Cons(D') \vdash \neg Just(\delta)\}$[9]. Then, we have the following result.

Theorem 3. *Let $\Delta = (D, W)$ be a default theory and let E be a set of formulas. Then, E is an extension of Δ iff $E = Th(W \cup Cons(D') \cup E')$ for some $D' \subseteq D$ such that (i) D' is grounded in W, (ii) D' is protected in D and (iii) $\Delta \ominus D'$ has extension E'.*

For general default theories $\Delta \ominus D'$, Condition *(iii)* is implementable by weak odd-, even- or well-orderedness. In this case, Theorem 3 furnishes a characterization of

[8] The approaches in [5, 10] do not apply to so-called *weak* semi-normal theories (allowing for default rules without justifications) that have been shown to be equivalent to general default theories in [8].

[9] The purpose of $\overline{D'}$ is to eliminate defaults with inconsistent justifications in $\Delta \ominus D'$.

extensions in terms of blocking and supporting sets (due to Condition *(ii)* and *(iii)*). In case $\Delta \ominus D'$ is semi-normal, the methods in [5, 10] could work just as fine (except for their syntax-dependency and the more complicated data structures needed). The test is even trivial in case $\Delta \ominus D'$ is normal or non-conflicting. This demonstrates the modularity of our approach as regards verifying Condition *(iii)*, that is, existence of extensions of $\Delta \ominus D'$.

For query-answering, it is important to notice that the set D' in Theorem 3 needs not to be maximal. In fact, D' can be any grounded subset of the generating default rules $GD(D, E)$ of extension E. This attributes to D' the character of an extension-dependent *default proof*: Given the set of generating default rules $GD(D, E)$ for some extension E, a default proof of some formula φ is simply a grounded set $D' \subseteq GD(D, E)$ such that $W \cup Cons(D') \vdash \varphi$. In such a predetermined setting, we do neither have to care about the consistent application of the rules in D' (this is assured by $D' \subseteq GD(D, E)$) nor (trivially) about the existence of an encompassing extension. Both issues, addressed by *(ii)* and *(iii)* in Theorem 3, are however of crucial importance, whenever there is no such extension at hand.

For example, in Default theory (1), the set $D' = \{\delta_3, \delta_1\}$ may serve as a default proof for f; it satisfies conditions *(i)* and *(ii)* because it is grounded and protected wrt (1). (The latter is easily verifiable in (2) and (3).) For showing that f belongs to an existing extension, it is now sufficient to demonstrate that $\Delta \ominus D'$ has an extension, notably *without* computing it. We get a non-conflicting theory

$$\Delta \ominus D' \quad = \quad (D \setminus (\{\delta_3, \delta_1\} \cup \{\delta_4\}), W \cup \{f, b\}) \quad = \quad (\{\delta_2\}, \{p, ab_p, f, b\})$$

which has obviously an extension, due to an empty block graph. Hence, we have shown that f is a default conclusion of (1) without computing the corresponding extension. The next corollary makes these ideas precise for query-answering:

Corollary 1. *Let $\Delta = (D, W)$ be a default theory and let φ be a formula. Then, $\varphi \in E$ for some extension E of Δ iff $W \cup Cons(D') \vdash \varphi$ for some $D' \subseteq D$ such that (i) D' is grounded in W, (ii) D' is protected in D and (iii) $\Delta \ominus D'$ has an extension.*

Query-answering thus boils down to finding an appropriate set of default rules that is sufficient for deriving the query and for protecting itself against possible threats, provided that its "application" preserves an encompassing extension.

6 Default Proofs for Query-Answering

After the compilation phase, we are able to decide by examining the block graph whether a default theory is non-conflicting, well-ordered, etc. or neither of them. In analogy to these classes of default theories (and for conceptual clarity) we give an incremental definition of a default proof. We start with default proofs for the simplest case of non-conflicting default theories.

Definition 6 (Pure default proof [10]). *Let $\Delta = (D, W)$ be a default theory and φ a formula. A set of default rules $P_0 \subseteq D$ is a pure default proof for φ from Δ iff*

[10] Observe that although pure default proofs ignore justifications, they are still non-monotonic because they may be invalidated after augmenting the underlying non-conflicting theory.

P1 $W \cup Cons(P_0) \vdash \varphi$,
P2 P_0 *is grounded in* W.

The problem of deciding whether a set P_0 is a pure default proof is co-NP-complete [13].

Actually, one may wonder whether we must stipulate an additional consistency condition in Definition 6, viz.

P3 $W \cup Cons(P_0) \not\vdash \neg Just(\delta)$ for all $\delta \in P_0$,

because this is central for applying default rules. But here, we concentrate on the very simple class of non-conflicting default theories, which ensures that **P3** holds anyway. Otherwise there would be at least one blocking set and the default theory would be conflicting. As a nice consequence, pure default proofs correspond to extension-dependent default proofs but without necessitating such an extension.

For example, let Δ'' be the default theory obtained from the one in (1) by leaving out $\frac{b:\neg ab_b}{f}$. Then, Δ'' is non-conflicting and $P_0 = \{\frac{p:\neg ab_p}{\neg f}\}$ is a pure default proof for $\neg f$. This proof is found without consistency checking nor any measures guaranteeing the existence of an encompassing extension.

We have the following result for non-conflicting default theories:

Theorem 4. *Let Δ be a non-conflicting default theory and φ a formula. Then $\varphi \in E$ for an extension E of Δ iff there is a pure default proof for φ from Δ.*

In case we have a conflicting yet well-ordered (or even) default theory, we have to take supporting sets into account, because then it is necessary to protect the constituent default rules of a default proof: Whenever we add a default rule to a partially constructed default proof, we have to make sure that there is some supporting set for this default (if necessary). Clearly, this must be done for the default rules in the supporting sets, too. We thus recursively add supporting sets to default proofs: Let $\Delta = (D, W)$ be a default theory and $\delta \in D$ a default rule. A set of default rules $C \subseteq D$ is a *complete support of* δ *in* Δ iff C is a minimal set such that for each $\delta' \in C \cup \{\delta\}$ there is some supporting set $S' \in \mathcal{S}_\Delta(\delta')$ with $S' \subseteq C$.

This leads us to the concept of supported default proofs:

Definition 7 (Supported default proof). *Let $\Delta = (D, W)$ be a default theory and φ a formula. A set of default rules $P \subseteq D$ is a supported default proof for φ from Δ iff $P = P_0 \cup C$ with $C = \bigcup_{\delta \in P_0} C_\delta$ such that*

SP1 *P_0 is a pure default proof for φ from Δ,*
SP2 *C_δ is a complete support of δ for each $\delta \in P_0$,*
SP3 *P contains no blocking set for each $\delta \in P$.*

In the full paper, we show that deciding whether a pure default proof is a supported default proof can be done within Σ_2^P of the polynomial hierarchy.

Condition **SP2** ensures that there is a supporting set for each default in P, if necessary. Condition **SP3** prevents P from blocking some of its own members; thus **SP2** and **SP3** together imply that P is a protected set of default rules. Therefore, **P3** is also valid for supported default proofs, because otherwise there would be a blocking set in

P, which is a contradiction to **SP3**. As an important consequence, we do not need any global consistency checks when querying even (or well-ordered) default theories.

Similar checks are done locally during the determination of supporting sets. For obtaining these sets during the query-answering phase, we can draw on the block graph for restricting the search space. To find out the blocking sets for a default δ, we only have to take the predecessors of δ in the block graph into account, which form usually a limited subset of all default rules. The consistent application of δ is then guaranteed by determining one supporting set for δ among a fixed subset of all default rules, namely the pre-predecessors of δ. We may also think of different heuristics for a more elaborated (re)construction of blocking sets, like caching selected blocking sets retained during the compilation phases.

For illustration, let us return to even Default theory (1). We have already seen that $P_0 = \{\delta_4\} = \{\frac{p\,:\,\neg ab_p}{\neg f}\}$ is a pure default proof for $\neg f$, that is, **SP1** holds. Because Theory (1) is conflicting, we have to warrant a complete support for each default in P_0. First of all, it is important to note that our approach leaves plenty of room for accomplishing this task, depending on whether and if so how many blocking sets were maintained from the compilation phase: In case we have kept $\mathcal{S}_\Delta(\delta_4)$, we can directly choose one of its members, yielding $C_{\delta_4} = \{\delta_4\}$. If not, we first verify whether it is actually necessary to generate a support for δ_4 by checking whether δ_4 has predecessors in the block graph. Since this is true in our example (δ_4 has predecessors δ_1 and δ_3), we must (re)generate one member of $\mathcal{S}_\Delta(\delta_4)$. This can be done by looking at the predecessors of δ_1 and δ_3 in the block graph, which immediately yields $\{\delta_4\}$ as the only supporting set; no matter whether or not $\mathcal{B}_\Delta(\delta_1)$ and $\mathcal{B}_\Delta(\delta_3)$ are given explicitly. In any case, we thus get $C = C_{\delta_4} = \{\delta_4\}$, so that we have

$$P = P_0 \cup C = \left\{\frac{p\,:\,\neg ab_p}{\neg f}\right\}.$$

This establishes Condition **SP2** for P. For verifying **SP3** it is sufficient to observe that no members of P are connected in the block graph. Hence P is a supported default proof for $\neg f$ from Theory (1). Given the block graph, this proof is found without any consistency checks and no measures guaranteeing the existence of an encompassing extension.

Similar arguments show that $\{\frac{p\,:\,b}{b}, \frac{b\,:\,\neg ab_b}{f}\}$ and $\{\frac{p\,:\,b}{b}, \frac{b\,:\,w}{w}\}$ are supported default proofs for f and w from Theory (1), respectively. It is important to note that for establishing this, we only had to warrant a support for $\frac{b\,:\,\neg ab_b}{f}$, since it is the only one among all involved default rules, having a predecessor in the block graph. As with $\frac{p\,:\,\neg ab_p}{\neg f}$ above, however, $\frac{b\,:\,\neg ab_b}{f}$ supports itself, so that no other defaults had to be added to the pure default proof obtained in the first case. In the second one, the default proof is even found without any search for supporting sets (nor any consistency check). This illustrates nicely how query-answering draws upon the information furnished by the block graph in order to concentrate efforts on the ultimately necessary defaults only.

As a result, we get that supported default proofs furnish a sound and complete concept for querying even default theories:

Theorem 5. *Let $\Delta = (D, W)$ be an even default theory and φ a formula. Then $\varphi \in E$ for an extension E of Δ iff there is a supported default proof for φ from Δ.*

The stipulation of an even default theory can be replaced by any other adequate condition guaranteeing the existence of extensions.

In the general case, dealing with arbitrary default theories, we must additionally guarantee that there is an encompassing extension, whose generating default rules comprise a given (supported) default proof. As always, this should be accomplished without actually computing such an extension. For this, we proceed as follows. After computing a supported default proof P from a theory Δ, we check whether $\Delta \ominus P$ has an extension. This is reflected by Condition **DP2** of the following definition. The resulting default proof is then called a general default proof:

Definition 8 (General default proof). *Let $\Delta = (D, W)$ be a default theory and φ a formula. A (general) default proof for φ from Δ is a set of default rules $P \subseteq D$ such that*

DP1 *P is a supported default proof for φ from Δ and*
DP2 *$\Delta \ominus P$ is even (well-ordered, or non-conflicting).*

Observe that **DP2** can be verified by means of the block graph in polynomial time. As above, the stipulation of even- or well-ordered- or even non-conflicting-ness can be replaced by any other condition guaranteeing the existence of extensions. This establishes further evidence of the modularity of our approach.

For illustration, let us render Default theory (1) a non-even, that is an *odd*, theory by adding self-circular default rule $\delta_5 = \frac{:f}{\neg f}$. We refer to the resulting default theory as Δ'. The corresponding block graph is given in Figure 2. This makes us add (δ_5, δ_5), (δ_4, δ_5)

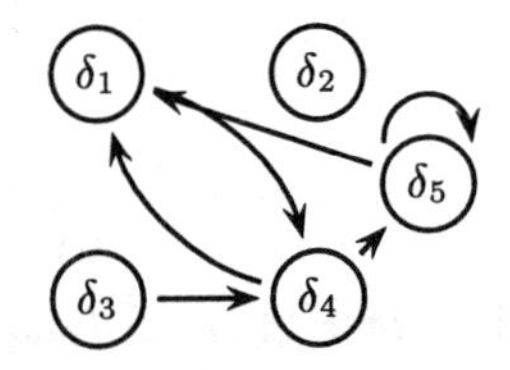

Fig. 2. Block graph of Δ'.

and (δ_5, δ_1) to the block graph in Figure 2; the rest of the graph remains unchanged. Δ' has a single extension: $Th(W \cup \{b, w, \neg f\})$. Now, let us reconsider in turn the previous (supported) default proofs, in the light of this change.

We start with $P = \{\delta_4\} = \{\frac{p\,:\,\neg ab_p}{\neg f}\}$. Since the set of predecessors of δ_4 remains unchanged, P is still a supported default proof; thus establishing **DP1**. To verify **DP2** for P, we have to construct the block graph $\Gamma_{\Delta' \ominus P}$ of $\Delta' \ominus P$ for verifying its even- or well-orderedness. By the monotonicity property expressed in Lemma 1, $\Gamma_{\Delta' \ominus P}$ is obtained from $\Gamma_{\Delta'}$ by simply deleting vertices and arcs. First, we delete in $\Gamma_{\Delta'}$ all defaults in $P = \{\delta_4\}$ along with all adjacent arcs. The same is done with δ_1 and δ_5 because $W \cup Cons(\{\delta_4\}) \vdash \neg Just(\delta_i)$ for $i = 1, 5$ (cf. (4)). As a result, we obtain for $\Gamma_{\Delta' \ominus P}$ a graph with vertices δ_2, δ_3 and no arcs, which implies that $\Delta' \ominus P$ is even non-conflicting. This shows that $\Delta' \ominus P$ has an extension and hence that P is a general default proof for $\neg f$ from Δ'.

Next, consider $P' = \{\delta_3, \delta_1\} = \{\frac{p\,:\,b}{b}, \frac{b\,:\,\neg ab_b}{f}\}$, the supported proof of f from (1), in the light of additional default rule $\delta_5 = \frac{:f}{\neg f}$. Unlike above, we do now encounter an additional predecessor of δ_1, namely, δ_5. Investigating the pre-predecessors of δ_5 yields two candidates for forming supporting sets: δ_4 and δ_5. Both of them can, however, be ruled out immediately, since their addition to P' would violate **SP3**. We see that their common consequent $\neg f$ is inconsistent with those in P'. Also, the empty supporting set must not be taken into account, because δ has predecessors (see full paper). Hence there is no way to construct a supporting set for δ_1, and so $\{\delta_3, \delta_1\}$ is no supported default proof, as reflected by the fact that Δ' has no extension containing f.

Finally, let us consider $P'' = \{\delta_3, \delta_2\} = \{\frac{p:b}{b}, \frac{b:w}{w}\}$, the supported proof of w from (1), in the light of additional default rule $\delta_5 = \frac{:f}{\neg f}$. Although the absence of predecessors for both default rules in $\Gamma_{\Delta'}$ does not urge us to determine (direct) supporting sets for them, we do actually need an indirect support for guaranteeing an encompassing extension. To see this, consider the block graph of $\Delta' \ominus P''$, given in Figure 3. We

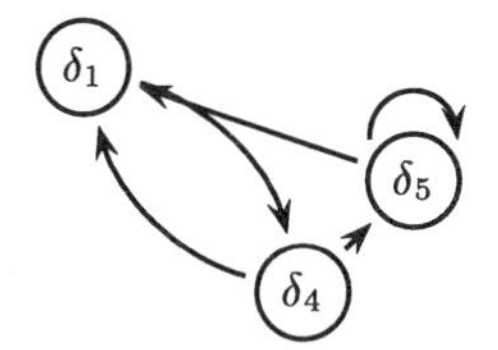

Fig. 3. Block graph of $\Delta' \ominus P''$.

encounter an odd cycle in $\Gamma_{\Delta' \ominus P''}$ which has an arc entering the odd cycle, namely (δ_4, δ_5), indicating that δ_5 can be blocked by δ_4. In fact, by adding δ_4 as an indirect support to P'', we can eliminate the odd cycle in the block graph and thus guarantee an encompassing extension. In more detail, we must now verify that $P'' \cup \{\delta_4\}$ satisfies **DP1** and **DP2**. The fact that $P'' \cup \{\delta_4\}$ is a supported default proof is established as shown above. For **DP2**, we must inspect the block graph of $\Delta' \ominus (P'' \cup \{\delta_4\})$, which turns out to be the empty graph. In all, $P'' \cup \{\delta_4\}$ is thus a valid default proof for w from Δ'.

Condition **DP2** is thus verified by first eliminating the defaults involved in the default proof from the block graph and then checking whether the resulting block graph is even, acyclic, or arcless, all of which is doable in polynomial time. In order to apply these criteria, however, we might have to add additional rules eliminating odd cycles.

We have the following result in the general case.

Theorem 6. *Let $\Delta = (D, W)$ be a (general) default theory and φ a formula. Then $\varphi \in E$ for an extension E of Δ iff there is a (general) default proof for φ from Δ.*

We have seen that the formation of default proofs benefits considerably from the usage of block graphs. As a major consequence, we may restrict our attention to ultimately necessary default rules; thus avoiding the construction of entire extensions. In fact, we have seen that irrelevant default rules, like $\frac{b:w}{w}$ for answering queries f or $\neg f$, are always ignored, since they are not related to the constituent rules of the respective proofs in the block graph. Contrariwise, we may ignore default rules $\frac{b:\neg ab_b}{f}$ and $\frac{p:\neg ab_p}{\neg f}$, when applying $\frac{b:w}{w}$, unless they must be called in for blocking extension-menacing rules, like $\frac{:f}{\neg f}$. In either case, both the non-interaction of $\frac{b:\neg ab_b}{f}$ and $\frac{p:\neg ab_p}{\neg f}$ with $\frac{b:w}{w}$ and the interaction between $\frac{p:\neg ab_p}{\neg f}$ and $\frac{:f}{\neg f}$ are read off the block graph.

As a result, all previous default proofs consist of true subsets of the generating defaults. Of course in the worst case, there are default theories and queries, for which we have to consider all generating defaults. But this is then arguably a matter of fact and unavoidable.

7 Discussion

We have introduced new concepts on default logic leading to alternative characterizations of default proofs and their encompassing extensions. The central role among these concepts is played by that of a block graph. First, it allows us to determine the existence of extensions in polynomial time by means of simple graph operations. Second, it tells us exactly which default rules must be considered for applying a default rule. The effort put into the computation of the block graph thus pays off whenever it allows us to validate default proofs by appeal to rather small subsets of the overall set of default rules.

We have presented our concepts in the context of query-answering. The resulting characterization of default proofs for queries is (to the best of our knowledge) the first one that does not appeal to the computation of entire extensions. For this, we have divided query-answering in Reiter's default logic in an off-line and an on-line process: We start with a compilation phase which results in the block graph Γ_Δ of a default theory Δ. Γ_Δ contains the essential information about the blocking relations between the default rules in Δ and is quadratic in the number of defaults. Unlike [7, 14], our approach does thus not suffer from exponential space complexity. The subsequent query-answering phase aims at finding a default proof P such that $\Delta \ominus P$ possesses an extension. This gives us a default proof which contains only the ultimately necessary defaults. To be more precise, a default rule belongs to P only (i) if it contributes to the derivation of the query or if it is needed (iia) for supporting a constituent rule of the proof or (iib) for supporting an encompassing extension. While the former is fixed by the standard inferential relation, the two latter are determined by the block graph. For delineating such a proof, we can draw on Γ_Δ for (re)computing the blocking and supporting sets of its constituent rules. Blocking sets are found among the direct predecessors of a rule, while the search for its supporting sets can be restricted to its pre-predecessors. This is clearly the more efficient the sparser the block graph. A default rule neither belonging to the actual proof nor being related to it via a path in the block graph must thus never be taken into account during query-answering (provided that an encompassing extension exists). This is what makes our approach different from extension-oriented ones [7, 14, 9, 3].

An important underlying problem is that of guaranteeing extensions of Δ or its derivate $\Delta \ominus P$. We addressed this problem by providing a range of criteria, each of which can be read off the block graph in polynomial time. However, our investment into the construction of a block graph does not only provide us with fast checks for encompassing extensions, which is also a great benefit in view of repetitive queries, but it furnishes moreover criteria that go beyond existing ones: We show in the full paper (i) that every default theory satisfying the criterion proposed by Papadimitriou and Sideri in [10] (which is itself a generalization of the one in [4]) is an even theory *but* not vice versa and (ii) that default theories satisfying the criterion proposed by Cholewiński in

[2] are orthogonal to even theories. Moreover, our criteria are fully syntax-independant which is not the case with any of the other approaches. Also, we can treat general default theories, while [4, 10] apply to semi-normal default theories only[11] and [2] imposes strong conditions on the occorrence of propositional variables. For fairness, we draw the readers attention to the fact that all initial analysis taken by the aforementioned approaches are polynomial, unlike the computation of the block graph.

Finally, it is important to note that our query-answering approach is modular in testing the existence of extensions. Depending on the different kinds of default theories (general, semi-normal or normal), we are able to choose different criteria for this test.

References

1. G. Brewka. *Nonmonotonic Reasoning: Logical Foundations of Commonsense*. Cambridge University Press, Cambridge, 1991.
2. P. Cholewiński. Reasoning with stratified default theories. In *Proc. Third International Conf. on Logic Programming and Nonmonotonic Reasoning*, 1995.
3. P. Cholewiński, V. Marek, and M. Truszczyński. Default reasoning system DeReS. In *Proc. Fifth International Conf. on the Principles of Knowledge Representation and Reasoning*. Morgan Kaufmann, 1996.
4. D. Etherington. *Reasoning with Incomplete Information*. Research Notes in AI, Pitman.
5. D. Etherington. Formalizing nonmonotonic reasoning. *Artificial Intelligence*, 31:41–85, 1987.
6. G. Gottlob. Complexity results for nonmonotonic logics. *J. Logic and Computation*, 2(3):397–425, June 1992.
7. U. Junker and K. Konolige. Computing the extensions of autoepistemic and default logic with a TMS. In *Proc. AAAI National Conf. on Artificial Intelligence*, 1990.
8. W. Marek and M. Truszczyński. *Nonmonotonic logic: context-dependent reasoning*. Artifical Intelligence. Springer, 1993.
9. I. Niemelä. Towards efficient default reasoning. In *Proc. International Joint Conf. on Artificial Intelligence*, 312–318. Morgan Kaufmann, 1995.
10. C. Papadimitriou and M. Sideri. Default theories that always have extensions. *Artificial Intelligence*, 69:347–357, 1994.
11. R. Reiter. A logic for default reasoning. *Artificial Intelligence*, 13(1-2):81–132, 1980.
12. T. Schaub. A new methodology for query-answering in default logics via structure-oriented theorem proving. *J. Automated Reasoning*, 15(1):95–165, 1995.
13. G. Schwarz and M. Truszczyński. Subnormal modal logics for knowledge representation. In *Proc. AAAI National Conf. on Artificial Intelligence*, 438–443. Morgan Kaufmann, 1993.
14. C. Schwind and V. Risch. Tableau-based characterization and theorem proving for default logic. *J. Automated Reasoning*, 13:223–242, 1994.

[11] Note that general default theories are reducible to *weak* semi-normal theories only [8], to which [4, 10] are inapplicable.

Towards State Update Axioms: Reifying Successor State Axioms

Michael Thielscher*

Dresden University of Technology
mit@pikas.inf.tu-dresden.de

Abstract. Successor state axioms are an optimal solution to the famous Frame Problem in reasoning about actions—but only as far as its representational aspect is concerned. We show how by gradually applying the principle of reification to these axioms, one can achieve gradual improvement regarding the inferential aspect without losing the representational merits. The resulting concept of *state update axioms* constitutes a novel version of what is known as the Fluent Calculus. We illustrate that under the provision that actions have no so-called open effects, any Situation Calculus specification can be transformed into an essentially equivalent Fluent Calculus specification, in which at the same time the representational and the inferential aspect of the Frame Problem are addressed. This alternative access to the Fluent Calculus both clarifies its role in relation to the most popular axiomatization paradigm and should help to enhance its acceptance.

1 Introduction

For a long time, the Fluent Calculus, introduced in [7] and so christened in [3], has been viewed exclusively as a close relative of approaches to the Frame Problem [12] which appeal to non-classical logics, namely, linearized versions of, respectively, the connection method [1, 2] and Gentzen's sequent calculus [11]. The affinity of the Fluent Calculus and these two formalisms has been emphasized by several formal comparison results. In [5], for example, the three approaches have been proved to deliver equivalent solutions to a resource-sensitive variant of STRIPS planning [4].

Yet the Fluent Calculus possesses a feature by which it stands out against the two other frameworks: It stays entirely within classical logic. In this setting the Fluent Calculus constitutes a successful attempt to address the Frame Problem as regards both the representational aspect (since no effect axiom or any other axiom needs to mention non-effects) and, at the same time, the inferential aspect (since carrying over persistent fluents from one situation to the next does not require separate deduction steps for each). Contrary to popular opinion, all this is achieved without relying on complete knowledge of the initial or any other situation. Nonetheless the Fluent Calculus has not yet received as much attention

*on leave from Darmstadt University of Technology

in the scientific community as, say, the Situation Calculus. One reason might be that, due to its heritage, the relation to the mainstream calculi, and in particular to the Situation Calculus, has not yet been convincingly elaborated.

The purpose of this paper is to present an alternative approach to the Fluent Calculus, where we start off from the Situation Calculus in the version where successor state axioms are used as means to solve the representational aspect of the Frame Problem [13]. We illustrate how the Fluent Calculus can be viewed as the result of gradually improving this approach in view of the inferential aspect but without losing its representational merits. The key is to gradually apply the principle of reification, which means to use terms instead of atoms as the formal denotation of statements. Along the path leading from successor state axioms to the Fluent Calculus lies an intermediate approach, namely, the alternative formulation of successor state axioms described by [9], in which atomic fluent formulas are reified. This alternative design inherits the representational advantages and additionally addresses the inferential Frame Problem. Yet it does so only under the severe restriction that complete knowledge of the values of the relevant fluents in the initial situation is available. The Fluent Calculus can then be viewed as a further improvement in that it overcomes this restriction by carrying farther the principle of reification to conjunctions of fluents. In the following section we illustrate by means of examples how successor state axioms can thus be reified to what we call *state update axioms*. In Section 3 we then present a fully mechanic method to derive state update axioms from effect specifications with arbitrary first-order condition. One restriction turns out necessary for this method to be correct, namely, that actions do not have so-called *open* effects.[1] In Section 4, we will briefly show how to design state update axioms for actions with such effects.

Viewed in the way we pursue in this paper, the Fluent Calculus presents itself as the result of a successful attempt to cope with the inferential Frame Problem, starting off from successor state axioms as a solution to the representational aspect. Our hope is that this alternative access clarifies the role of this axiomatization paradigm in relation to the most popular approach and helps enhancing its acceptance. Following the new motivation it should become clearer that the Fluent Calculus provides an expressive axiomatization technique, in the setting of classical logic, which altogether avoids non-effect axioms and at the same time successfully copes with the inferential aspect of the Frame Problem.

2 From Situation Calculus to Fluent Calculus

2.1 From Successor State Axioms (I) ...

Reasoning about actions is inherently concerned with change: Properties expire, objects come into being and cease to exist, true statements about the state

[1] This concept is best explained by an example. This axiom specifies an open effect: $\forall x, y, s.\ Bomb(x) \wedge Nearby(x, y, s) \supset Destroyed(y, Do(Explodes(x), s))$. Even after instantiating the action expression $Explodes(x)$ and the situation term s, the effect literal still carries a variable, y, so that the action may have infinitely many effects.

of affairs at some time may be entirely wrong at another time. The first and fundamental challenge of formalizing reasoning about actions is therefore to account for the fact that most properties in the real world possess just a limited period of validity. This unstable nature of properties which vary in the course of time has led to calling them "fluents." In order to account for fluents changing their truth values in the course of time as consequences of actions, the Situation Calculus paradigm [12] is to attach a situation argument to each fluent, thus limiting its range of validity to a specific situation. The performance of an action then brings about a new situation in which certain fluents may no longer hold.

As an example which will be used throughout the paper, we will formalize the reasoning that led to the resolution of the following little mystery:

> A reliable witness reported that the murderer poured some milk into a cup of tea before offering it to his aunt. The old lady took a drink or two and then she suddenly fell into the armchair and died an instant later, by poisoning as has been diagnosed afterwards. According to the witness, the nephew had no opportunity to poison the tea beforehand. This proves that it was the milk which was poisoned and by which the victim was murdered.

The conclusion in this story is obviously based on some general commonsense knowledge of poisoned substances and the way they may affect people's health. To begin with, let us formalize by means of the Situation Calculus the relevant piece of knowledge that mixing a poisoned substance into another one causes the latter to be poisoned as well. To this end, we use the binary predicate $Poisoned(x, s)$ representing the fact that x is poisoned in situation s, the action term $Mix(p, x, y)$ denoting the action carried out by agent p of mixing x into y, and the binary function $Do(a, s)$ which denotes the situation to which leads the performance of action a in situation s.[2] With this signature and its semantics the following axiom formalizes the fact that if x is poisoned in situation s then y, too, is poisoned in the situation that obtains when someone mixes x into y:

$$Poisoned(x, s) \supset Poisoned(y, Do(Mix(p, x, y), s)) \tag{1}$$

The second piece of commonsense knowledge relevant to our example concerns the effect of drinking poisoned liquids. Let $Alive(x, s)$ represent the property of x being alive in situation s, and let the action term $Drink(p, x)$ denote that p drinks x. Then the following axiom encodes the fact that if x is poisoned then person p ceases to being among the livings after she had drunk x:

$$Alive(p, s) \wedge Poisoned(x, s) \supset \neg Alive(p, Do(Drink(p, x), s)) \tag{2}$$

These two effect axioms, however, do not suffice to solve the mystery due to the Frame Problem, which has been uncovered as early as in [12]. To see why,

[2] A word on the notation: Predicate and function symbols, including constants, start with a capital letter whereas variables are in lower case, sometimes with sub- or superscripts. Free variables in formulas are assumed universally quantified.

let S_0 be a constant by which we denote the initial situation, and consider the assertion,

$$\neg Poisoned(Tea, S_0) \wedge Alive(Nephew, S_0) \wedge Alive(Aunt, S_0) \quad (3)$$

Even with $Poisoned(Milk, S_0)$ added, $\neg Alive(Aunt, S_2)$ does not yet follow (where $S_2 = Do(Drink(Aunt, Tea), Do(Mix(Nephew, Milk, Tea), S_0)))$, because $Alive(Aunt, Do(Mix(Nephew, Milk, Tea), S_0))$ is needed for axiom (2) to apply but cannot be concluded. In order to obtain this and other intuitively expected conclusions, a number of non-effect axioms (or "frame axioms") need to be supplied, like the following, which says that people survive the mixing of substances:

$$Alive(x, s) \supset Alive(x, Do(Mix(p, y, z), s)) \quad (4)$$

Now, the Frame Problem is concerned with the problems that arise from the apparent need for non-effect axioms like (4). Actually there are two aspects of this famous problem: The *representational* Frame Problem is concerned with the proliferation of all the many frame axioms. The *inferential* Frame Problem describes the computational difficulties raised by the presence of many non-effect axioms when it comes to making inferences on the basis of an axiomatization: To derive the consequences of a sequence of actions it is necessary to carry, one-by-one and almost all the time using non-effect axioms, each property through each intermediate situation.

With regard to the representational aspect of the Frame Problem, successor state axioms [13] provide a solution which is optimal in a certain sense, namely, in that it requires no extra frame axioms at all. The key idea is to combine, in a clear elaborated fashion, several effect axioms into a single one. The result, more complex than simple effect axioms like (1) and (2) but still mentioning solely effects, is designed in such a clever way that it implicitly contains sufficient information also about non-changes of fluents.

The procedure by which these axioms are set up is the following. Suppose $F(\boldsymbol{x})$ is among the fluents one is interested in. On the assumption that a fixed, finite set of actions is considered relevant, it should be possible to specify with a single formula $\gamma_F^+(\boldsymbol{x}, a, s)$ all circumstances by which $F(\boldsymbol{x})$ would be caused to become true. That is to say, $\gamma_F^+(\boldsymbol{x}, a, s)$ describes all actions a and conditions relative to situation s so that $F(\boldsymbol{x})$ is a positive effect of performing a in s. For example, among the actions we considered above there is one, and only one, by which the fluent $Poisoned(x)$ is made true, namely, mixing some poisonous y into x. Hence an adequate definition of $\gamma_{Poisoned}^+(x, a, s)$ is the formula $\exists p, y[a = Mix(p, y, x) \wedge Poisoned(y, s)]$.

A dual formula, $\gamma_F^-(\boldsymbol{x}, a, s)$, defines the circumstances by which fluent $F(\boldsymbol{x})$ is caused to become false. In our example we consider no way to 'decontaminate' a substance, which is why $\gamma_{Poisoned}^-(x, a, s)$ should be equated with a logical contradiction. For our second fluent, $Alive(x)$, the situation is just the other way round: While $\gamma_{Alive}^+(x, a, s)$ is false for any instance, the appropriate definition of $\gamma_{Alive}^-(x, a, s)$ is $\exists y[a = Drink(x, y) \wedge Alive(x, s) \wedge Poisoned(y, s)]$.

On the basis of suitable definitions for both γ_F^+ and γ_F^-, a complete account can be given of how the truth value of fluent F in a new situation depends on the old one, namely,

$$F(\boldsymbol{x}, Do(a,s)) \equiv \gamma_F^+(\boldsymbol{x}, a, s) \vee [F(\boldsymbol{x}, s) \wedge \neg \gamma_F^-(\boldsymbol{x}, a, s)] \tag{5}$$

This is the general form of *successor state axioms*.[3] It says that the fluent F holds in a new situation if, and only if, it is either a positive effect of the action being performed, or it was already true and the circumstances were not such that the fluent had to become false. Notice that both γ^+ and γ^- talk exclusively about effects (positive and negative), not at all about non-effects. Nonetheless, by virtue of being bi-conditional, a successor state axiom implicitly contains all the information needed to entail any non-change of the fluent in question. For whenever neither $\gamma_F^+(\boldsymbol{x}, a, s)$ nor $\gamma_F^-(\boldsymbol{x}, a, s)$ is true, then (5) rewrites to the simple equivalence $F(\boldsymbol{x}, Do(a,s)) \equiv F(\boldsymbol{x}, s)$.

The two successor state axioms for our example domain, given the respective formulas γ from above, are

$$\begin{aligned} Poisoned(x, Do(a,s)) \equiv\ & \exists p, y\, [\, a = Mix(p,y,x) \wedge Poisoned(y,s)\,] \\ & \vee\ Poisoned(x,s) \end{aligned} \tag{6}$$

and

$$\begin{aligned} & Alive(x, Do(a,s)) \equiv \\ & \quad Alive(x,s) \wedge \neg\exists y\, [\, a = Drink(x,y) \wedge Alive(x,s) \wedge Poisoned(y,s)\,] \end{aligned} \tag{7}$$

The latter, for instance, suffices to conclude that $Alive(Aunt, S_0)$ is not affected by the action $Mix(Nephew, Milk, Tea)$—assuming "unique names" for actions, i.e., $Mix(p', x', y') \neq Drink(x,y)$. Thus we can spare the frame axiom (4).

By specifying the effects of actions in form of successor state axioms it is possible to avoid frame axioms altogether. These axioms thus provide us with an in a certain sense optimal solution to the Frame Problem, as far as the representational aspect is concerned.

2.2 ... via Successor State Axioms (II) ...

While successor state axioms are a good way to overcome the representational Frame Problem since no frame axioms at all are required, the inferential aspect is fully present. In order to derive which fluents hold and which do not after a sequence of actions, it is still necessary to carry, one-by-one, each fluent through each intermediate situation by separate instances of successor state axioms. In this respect nothing seems gained by incorporating knowledge of non-effects in complex effect axioms instead of using explicit frame axioms.

---

[3] For the sake of clarity we ignore the concept of action precondition in this paper, as it is irrelevant for our discussion (see Section 4).

However, it has been shown in [9] that by formulating successor state axioms in a way that is somehow dual to the scheme (5), the inferential aspect can be addressed at least to a certain extent. Central to this alternative is the representation technique of reification. It means that properties like $Poisoned(x)$ are formally modeled as terms, in other words as objects, in logical axiomatizations. This allows for a more flexible handling of these properties within first-order logic. Let, to this end, $Holds(f, s)$ be a binary predicate representing the fact that in situation s holds the fluent f, now formally a term but still meaning a proposition.

The key to the alternative form of successor state axioms is to devise one for each action, and not for each fluent, which gives a complete account of the positive and negative effects of that action. Suppose $A(\boldsymbol{x})$ is an action, then it should be possible to specify with a single formula $\delta_A^+(\boldsymbol{x}, f, s)$ the necessary and sufficient conditions on f and s so that f is a positive effect of performing $A(\boldsymbol{x})$ in s. In our running example, the appropriate definition of $\delta_{Mix}^+(p, x, y, f, s)$, say, is $[f = Poisoned(y, s)] \wedge Holds(Poisoned(x), s)$, while $\delta_{Drink}^+(p, x, f, s)$ should be equated with a logical contradiction since $Drink(p, x)$ has no relevant positive effect. A dual formula, $\delta_A^-(\boldsymbol{x}, f, s)$, defines the necessary and sufficient conditions on f and s so that f is a negative effect of performing $A(\boldsymbol{x})$ in s. For instance, $\delta_{Mix}^-(p, x, y, f, s)$ should be false in any case, while $\delta_{Drink}^-(p, x, f, s)$ is suitably described by $[f = Alive(p)] \wedge Holds(Alive(p), s) \wedge Holds(Poisoned(x), s)$.

On the basis of δ_A^+ and δ_A^-, a complete account can be given of which fluents hold in situations reached by performing $A(\boldsymbol{x})$, namely,

$$Holds(f, Do(A(\boldsymbol{x}), s)) \equiv \delta_A^+(\boldsymbol{x}, f, s) \vee [\, Holds(f, s) \wedge \neg \delta_A^-(\boldsymbol{x}, f, s)\,] \tag{8}$$

That is to say, the fluents which hold after performing the action $A(\boldsymbol{x})$ are exactly those which are among the positive effects or which held before and are not among the negative effects. The reader may contrast this scheme with (5) and in particular observe the reversed roles of fluents and actions.[4]

Given the formulas $\delta_{Mix}^+(p, x, y, f, s)$, $\delta_{Mix}^-(p, x, y, f, s)$, $\delta_{Drink}^+(p, x, f, s)$, and $\delta_{Drink}^+(p, x, f, s)$, respectively, from above, we thus obtain these two successor state axioms of type (II):

$$\begin{aligned} Holds(f, Do(Mix(p, x, y), s)) \equiv\ & f = Poisoned(y) \wedge Holds(Poisoned(x), s) \\ & \vee\ Holds(f, s) \end{aligned} \tag{9}$$

and

$$\begin{aligned} & Holds(f, Do(Drink(p, x), s)) \equiv \\ & \qquad Holds(f, s) \wedge \neg\, [\, f = Alive(p) \wedge Holds(Alive(p), s) \\ & \qquad\qquad\qquad\qquad \wedge Holds(Poisoned(x), s)\,] \end{aligned} \tag{10}$$

Notice that as before non-effects are not explicitly mentioned and no additional frame axioms are required, so the representational aspect of the Frame Problem

[4] Much like [13] roots in the axiomatization technique of [6], the foundations for the alternative form of successor state axioms were laid in [10].

is addressed with the alternative notion of successor state axioms just as well. The inferential advantage of the alternative design shows if we represent the collection of fluents that are true in a situation s by equating the atomic formula $Holds(f, s)$ with the conditions on f to hold in s. The following formula, for instance, constitutes a suitable description of the initial situation in our example:

$$\begin{aligned}&Holds(f, S_0) \equiv \\ &\qquad f = Alive(Nephew) \vee f = Alive(Aunt) \vee f = Poisoned(Milk)\end{aligned} \tag{11}$$

The crucial feature of this formula is that the situation argument, S_0, occurs only once. With this representational trick it becomes possible to obtain a complete description of a successor situation in one go, that is, by singular application of a successor state axiom. To see why, consider the axiom which specifies the effects of mixing, (9). If we substitute p, x, and y by $Nephew$, $Milk$, and Tea, respectively, and s by S_0, then we can replace the sub-formula $Holds(f, S_0)$ of the resulting instance by the equivalent disjunction as given in axiom (11). So doing yields the formula,

$$\begin{aligned}&Holds(f, Do(Mix(Nephew, Milk, Tea), S_0)) \equiv \\ &\qquad f = Poisoned(Tea) \wedge Holds(Poisoned(Milk), S_0) \\ &\qquad \vee f = Alive(Nephew) \vee f = Alive(Aunt) \vee f = Poisoned(Milk)\end{aligned}$$

which all at once provides a complete description of the successor situation. Given suitable axioms for equality, the above can be simplified, with the aid of (11), to

$$\begin{aligned}&Holds(f, Do(Mix(Nephew, Milk, Tea), S_0)) \equiv \\ &\qquad f = Poisoned(Tea) \vee f = Alive(Nephew) \\ &\qquad \vee f = Alive(Aunt) \vee f = Poisoned(Milk)\end{aligned}$$

The reader may verify that we can likewise infer the result of $Drink(Aunt, Tea)$ in the new situation by applying the appropriate instance of successor state axiom (10), which, after simplification, yields

$$\begin{aligned}&Holds(f, Do(Drink(Aunt, Tea), Do(Mix(Nephew, Milk, Tea), S_0))) \equiv \\ &\qquad f = Poisoned(Tea) \vee f = Alive(Nephew) \vee f = Poisoned(Milk)\end{aligned}$$

At first glance it seems that the alternative design of successor state axioms provides an overall satisfactory solution to both aspects of the Frame Problem. No frame axioms at all are needed, and one instance of a single successor state axiom suffices to carry over to the next situation all unchanged fluents. However, the proposed method of inference relies on the very strong assumption that we can supply a complete account of what does and what does not hold in the initial situation. Formula (11) provides such a complete specification, because it says that any fluent is necessarily false in S_0 which does not occur to the right of the equivalence symbol. Unfortunately it is impossible to formulate partial knowledge of the initial state of affairs in a similarly advantageous fashion. Of course one can start with an incomplete specification like, for instance,

$$Holds(f, S_0) \subset [f = Alive(Nephew) \vee f = Alive(Aunt)] \wedge f \neq Poisoned(Tea)$$

which mirrors the incomplete description we used earlier (c.f. formula (3)). But then the elegant inference step from above, where we have simply replaced a sub-formula by an equivalent, is no longer feasible. In this case one is in no way better off with the alternative notion of successor state axioms; again separate instances need to be applied, one for each fluent, in order to deduce what holds in a successor situation.

2.3 ... to State Update Axioms

So far we have used reification to denote single properties by terms. The 'meta'-predicate *Holds* has been introduced which relates a reified fluent to a situation term, thus indicating whether the corresponding property is true in the associated situation. When formalizing collected information about a particular situation S as to which fluents are known to hold in it, the various corresponding atoms $Holds(f_i, S)$ are conjuncted using the standard logical connectives. We have seen how the inferential aspect of the Frame Problem is addressed if this is carried out in a certain way, namely, by equating $Holds(f, s)$ with some suitable formula Ψ. The effects of an action a can then be specified in terms of how Ψ modifies to some formula Ψ' such that $Holds(f, Do(a, s)) \equiv \Psi'$. We have also seen, however, that this representation technique is still not sufficiently flexible in that it is impossible to construct a first-order formula Ψ so that $Holds(f, S_0) \equiv \Psi$ provides a correct incomplete specification of S_0. Yet it is possible to circumvent this drawback by carrying farther the principle of reification, to the extent that not only single fluents but also their conjunctions are formally treated as terms. Required to this end is a binary function which to a certain extent reifies the logical conjunction. This function shall be denoted by the symbol "$\circ$" and written in infix notation, so that, for instance, the term $Alive(Nephew) \circ Poisoned(Milk)$ is the reified version of $Alive(Nephew) \wedge Poisoned(Milk)$. The use of the function "$\circ$" is the characteristic feature of axiomatizations which follow the paradigm of Fluent Calculus.

The union of all relevant fluents that hold in a situation is called the *state* (of the world) in that situation. Recall that a situation is characterized by the sequence of actions that led to it. While the world possibly exhibits the very same state in different situations,[5] the world is in a unique state in each situation. A function denoted by $State(s)$ shall relate situations s to the corresponding states, which are reified collections of fluents.

Modeling entire states as terms allows the use of variables to express mere partial information about a situation. The following, for instance, is a correct

---

[5] If, for example, the tea was already poisoned initially, then the state of the world prior to and after $Mix(Nephew, Milk, Tea)$ would have been the same—in terms of which of the two liquids are poisoned and who of our protagonists is alive.

incomplete account of the initial situation S_0 in our mystery story (c.f. (3)):

$$\exists z \, [\, State(S_0) = Alive(Nephew) \circ Alive(Aunt) \circ z \\ \wedge \forall z'. \, z \neq Poisoned(Tea) \circ z' \,] \tag{12}$$

That is to say, of the initial state it is known that both $Alive(Nephew)$ and $Alive(Aunt)$ are true and that possibly some other facts z hold, too—with the restriction that z must not include $Poisoned(Tea)$, of which we know it is false.

The binary function " $\circ$ " needs to inherit from the logical conjunction an important property. Namely, the order is irrelevant in which conjuncts are given. Formally, order ignorance is ensured by stipulating associativity and commutativity, that is, $\forall x, y, z. \, (x \circ y) \circ z = x \circ (y \circ z)$ and $\forall x, y. \, x \circ y = y \circ x$. It is convenient to also reify the empty conjunction, a logical tautology, by a constant usually denoted $\emptyset$ and which satisfies $\forall x. \, x \circ \emptyset = x$. The three equational axioms, jointly abbreviated AC1, in conjunction with the standard axioms of equality entail the equivalence of two state terms whenever they are built up from an identical collection of reified fluents.[6] In addition, denials of equalities, such as in the second part of formula (12), need to be derivable. This requires an extension of the standard assumption of "unique names" for fluents to uniqueness of states, denoted by $EUNA$ (see, e.g., [8, 14]).

The assertion that some fluent f holds (resp. does not hold) in some situation s can now be formalized by $\exists z. \, State(s) = f \circ z$ (resp. $\forall z. \, State(s) \neq f \circ z$). This allows to reintroduce the *Holds* predicate, now, however, not as a primitive notion but as a derived concept:

$$Holds(f, s) \equiv \exists z. \, State(s) = f \circ z \tag{13}$$

In this way, any Situation Calculus assertion about situations can be directly transferred to a formula of the Fluent Calculus. For instance, the (quite arbitrary) Situation Calculus formula $\exists x. \, Poisoned(x, S_0) \vee \neg Alive(Aunt, S_0)$ reads $\exists x. \, Holds(Poisoned(x), S_0) \vee \neg Holds(Alive(Aunt), S_0)$ in the Fluent Calculus. We will use the notation $HOLDS(\Psi)$ to denote the formula that results from transforming a Situation Calculus formula Ψ into the reified version using the *Holds* predicate.

Knowledge of effects of actions is formalized in terms of specifying how a current state modifies when moving on to a next situation. The universal form of what we call *state update axiom* is

$$\Delta(s) \supset \Gamma[State(Do(A, s)), State(s)] \tag{14}$$

where $\Delta(s)$ states conditions on s, or rather on the corresponding state, under which the successor state is obtained by modifying the current state according to Γ. Typically, condition $\Delta(s)$ is a compound formula consisting of $Holds(f, s)$

---

[6] The reader may wonder why function " $\circ$ " is not expected to be idempotent, i.e., $\forall x. \, x \circ x = x$, which is yet another property of logical conjunction. The (subtle) reason for this is given below.

atoms, as defined with the foundational axiom (13). The component Γ defines the way the state in situation s modifies according to the effects of the action under consideration. Actions may initiate and terminate properties. We will discuss the designing of Γ for these two cases in turn.

If an action has a positive effect, then the fluent which becomes true simply needs to be coupled onto the state term. An example is the following axiomatization of the (conditional) effect of mixing a liquid into a second one:

$$\begin{aligned} & Holds(Poisoned(x), s) \wedge \neg Holds(Poisoned(y), s) \supset \\ & \qquad State(Do(Mix(p,x,y),s)) = State(s) \circ Poisoned(y) \\ & \neg Holds(Poisoned(x), s) \vee Holds(Poisoned(y), s) \supset \\ & \qquad State(Do(Mix(p,x,y),s)) = State(s) \end{aligned} \tag{15}$$

That is to say, if x is poisoned and y is not, then the new state is obtained from the predecessor just by adding the fluent $Poisoned(y)$, else nothing changes at all and so the two states are identical. Notice that neither of the two state update axioms mentions any non-effects.

If we substitute, in the two axioms (15), p, x, and y by *Nephew*, *Milk*, and *Tea*, respectively, and s by S_0, then we can replace the term $State(S_0)$ in both resulting instances by the equal term as given in axiom (12). So doing yields,

$$\begin{aligned} \exists z\, [\ & Holds(Poisoned(Milk), S_0) \wedge \neg Holds(Poisoned(Tea), S_0) \supset \\ & \qquad State(Do(Mix(Nephew, Milk, Tea), S_0)) \\ & \qquad\quad = Alive(Nephew) \circ Alive(Aunt) \circ z \circ Poisoned(Tea) \\ & \wedge \neg Holds(Poisoned(Milk), S_0) \vee Holds(Poisoned(Tea), S_0) \supset \\ & \qquad State(Do(Mix(Nephew, Milk, Tea), S_0)) \\ & \qquad\quad = Alive(Nephew) \circ Alive(Aunt) \circ z \\ & \wedge \forall z'.\, z \neq Poisoned(Tea) \circ z' \ ] \end{aligned}$$

which implies, using the abbreviation $S_1 = Do(Mix(Nephew, Milk, Tea), S_0)$ and the correspondence (13) along with axioms for equality and assertion (12),

$$\begin{aligned} \exists z\, [\ & Holds(Poisoned(Milk), S_0) \supset \\ & \qquad State(S_1) = Alive(Nephew) \circ Alive(Aunt) \\ & \qquad\qquad \circ Poisoned(Milk) \circ Poisoned(Tea) \circ z \\ & \wedge \neg Holds(Poisoned(Milk), S_0) \supset \\ & \qquad State(S_1) = Alive(Nephew) \circ Alive(Aunt) \circ z \\ & \qquad \wedge \neg Holds(Poisoned(Tea), S_1) \ ] \end{aligned}$$

In this way we have obtained from an incomplete initial specification a still partial description of the successor state, which includes the unaffected fluents $Alive(Nephew)$ and $Alive(Aunt)$. These properties thus survived the application of the effects axioms without the need to be carried over, one-by-one, by separate application of axioms.

If an action has a negative effect, then the fluent f which becomes false needs to be withdrawn from the current state $State(s)$. The schematic equation $State(Do(A, s)) \circ f = State(s)$ serves this purpose. Incidentally, this scheme is the sole reason for not stipulating that "$\circ$" be idempotent. For otherwise the equation $State(Do(A, s)) \circ f = State(s)$ would be satisfied if $State(Do(A, s))$ contained f. Hence this equation would not guarantee that f becomes false. Vital for our scheme is also to ensure that state terms do not contain any fluent twice or more, i.e.,

$$\forall s, x, z.\ State(s) = x \circ x \circ z \supset x = \emptyset \tag{16}$$

These preparatory remarks lead us to the following axiomatization of the (conditional) effect of drinking:

$$\begin{aligned} & Holds(Alive(p) \circ Poisoned(x), s) \supset \\ & \qquad State(Do(Drink(p, x), s)) \circ Alive(p) = State(s) \\ & \neg Holds(Alive(p), s) \vee \neg Holds(Poisoned(x), s) \supset \\ & \qquad State(Do(Drink(p, x), s)) = State(s) \end{aligned} \tag{17}$$

That is to say, if p is alive and x is poisoned, then the new state is obtained from the predecessor just by terminating $Alive(p)$, else nothing changes at all.[7]

Applying the two axioms (17) to what we have derived about the state in situation S_1 yields, setting $S_2 = Do(Drink(Aunt, Tea), S_1)$ and performing straightforward simplifications,

$$\begin{aligned} \exists z\ [\ & Holds(Poisoned(Milk), S_0) \supset \\ & \qquad State(S_2) \circ Alive(Aunt) = Alive(Nephew) \circ Alive(Aunt) \\ & \qquad\qquad \circ Poisoned(Milk) \circ Poisoned(Tea) \circ z \\ \wedge\ & \neg Holds(Poisoned(Milk), S_0) \supset \\ & \qquad State(S_2) = Alive(Nephew) \circ Alive(Aunt) \circ z\ ] \end{aligned}$$

This partial description[8] of the successor state again includes every persistent fluent without having applied separate deduction steps for each. The Fluent Calculus thus provides a solution to both the representational and the inferential aspect of the Frame Problem which is capable of dealing with incomplete knowledge about states.

[7] Actions may of course have both positive and negative effects at the same time, in which case the component Γ of a state update axiom combines the schemes for initiating and terminating fluents. This general case is dealt with in Section 3.

[8] which by the way, since $State(S_2) = Alive(Nephew) \circ Alive(Aunt) \circ z$ implies that $Holds(Alive(Aunt), S_2)$, leads directly to the resolution of the murder mystery: Along with the statement of the witness, $\neg Holds(Alive(Aunt), S_2)$, the formula above logically entails the explanation that $Holds(Poisoned(Milk), S_0)$.

3 The General Method

Having illustrated the design and use of state update axioms by example, in this section we will present a general, fully mechanic procedure by which is generated a suitable set of state update axioms from a given collection of Situation Calculus effect axioms, like (1) and (2). As indicated in the introduction, we will only consider actions without open effects (c.f. Footnote 1). This is reflected in the assumption that each positive effect specification be of the following form, where A denotes an action and F a fluent:

$$\varepsilon^{+}_{A,F}(\boldsymbol{x}, s) \supset F(\boldsymbol{y}, Do(A(\boldsymbol{x}), s)) \tag{18}$$

Here, ε is a first-order formula whose free variables are among $\boldsymbol{x}, s$; and $\boldsymbol{y}$ contains only variables from $\boldsymbol{x}$. Notice that it is the very last restriction which ensures that the effect specification does not describe what is called an open effect: Except for the situation term, all arguments of the effect F are bound by the action term $A(\boldsymbol{x})$. Likewise, negative effect specifications are of the form

$$\varepsilon^{-}_{A,F}(\boldsymbol{x}, s) \supset \neg F(\boldsymbol{y}, Do(A(\boldsymbol{x}), s)) \tag{19}$$

where again ε is a first-order formula whose free variables are among $\boldsymbol{x}, s$ and where $\boldsymbol{y}$ contains only variables from $\boldsymbol{x}$.[9] We assume that a given set $\mathcal{E}$ of effect axioms is consistent in that for all A and F the unique names assumption entails $\neg\exists \boldsymbol{x}, s\, [\varepsilon^{+}_{A,F}(\boldsymbol{x}, s) \wedge \varepsilon^{-}_{A,F}(\boldsymbol{x}, s)]$.

Fundamental for any attempt to solve the Frame Problem is the assumption that a given set of effect axioms is *complete* in the sense that it specifies all relevant effects of all involved actions.[10] Our concern, therefore, is to design state update axioms for a given set of effect specifications which suitably reflect the completeness assumption. The following instance of scheme (14) is the general form of state update axioms for deterministic actions with only direct effects:

$$\Delta(s) \supset State(Do(A, s)) \circ \vartheta^{-} = State(s) \circ \vartheta^{+}$$

where ϑ^{-} are the negative effects and ϑ^{+} the positive effects, respectively, of action A under condition $\Delta(s)$. The main challenge for the design of these state update axioms is to make sure that condition Δ is strong enough for the equation in the consequent to be sound. Neither must ϑ^{+} include a fluent that already holds in situation s (for this would contradict the foundational axiom about multiple occurrences, (16)), nor should ϑ^{-} specify a negative effect that is already false in s (for then *EUNA* implies that the equation be false). This is the motivation behind step 1 and 2 of the procedure below. The final and main step 3 reflects the fact that actions with conditional effects require more than one state update axiom, each applying in different contexts:

[9] Our two effect axioms at the beginning of Section 2.1 fit this scheme, namely, by equating $\varepsilon^{+}_{Mix,Poisoned}(p, x, y, s)$ with $Poisoned(x, s)$ and $\varepsilon^{-}_{Drink,Alive}(p, x, s)$ with $Alive(p, s) \wedge Poisoned(x, s)$.

[10] If actions have additional, indirect effects, then this gives rise to the so-called Ramification Problem; see Section 4.

1. Rewrite to $\varepsilon^+_{A,F}(\boldsymbol{x}, s) \wedge \neg F(\boldsymbol{y}, s) \supset F(\boldsymbol{y}, Do(A(\boldsymbol{x}), s))$ each positive effect axiom of the form (18).
2. Similarly, rewrite to $\varepsilon^-_{A,F}(\boldsymbol{x}, s) \wedge F(\boldsymbol{y}, s) \supset \neg F(\boldsymbol{y}, Do(A(\boldsymbol{x}), s))$ each negative effect axiom of the form (19).
3. For each action A, let the following $n \geq 0$ axioms be all effect axioms thus rewritten (positive and negative) concerning A:

$$\begin{aligned}
&\varepsilon_1(\boldsymbol{x}, s) \supset F_1(\boldsymbol{y}_1, Do(A(\boldsymbol{x}), s)), \ \ldots, \ \varepsilon_m(\boldsymbol{x}, s) \supset F_m(\boldsymbol{y}_m, Do(A(\boldsymbol{x}), s)) \\
&\varepsilon_{m+1}(\boldsymbol{x}, s) \supset \neg F_{m+1}(\boldsymbol{y}_{m+1}, Do(A(\boldsymbol{x}), s)), \ \ldots, \\
&\qquad\qquad\qquad\qquad\qquad\qquad \varepsilon_n(\boldsymbol{x}, s) \supset \neg F_n(\boldsymbol{y}_n, Do(A(\boldsymbol{x}), s))
\end{aligned}$$

Then, for any pair of subsets $\mathcal{I}_+ \subseteq \{1, \ldots, m\}$, $\mathcal{I}_- \subseteq \{m+1, \ldots, n\}$ (including the empty ones) introduce the following state update axiom:

$$\begin{aligned}
&\textstyle\bigwedge_{i \in \mathcal{I}^+ \cup \mathcal{I}^-} HOLDS(\varepsilon_i(\boldsymbol{x}, s)) \wedge \bigwedge_{j \notin \mathcal{I}^+ \cup \mathcal{I}^-} HOLDS(\neg\varepsilon_j(\boldsymbol{x}, s)) \\
&\qquad \supset State(Do(A(\boldsymbol{x}), s)) \circ \vartheta^{\mathcal{I}_-} = State(s) \circ \vartheta^{\mathcal{I}_+}
\end{aligned} \tag{20}$$

where $\vartheta^{\mathcal{I}_-}$ is the term $F_1 \circ \ldots \circ F_k$ if $\{F_1, \ldots, F_k\} = \{F_i(\boldsymbol{y}_i) : i \in \mathcal{I}^-\}$ and, similarly, $\vartheta^{\mathcal{I}_+}$ is the term $F_1 \circ \ldots \circ F_k$ if $\{F_1, \ldots, F_k\} = \{F_i(\boldsymbol{y}_i) : i \in \mathcal{I}^+\}$.[11]

Step 3 blindly considers all combinations of positive and negative effects. Some of the state update axiom thus obtained may have inconsistent antecedent, in which case they can be removed. To illustrate the interaction of context-dependent positive and negative effects, let us apply our procedure to these two effect axioms:

$$\begin{aligned}
Loaded(s) &\supset Dead(Do(Shoot, s)) \\
true &\supset \neg Loaded(Do(Shoot, s))
\end{aligned}$$

After rewriting according to steps 1 and 2, step 3 produces four state update axioms, viz.

$$\begin{aligned}
&\neg\,[\,Holds(Loaded, s) \wedge \neg Holds(Dead, s)\,] \wedge \neg\,[\,true \wedge Holds(Loaded, s)\,] \\
&\qquad \supset State(Do(Shoot, s)) \circ \emptyset = State(s) \circ \emptyset \\
&\neg\,[\,Holds(Loaded, s) \wedge \neg Holds(Dead, s)\,] \wedge true \wedge Holds(Loaded, s) \\
&\qquad \supset State(Do(Shoot, s)) \circ Loaded = State(s) \circ \emptyset \\
&Holds(Loaded, s) \wedge \neg Holds(Dead, s) \wedge \neg\,[\,true \wedge Holds(Loaded, s)\,] \\
&\qquad \supset State(Do(Shoot, s)) \circ \emptyset = State(s) \circ Dead \\
&Holds(Loaded, s) \wedge \neg Holds(Dead, s) \wedge true \wedge Holds(Loaded, s) \\
&\qquad \supset State(Do(Shoot, s)) \circ Loaded = State(s) \circ Dead
\end{aligned}$$

Logical simplification of the premises of the topmost two axioms yields

$$\begin{aligned}
\neg Holds(Loaded, s) &\supset State(Do(Shoot, s)) = State(s) \\
Holds(Dead, s) \wedge Holds(Loaded, s) &\supset State(Do(Shoot, s)) \circ Loaded = State(s)
\end{aligned}$$

[11] Thus $\vartheta^{\mathcal{I}_-}$ contains the negative effects and $\vartheta^{\mathcal{I}_+}$ the positive effects specified in the update axiom. If either set is empty then the respective term is the unit element, $\emptyset$.

The third axiom can be abandoned because of an inconsistent antecedent, while the fourth axiom simplifies to

$$\begin{array}{l} \mathit{Holds}(\mathit{Loaded}, s) \wedge \neg \mathit{Holds}(\mathit{Dead}, s) \supset \\ \qquad \mathit{State}(\mathit{Do}(\mathit{Shoot}, s)) \circ \mathit{Loaded} = \mathit{State}(s) \circ \mathit{Dead} \end{array}$$

(The interested reader may verify that applying the general procedure to our effect axioms (1) and (2) yields four axioms which, after straightforward simplification, turn out to be (15) and (17), respectively.)

The following primary theorem for the Fluent Calculus shows that the resulting set of state update axioms correctly reflects the effect axioms if the fundamental completeness assumption is made.

Theorem 1. *Consider a finite set $\mathcal{E}$ of effect axioms which complies with the assumption of consistency, and let SUA be the set of state update axioms generated from $\mathcal{E}$. Suppose $\mathcal{M}$ is a model of $SUA \cup \{(13), (16)\} \cup EUNA$,*[12] *and consider a fluent term $F(\tau)$, an action term $A(\rho)$, and a situation term σ. Then $\mathcal{M} \models Holds(F(\tau), Do(A(\rho), \sigma))$ iff*

1. *$\mathcal{M} \models \varepsilon^{+}_{A,F}(\rho, \sigma)$, for the instance $\varepsilon^{+}_{A,F}(\rho, \sigma) \supset F(\tau, Do(A(\rho), \sigma))$ of some axiom in $\mathcal{E}$;*
2. *or $\mathcal{M} \models Holds(F(\tau), \sigma)$ and no instance $\varepsilon^{-}_{A,F}(\rho, \sigma) \supset \neg F(\tau, Do(A(\rho), \sigma))$ of an axiom in $\mathcal{E}$ exists such that $\mathcal{M} \models \varepsilon^{-}_{A,F}(\rho, \sigma)$.*

4 Conclusion

We have pursued a new motivation for the Fluent Calculus, namely, as the outcome of applying the principle of reification to successor state axioms. The resulting concept of state update axioms copes with the inferential Frame Problem without losing the solution to the representational aspect. We have shown how, much like in [13], a suitable collection of these axioms can be automatically derived from a complete (wrt. the relevant fluents and actions) set of single effect axioms, provided actions have no open effects. Since state update axioms cover the entire change an action causes in order to solve the inferential aspect of the Frame Problem, their number is, in the worst case, exponentially larger than the number of single effect axioms. This is perfectly acceptable since actions are viewed as having very few effects compared to the overall number of fluents.

Open effects can only be implicitly described in state update axioms. An example is the following axiom (c.f. Footnote 1):

$$\begin{array}{l} \mathit{Bomb}(x) \supset \\ \quad \left[\forall f, y \left[\begin{array}{l} f = \mathit{Destroyed}(y) \wedge \mathit{Holds}(\mathit{Nearby}(x,y), s) \wedge \neg \mathit{Holds}(f, s) \\ \qquad \equiv \exists z.\, w = f \circ z \end{array} \right] \right. \\ \qquad \left. \supset \mathit{State}(\mathit{Do}(\mathit{Explodes}(x), s)) = z \circ w \right] \end{array}$$

[12] Recall that $EUNA$, the extended unique names assumption, axiomatizes equality and inequality of terms with the function "$\circ$".

in which w, the positive effects of the action, is defined rather than explicitly given. It lies in the nature of open effects that a suitable state update axiom can only implicitly describe the required update and so does no longer solve the inferential Frame Problem (though it still covers the representational aspect).

The problem of action preconditions has been ignored for the sake of clarity. Their dealing with requires no special treatment in the Fluent Calculus since each Situation Calculus assertion about what holds in a situation corresponds directly to a Fluent Calculus assertion via the fundamental relation (13).

The basic Fluent Calculus as investigated in this paper assumes state update axioms to describe all effects of an action. The solution to the Ramification Problem of [14], and in particular its axiomatization in the Fluent Calculus, furnishes a ready approach for elaborating the ideas developed in the present paper so as to deal with additional, indirect effects of actions.

The version of the Fluent Calculus we arrived at in this paper differs considerably from its roots [7], e.g. in that it exploits the full expressive power of first-order logic. In so doing it is much closer to the variant introduced in [14], but still novel is the notion of state update axioms. In particular the new function $State(s)$ seems to lend more elegance to effect specifications and at the same time emphasizes the relation to the Situation Calculus.

References

1. W. Bibel. A deductive solution for plan generation. *New Generation Computing*, 4:115–132, 1986.
2. W. Bibel. Let's plan it deductively! In M. Pollack, editor, *Proceedings of IJCAI*, pages 1549–1562, Nagoya, Japan, August 1997. Also to appear in *Artif. Intell.*
3. S.-E. Bornscheuer and M. Thielscher. Explicit and implicit indeterminism: Reasoning about uncertain and contradictory specifications of dynamic systems. *J. of Logic Programming*, 31(1–3):119–155, 1997.
4. R. E. Fikes and N. J. Nilsson. STRIPS: A new approach to the application of theorem proving to problem solving. *Artif. Intell.*, 2:189–208, 1971.
5. G. Große, S. Hölldobler, and J. Schneeberger. Linear Deductive Planning. *J. of Logic and Computation*, 6(2):233–262, 1996.
6. A. R. Haas. The case for domain-specific frame axioms. In F. M. Brown, editor, *The Frame Problem in Artificial Intelligence*, pages 343–348, Los Altos, CA, 1987.
7. S. Hölldobler and J. Schneeberger. A new deductive approach to planning. *New Generation Computing*, 8:225–244, 1990.
8. S. Hölldobler and M. Thielscher. Computing change and specificity with equational logic programs. *Annals of Mathematics and Artif. Intell.*, 14(1):99–133, 1995.
9. H. Khalil. Formalizing the effects of actions in first order logic. Master's thesis, Dept. of Computer Science, Darmstadt University of Technology, 1996.
10. R. Kowalski. *Logic for Problem Solving*. Elsevier, 1979.
11. M. Masseron, C. Tollu, and J. Vauzielles. Generating plans in linear logic I. Actions as proofs. *J. of Theoretical Computer Science*, 113:349–370, 1993.
12. J. McCarthy and P. J. Hayes. Some philosophical problems from the standpoint of artificial intelligence. *Machine Intell.*, 4:463–502, 1969.
13. R. Reiter. The frame problem in the situation calculus: A simple solution (sometimes) and a completeness result for goal regression. In V. Lifschitz, editor, *Artificial Intelligence and Mathematical Theory of Computation*, pages 359–380, 1991.

14. M. Thielscher. Ramification and causality. *Artif. Intell.*, 89(1–2):317–364, 1997.

A Mechanised Proof System for Relation Algebra using Display Logic

Jeremy E. Dawson * and Rajeev Goré **

Department of Computer Science and Automated Reasoning Project,
Australian National University, Canberra, Australia
jeremy@arp.anu.edu.au, rpg@arp.anu.edu.au

Abstract. We describe an implementation of the Display Logic calculus for relation algebra as an Isabelle theory. Our implementation is the first mechanisation of any display calculus. The inference rules of Display Logic are coded directly as Isabelle theorems, thereby guaranteeing the correctness of all derivations. Our implementation generalises easily to handle other display calculi. It also provides a useful interactive proof assistant for relation algebras.
We describe various tactics and derived rules developed for simplifying proof search, including an automatic cut-elimination procedure, and example theorems proved using Isabelle. We show how some relation algebraic theorems proved using our system can be put in the form of structural rules of Display Logic, facilitating later re-use. We then show how the implementation can be used to prove results comparing alternative formalizations of relation algebra from a proof-theoretic perspective.

Keywords: proof systems for relation algebra, non-classical logics, automated deduction, display logic, description logics

1 Introduction

Relation algebras are extensions of Boolean algebras; whereas Boolean algebras model subsets of a given set, relation algebras model binary relations on a given set. Thus relation algebras have relational operations such as composition and converse. As each relation is itself a set (of pairs), relation algebras also have the Boolean operations such as intersection (conjunction) and complement (negation). Relation algebras form the basis of relational databases and of the specification and proof of correctness of programs. Recently, relation algebras and their extensions, Peirce algebras, have also been shown to from the basis of description logics [5]. Just as Boolean algebras can be studied as a logical system (classical propositional logic), relation algebras can also be studied in a logical, rather than an algebraic, fashion. In particular, relation algebras can be formulated using Display Logic [12], and in several other ways [16].

Display Logic [1] is a syntactic proof system for non-classical logic, based on the Gentzen sequent calculus [9]. Its advantages include a generic cut-elimination

* Supported by the Defence Science and Technology Organization
** Supported by an Australian Research Council QEII Fellowship

theorem, which applies whenever the rules for the display calculus satisfy certain conditions. It is a significant general logical formalism, applicable to many logics; its mechanisation is therefore an important challenge for automated reasoning.

In this paper we describe an implementation of $\boldsymbol{\delta}\mathbf{RA}$ [12], a display logic formulation of relation algebras, using the Isabelle theorem prover. In this implementation the rules of $\boldsymbol{\delta}\mathbf{RA}$ form the axioms of an Isabelle object logic – $\boldsymbol{\delta}\mathbf{RA}$ is not built upon one of the standard Isabelle object logics (as is the case, for example, with RALL [17]). The ease with which this display calculus can be implemented in Isabelle highlights the generality of both Isabelle and Display Logic. We demonstrate how Isabelle can be used to show the relationship between $\boldsymbol{\delta}\mathbf{RA}$ and two other formalizations of relation algebras.

1.1 Other mechanised proof systems for relation algebras

RALL [17] is a theorem proving system for relation algebra, based on Isabelle. It uses the atomicity of relation algebras; every relation algebra can be embedded in an atomic relation algebra. Although RALL provides automated proof search, this feature is heavily dependent on the atomization of the relation algebra. It is still not clear to us to what extent RALL is applicable to relation algebras which are not themselves atomic. RALL is built upon the `HOL` Isabelle theory, whereas our implementation of $\boldsymbol{\delta}\mathbf{RA}$ is built directly upon Isabelle's metalogic.

RALF [2] is a graphically oriented relation algebra formula manipulation system and proof assistant. It contains a large number of hard-coded transformation rules, and the super-user can add others. However the rules do not form a formal calculus. Thus it does not demonstrate that the results it derives follow from any formalization of relation algebra. In fact it may be seen as complementing a system such as that described in this paper since an interesting avenue for further work would be to obtain the transformation rules of RALF as $\boldsymbol{\delta}\mathbf{RA}$ derived rules, giving a rigrorous basis to RALF and a graphical front-end to $\boldsymbol{\delta}\mathbf{RA}$.

2 Relation Algebras

A **relation** on a set U is a set of ordered pairs (a, b) of elements of U, ie a subset of $U \times U$. Since a relation is itself a set (of ordered pairs), the operations on relations include the operations of Boolean algebras. Other relational operations are (as defined in [12]) **composition** $R \circ S = \{(a, b) \mid \exists c.\ (a, c) \in R \text{ and } (c, b) \in S\}$, its **identity** $\{(a, a) \mid a \in U\}$, and **converse** $\smile R = \{(a, b) \mid (b, a) \in R\}$.

Relation algebras are an abstraction of the notion of binary relations on a set U; reference to the actual elements of the set is abstracted away, leaving the relations themselves as the objects under consideration. Chin & Tarski [6] give a finite equational axiomatization of relation algebras. The following definition, equivalent to that in [6], is taken from [5] and [12].

A relation-type algebra $\mathcal{R} = \langle R, \vee, \wedge, \neg, \bot, \top, \circ, \smile, \mathbf{1} \rangle$ (ie, a free algebra on the set of variables R with binary operators $\vee$, $\wedge$ and $\circ$, unary operators $\neg$ and $\smile$, and constants $\bot$, $\top$ and $\mathbf{1}$) is a **relation algebra** if it satisfies the following axioms for each $r, s, t \in R$

(RA1) $(R, \vee, \wedge, \neg, \bot, \top)$ is a Boolean algebra

(RA2) $r \circ (s \circ t) = (r \circ s) \circ t$ (RA3) $r \circ \mathbf{1} = r = \mathbf{1} \circ r$

(RA4) $\smile\smile r = r$ (RA5) $(r \vee s) \circ t = (r \circ t) \vee (s \circ t)$

(RA6) $\smile (r \vee s) = (\smile r) \vee (\smile s)$ (RA7) $\smile (r \circ s) = (\smile s) \circ (\smile r)$

(RA8) $\smile r \circ \neg(r \circ s) \leq \neg s$

Further operators can be defined: $A \leq B$, as used in (RA8), means $(A \wedge B) = A$. Duals of $\mathbf{1}$ and $\circ$ are defined: $\mathbf{0}$ is $\neg\mathbf{1}$, and $A + B$ is $\neg(\neg A \circ \neg B)$. We call this theory of relation algebras **RA**. A *proper* relation algebra consists of a set of relations on a set; note that not all relation algebras are isomorphic to a proper relation algebra ([16], p.444-5).

3 Relation Algebras in Display Logic

A number of different logical systems can be formulated using the method, or style, of Display Logic [1]. These include several normal modal logics [21], and intuitionistic logic [11]. Display Logic resembles the Gentzen sequent calculus **LK**, but with significant differences. For example, the rules for introducing the connective '$\vee$' (on the right) in **LK** and Display Logic are

$$\frac{\Gamma \vdash \Delta, P, Q}{\Gamma \vdash \Delta, P \vee Q}(\mathbf{LK}\text{-} \vdash \vee) \qquad \text{and} \qquad \frac{X \vdash P, Q}{X \vdash P \vee Q}(\text{DL-} \vdash \vee)$$

Whereas, in **LK**, Γ and Δ denote comma-separated lists of formulae, in Display Logic, X denotes a Display Logic structure, which can involve several *structural* operators (one of which is ','). (See [12] or [7] for a full explanation of *structures* and *formulae*). In Display Logic, unlike in **LK**, the introduced formula (here, $P \vee Q$) stands by itself on one side of the turnstile. However there are also rules (the "display postulates")which effectively allow moving formulae from one side to the other. In [12] there is a Display Logic formulation of relation algebras, called **δRA** whose structural operators are ',' ';' '$*$' '$\bullet$' 'I' 'E' . As in **LK**, the use of ',' to stand for either '$\wedge$' or '$\vee$' (depending on the position) reflects the duality between them. Likewise in **δRA** ';' stands for either '$\circ$' or '$+$', whose duality was noted above. Each structural operator stands for one, or two (depending on the position), formula operators. Thus I stands for truth or falsity, E for the identity relation or its complement, $*$ for Boolean negation and $\bullet$ for the relational converse. The rules of **δRA** are sound and complete with respect to the equational axiomatization of Chin & Tarski [12].

Like **LK**, Display Logic is suited to backwards search for proof of a sequent, provided that a sequent is provable without using the cut rule. In terms of backwards proof, the logical introduction rules eliminate the logical operators and replace them with corresponding structural operators. The cut-elimination theorem of **LK** applies in all Display Logics, provided that the rules satisfy certain conditions, which are relatively easily checked ([1]). Indeed the procedure for eliminating cuts from a proof has been automated, see §5.4.

Unlike **LK**, Display Logic has bi-directional rules, the display postulates. Therefore any proof search technique must avoid a naive application of these *ad infinitum*. Although this is a difficulty in developing a purely automatic proof search technique, this is not insurmountable, and requires further work. For example, the techniques described below are sufficient to obtain a decision procedure for the classical propositional calculus (see [7], §2.6.1 for details). In any case it is known that the theory of relation algebras is undecidable (see [16], p.432).

Note that although Display Logic is suited to backwards proof, we will always write a proof tree, or a rule, with the premises at the top and the conclusion at the bottom (except that a double line in a rule means that it is bi-directional), and we will refer to the order of steps in a proof as though it is done in a forward direction, from premises to conclusion. As an alternative to writing a rule with one premise $\mathcal{P}$ and a conclusion $\mathcal{C}$ separated by a horizontal line, we often write $\mathcal{P} \Longrightarrow \mathcal{C}$, and for a bi-directional rule we often write $\mathcal{P} \iff \mathcal{C}$. For a full explanation of Display Logic and $\boldsymbol{\delta}$**RA** see [12].

4 Isabelle

Isabelle is an interactive computer-based proof system, described in [18]. Its capabilities include higher-order unification and term rewriting. It is written in Standard ML; when it is running, the user can interact with it by entering further ML commands, and can program complex proof sequences in ML. Isabelle provides a number of basic proof steps for backwards proof (*tactics*), as well as *tacticals* for combining these. Isabelle also supports forward proof.

It has a simple meta-logic, an intuitionistic higher-order logic. The user then has to augment this by defining an object-level logic, though normally one would use one of several which are packaged with the Isabelle distribution [19].

Thus a logical system such as $\boldsymbol{\delta}$**RA** can be implemented in Isabelle so that the only proof rules available are those of $\boldsymbol{\delta}$**RA**. This contrasts with an implementation in other proof systems, where the $\boldsymbol{\delta}$**RA** rules would be added to the rules of the logic on which that proof system was based. (Of course, systems can be implemented in this way in Isabelle too; thus RALL [17] is implemented on top of the `HOL` Isabelle theory).

Isabelle theorems (including the axioms of the theory) may but need not involve Isabelle's metalogical operators. In the Isabelle implementation of Display Logic, an Isabelle theorem (of type `thm`) may be either a simple sequent of $\boldsymbol{\delta}$**RA**, or a sequent *rule*. Whereas a sequent of $\boldsymbol{\delta}$**RA** becomes an Isabelle theorem of the form $X \vdash Y$, a sequent rule becomes an Isabelle theorem containing also the operators ==> (implication) or == (equality). For example, the $(\vdash \vee)$ rule (shown in §3) appears as

```
"$?X |- ?A, ?B ==> $?X |- ?A v ?B"
```

in Isabelle. (The '?' denotes a variable which can be instantiated, and the '$' denotes a structural variable or constant.) The variant $A \vdash A$ of the (id) rule (see §5.4) appears as `"?A |- ?A"`. The bi-directional `dcp` rule, shown in Fig. 1, appears as `"$?X |- $?Y == * $?Y |- * $?X"`.

Isabelle can be used to show the equivalence of $\delta\mathbf{RA}$ to the various other systems; in most cases this serves only to confirm proofs appearing in the literature, and in many cases it can only provide proofs of the various inductive steps of a proof that requires structural induction. However it shows that $\delta\mathbf{RA}$ is effective for proving the axioms or rules of the other systems.

5 Implementation in Isabelle

The syntax of $\delta\mathbf{RA}$ has been implemented in Isabelle. The source files are presently available at `http://arp.anu.edu.au:80/~jeremy/files`, and further details may be found in [7]. The syntax distinguishes structure and formula variables; externally, at the user level, a structure variable is preceded by '$'. The method used for this was taken directly from the Isabelle theory `LK` ([19], Ch 6).

The $\delta\mathbf{RA}$ rules have been implemented in Isabelle. Some particularly useful derived rules are shown in Fig. 1. It will be noticed that the two unary structural operators commute, and both distribute (in a sense) over both the binary structural operators. These distributive rules are used, for example, in §7.1. Further, there is a one-directional distributive rule for ';' over ','. The rules `blcEa` and `blcEs`, in particular, are elegant results which are much easier to prove for the case of a proper relation algebra.

$$\begin{array}{c} X \vdash **Y \\ \hline\hline X \vdash Y \end{array}\ (\texttt{rssS}) \qquad \begin{array}{c} X \vdash Y \\ \hline\hline *Y \vdash *X \end{array}\ (\texttt{dcp}) \qquad \begin{array}{c} X \vdash \bullet\bullet Y \\ \hline\hline X \vdash Y \end{array}\ (\texttt{rbbS}) \qquad \begin{array}{c} X \vdash Y \\ \hline\hline \bullet X \vdash \bullet Y \end{array}\ (\texttt{blbl})$$

$$\begin{array}{c} *(X,Y) \vdash Z \\ \hline\hline *Y, *X \vdash Z \end{array}\ (\texttt{stcdista}) \qquad \begin{array}{c} *(X;Y) \vdash Z \\ \hline\hline *X; *Y \vdash Z \end{array}\ (\texttt{stscdista})$$

$$\begin{array}{c} \bullet(X;Y) \vdash Z \\ \hline\hline \bullet Y; \bullet X \vdash Z \end{array}\ (\texttt{blscdista}) \qquad \begin{array}{c} (\bullet X, \bullet Y) \vdash Z \\ \hline\hline \bullet(X,Y) \vdash Z \end{array}\ (\texttt{blcdista})$$

$$\frac{(X;Y),(X;Z) \vdash W}{X;(Y,Z) \vdash W}\ (\texttt{sccldista}) \qquad \frac{(X;Z),(Y;Z) \vdash W}{(X,Y);Z \vdash W}\ (\texttt{sccrdista})$$

$$\begin{array}{c} X;Y \vdash Z \\ \hline\hline \bullet Y; \bullet X \vdash \bullet Z \end{array}\ (\texttt{tagae}) \qquad \begin{array}{c} \bullet * X \vdash Y \\ \hline\hline * \bullet X \vdash Y \end{array}\ (\texttt{bsA}) \qquad \begin{array}{c} E \vdash X \\ \hline\hline *E \vdash X \end{array}\ (\texttt{sEa}) \qquad \begin{array}{c} I \vdash X \\ \hline\hline *I \vdash X \end{array}\ (\texttt{sIa})$$

$$\begin{array}{c} \bullet X, E \vdash Y \\ \hline\hline X, E \vdash Y \end{array}\ (\texttt{blcEa}) \qquad \begin{array}{c} Y \vdash \bullet X, E \\ \hline\hline Y \vdash X, E \end{array}\ (\texttt{blcEs})$$

$$\frac{A \vdash X \quad B \vdash X}{A \vee B \vdash X}\ (\texttt{orA}) \qquad \frac{X \vdash A \quad X \vdash B}{X \vdash A \wedge B}\ (\texttt{andS})$$

Fig. 1. Selected Derived Rules

Of the δ**RA** logical introduction rules, some can be applied always (in a backwards proof) when the relevant formula is displayed. Others, namely ($\vdash \wedge$), ($\vee \vdash$), ($\vdash \circ$) and ($+ \vdash$), are subject to the "constraint" that the structure on the other side of $\vdash$ be of a particular form. The rules `andS` and `orA` are equivalent to ($\vdash \wedge$) and ($\vee \vdash$). `andS` and `orA` can be applied always when the relevant formula is displayed. Unfortunately the introduction rules for '$\circ$' and '$+$' have no such equivalents. Thus the introduction of '$\circ$' on the right of $\vdash$ is always subject to the "constraint" that the structure on the left of $\vdash$ be of a particular form; similarly for introducing '$+$' on the left.

As in classical sequent calculus, the *modus operandi* for proving a sequent is to start the proof from the bottom, ie, from the goal, and work "backwards". The first step is, where possible, to remove all the formula (logical) operators, using the logical introduction rules. In Display Logic (unlike classical sequent calculus) this requires that the formula which is to be "broken up" must first be displayed. As any formula can be displayed, this is not a difficulty, except in that it is tedious to select display postulates one-by-one. The tactics described in the next two subsections were first written to help in this style of proof.

5.1 Display Tactics

The functions described below were written to help in the process of displaying a chosen substructure.

`disp_tac : string -> int -> tactic` replaces a subgoal by one in which a selected structure is displayed. The structure selected is indicated by a string argument, of which the first character is '`|`' or '`-`', to indicate the left or right side of the turnstile. Remaining characters are

'`l`' or '`r`', to indicate the left or right operand of the ',' operator
'`L`' or '`R`', to indicate the left or right operand of the ';' operator
'`*`' or '`@`', to indicate the operand of the '$*$' or the '$\bullet$' operator

For example, in the (sub)goal $R \vdash ((*T; \bullet * S), \bullet * Q); Y$ the string `"-L"` denotes the sub-structure $((*T; \bullet * S), \bullet * Q)$. Thus `disp_tac "-L"` performs the first (bottom) step of the proof example shown below.

`fdisp : string -> thm -> thm` is the corresponding function for forward proof, displaying the selected part of the conclusion of the given theorem.

`d_b_tac : string -> (int -> tactic) -> int -> tactic`

`d_b_tac` *str tacfn sg* takes subgoal *sg*, displays the part selected by *str* (as for `disp_tac`) and applies *tacfn* to the transformed subgoal. It then applies the reverse of the display steps to each resulting subgoal. This requires that the subgoals produced by *tacfn* are in a suitable form to permit the display steps used to be reversed. This will be the case if *tacfn* only affects the displayed sub-structure, and leaves the rest of the sequent alone. This was the purpose of deriving the rules `andS` and `orA` (which satisfy this requirement) from `ands` and `ora` (which do not).

For example, if we start with the subgoal $R \vdash ((*T; \bullet * S), \bullet * Q); Y$ then `d_b_tac "-L" (rtac mrs)` applies the monotonicity rule ($\vdash M$) (`mrs`) to the sub-structure $(*T; \bullet * S), \bullet * Q$ by the following steps (read upwards)

$$\cfrac{\cfrac{\cfrac{R \vdash (*T; \bullet * S); Y}{R; * \bullet Y \vdash *T; \bullet * S} \text{(reverse of disp\_tac "-L")}}{R; * \bullet Y \vdash (*T; \bullet * S), \bullet * Q} \text{(mrs)}}{R \vdash ((*T; \bullet * S), \bullet * Q); Y} \text{(disp\_tac "-L")}$$

`fdispfun : string -> (thm -> thm) -> thm -> thm`
is a corresponding operation for forward proof. `fdispfun` *str thfn th* displays the part of theorem *th* indicated by string *str*, applies transformation *thfn* to the result, then reverses the display steps used.

5.2 Search-and-Replace Tactics

Tactics were written to display all structures in a sequent in turn, and apply any appropriate transformations (selected from a given set) to the structure. These can be used to rewrite all substructures of a particular form, for example, to rewrite all occurrences of "$X, (Y, Z)$" to "$(X, Y), Z$".

`glob_tac :`
`(term -> int -> (int -> tactic) list) -> int -> tactic`
The first argument of `glob_tac` is a function *actfn*, which is called as *actfn tm asp*, where *tm* is a subterm, and *asp* has the value 0 or 1 to indicate that *tm* is displayed as the antecedent or succedent part of a sequent. *actfn tm asp* should return either `[]` or `[`*tacf*`]`, where, when *sg* is a subgoal number, *tacf sg* is a tactic. When *tm* is displayed as the antecedent (if $asp = 0$) or succedent (if $asp = 1$), *tacf* should be a tactic which changes only *tm*.
Then `glob_tac` *actfn sg* displays in turn every subterm of subgoal *sg*, and applies the tactic, if any, given by *actfn tm asp*, (where *tm* is the subterm displayed on the side indicated by *asp*) to the sequent thus produced. All the display steps are eventually reversed, in the same way as for `d_b_tac`.
`glob_tac` uses a top-down strategy in that it first tests the whole structure with the function *actfn*, then, recursively, each sub-structure. If *actfn* returns a tactic function (ie, not the empty list) then the changed structure is tested again.

`bup_tac` is like `glob_tac`, but uses a bottom-up strategy, displaying and testing the smallest sub-structures first.

`fgl_fun, fbl_fun :`
`(term -> int -> (thm -> thm) list) -> thm -> thm`
These functions are like `glob_tac` and `bup_tac`, but for doing forward proof. Their first argument is again a function *actfn*, called as *actfn tm asp*. Here *actfn tm asp* should return either `[]` or `[`*thtr*`]`, where *thtr* is a theorem transformation, ie a function of type `thm -> thm`, which changes only the term *tm* displayed on the side indicated by *asp*.

The first implementation of these tactics and functions was described in [7], §2.5.3. A later implementation gives the user great flexibility in programming the search strategy, enabling him/her to specify (for example) whether to traverse the structure top-down or bottom-up, whether or not to attempt further changes to a rewritten substructure, whether or not to repeat the whole procedure when any substructure is rewritten, etc. This implementation is described in [8, §4.4].

5.3 Application of the Search-and-Replace Tactics

We now describe some uses of the search-and-replace tactics. Provided the function *actfn*, as described in §5.2, returns an action which changes only the displayed subterm *tm*, these tactics perform local "rewrites" on subexpressions.

Examples of the use of these tactics are (1) to eliminate $**$ and $\bullet\bullet$, (2) to use the various derived distributive laws to push $*$ and $\bullet$ down, and (3) to eliminate logical operators, where possible, using a set of logical introduction rules including `andS` and `orA` (see Fig. 1) – this is only for backward proof.

As these examples illustrate, we can achieve the power of a procedure which rewrites using equalities or equivalences, such as `rewrite_tac` and `rewrite_rule` in Isabelle and `REWRITE_TAC` and `REWRITE_RULE` in HOL, even though the δ**RA** formalization does not contain any general rewriting facility. However δ**RA** requires more than simple rewriting due to the logical introduction rules with "constraints".

The fact that `pr_intr` is available only for backward proof illustrates that these tactics can be used with uni-directional rules, such as the logical introduction rules, as well as with bi-directional rules. This useful aspect of the tactics contrasts with the more typical rewriting tactics, such as in Isabelle and HOL, where the rules used to rewrite an expression *must* be equalities.

In [12], §4, a function τ which translates structures to formulae – for example, to convert $B; \bullet A \vdash \bullet D, *C$ to $B \circ \smile A \vdash \smile D \vee \neg C$ – is crucial in the proof of soundness of δ**RA**. Its implementation in a forward proof is another example of the use of a search-and-replace function. For this we can use some of the logical introduction rules, in their original form (as given in [12]), using `fbl_fun`. Note that this must be done bottom up, ie, on the smallest sub-structures first, because only at the bottom are the sub-structures also formulae.

5.4 Other Tactics and Methods

idf_tac In the identity rule (id) $p \vdash p$, p stands for a primitive proposition, not an arbitrary formula. It is, however, true that $A \vdash A$ for any formula A; this is proved by induction (see [12], Lemma 2). The restriction to primitive propositions is not reflected in the Isabelle implementation, which contains the rule $A \vdash A$, where A is a variable standing for any formula. However the tactic `idf_tac` will convert an identity subgoal such as $q \circ (\smile p \vee \neg r) \vdash q \circ (\smile p \vee \neg r)$ into three separate subgoals $\quad q \vdash q \quad p \vdash p \quad r \vdash r$. The tactic can therefore provide a proof, from the rule $p \vdash p$, of any given instance of the general theorem (which is provable only by induction) that $A \vdash A$ for any formula A.

This tactic uses a set of six derived results (one for each logical operator, of which two examples are shown).

$$\frac{A_1 \vdash A \quad B_1 \vdash B}{A_1 \circ B_1 \vdash A \circ B}\,(\texttt{cong\_comp}) \qquad \frac{A \vdash A_1}{\neg A_1 \vdash \neg A}\,(\texttt{cong\_not})$$

Structural operators from logical ones It is also possible to implement the function τ in a backwards proof. We can "create" logical connectives (as we read a proof upwards); this can on occasions help in managing a complex proof. The tactic `tau_tac` does this. It uses variants of the logical introduction rules where all the premises are "structure-free" and "connective-free", ie of the form $A \vdash X$ or $X \vdash A$. Structural or logical operators appear only in their conclusions.

The inverse of τ is implemented for forward proof. This use of the search-and-replace method can only be done top down (using `fgl_fun`), and it uses the cut rule. Not all formula operators will necessarily be changed to structural ones: for example, '$\circ$' or '$\wedge$' in a succedent position and '$+$' or '$\vee$' in an antecedent position cannot be converted to structural operators.

Flipping theorems It may be observed in Fig. 1 of [12] that many of the rules in the two columns are symmetric about the turnstile, for example

$$(A;\vdash)\,\frac{X;(Y;Z) \vdash W}{(X;Y);Z \vdash W}\,(\texttt{arA}) \qquad (\vdash A;)\,\frac{W \vdash X;(Y;Z)}{W \vdash (X;Y);Z}\,(\texttt{arS})$$

In fact most rules in the right-hand column of that figure could be derived from the corresponding rules in the left-hand column by "flipping" them about the turnstile, ie, interchanging the parts before and after $\vdash$. Likewise, "flipped" versions were derived for those rules in Fig. 1 which use only structural operators. The procedure for doing this has been formalized in the theorem transformation functions `flip_st_p` and `flip_st_c`, of type `thm -> thm`. These use the theorem `dcp` (see Fig. 1) to swap the sides of each sequent (by putting a $*$ in front of each side), use the distributive laws to push the $*$ down as far as possible, and rename variables to absorb the $*$. The transformations require the rules to be in structural form (see §6.1 below). This applies generally in display logics – having proved one theorem or rule in structural form, you then get one "free", with each part moved to the other side of the turnstile. The theorem transformations `flip_bl_p` and `flip_bl_c` do the same thing, but with $\bullet$ instead of $*$. These transform a theorem to a corresponding one in terms of the converse relations.

Automated cut-elimination Belnap's original proof of cut-elimination is operational; see the proof sketch in [12, Appendix]. It was implemented as follows.

- The rules used in a proof are stored in the nodes of a tree using an appropriate ML datatype, where each node represents a single proof step. Thus, given the original sequent, we can reproduce the original proof.
- Functions were written in ML to turn an Isabelle proof, using the various compound tactics described in §5.2, into a proof consisting of single steps, and then to represent that proof as a tree, as above.

- Functions were written to allow us to rearrange the nodes of such proof trees to push a cut upwards until the cut-formula becomes principal in both premises of that cut (either due to an introduction rule or an axiom $p \vdash p$).
- Such a cut is either eliminated completely or is converted to cuts on smaller formulae using a look-up table that gives a recipe of how to do so for each logical connective. Display logic gurantees that such a recipe always exists.
- This procedure is applied recursively to remove all cuts from the proof tree.

Full details are given in [7, Ch. 5].

One novel use of this procedure would be to convert each proof of Chin and Tarski [6] into a **δRA** proof as follows. Simply begin to convert the proof by following the text. Sooner or later, the text "stores" the current reasoning as a lemma. At this point, we let the converted text constitute a premise of cut. We then continue to convert the text, until the lemma is used. At this point we insert a cut in the converted proof. Now applying the automated cut-elimination procedure will give a new, purely cut-free proof. Often, this proof is shorter, although, in general, it will be exponentially longer.

6 Results from the δRA implementation

6.1 Theorems proved using Isabelle

We give some examples of theorems and derived rules that have been proved in **δRA**, using Isabelle. They show how some relation algebraic results can be turned into structural derived rules, ie, rules in which all the operators are structural, and all the variables denote arbitrary structures. This (together with the display postulates) enables them to be reused in doing proofs in structural form.

Chin & Tarski [6] give a number of results whose proof is far from simple. For example, their Theorem 2.7 and the corresponding structural rule are

$$(a \circ b) \wedge c \vdash a \circ ((\smile a \circ c) \wedge b) \qquad \frac{X; ((\bullet X; Z), Y) \vdash W}{(X; Y), Z \vdash W}$$

Chin & Tarski [6] Theorem 2.11 and its corresponding structural rule are

$$(r \circ s) \wedge (t \circ u) \vdash (r \circ ((\smile r \circ t) \wedge (s \circ \smile u))) \circ u \quad \frac{(U; ((\bullet U; W), (V; \bullet X))); X \vdash Z}{(U; V), (W; X) \vdash Z}$$

The Dedekind rule ([17], §5.3; [2], §2.1) and its corresponding structural rule are

$$(q \circ r) \wedge s \vdash (q \wedge (s \circ \smile r)) \circ (r \wedge (\smile q \circ s)) \qquad \frac{(X, (Z; \bullet Y)); (Y, (\bullet X; Z)) \vdash W}{(X; Y), Z \vdash W}$$

All these results were proved in **δRA**, using Isabelle.

Some interesting theorems were useful in various areas of [7]. Again, they can be proved in the form of structural rules. These show an interesting similarity between the Boolean and relational "times" operators, ',' and ';', and their identities, I and E. Firstly, we have

$$\frac{((X;Y),E);Z \vdash W}{((X,\bullet Y);I),Z \vdash W}\,(\texttt{thab})$$

Next we have two results which can be proved from it.

$$\frac{(X,E);(Y,Z) \vdash W}{((X,E);Y),Z \vdash W}\,(\texttt{th78sr}) \qquad \frac{(X;I),(Y;Z) \vdash W}{((X;I),Y);Z \vdash W}\,(\texttt{th56sr})$$

These results have the following corollaries, giving examples of cases where the Boolean and relational products of two quantities are equal.

$$\frac{(X,E);(E,Z) \vdash W}{(X,E),(E,Z) \vdash W}\,(\texttt{cor78i}) \qquad \frac{(X;I),(I;Z) \vdash W}{(X;I);(I;Z) \vdash W}\,(\texttt{cor56i})$$

6.2 Proving the RA axioms

Using Isabelle we can mechanize the formal proofs of completeness of $\boldsymbol{\delta}$**RA** with respect to **RA**. Here, we sketch how this is done. We show, informally, how any result provable in **RA** can be proved in $\boldsymbol{\delta}$**RA**. This completeness result is weaker than Theorem 6 of [12] in that we only show how to derive, in $\boldsymbol{\delta}$**RA**, any **RA**-equation (expressed in terms of *formulae*); by contrast, [12] also deals with $\boldsymbol{\delta}$**RA**-sequents involving *structures* whose translation into a formula inequality is valid in **RA**.

In [12] a (Lindenbaum) relation algebra is constructed from $\boldsymbol{\delta}$**RA**. Here we consider the proof system which corresponds to **RA**. By the completeness theorem of Birkhoff [3], discussed in [14], the proof rules for **RA** consist only of substituting terms for variables in the relevant equations (defining a Boolean algebra and (RA2) to (RA8)), and replacing equals by equals.

Firstly, following [12], Theorem 6, we let $A = B$ mean $A \vdash B$ and $B \vdash A$; more formally, we introduce the following rules

$$\frac{A \vdash B \quad B \vdash A}{A = B}(\texttt{eqI}) \qquad \frac{A = B}{A \vdash B}(\texttt{eqD1}) \qquad \frac{A = B}{B \vdash A}(\texttt{eqD2})$$

This equality was easily proved to be reflexive, symmetric and transitive. We define $A \leq B$ as $(A \wedge B) = A$, as in **RA**. It is then proved that $A \leq B \iff A \vdash B$.

Then the **RA** axioms can be proved; (RA2) to (RA8) are proved using Isabelle. Axiom (RA1), which says that the logical operators form a Boolean algebra, follows from the fact that $\boldsymbol{\delta}$**RA** contains classical propositional logic.

In using the **RA** axioms we would implicitly rely on the equality operator $=$ being symmetric and transitive, which we proved, as indicated above. The proof system for **RA** also uses the fact, as noted above, that equality permits substitution of equals. That is, suppose $C[A]$ and $C[B]$ are formulae such that $C[B]$ is obtained from $C[A]$ by replacing some occurrences of A by B. If we now prove in **RA** that $A = B$ and $C[A] = D$ then we could deduce $C[B] = D$ by substituting B for A.

To show that this can be done in $\boldsymbol{\delta}$**RA** we have to use induction on the structure of $C[\cdot]$. Each induction step uses one of six derived results, of which two are shown below.

$$\frac{A_1 = A \quad B_1 = B}{A_1 \wedge B_1 = A \wedge B}\,(\texttt{eqc\_and}) \qquad \frac{A_1 = A}{\smile A_1 = \smile A}\,(\texttt{eqc\_conv})$$

Alternatively, any given instance of this result could be proved with the help of the tactic `idf_tac`, described in §5.4. Details are given in [7].

7 Comparing Different Logical Calculi for RA Using δRA

In this section we describe how we used the implementation of **δRA** in Isabelle to demonstrate the soundness of some other logical systems for relation algebra. (As described in §6.2, we did the same for **RA**; as we use **RA** as a "reference point" we treated this as a demonstration of the completeness of **δRA**).

7.1 Gordeev's term rewriting system for RA

Gordeev [10] has given a system **NRA** of rules for rewriting relational equations of the form $F = \top$ (where F is a relational formula and $\top$ is the unique largest element). A derivation of F in **NRA** is a sequence of permitted rewrites ("reductions") $F \longrightarrow \cdots \longrightarrow \top$. Thus **NRA** proves equations of the form $F = \top$. However it is known that every Boolean combination of equations is equivalent to an equation of the form $F = \top$ (see [16], pp. 435–6, pp. 439–440). Gordeev has shown that F is derivable in **NRA** iff $F = \top$ is derivable in **RA**.

It is expected that details of **NRA** will be published in due course; the actual rules are not pertinent here, since we simply outline how Isabelle was used as the basis of a proof of the soundness of **NRA**.

NRA, uses "literals" (L) and "positive terms" (A, B, C, D). Literals are of the shape P, $\neg P$, $\smile P$, $\neg \smile P$, $\top$, $\bot$, $\mathbf{0}$ or $\mathbf{1}$, for any given primitive relation symbol P. Positive terms are built up from literals using the binary operators. Any **RA** term is reducible to a positive term by the distributive rules which show how the unary operators distribute over the binary operators (see Fig. 1 for the structural counterparts).

In **NRA** there are 12 rewriting rules (NRA1)-(NRA12), but, given a formula F, some may only be used to rewrite F or a conjunct of F. The permitted rewrites, or "reductions", are

- $F[G'] \longrightarrow F[G'']$, where $G' \longrightarrow G''$ is an instance of any rule (NRAi) except for (NRA9) and (NRA10)
- $G_1 \wedge G_2 \wedge \ldots \wedge G'_j \wedge \ldots \wedge G_k \longrightarrow G_1 \wedge G_2 \wedge \ldots \wedge G''_j \wedge \ldots \wedge G_k$ where $1 \leq j \leq k$ and $G'_j \longrightarrow G''_j$ is an instance of (NRA9) or (NRA10)

We take it that $F[\cdot]$ above must be such that G is a 'sub-positive-term' of $F[G]$, ie, that G is not a literal properly contained in a larger literal within $F[G]$ (for example, $F[G]$ may not be $\neg G \vee H$). This point was not spelt out in the brief communication [10], and we realized the need for it only in doing the proof in Isabelle. This gives an example of the value of a computer-based theorem prover in ensuring such details are not overlooked.

The following lemma explains the **NRA** reduction step in terms of **δRA**.

Lemma 1. *Whenever $F' \longrightarrow F''$ in* **NRA***, then* $\top \vdash F'' \Longrightarrow \top \vdash F'$ *in* $\boldsymbol{\delta}$**RA**.

Proof. For rules other than (NRA9) and (NRA10), we show that $F'' \vdash F'$ in $\boldsymbol{\delta}$**RA**. Let $F' = F[G']$ and $F'' = F[G'']$, where the **NRA** rule is $G' \longrightarrow G''$. For each rule the $\boldsymbol{\delta}$**RA** theorem $G'' \vdash G'$ was proved in Isabelle.

From $G'' \vdash G'$ we show $F[G''] \vdash F[G']$ by induction on the structure of $F[\cdot]$. As $F[\cdot]$ uses only literals and the binary operators, the inductive steps rely on the four $\boldsymbol{\delta}$**RA**-theorems `cong_or`, `cong_and`, `cong_comp` and `cong_rs` (see §5.4), and the identity rule $A \vdash A$.

Since $F'' \vdash F'$, if we assume $\top \vdash F''$, then the (cut) rule gives $\top \vdash F'$.

We now look at rules (NRA9) and (NRA10). For these rules, where the **NRA** rule is $G' \longrightarrow G''$, we have proved the $\boldsymbol{\delta}$**RA** derived rule $\top \vdash G'' \Longrightarrow \top \vdash G'$ in Isabelle. So from this we need to show $\top \vdash F[G''] \Longrightarrow \top \vdash F[G']$, where G' is a conjunct of $F[G']$. This is shown by induction on the number of conjuncts in $F[\cdot]$. The inductive step uses the following $\boldsymbol{\delta}$**RA**-theorem.

$$\frac{\top \vdash G1'' \Longrightarrow \top \vdash G1' \qquad \top \vdash G2'' \Longrightarrow \top \vdash G2'}{\top \vdash G1'' \wedge G2'' \Longrightarrow \top \vdash G1' \wedge G2'} \; (\texttt{G910meth})$$

□

`G910meth` is a theorem whose use of Isabelle's metalogic is more complex than others seen so far, with nested occurrences of the metalogical operator $\Longrightarrow$.

Lemma 1 implies that for every **NRA**-derivation of F there is a $\boldsymbol{\delta}$**RA**-derivation of $\top \vdash F$. Thus **NRA** is sound with respect to **RA**.

7.2 A point-variable sequent calculus

In [7], Ch. 4, we described the sequent calculus system of Maddux [15], $\mathcal{M}$, for relation algebras, and showed how its sequents and proofs correspond to sequents and proofs in $\boldsymbol{\delta}$**RA**. We characterized the sequents in $\mathcal{M}$ which can be translated into $\boldsymbol{\delta}$**RA**by describing the pattern of relation variables and point variables in the $\mathcal{M}$-sequent as a graph, and showing that whether or not an $\mathcal{M}$-sequent can be translated into $\boldsymbol{\delta}$**RA** depends on the shape of the graph.

As a sequent in $\mathcal{M}$ can often be translated to any of several sequents in $\boldsymbol{\delta}$**RA**, we showed that these several sequents are equivalent in $\boldsymbol{\delta}$**RA**.

The result relating proofs in the two systems was that if an $\mathcal{M}$-sequent $\mathcal{S}$ can be translated into a $\boldsymbol{\delta}$**RA**-sequent $\mathcal{S}'$, then any $\mathcal{M}$-proof of $\mathcal{S}$ can be translated into a $\boldsymbol{\delta}$**RA**-proof of $\mathcal{S}'$, provided that the $\mathcal{M}$-proof uses at most four point variables. This required showing how such a proof in $\mathcal{M}$ can be divided into parts each of which could be turned into a part of a proof in $\boldsymbol{\delta}$**RA**.

As not all $\mathcal{M}$-sequents containing at most four point variables can be translated into $\boldsymbol{\delta}$**RA**, not all the intermediate stages of such a proof could be translated into $\boldsymbol{\delta}$**RA**, that is, not every individual step of the $\mathcal{M}$-proof could be converted into a corresponding step of a proof in $\boldsymbol{\delta}$**RA**. Rather, a larger portion of the $\mathcal{M}$-proof had to be treated as a unit; we showed that we could always re-arrange the $\mathcal{M}$-proof to get these portions of a particular form, which conformed to a theorem that we had proved in Isabelle for $\boldsymbol{\delta}$**RA**.

This result provided a constructive proof of one direction of the result of Maddux ([15], Thm. 6) that the class of relation algebras is characterized by the $\mathcal{M}$-sequents provable using four variables.

8 Conclusion

We described an implementation, in Isabelle, of the display calculus $\boldsymbol{\delta}$**RA** for relation algebras. This is the first implementation of any display calculus. We intend to generalise it to other display calculi in future work. Although a $\boldsymbol{\delta}$**RA** proof involved frequent and tedious use of the display postulates, we provided a number of tactics and other functions to make this aspect of constructing proofs easier. We also provided tactics to search for certain patterns and replace them wherever they are found, giving the power of a rewrite rule in a term rewriting system. Most, but not all, of the rules for introducing logical connectives could be used in this manner. Certain other tactics and functions, such as those which transform between structural and logical forms of sequents, have proved convenient in proofs on occasions. There is scope for further work to devise more automatic techniques than those described here.

The use of display postulates to "display" a subterm resembles the use of window inference [20],[13] to "focus" upon a subterm. Further work is needed to connect these two modes of reasoning.

Some useful derived rules were proved, for example those showing how the unary operators distribute over the binary operators. We showed how the $\boldsymbol{\delta}$**RA** system can justify inferences in the (axiomatic) theory of relation algebras. Although this result appears as Theorem 6 of [12], the published proof omits most of the detail. The detail is provided in the Isabelle proofs, giving an example of the value of a mechanized theorem prover to confirm the details of proofs which are too voluminous to publish or even too tedious to check by hand.

Some important (and difficult) theorems from the literature were also proved, for example some from [6]. Other interesting theorems were discovered and proved, some of which were needed for the work outlined in §7.2.

In §7.1 we described an equational theory due to Gordeev for relation algebras, and showed how its inferences can be justified in $\boldsymbol{\delta}$**RA**. As mentioned there, we obtained a corresponding result for the sequent calculus system of Maddux. This showed how Isabelle could be used to demonstrate the relationship between different logical systems for relation algebra.

Theorems 3.6 and 3.8 of [5] show that Peirce algebras can be embedded inside relation algebras in various ways. The authors also show that Peirce algebras form the basis of most of the common Knowledge Representation languages like KL-ONE [4]. Minor modifications of our system should give mechanised proof systems for these logics.

References

1. Nuel D. Belnap, Display Logic, Journal of Philosophical Logic 11 (1982), 375–417.
2. Rudolf Berghammer & Claudia Hattensperger, Computer-Aided Manipulation of Relational Expressions and Formulae Using RALF, preprint.
3. Garrett Birkhoff, On the Structure of Abstract Algebras, Proc. Cambridge Phil. Soc. 31 (1935), 433-454.
4. R.J. Brachman & J.G. Schmolze, An overview of the KL-ONE knowledge representation system, Cognitive Science 9(2) (1985), 171–216.
5. Chris Brink, Katarina Britz & Renate A. Schmidt, Peirce Algebras, Formal Aspects of Computing 6 (1994), 339–358.
6. Louise H. Chin & Alfred Tarski, Distributive and Modular Laws in the Arithmetic of Relation Algebras, University of California Publications in Mathematics, New Series, I (1943–1951), 341–384.
7. Jeremy E. Dawson, Mechanised Proof Systems for Relation Algebras, Grad. Dip. Sci. sub-thesis, Dept of Computer Science, Australian National University. Available at `http://arp.anu.edu.au:80/~jeremy/thesis.dvi`
8. Jeremy E. Dawson, Simulating Term-Rewriting in LPF and in Display Logic, submitted. Available at `http://arp.anu.edu.au:80/~jeremy/rewr/rewr.dvi`
9. Jean H. Gallier, Logic for Computer Science : Foundations of Automatic Theorem Proving, Harper & Row, New York, 1986.
10. Lev Gordeev, personal communication.
11. Rajeev Goré, Intuitionistic Logic Redisplayed, Automated Reasoning Project TR-ARP-1-95, ANU, 1995.
12. Rajeev Goré, Cut-free Display Calculi for Relation Algebras, Computer Science Logic, Lecture Notes in Computer Science 1249 (1997), 198–210. (or see `http://arp.anu.edu.au/~rpg/publications.html`).
13. Jim Grundy, Transformational Hierarchical Reasoning, The Computer Journal 39 (1996), 291–302.
14. Gérard Huet & Derek C. Oppen, Equations and Rewrite Rules – A Survey, in Formal Languages: Perspectives and Open Problems, R.V. Book (ed), Academic Press (1980), 349–405.
15. Roger D. Maddux, A Sequent Calculus for Relation Algebras, Annals of Pure and Applied Logic 25 (1983), 73–101.
16. Roger D. Maddux, The Origin of Relation Algebras in the Development and Axiomatization of the Calculus of Relations, Studia Logica 50 (1991), 421–455.
17. David von Oheimb & Thomas F. Gritzner, RALL: Machine-supported proofs for Relation Algebra, Proceedings of CADE-14, Lecture Notes in Computer Science 1249 (1997), 380–394.
18. Lawrence C. Paulson, The Isabelle Reference Manual, Computer Laboratory, University of Cambridge, 1995.
19. Lawrence C. Paulson, Isabelle's Object-Logics, Computer Laboratory, University of Cambridge, 1995.
20. Peter J. Robinson & John Staples, Formalizing a Hierarchical Structure of Practical Mathematical Reasoning, J. Logic & Computation, 3 (1993), 47–61.
21. Heinrich Wansing, Sequent Calculi for Normal Modal Propositional Logics, Journal of Logic and Computation 4 (1994), 124–142.

Relative Similarity Logics are Decidable: Reduction to FO^2 with Equality*

Stéphane Demri and Beata Konikowska

[1] Laboratoire LEIBNIZ - C.N.R.S., Grenoble, France
[2] Institute of Computer Science, Polish Academy of Sciences, Warszawa, Poland

Abstract. We show the decidability of the satisfiability problem for relative similarity logics that allow classification of objects in presence of incomplete information. As a side-effect, we obtain a finite model property for such similarity logics. The proof technique consists of reductions into the satisfiability problem for the decidable fragment FO^2 with equality from classical logic. Although the reductions stem from the standard translation from modal logic into classical logic, our original approach (for instance handling nominals for atomic properties and decomposition in terms of components encoded in the reduction) can be generalized to a larger class of relative logics, opening ground for further investigations.

1 Introduction

Background. Classification of objects in presence of incomplete information has been long recognized as an issue of concern for various AI problems that deal with commonsense knowledge as well as scientific and engineering knowledge (expert systems, image recognition, knowledge bases and so on). Similarity -sometimes termed "weak equivalence"- provides a basic tool each time when we classify objects with respect to their properties. There exist several formal systems capturing the notion of similarity from the logical viewpoint [Vak91a,Vak91b]. In the present paper we base on the formalization given in [Kon97], where, contrary to [Vak91a,Vak91b], similarity is treated as a relative notion. More precisely, in [Kon97] similarity is defined as a reflexive and symmetric binary relation sim_P, parametrized by the set P of properties with respect to which the objects are classified as either similar or dissimilar. Thus, instead of a single similarity relation we have a whole family $(sim_P)_{P \subseteq PROP}$, where $PROP$ is the set of all the properties considered in a given system. When talking about similarity or equivalence it is natural to talk about *lower* and *upper approximation* $L(sim_P)A$, $U(sim_P)A$ of a given set A of objects with respect to the similarity sim_P. The above operations stem from rough set theory [Paw81], with $L(sim_P)A$ being the set of all objects in A which are not similar (in the sense of sim_P) to any object outside A and $U(sim_P)A$ - the set of all objects of the universe which are similar to some object in A. Thus, the above operations could be considered

* This work has been partially supported by the Polish-French Project "Rough-set based reasoning with incomplete information: some aspects of mechanization", ♯7004.

as the operations of taking "interior" and "closure" of the set A with respect to similarity sim_P. However, the analogy is not complete, since similarity is not transitive, and hence the above operations are not idempotent.

Practical importance of the approximation operations is quite obvious: if we can distinguish objects only up to similarity, then when looking for objects belonging to some set A we should take those in $L(sim_P)A$, if we want to consider only the objects sure to belong to A, and those in $U(sim_P)A$ if our aim is not to overlook any object which might possibly belong to A.

Our objectives. The formal system introduced in [Kon97] features the above operations, which generate a family of interdependent relative modalities. The resulting polymodal logic is equipped with a complete deduction system. However, from the viewpoint of any practical applications of the similarity logic in the area of Artificial Intelligence mentioned above an issue of great importance is whether the logic is decidable. A positive answer to this question might provide not only a decision procedure, but also a better understanding of the logical analysis of similarity. These are the objectives of the present paper. Up to now, the question of decidability has been open, which is hardly surprising in view of the high expressive power of the logic. Indeed: its language admits implicitly the universal modal operator, and nominals for atomic propositions as well for atomic properties; in addition, the modal operators are interdependent. Nominals (or names) are used in numerous non-classical logics with various motivations (see e.g. [Orło84a,PT91,Bla93,Kon97]) and they usually greatly increase the expressive power of the logics (causing additional difficulties with proving (un)decidability -see e.g. [PT91]). Furthermore, since finite submodels can be captured in the language up to an isomorphism (which is yet another evidence of the expressive power of similarity logics), there is no hope of proving decidability by showing a finite model property for a class of models including strictly the class of standard models with a bound on the model's size (see e.g. [Vak91c,Bal97]). On the other hand, the intersection operator, which is implicitly present in the interpretation of the modal terms, is known to behave badly for filtration-like constructions.

Our contribution. We prove that the logic defined in [Kon97] together with some of its variants is decidable by translating it to a decidable fragment of first-order logic: the two-variable fragment FO^2 containing equality, but no function symbols (see e.g. [Mor75]). Although there are known methods of handling the universal modal operator, the Boolean operations for modal terms and nominals for atomic propositions in order to translate them into FO^2 with equality (see for example the survey papers [Ben98,Var97]), the extra features of the similarity logics require some significant extra work in order to be also translated to such a fragment. This is achieved in the present paper. Unlike the Boolean Modal Logic BML [GP90], for which decidability can be proved via the finite model property for a class of models, reduction of satisfiability for the similarity logics to FO^2 with equality is the only known decidability proof we are aware of, and therefore we solve an open problem here. As a side-effect, we prove the finite model property. More importantly, the novelty of our approach allows us to generalize the translation to a large class of relative modal logics.

Plan of the paper. The paper is structured as follows. In Section 2 the relative similarity logics we deal with in the paper are defined, and some results about their expressive power and complexity are stated. In Section 3, we define the translation of the main relative similarity logic $\mathcal{L}$ into FO^2 with equality, and show its faithfulness. Decidability and finite model property for $\mathcal{L}$ are obtained partly by considering the analogous properties of the fragment FO^2 with equality. In Section 4, we investigate some variants of $\mathcal{L}$, and show their decidability and the finite model property. Section 5 concludes the paper by providing some generalizations of the results proved in the preceding Sections, and stating what is known about the computational complexity of $\mathcal{L}$-satisfiability. In addition, several examples of formula translations are given.

2 Similarity logics

2.1 Information systems and similarity

The information systems that proposed for representation of knowledge are the foundational structures, on which the semantics of the relative similarity logic is based. An *information system* S is defined as a pair $\langle ENT, PROP\rangle$ where ENT is a non-empty set of *entities* (also called *objects*) and $PROP$ is a non-empty set of *properties* (also called *attributes*) -see e.g. [Paw81]. Each property $prop$ is a mapping $ENT \rightarrow \mathcal{P}(Val_{prop}) \setminus \emptyset$ and Val_{prop} is the set of *values* of the property $prop$ -see e.g. [OP84]. In that setting, two entities e_1, e_2 are said to be *similar with respect to some set $P \subseteq PROP$ of properties* (in short $e_1\ sim_P\ e_2$) iff for any $prop \in P$, $prop(e_1) \cap prop(e_2) \neq \emptyset$. The polymodal frames of the relative similarity logics are isomorphic to structures of the form $(ENT, PROP, (sim_P)_{P \subseteq PROP})$. Other relationships between entities can be found in the literature -see e.g. [FdCO84,Orło84b]. For instance, two entities e_1, e_2 are said to be *negatively similar (resp. indiscernible) with respect to some set $P \subseteq PROP$ of properties* (in short $e_1\ nsim_P\ e_2$ -resp. $e_1\ ind_P\ e_2$) iff for any $prop \in P$, $-prop(e_1) \cap -prop(e_2) \neq \emptyset$ - resp. $prop(e_1) = prop(e_2)$.

The family $(sim_P)_{P \subseteq PROP}$ of similarity relations stemming from some information system $S = \langle ENT, PROP\rangle$ induces certain approximations of subsets of entities in S. Indeed, let $L(sim_P)X$ (resp. $U(sim_P)X$) be the lower (resp. upper) sim_P-approximation of the set X of entities defined as follows:

- $L(sim_P)X \stackrel{\text{def}}{=} \{e \in ENT : \forall\, e' \in ENT,\ (e, e') \in sim_P \text{ implies } e' \in X\}$;
- $U(sim_P)X \stackrel{\text{def}}{=} \{e \in ENT : \exists\, e' \in ENT,\ (e, e') \in sim_P \text{ and } e' \in X\}$.

Obviously $L(sim_P)X \subseteq X \subseteq U(sim_P)X$ and $L(sim_P)X = ENT \setminus U(sim_P)(ENT \setminus X)$. These approximations are rather crucial in rough set theory since they allow to classify objects in presence of incomplete information. That is why, the semantics of modal operators in the relative similarity logics shall use these approximations as modal operations. We invite the reader to consult [Oe97] for examples of rough set analysis of incomplete information.

2.2 Syntax and semantics

The set of primitive symbols of the polymodal language $\mathtt{L}$ is composed of

- a set $\mathtt{VARSE} = \{\mathtt{E}_1, \mathtt{E}_2, \ldots\}$ of variables representing sets of entities,
- a set $\mathtt{VARE} = \{\mathtt{x}_1, \mathtt{x}_2, \ldots\}$ of variables representing individual entities,
- symbols for the classical connectives $\neg$, $\wedge$ (negation and conjunction), and
- a countably infinite set $\{[\mathtt{A}] : \mathtt{A} \in \mathtt{TERM}\}$ of unary modal operators where the set $\mathtt{TERM}$ of terms is the smallest set containing
  - the constant $\mathtt{0}$ representing the empty set of properties,
  - a countably infinite set $\mathtt{VARP} = \{\mathtt{p}_1, \mathtt{p}_2, \ldots\}$ of variables representing individual properties,
  - a countably infinite set $\mathtt{VARSP} = \{\mathtt{P}_1, \mathtt{P}_2, \ldots\}$ of variables representing sets of properties,

  and closed under the Boolean operators $\cap, \cup, -$.

The formation rules of the set $\mathtt{FORM}$ of formulae are those of the classical propositional calculus plus the rule: if $\mathtt{F} \in \mathtt{FORM}$ and $\mathtt{A} \in \mathtt{TERM}$, then $[\mathtt{A}]\mathtt{F} \in \mathtt{FORM}$. We use the connectives $\vee, \Rightarrow, \Leftrightarrow$, $\langle \mathtt{A} \rangle$ as abbreviations with their standard meanings. For any syntactic category $\mathtt{X}$ and any syntactic object $\mathtt{O}$, we write $\mathtt{X}(\mathtt{O})$ to denote the set of those elements of $\mathtt{X}$ that occur in $\mathtt{O}$. Moreover, for any syntactic object $\mathtt{O}$, we write $|\mathtt{O}|$ to denote its *length* (or *size*), that is the number of symbol occurrences in $\mathtt{O}$. As usual, $sub(\mathtt{F})$ denotes the set of *subformulae* of the formula $\mathtt{F}$ (including $\mathtt{F}$ itself).

Definition 1. *A* $\mathtt{TERM}$-interpretation v *is a map* $v : \mathtt{TERM} \to \mathcal{P}(PROP)$ *such that* $PROP$ *is a non-empty set and for any* $\mathtt{A}_1, \mathtt{A}_2 \in \mathtt{TERM}$,

- *if* $\mathtt{A}_1, \mathtt{A}_2 \in \mathtt{VARP}$ *and* $\mathtt{A}_1 \neq \mathtt{A}_2$, *then* $v(\mathtt{A}_1) \neq v(\mathtt{A}_2)$,
- *if* $\mathtt{A}_1 \in \mathtt{VARP}$, *then* $v(\mathtt{A}_1)$ *is a singleton, i.e.* $v(\mathtt{A}_1) = \{prop\}$ *for some* $prop \in PROP$,
- $v(\mathtt{0}) = \emptyset$, $v(\mathtt{A}_1 \cap \mathtt{A}_2) = v(\mathtt{A}_1) \cap v(\mathtt{A}_2)$, $v(\mathtt{A}_1 \cup \mathtt{A}_2) = v(\mathtt{A}_1) \cup v(\mathtt{A}_2)$,
- $v(-\mathtt{A}_1) = PROP \setminus v(\mathtt{A}_1)$.

For any $\mathtt{A}, \mathtt{B} \in \mathtt{TERM}$, we write $\mathtt{A} \equiv \mathtt{0}$ (resp. $\mathtt{A} \equiv \mathtt{B}$) when for any $\mathtt{TERM}$-interpretation v, $v(\mathtt{A}) = \emptyset$ (resp. $v(\mathtt{A}) = v(\mathtt{B})$).

Definition 2. *A* model $\mathcal{U}$ *is a structure* $\mathcal{U} = (ENT, PROP, (sim_P)_{P \subseteq PROP}, v)$ *where* ENT *and* $PROP$ *are non-empty sets and* $(sim_P)_{P \subseteq PROP}$ *is a family of binary relations over* ENT *such that*

- *for any* $\emptyset \neq P \subseteq PROP$, sim_P *is reflexive and symmetric,*
- *for any* $P, P' \subseteq PROP$, $sim_{P \cup P'} = sim_P \cap sim_{P'}$ *and* $sim_\emptyset = ENT \times ENT$.

Moreover, v *is a mapping* $v : \mathtt{VARE} \cup \mathtt{VARSE} \cup \mathtt{TERM} \to \mathcal{P}(ENT) \cup \mathcal{P}(PROP)$ *such that* $v(\mathtt{E}) \subseteq ENT$ *for any* $\mathtt{E} \in \mathtt{VARSE}$, $v(\mathtt{x}) = \{e\}$, *where* $e \in ENT$ *for any* $\mathtt{x} \in \mathtt{VARE}$ *and the restriction of* v *to* $\mathtt{TERM}$ *is a* $\mathtt{TERM}$*-interpretation.*

Since the set of nominals for properties is countably infinite, and any two different nominals are interpreted by different properties, each model has an infinite set of properties. Let $\mathcal{U} = (ENT, PROP, (sim_P)_{P \subseteq PROP}, v)$ be a model. As usual, we say that a formula F is *satisfied by an entity* $e \in ENT$ *in* $\mathcal{U}$ (written $\mathcal{U}, e \models \mathtt{F}$) if the following conditions are satisfied.

- $\mathcal{U}, e \models \mathtt{x}$ iff $\{e\} = v(\mathtt{x})$; $\mathcal{U}, e \models \mathtt{E}$ iff $e \in v(\mathtt{E})$;
- $\mathcal{U}, e \models \neg\mathtt{F}$ iff not $\mathcal{U}, e \models \mathtt{F}$; $\mathcal{U}, e \models \mathtt{F} \wedge \mathtt{G}$ iff $\mathcal{U}, e \models \mathtt{F}$ and $\mathcal{U}, e \models \mathtt{G}$;
- $\mathcal{U}, e \models [\mathtt{A}]\mathtt{F}$ iff for any $e' \in sim_{v(\mathtt{A})}(e)$, $\mathcal{U}, e' \models \mathtt{F}$.

A formula F is *true* in a model $\mathcal{U}$ (written $\mathcal{U} \models \mathtt{F}$) iff for any $e \in ENT$, $\mathcal{U}, e \models \mathtt{F}$ - or, equivalently, iff for some $e \in ENT$, $\mathcal{U}, e \models [\mathtt{0}]\mathtt{F}$. A formula F is said to be *valid* iff F is true in all models. A formula F is said to be *satisfiable* iff $\neg\mathtt{F}$ is not valid. The similarity logic $\mathcal{L}$ is said to have the *finite model property* iff every satisfiable formula is satisfied in some model $\mathcal{U} = (ENT, PROP, (sim_P)_{P \subseteq PROP}, v)$ with a finite set ENT such that, for any $P \subseteq PROP$, $sim_P = sim_{P \cap P_0}$, where $P_0 \subseteq PROP$ is finite and nonempty (P_0 is called the *relevant part* of $PROP$ in $\mathcal{U}$). Consequently, if $\mathcal{L}$ has the finite model property, then every satisfiable formula has a model $(ENT, PROP, (sim_P)_{P \subseteq PROP}, v)$ such that for any $\emptyset \neq P \subseteq PROP$, $sim_P = \bigcap_{x \in P} sim_{\{x\}}$.

The similarity logic defined in [Kon97] is not exactly the logic $\mathcal{L}$ defined above, since in [Kon97] the set of properties was supposed to be fixed, and constants representing properties were used instead of variables. For any set X, we write $\mathcal{L}_X$ to denote the logic that differs from $\mathcal{L}$ in the following points: (1) the set of properties $PROP$ is fixed in all the models and equals X, (2) VARP and X have the same cardinality. In various places in the paper, we implicitly use the facts that satisfiability is insensitive to the renaming of any sort of variables, and that any two models isomorphic in the standard sense satisfy the same set of formulae. Moreover, for the logics $\mathcal{L}_X$, as far as satisfiability is concerned, it is irrelevant whether we fix the interpretation of each nominal for the properties.

2.3 Expressive power and complexity lower bound

Since the language of the relative similarity logic $\mathcal{L}$ contains nominals, the universal modal operator and a family of standard modal operators, its expressive power is quite high. In Proposition 1 below, we shall state a counterpart of Corollary 4.17 in [GG93] (see also Theorem 2.8 in [PT91]) saying that finite submodels can be captured in the language up to isomorphism. In Proposition 1 below, we show that for any finite structure $\mathcal{S}$ there is a formula $\mathtt{F}_{\mathcal{S}}$ such that a model $\mathcal{U}$ satisfies $\mathtt{F}_{\mathcal{S}}$ iff $\mathcal{S}$ is a substructure of $\mathcal{U}$ up to isomorphism. Although this shows that the expressive power of the logic is high, it has a very unpleasant consequence: there is no hope of characterizing $\mathcal{L}$-satisfiability by a class of *finite non-standard* models the way it is done in [Vak91c,Bal97]. It means for instance that proving the finite model property of $\mathcal{L}$ by a standard filtration-like technique becomes highly improbable since $\mathcal{L}$ has implicitly the intersection operator in the language.

In Proposition 1 below, the structure $\mathcal{S}$ encodes a finite part of some model. The set $\{1,\ldots,n\}$ should be understood as a finite set of entities, and $\{1,\ldots,l\}$ as a finite set of properties. Moreover, only a finite set $\{\mathtt{E}_1,\ldots,\mathtt{E}_k\}$ of atomic propositions is taken into account. For $i \in \{1,\ldots,n\}$ and $j \in \{1,\ldots,k\}$, $j \in v'(i)$ is to mean that $\mathtt{E}_j$ is satisfied by i.

Proposition 1. *Let $\mathcal{S} = \langle \{1,\ldots,n\}, \{1,\ldots,l\}, (R(P))_{P\subseteq\{1,\ldots,l\}}, v'\rangle$ be a structure such that each $R(P)$ is a reflexive and symmetric relation, $R(\emptyset)$ is the universal relation, for any $P, P' \subseteq \{1,\ldots,l\}$, $R(P \cup P') = R(P) \cap R(P')$ and v' is a mapping $\{1,\ldots,n\} \rightarrow \mathcal{P}(\{1,\ldots,k\})$ for some $k \geq 1$. Then, there is formula $\mathtt{F}_{\mathcal{S}}$ such that for any $\mathcal{L}$-model $\mathcal{U}$, $\mathcal{U} \models \mathtt{F}_{\mathcal{S}}$ iff there is an 1-1 mapping $\Psi_1 : \{1,\ldots,n\} \rightarrow ENT$ and an injective mapping $\Psi_2 : \{1,\ldots,l\} \rightarrow PROP$ with the following properties*

- *for any $i \in \{1,\ldots,k\}$, $v(\mathtt{E}_i) = \{\Psi_1(s) : i \in v'(s)\}$;*
- *for any $P \subseteq PROP$ such that there is $P' \subseteq \{1,\ldots,l\}$ verifying $\{\Psi_2(i) : i \in P'\} = P$, we have $sim_P = R(P')$.*

Proof. The formula $\mathtt{F}_{\mathcal{S}}$ is the conjunction of the following formulae.

1. $[0](\mathtt{x}_1 \vee \ldots \vee \mathtt{x}_n) \Leftrightarrow [0] \bigwedge_{1\leq i<j\leq n} \neg(\mathtt{x}_i \wedge \mathtt{x}_j)$;
2. $[0](\bigwedge_{i\in\{1,\ldots,n\}}(\mathtt{x}_i \Rightarrow (\bigwedge_{u\in\{1,\ldots,k\}} s_u\mathtt{E}_u)))$ where s_u is the empty string if $u \in v(i)$, otherwise $s_u \stackrel{\text{def}}{=} \neg$;
3. for any $\{i_1,\ldots,i_q\} \subseteq \{1,\ldots,l\}$ and all $i \in \{1,\ldots,n\}$,

$$[0](\mathtt{x}_i \Rightarrow ((\bigwedge_{j\in R_{\{i_1,\ldots,i_q\}}(i)} \langle \mathtt{p}_{i_1} \cup \ldots \cup \mathtt{p}_{i_q}\rangle \mathtt{x}_j) \wedge (\bigwedge_{j\notin R_{\{i_1,\ldots,i_q\}}(i)} \neg\langle \mathtt{p}_{i_1} \cup \ldots \cup \mathtt{p}_{i_q}\rangle \mathtt{x}_j))$$

Before establishing decidability of $\mathcal{L}$-satisfiability, one can provide a lower bound for the complexity of this problem using [Hem96].

Proposition 2. *$\mathcal{L}$-satisfiability is* **EXPTIME**-*hard.*

When no nominals for properties and entities are allowed satisfiability can be shown to be in **EXPTIME** [Dem98].

3 Translation from $\mathcal{L}$ into FO^2 with equality

3.1 A known decidable fragment of classical logic

Consistently with the general convention, by FO^2 we mean a fragment of first-order logic (FOL for short) without equality or function symbols using only 2 variables (denoted by $\mathtt{y}_0$ and $\mathtt{y}_1$ in the sequel). We shall translate the similarity logics into a slight extension of FO^2 obtained by augmenting the language with identity. Actually, we shall restrict ourselves to the following vocabulary:

- a countable set $\{\mathtt{P}_i : i \in \omega\} \cup \{\mathtt{Q}_i : i \in \omega\}$ of unary predicate symbols,
- a countable set $\{\mathtt{R}_{i,j} : i, j \in \omega\}$ of binary predicate symbols,

– the symbol = (interpreted as identity).

In what follows, by a first-order formula we mean a formula belonging to just this fragment of FOL (written FO^2[=] in the sequel). As usual, a first-order structure $\mathcal{M}$ (restricted to this fragment) is a pair $\langle D, m\rangle$ such that D is a non-empty set and m is an interpretation function with $m(\mathtt{P}_i) \cup m(\mathtt{Q}_i) \subseteq D$ for $i \in \omega$, $m(\mathtt{R}_{i,j}) \subseteq D \times D$ for $i, j \in \omega$ and $m(=) \stackrel{\text{def}}{=} \{\langle a, a\rangle : a \in D\}$. As usual, a *valuation* $v_{\mathcal{M}}$ for $\mathcal{M}$ is a mapping $v_{\mathcal{M}} : \{\mathtt{y}_0, \mathtt{y}_1\} \to D$. We write $\mathcal{M}, v_{\mathcal{M}} \models \mathtt{F}$ to denote that F is satisfied in $\mathcal{M}$ under $v_{\mathcal{M}}$, and omit $v_{\mathcal{M}}$ when F is closed. It is known that FO^2[=] has the finite model property, FO^2[=]-satisfiability is decidable [Mor75] and **NEXPTIME**-complete [Lew80,GKV97]. Actually, F is FO^2[=]-satisfiable iff F has a model of size $2^{c\times|\mathtt{F}|}$ for some fixed $c > 0$ [GKV97].

3.2 Normal forms

Let $\mathtt{F} \in \mathtt{FORM}$ be such that[1] $\mathtt{VARP}(\mathtt{F}) = \{\mathtt{p}_1, \ldots, \mathtt{p}_l\}$ and $\mathtt{VARSP}(\mathtt{F}) = \{\mathtt{P}_1, \ldots, \mathtt{P}_n\}$. In the rest of this section, we assume that $n \geq 1$ and $l \geq 1$. The degenerate cases make no additional difficulties and they are treated in a separate section. For any integer $k \in \{0, \ldots, 2^n - 1\}$, by $\mathtt{B}_k$ we denote the term

$$\mathtt{A}_1 \cap \ldots \cap \mathtt{A}_n$$

where, for any $s \in \{1, \ldots, n\}$, $\mathtt{A}_s = \mathtt{P}_s$ if $bit_s(k) = 0$, and $\mathtt{A}_s = -\mathtt{P}_s$ otherwise, with $bit_s(k)$ denoting the sth bit in the binary representation of k. For any integer $k \in \{0, \ldots, 2^n - 1\}$, we denote

$$\mathtt{A}_{k,0} \stackrel{\text{def}}{=} \mathtt{B}_k \cap -\mathtt{p}_1 \cap \ldots \cap -\mathtt{p}_l$$

Finally, for any $\langle k, k'\rangle \in \{0, \ldots, 2^n - 1\} \times \{1, \ldots, l\}$, we denote $\mathtt{A}_{k,k'} \stackrel{\text{def}}{=} \mathtt{B}_k \cap \mathtt{p}_{k'}$. For any TERM-interpretation $v : \mathtt{TERM} \to \mathcal{P}(PROP)$, the family

$$\{v(\mathtt{A}_{k,k'}) : \langle k, k'\rangle \in \{0, \ldots, 2^n - 1\} \times \{0, \ldots, l\}\}$$

is a partition of $PROP$. Moreover, for any term $\mathtt{A} \in \mathtt{TERM}(\mathtt{F})$, either $\mathtt{A} \equiv 0$ or there is a unique non-empty set $\{\mathtt{A}_{k_1,k'_1}, \ldots, \mathtt{A}_{k_u,k'_u}\}$ such that $\mathtt{A} \equiv \mathtt{A}_{k_1,k'_1} \cup \ldots \cup \mathtt{A}_{k_u,k'_u}$. The normal form of A, written $N(\mathtt{A})$, is either 0 or $\mathtt{A}_{k_1,k'_1} \cup \ldots \cup \mathtt{A}_{k_u,k'_u}$ according to the two cases above. Such a decomposition, introduced in [Kon97], generalizes with nominals the canonical disjunctive normal form for the propositional calculus. $N(\mathtt{A})$ can be computed by an effective procedure.

For any $k' \in \{1, \ldots, l\}$, we write $occ_{k'}$ to denote the set

$$\{k \in \{0, \ldots, 2^n - 1\} : \exists \mathtt{A} \in \mathtt{TERM}(\mathtt{F}),\ N(\mathtt{A}) = \ldots \cup \mathtt{A}_{k,k'} \cup \ldots\}$$

[1] Without any loss of generality we can assume that if l (resp. n) nominals for properties (resp. for entities) occur in F they are precisely the l (resp. n) first in the enumeration of VARP (resp. VARE) since satisfiability is not sensitive to the renaming of variables.

Informally, $occ_{k'}$ is the set of indices k such that $\mathtt{A}_{k,k'}$ occurs in the *normal form* of some element of TERM(F). We write $setocc_{k'}$ to denote the set

$$\{X \subseteq occ_{k'} : card(occ_{k'}) - 1 \leq card(X) \leq 2^n - 1\}$$

The definition of $setocc_{k'}$ is motivated by the fact that for any TERM-interpretation v, there is only one $k \in \{0, \ldots, 2^n - 1\}$ such that $v(\mathtt{A}_{k,k'}) \neq \emptyset$, and for this very k, $v(\mathtt{A}_{k,k'}) = v(\mathtt{p}_{k'})$. For each $X \in setocc_{k'}$ in turn, in the forthcoming constructions we shall enforce $v(\mathtt{A}_{k,k'}) = \emptyset$ for any $k \in X$.

3.3 The translation

In this section, we define an extension to $\mathcal{L}$ of the translation ST defined in [Ben83] of modal formulae into a first-order language containing a binary predicate, a countable set of unary predicate symbols and two individual variables (due to a smart recycling of the variables). Our translation of the nominals for entities is similar to the translation of nominals in [GG93]. However, we take into account the decomposition of terms into components in order to obtain a faithful translation. The translation of nominals for atomic properties is a twofold one: we take it into account both in defining the normal form of terms, and in the generalized disjunction defining the translation T below.

Let $\mathtt{F} \in \mathtt{FORM}$ be such that $\mathtt{VARP(F)} = \{\mathtt{p}_1, \ldots, \mathtt{p}_l\}$, $\mathtt{VARSP(F)} = \{\mathtt{P}_1, \ldots, \mathtt{P}_n\}$ and $\mathtt{VARE(F)} = \{\mathtt{x}_1, \ldots, \mathtt{x}_q\}$. Before defining ST' - the mapping translating $\mathcal{L}$-formulae into FO^2-formulae - let us state what are the main features we intend that mapping to have. Analogously to ST, ST' encodes the quantification in the interpretation of [A] into the language of FO^2 by using the standard universal quantifier $\forall$ and by introducing a binary predicate symbol $\mathtt{R}_\mathtt{A}$ for each $\mathtt{A} \in \mathtt{TERM}$. However, this is not exactly the way ST' is defined. Actually, to each component $\mathtt{A}_{k,k'}$ we associate the predicate symbol $\mathtt{R}_{k,k'}$. The main idea of ST' is therefore to treat components as constants, which means that the translation of [A]G is uniquely determined by the components (if any) of the normal form of A. Then, the conditions on the $\mathcal{L}$-models justify why a modal operator indexed by the *union* of components is translated into a formula involving a *conjunction* of atomic formulae. Let ST' be defined as follows (ST' is actually parametrized by F and $i \in \{0, 1\}$):

(1) $ST'(\mathtt{E}_j, \mathtt{y}_i) \stackrel{\text{def}}{=} \mathtt{P}_j(\mathtt{y}_i)$; $ST'(\mathtt{x}_j, \mathtt{y}_i) \stackrel{\text{def}}{=} \mathtt{Q}_j(\mathtt{y}_i)$;
(2) $ST'(\neg\mathtt{G}, \mathtt{y}_i) \stackrel{\text{def}}{=} \neg ST'(\mathtt{G}, \mathtt{y}_i)$; $ST'(\mathtt{G} \wedge \mathtt{H}, \mathtt{y}_i) \stackrel{\text{def}}{=} ST'(\mathtt{G}, \mathtt{y}_i) \wedge ST'(\mathtt{H}, \mathtt{y}_i)$;
(3)

$$ST'([\mathtt{A}]\mathtt{G}, \mathtt{y}_i) \stackrel{\text{def}}{=} \begin{cases} \forall\, \mathtt{y}_0\ ST'(\mathtt{G}, \mathtt{y}_0) \text{ if } N(\mathtt{A}) = 0 \\ \forall\, \mathtt{y}_{1-i}(\mathtt{R}_{k_1,k'_1}(\mathtt{y}_i, \mathtt{y}_{1-i}) \wedge \ldots \wedge \mathtt{R}_{k_u,k'_u}(\mathtt{y}_i, \mathtt{y}_{1-i})) \Rightarrow ST'(\mathtt{G}, \mathtt{y}_{1-i}) \\ \text{if } N(\mathtt{A}) = \mathtt{A}_{k_1,k'_1} \cup \ldots \cup \mathtt{A}_{k_u,k'_u} \end{cases}$$

By adopting the standard definition $\langle\mathtt{A}\rangle\mathtt{G} \stackrel{\text{def}}{=} \neg[\mathtt{A}]\neg\mathtt{G}$, ST' can be easily defined for $\langle\mathtt{A}\rangle\mathtt{G}$: the existential quantification is involved instead of universal one.

Let $\mathtt{G}_0$ be a first-order formula (in FO^2) expressing the fact that, for any $\langle k,k'\rangle \in \{0,\ldots,2^n-1\}\times\{0,\ldots,l\}$, $\mathtt{R}_{k,k'}$, is interpreted as a reflexive and symmetric binary relation. Let $\mathtt{G}_1$ be a first-order formula expressing the fact that, for any $i \in \{1,\ldots,q\}$, $\mathtt{Q}_i$ is interpreted[2] as a singleton, e.g.

$$\bigwedge_{i=1}^{q} \exists\, \mathtt{y}_0\ (\mathtt{Q}_i(\mathtt{y}_0) \wedge \forall\, \mathtt{y}_1\ \neg \mathtt{y}_0{=}\mathtt{y}_1 \Rightarrow \neg \mathtt{Q}_i(\mathtt{y}_1))$$

In the case when $\mathtt{VARE(F)} = \emptyset$, $\mathtt{G}_1 \stackrel{\text{def}}{=} \forall \mathtt{y}_0\ \mathtt{y}_0 = \mathtt{y}_0$. Let $\mathtt{T}_1(\mathtt{F})$ be the first-order formula (in FO^2[=]) defined by

$$\mathtt{T}_1(\mathtt{F}) \stackrel{\text{def}}{=} \mathtt{G}_0 \wedge \mathtt{G}_1 \wedge \exists\, \mathtt{y}_0\ ST'(\mathtt{F},\mathtt{y}_0)$$

The translation is not quite finished yet. Indeed, although at least one of the components $\mathtt{p}_1 \cap \mathtt{P}_1$ or $\mathtt{p}_1 \cap -\mathtt{P}_1$ is interpreted by the empty set of properties, this fact is not taken into account in ST' (considering e.g. $n=1$). This is a serious gap since at least one of the predicate symbols $\mathtt{R}_{0,1}$ or $\mathtt{R}_{1,1}$ should be interpreted as the universal relation. The forthcoming developments provide an answer to this technical problem.

Let $\mathtt{G}$ be a first-order formula, $k' \in \{1,\ldots,l\}$ and $X_{k'} \in setocc_{k'}$. We write $\mathtt{G}[k',X_{k'}]$ to denote the first-order formula obtained from $\mathtt{G}$ by substituting:

- every occurrence of $\mathtt{R}_{k,k'}(\mathtt{z}_1,\mathtt{z}_2) \Rightarrow \mathtt{H}$ with $\mathtt{H}$ if $k \in X_{k'}$,
- every occurrence of $\mathtt{F}' \wedge \mathtt{R}_{k,k'}(\mathtt{z}_1,\mathtt{z}_2) \wedge \mathtt{F}'' \Rightarrow \mathtt{H}$ with $\mathtt{F}' \wedge \mathtt{F}'' \Rightarrow \mathtt{H}$ if $k \in X_{k'}$ -the degenerate cases are omitted here-

(this rewriting procedure is confluent and always terminates). ¿From a semantical viewpoint, the substitution is equivalent[3] to satisfaction of the condition $(k \in X_{k'})$: $\forall \mathtt{z}_1,\mathtt{z}_2,\ \mathtt{R}_{k,k'}(\mathtt{z}_1,\mathtt{z}_2)$. For $\langle X_1,\ldots,X_l\rangle \in setocc_1 \times \ldots \times setocc_l$, we write $\mathtt{G}[X_1,\ldots,X_l]$ to denote the first-order formula $\mathtt{G}[1,X_1][2,X_2]\ldots[l,X_l]$. Observe that for any permutation σ on $\{1,\ldots,l\}$,

$$\mathtt{G}[\sigma(1),X_{\sigma(1)}][\sigma(2),X_{\sigma(2)}]\ldots[\sigma(l),X_{\sigma(l)}] = \mathtt{G}[1,X_1][2,X_2]\ldots[l,X_l]$$

Let $\mathtt{T}(\mathtt{F})$ be the formula

$$\mathtt{T}(\mathtt{F}) \stackrel{\text{def}}{=} \bigvee \{\mathtt{T}_1(\mathtt{F})[X_1,\ldots,X_l] : \langle X_1,\ldots,X_l\rangle \in setocc_1 \times \ldots \times setocc_l\}$$

Observe that $\mathtt{T}$ is exponential-time in $|\mathtt{F}|$ and the size of the formula obtained by translation may increase exponentially. It is however, not clear whether there

[2] Let FO$^2[\exists^{=1}]$ be FO^2 augmented with the existential quantifier $\exists^{=1}$ meaning "there exists exactly one". FO$^2[\exists^{=1}]$-satisfiability has been proved to be in **NEXPTIME** (see e.g. [PST97]). By defining $\mathtt{G}_1$ by $\mathtt{G}_1 \stackrel{\text{def}}{=} \bigwedge_{i=1}^{q} \exists^{=1}\ \mathtt{y}_0\ \mathtt{Q}_i(\mathtt{y}_0)$ we are able to prove decidability of $\mathcal{L}$-satisfiability via a translation into FO$^2[\exists^{=1}]$.

[3] Another solution consists in defining $\mathtt{G}[k',X_{k'}]$ as the formula $(\bigwedge_{k\in X_{k'}} \forall \mathtt{y}_0,\mathtt{y}_1,\ \mathtt{R}_{k,k'}(\mathtt{y}_0,\mathtt{y}_1)) \wedge \mathtt{G}$.

exists a tighter translation that characterizes more accurately the complexity class of $\mathcal{L}$-satisfiability. Observe also that $\mathtt{T(F)}$ is classically equivalent to

$$\mathtt{G_0} \wedge \mathtt{G_1} \wedge \exists \mathtt{y_0} \bigvee \{ST'(\mathtt{F}, \mathtt{y_0})[X_1, \ldots, X_l] : \langle X_1, \ldots, X_l \rangle \in setocc_1 \times \ldots \times setocc_l\}$$

Example 1. Let $\mathtt{F}$ be the formula $\langle \mathtt{p_1} \cup \mathtt{p_2} \rangle \neg \mathtt{E_1} \wedge [\mathtt{P_1}]\mathtt{E_1}$. Then $\mathtt{F}$ is $\mathcal{L}$-satisfiable, and the translation of $\mathtt{F}$ is the disjunction of the following formulae:

1. $\mathtt{G_0} \wedge \mathtt{G_1} \wedge \exists \mathtt{y_0}\ (\exists \mathtt{y_1}\ \mathtt{R_{1,1}(y_0, y_1)} \wedge \mathtt{R_{1,2}(y_0, y_1)} \wedge \neg \mathtt{P_1(y_1)}) \wedge (\forall \mathtt{y_1}\ \mathtt{R_{0,0}(y_0, y_1)} \Rightarrow \mathtt{P_1(y_1)})$
2. $\mathtt{G_0} \wedge \mathtt{G_1} \wedge \exists \mathtt{y_0}\ (\exists \mathtt{y_1}\ \mathtt{R_{0,1}(y_0, y_1)} \wedge \mathtt{R_{1,2}(y_0, y_1)} \wedge \neg \mathtt{P_1(y_1)}) \wedge (\forall \mathtt{y_1}\ \mathtt{R_{0,0}(y_0, y_1)} \wedge \mathtt{R_{0,1}(y_0, y_1)} \Rightarrow \mathtt{P_1(y_1)})$
3. $\mathtt{G_0} \wedge \mathtt{G_1} \wedge \exists \mathtt{y_0}\ (\exists \mathtt{y_1}\ \mathtt{R_{1,1}(y_0, y_1)} \wedge \mathtt{R_{0,2}(y_0, y_1)} \wedge \neg \mathtt{P_1(y_1)}) \wedge (\forall \mathtt{y_1}\ \mathtt{R_{0,0}(y_0, y_1)} \wedge \mathtt{R_{0,2}(y_0, y_1)} \Rightarrow \mathtt{P_1(y_1)})$
4. $\mathtt{G_0} \wedge \mathtt{G_1} \wedge \exists \mathtt{y_0}\ (\exists \mathtt{y_1}\ \mathtt{R_{0,1}(y_0, y_1)} \wedge \mathtt{R_{0,2}(y_0, y_1)} \wedge \neg \mathtt{P_1(y_1)}) \wedge (\forall \mathtt{y_1}\ \mathtt{R_{0,0}(y_0, y_1)} \wedge \mathtt{R_{0,1}(y_0, y_1)} \wedge \mathtt{R_{0,2}(y_0, y_1)} \Rightarrow \mathtt{P_1(y_1)})$

The translation takes into account that $N(\mathtt{p_1} \cup \mathtt{p_2}) = \mathtt{A_{0,1}} \cup \mathtt{A_{1,1}} \cup \mathtt{A_{0,2}} \cup \mathtt{A_{1,2}}$ and $N(\mathtt{P_1}) = \mathtt{A_{0,0}} \cup \mathtt{A_{0,1}} \cup \mathtt{A_{0,2}}$.

3.4 Faithfulness of the translation

The rest of this section is devoted to proving Proposition 3 below and stating certain corollaries (some being consequences of the proof of Proposition 3).

Proposition 3. *(1)* $\mathtt{F}$ *is* $\mathcal{L}$*-satisfiable iff (2)* $\mathtt{T(F)}$ *is first-order satisfiable.*

Proof. (1) implies (2). First assume $\mathcal{U}, e_0 \models \mathtt{F}$ for some model

$$\mathcal{U} = (ENT, PROP, (sim_P)_{P \subseteq PROP}, v)$$

and $e_0 \in ENT$ (this is the easier part of the proof). Let us define the following first-order structure $\mathcal{M} \stackrel{\text{def}}{=} \langle D, m \rangle$:

- $D \stackrel{\text{def}}{=} ENT$; for any $i \in \omega$, $m(\mathtt{Q}_i) \stackrel{\text{def}}{=} v(\mathtt{x}_i)$ and $m(\mathtt{P}_i) \stackrel{\text{def}}{=} v(\mathtt{E}_i)$;
- for any $\langle k, k' \rangle \in \{0, \ldots, 2^n - 1\} \times \{1, \ldots, l\}$, $m(\mathtt{R}_{k,k'}) \stackrel{\text{def}}{=} sim_{v(\mathtt{A}_{k,k'})}$ (for the other values of $\langle k, k' \rangle$ the interpretation of $\mathtt{R}_{k,k'}$ is not constrained).

Let $\langle i_1, \ldots, i_l \rangle \in \{0, \ldots, 2^n - 1\}^l$ be such that for any $k' \in \{1, \ldots, l\}$, $v(\mathtt{A}_{i_{k'},k'}) = v(\mathtt{p}_{k'})$. Such a sequence $\langle i_1, \ldots, i_l \rangle$ is unique. So, for any $k' \in \{1, \ldots, l\}$, $X_{k'} \stackrel{\text{def}}{=} occ_{k'} \setminus \{i_{k'}\}$. It is easy to show that $\mathcal{M} \models \mathtt{G_0} \wedge \mathtt{G_1}$ since $\mathcal{U}$ is a model. We claim that $\mathcal{M} \models \mathtt{T_1(F)}[X_1, \ldots, X_l]$, and therefore $\mathcal{M} \models \mathtt{T(F)}$. To prove such a result, let us show that for any $\mathtt{G} \in sub(\mathtt{F})$, $e \in ENT$, $i \in \{0, 1\}$, $\mathcal{U}, e \models \mathtt{G}$ iff $\mathcal{M}, v_{\mathcal{M}}[\mathtt{y}_i \leftarrow e] \models ST'(\mathtt{G}, \mathtt{y}_i)[X_1, \ldots, X_l]$. We write $v_{\mathcal{M}}[\mathtt{y}_i \leftarrow e]$ to denote a first-order valuation $v_{\mathcal{M}}$ such that $v_{\mathcal{M}}(\mathtt{y}_i) = e$. It entails $\mathcal{M} \models \exists \mathtt{y_0}\ ST'(\mathtt{G}, \mathtt{y_0})[X_1, \ldots, X_l]$. We omit the base case and the cases in the induction step when the outermost connective is Boolean. Here are the remaining cases.
Case 1: $\mathtt{G} = [\mathtt{A}]\mathtt{F_1}$ and $N(\mathtt{A}) = \mathtt{0}$

$\mathcal{U}, e \models [\mathtt{A}]\mathtt{F}_1$ iff for any $e' \in sim_{v(\mathtt{A})}(e)$, $\mathcal{U}, e' \models \mathtt{F}_1$
iff for any $e' \in ENT$, $\mathcal{U}, e' \models \mathtt{F}_1$
iff for any $e' \in D$, $\mathcal{M}$, $v_{\mathcal{M}}[\mathtt{y}_i \leftarrow e'] \models ST'(\mathtt{F}_1, \mathtt{y}_i)[X_1, \ldots, X_l]$
iff $\mathcal{M} \models \forall \mathtt{y}_0\ ST'(\mathtt{F}_1, \mathtt{y}_0)[X_1, \ldots, X_l]$
iff $\mathcal{M} \models ST'([\mathtt{A}]\mathtt{F}_1, \mathtt{y}_i)[X_1, \ldots, X_l]$

Case 2: $\mathtt{G} = [\mathtt{A}]\mathtt{F}_1$ and $N(\mathtt{A}) = \mathtt{A}_{k_1,k'_1} \cup \ldots \cup \mathtt{A}_{k_u,k'_u}$
First observe that for any $k' \in \{1, \ldots, l\}$ and $k \in X_{k'}$, $v(\mathtt{A}_{k,k'}) = \emptyset$.

$\mathcal{U}, e \models [\mathtt{A}]\mathtt{F}_1$ iff for any $e' \in \bigcap_{i \in \{1,\ldots,u\}} sim_{v(\mathtt{A}_{k_i,k'_i})}(e)$, $\mathcal{U}, e' \models \mathtt{F}_1$
iff for any $e' \in \bigcap_{i \in \{1,\ldots,u\}} sim_{v(\mathtt{A}_{k_i,k'_i})}(e)$,
$\mathcal{M}$, $v_{\mathcal{M}}[\mathtt{y}_{1-i} \leftarrow e'] \models ST'(\mathtt{F}_1, \mathtt{y}_{1-i})[X_1, \ldots, X_l]$
iff for any $e' \in \bigcap_{i \in \{1,\ldots,u\}} m(\mathtt{R}_{k_i,k'_i})(e)$,
$\mathcal{M}$, $v_{\mathcal{M}}[\mathtt{y}_{1-i} \leftarrow e'] \models ST'(\mathtt{F}_1, \mathtt{y}_{1-i})[X_1, \ldots, X_l]$
iff $\mathcal{M}$, $v_{\mathcal{M}}[\mathtt{y}_i \leftarrow e] \models \forall \mathtt{y}_{1-i}\ (\mathtt{R}_{k_1,k'_1}(\mathtt{y}_i, \mathtt{y}_{1-i}) \wedge \ldots \wedge \mathtt{R}_{k_u,k'_u}(\mathtt{y}_i, \mathtt{y}_{1-i})) \Rightarrow$
$ST'(\mathtt{F}_1, \mathtt{y}_{1-i})[X_1, \ldots, X_l]$
iff $\mathcal{M}$, $v_{\mathcal{M}}[\mathtt{y}_i \leftarrow e] \models \forall \mathtt{y}_{1-i}\ (\bigwedge_{1 \leq i \leq u,\ k_i \notin X_{k'_i}} \mathtt{R}_{k_i,k'_i}(\mathtt{y}_i, \mathtt{y}_{1-i})) \Rightarrow$
$ST'(\mathtt{F}_1, \mathtt{y}_{1-i})[X_1, \ldots, X_l]$
($\bigwedge_{1 \leq i \leq u,\ k_i \notin X_{k'_i}} \mathtt{R}_{k_i,k'_i}(\mathtt{y}_i, \mathtt{y}_{1-i})$ is $\top$ if the conjunction is empty)
iff $\mathcal{M}$, $v_{\mathcal{M}}[\mathtt{y}_i \leftarrow e] \models (\forall \mathtt{y}_{1-i}\ (\mathtt{R}_{k_1,k'_1}(\mathtt{y}_i, \mathtt{y}_{1-i}) \wedge \ldots \wedge \mathtt{R}_{k_u,k'_u}(\mathtt{y}_i, \mathtt{y}_{1-i})) \Rightarrow$
$ST'(\mathtt{F}_1, \mathtt{y}_{1-i}))[X_1, \ldots, X_l]$
iff $\mathcal{M} \models ST'([\mathtt{A}]\mathtt{F}_1, \mathtt{y}_i)[X_1, \ldots, X_l]$

In the previous line the substitution operation is performed only on $ST'(\mathtt{F}_1, \mathtt{y}_{1-i})$ whereas in the next line it is performed on the whole expression.

(2) implies (1). Omitted because of lack of space.

Corollary 1. *(1) The $\mathcal{L}$-satisfiability problem is decidable. (2) $\mathcal{L}$ has the finite model property. In particular, every $\mathcal{L}$-satisfiable formula* F *has a model such that $card(ENT) \leq 2^{2^{p(|\mathtt{F}|)}}$ for some fixed polynomial $p(n)$, and the cardinality of the relevant part of $\mathcal{U}$ is at most $2^n + l$, where $n = card(\mathtt{VARSP}(\mathtt{F}))$ and $l = card(\mathtt{VARP}(\mathtt{F}))$.*

As observed by one referee, formalizing concepts from similarity theory directly in first-order logic could be another alternative.

3.5 The degenerate cases

In the previous section we have assumed that $n \geq 1$ and $l \geq 1$. Now let us examine the remaining cases. If $l \geq 1$ and no variable for sets of properties occurs in the formula ($n = 0$), then we consider the following components: $\mathtt{A}_0 = -\mathtt{p}_1 \cap \ldots \cap -\mathtt{p}_l$ and $\mathtt{A}_{k'} = \mathtt{p}_{k'}$ for $k' \in \{1, \ldots, l\}$. Condition (3) in the definition of ST' becomes:

$$ST'([\mathtt{A}]\mathtt{G}, \mathtt{y}_i) \stackrel{\text{def}}{=} \begin{cases} \forall\ \mathtt{y}_0\ ST'(\mathtt{G}, \mathtt{y}_0) \text{ if } N(\mathtt{A}) = 0 \\ \forall\ \mathtt{y}_{1-i}(\mathtt{R}_{0,k'_1}(\mathtt{y}_i, \mathtt{y}_{1-i}) \wedge \ldots \wedge \mathtt{R}_{0,k'_u}(\mathtt{y}_i, \mathtt{y}_{1-i})) \Rightarrow ST'(\mathtt{G}, \mathtt{y}_{1-i}) \\ \text{if } N(\mathtt{A}) = \mathtt{A}_{k'_1} \cup \ldots \cup \mathtt{A}_{k'_u} \end{cases}$$

If $n \geq 1$ and no variable for individual properties occurs in the formula ($l = 0$), then Condition (3) in the definition of ST' becomes:

$$ST'([\mathtt{A}]\mathtt{G}, \mathtt{y}_i) \stackrel{\text{def}}{=} \begin{cases} \forall\, \mathtt{y}_0\; ST'(\mathtt{G}, \mathtt{y}_0) \text{ if } N(\mathtt{A}) = 0 \\ \forall\, \mathtt{y}_{1-i}(\mathtt{R}_{k_1,0}(\mathtt{y}_i, \mathtt{y}_{1-i}) \wedge \ldots \wedge \mathtt{R}_{k_u,0}(\mathtt{y}_i, \mathtt{y}_{1-i})) \Rightarrow ST'(\mathtt{G}, \mathtt{y}_{1-i}) \\ \text{if } N(\mathtt{A}) = \mathtt{B}_{k_1} \cup \ldots \cup \mathtt{B}_{k_u} \text{ (see Section 3.2)} \end{cases}$$

Moreover, $\mathtt{T}(\mathtt{F})$ is simply defined as $\mathtt{T}_1(\mathtt{F})$. In the case when $n = 0$, $l = 0$ and $\mathtt{TERM}(\mathtt{F}) \neq \emptyset$, by substituting every occurrence of $\mathtt{0}$ in $\mathtt{F}$ by $\mathtt{p}_1 \cap -\mathtt{p}_1$ we preserve $\mathcal{L}$-satisfiability and reduce the case to the previous one. Otherwise, $\mathtt{F}$ is a formula of the propositional calculus and therefore it poses no difficulty with respect to decidability.

4 Decidability results for variants of $\mathcal{L}$

4.1 Fixed finite set of properties

In this section we consider a finite set $PROP$ of properties, and show that $\mathcal{L}_{PROP}$ shares various features with $\mathcal{L}$. Actually $\mathcal{L}_{PROP}$ corresponds to the similarity logic with a fixed finite set of properties defined in [Kon97]. Without any loss of generality we can assume that $PROP = \{1, \ldots, \alpha\}$ for some $\alpha \geq 1$ and $\mathtt{VARP} = \{\mathtt{p}_1, \ldots, \mathtt{p}_\alpha\}$.

Let $\mathtt{F}$ be a $\mathcal{L}_{PROP}$-formula such that $\mathtt{VARP}(\mathtt{F}) = \{\mathtt{p}_{i_1}, \ldots, \mathtt{p}_{i_l}\}$ and $\mathtt{VARSP}(\mathtt{F}) = \{\mathtt{P}_{j_1}, \ldots, \mathtt{P}_{j_n}\}$. For any interpretation v possibly occurring in some $\mathcal{L}_{PROP}$-model and for any $\mathtt{A} \in \mathtt{TERM}(\mathtt{F})$, if $v(\mathtt{A}) = \{k_1, \ldots, k_s\}$, then $N_v(\mathtt{A}) \stackrel{\text{def}}{=} \mathtt{p}_{k_1} \cup \ldots \cup \mathtt{p}_{k_s}$ otherwise ($v(\mathtt{A}) = \emptyset$) $N_v(\mathtt{A}) \stackrel{\text{def}}{=} \mathtt{0}$. We write $v_{\mathtt{F}}$ (resp. $N_v(\mathtt{F})$) to denote the restriction of v to $\mathtt{TERM}(\mathtt{F})$ (resp. the formula obtained from $\mathtt{F}$ by substituting every occurrence of $\mathtt{A}$ by $N_v(\mathtt{A})$). Let $X_{\mathtt{F}}$ be the *finite* set

$$\{v_{\mathtt{F}} : v \text{ interpretation possibly occurring in some } \mathcal{L}_{PROP} \text{ -model}\}$$

Proposition 4. *Let* $\mathtt{F}$ *be a* $\mathcal{L}_{PROP}$*-formula. (1)* $\mathtt{F}$ *is* $\mathcal{L}_{PROP}$*-satisfiable iff (2)* $\bigvee_{v'' \in X_{\mathtt{F}}} N_{v''}(\mathtt{F})$ *is* $\mathcal{L}$*-satisfiable.*

Proof. (1) implies (2): Assume $\mathcal{U} = (ENT, PROP, (sim_P)_{P \subseteq PROP}, v), e_0 \models \mathtt{F}$ for some $e_0 \in ENT$. It is easy to check that $\mathcal{U}', e_0 \models N_v(\mathtt{F})$ where $\mathcal{U}'$ is defined from $\mathcal{U}$ by only replacing v by v' defined as follows: for any $i \in \{1, \ldots, \alpha\}$, $v'(\mathtt{p}_i) \stackrel{\text{def}}{=} \{i\}$. Let $\mathcal{U}'' = (ENT, \omega, (sim''_P)_{P \subseteq \omega}, v'')$ be an $\mathcal{L}$-model such that

- v' and v'' are identical for the common sublanguage,
- for any $i > \alpha$, $v''(\mathtt{p}_i) \stackrel{\text{def}}{=} \{i\}$, for any $P \subseteq \omega$, $sim''_P \stackrel{\text{def}}{=} sim_{P \cap PROP}$.

It is a routine task to check that $\mathcal{U}'', e_0 \models N_v(\mathtt{F})$ and therefore $\mathcal{U}'', e_0 \models \bigvee_{v'' \in X_{\mathtt{F}}} N_{v''}(\mathtt{F})$. Indeed, for any $\mathtt{A} \in \mathtt{TERM}(\mathtt{F})$, $sim_{v(\mathtt{A})} = sim''_{v''(N_v(\mathtt{A}))}$.

(2) implies (1): Now assume $\bigvee_{v'' \in X_{\mathtt{F}}} N_{v''}(\mathtt{F})$ is $\mathcal{L}$-satisfiable. There exist an $\mathcal{L}$-model $\mathcal{U}' = (ENT', PROP', (sim'_P)_{P \subseteq PROP'}, v')$, $e_0 \in ENT'$ and $v''_0 \in X_{\mathtt{F}}$

such that $\mathcal{U}', e_0 \models N_{v_0''}(\mathtt{F})$. By the proof of Proposition 3, we can assume that $\{u_1, \ldots, u_\alpha\}$ is a relevant part of $PROP'$ in $\mathcal{U}'$ such that for any $i \in \{1, \ldots, \alpha\}$, $v'(\mathtt{p}_i) = u_i$. Indeed, $PROP'$ is at least countable, $\mathtt{VARSP}(\bigvee_{v'' \in X_{\mathtt{F}}} N_{v''}(\mathtt{F})) = \emptyset$ and $card(\mathtt{VARP}(\bigvee_{v'' \in X_{\mathtt{F}}} N_{v''}(\mathtt{F}))) \leq \alpha$.

Let $\mathcal{U} = (ENT', PROP, (sim_P)_{P \subseteq PROP}, v)$ be the $\mathcal{L}_{PROP}$-model such that:

- v and v_0'' are identical for the common sublanguage;
- for any $P \subseteq PROP$, $sim_P \stackrel{\text{def}}{=} sim'_{\{u_i : i \in P\}}$.

It is a routine task to check that $\mathcal{U}, e_0 \models \mathtt{F}$ since for any $\mathtt{A} \in \mathtt{TERM}(\mathtt{F})$, $sim_{v(\mathtt{A})} = sim'_{v'(N_{v_0''}(\mathtt{A}))}$.

Corollary 2. *$\mathcal{L}_{PROP}$-satisfiability is decidable and $\mathcal{L}_{PROP}$ has the finite model property.*

Example 2. (Example 1 continued) Let $\mathtt{F}$ be the formula $\langle \mathtt{p}_1 \cup \mathtt{p}_2 \rangle \neg \mathtt{E}_1 \wedge [\mathtt{P}_1]\mathtt{E}_1$ for the logic $\mathcal{L}_{\{1,2\}}$. Then, $\mathtt{F}$ *is not* $\mathcal{L}_{\{1,2\}}$-satisfiable, although $\mathtt{F}$ *is* $\mathcal{L}$-satisfiable. The formula $\bigvee_{v' \in X_{\mathtt{F}}} N_{v'}(\mathtt{F})$ is the disjunction of the following formulae:

1. $(\langle \mathtt{p}_1 \cup \mathtt{p}_2 \rangle \neg \mathtt{E}_1 \wedge [0]\mathtt{E}_1) \vee (\langle \mathtt{p}_1 \cup \mathtt{p}_2 \rangle \neg \mathtt{E}_1 \wedge [\mathtt{p}_1]\mathtt{E}_1)$
2. $(\langle \mathtt{p}_1 \cup \mathtt{p}_2 \rangle \neg \mathtt{E}_1 \wedge [\mathtt{p}_2]\mathtt{E}_1) \vee (\langle \mathtt{p}_1 \cup \mathtt{p}_2 \rangle \neg \mathtt{E}_1 \wedge [\mathtt{p}_1 \cup \mathtt{p}_2]\mathtt{E}_1)$

4.2 Fixed infinite set of properties

In this section we consider some infinite set $PROP$ of properties, and show that $\mathcal{L}_{PROP}$ shares various features with $\mathcal{L}$. Actually, $\mathcal{L}_{PROP}$ corresponds to the similarity logic with a fixed infinite set of properties defined in [Kon97]. Without any loss of generality we can assume that $\omega \subseteq PROP$ (there is an injective map f from ω into $PROP$) and $\{\mathtt{p}_1, \mathtt{p}_2, \ldots\} \subseteq \mathtt{VARP}$.

Proposition 5. *Let* $\mathtt{F}$ *be a* $\mathcal{L}_{PROP}$*-formula. (1)* $\mathtt{F}$ *is* $\mathcal{L}_{PROP}$*-satisfiable iff (2)* $\mathtt{F}$ *is* $\mathcal{L}$*-satisfiable.*

Proof. Omitted because of lack of space.

Corollary 3. *$\mathcal{L}_{PROP}$-satisfiability is decidable and $\mathcal{L}_{PROP}$ has the finite model property.*

5 Concluding remarks

We have shown that the relative similarity logics $\mathcal{L}$ and $\mathcal{L}_X$ for some non-empty set X of properties have a *decidable* satisfiability problems. Moreover, we have also established that such logics have the finite model property. The decidability proof reduces satisfiability in our logic to satisfiability in $FO^2[=]$, a decidable fragment of classical logic [Mor75]. Although our reduction takes advantage of the standard translation ST [Ben83] of modal logic into classical logic, the novelty of our approach consists in the method of handling nominals for atomic properties and decomposition in terms of components *encoded* in the translation. The reduction into $FO^2[=]$ can be generalized to any relative logics provided,

1. the conditions[4] on the relations of the models can be expressed by a first-order formula involving at most two variables (see the definition of the formula G_0 in Section 3.3), and
2. the class of binary relations underlying the logic is closed under intersection.

For instance, if in the definition of $\mathcal{L}$ we replace reflexivity by *weak reflexivity*, then decidability and finite model property still hold true[5]. This is particularly interesting since weakly reflexive and symmetric modal frames represent exactly the *negative* similarity relations in information systems (see e.g. [Vak91a,DO96]). For the sake of comparison, the class of reflexive and symmetric modal frames represent precisely the *positive* similarity relation in information systems.

We have also shown that $\mathcal{L}$-satisfiability is **EXPTIME**-hard (by taking advantage of the general results from [Hem96]), and that the problem can be solved by a deterministic Turing machine in time $O(2^{2^{2^{p(n)}}})$ for some polynomial $p(n)$, where n is the length of the tested formula. Indeed: the translation process $\mathtt{T}$ is exponential in time in the length of the formula, $\mathtt{T}$ may increase exponentially the length of the formula and satisfiability for $FO^2[=]$ is in **NEXPTIME**. It is therefore an open problem to characterize more accurately the complexity class of $\mathcal{L}$-satisfiability. However, the translations we have established can already be used to mechanize the relative similarity logics by taking $FO^2[=]$ as the target logic and by using a theorem prover dedicated to it. We conjecture that more efficient methods might exist for the mechanization.

References

[Bal97] Ph. Balbiani. Axiomatization of logics based on Kripke models with relative accessibility relations. In *[Oe97]*, pages 553–578, 1997.

[Ben83] J. van Benthem. *Modal logic and classical logic.* Bibliopolis, 1983.

[Ben98] J. van Benthem. The range of modal logic - An essay in memory of George Gargov. *Journal of Applied Non-Classical Logics*, 1998. To appear.

[Bla93] P. Blackburn. Nominal tense logic. *Notre Dame Journal of Formal Logic*, 34(1):56–83, 1993.

[Dem98] S. Demri. A logic with relative knowledge operators. *Journal of Logic, Language and Information*, 1998. To appear.

[DO96] S. Demri and E. Orłowska. Informational representability: Abstract models versus concrete models. In D. Dubois and H. Prade, editors, *Linz Seminar on Fuzzy Sets, Logics and Artificial Intelligence, Linz, Austria.* Kluwer Academic Publishers, February 1996. To appear.

[FdCO84] L. Fariñas del Cerro and E. Orłowska. DAL - A logic for data analysis. In T. O'Shea, editor, *ECAI-6*, pages 285–294, September 1984.

[GG93] G. Gargov and V. Goranko. Modal logic with names. *Journal of Philosophical Logic*, 22(6):607–636, December 1993.

[4] If the conditions require more than two variables, the translation into $FO^k[=]$ for some $k > 2$ is also possible. For instance, the logic defined in [Orło84a] can be translated into $FO^3[=]$ by taking advantage of the results of the present paper.

[5] A binary relation R is *weakly reflexive* iff for any element x of some domain either $R(x) = \emptyset$ or $(x, x) \in R$.

[GKV97] E. Grädel, Ph. Kolaitis, and M. Vardi. On the decision problem for two-variable first-order logic. *Bulletin of Symbolic Logic*, 3(1):53–69, 1997.

[GP90] G. Gargov and S. Passy. A note on boolean modal logic. In P. Petkov, editor, *Summer School and Conference on Mathematical Logic '88*, pages 299–309. Plenum Press, 1990.

[Hem96] E. Hemaspaandra. The price of universality. *Notre Dame Journal of Formal Logic*, 37(2):173–203, 1996.

[Kon97] B. Konikowska. A logic for reasoning about relative similarity. *Studia Logica*, 58(1):185–226, 1997.

[Lew80] H. Lewis. Complexity results for classes of quanticational formulas. *Journal of Computer and System Sciences*, 21:317–353, 1980.

[Mor75] M. Mortimer. On language with two variables. *Zeit. für Math. Logik and Grund. der Math.*, 21:135–140, 1975.

[Oe97] E. Orlowska (ed.). *Incomplete Information: Rough Set Analysis*. Studies in Fuzziness and Soft Computing. Physica-Verlag, Heidelberg, 1997.

[OP84] E. Orłowska and Z. Pawlak. Representation of nondeterministic information. *Theoretical Computer Science*, 29:27–39, 1984.

[Orło84a] E. Orłowska. Logic of indiscernibility relations. In A. Skowron, editor, *5th Symposium on Computation Theory, Zaborów, Poland*, pages 177–186. LNCS 208, Springer-Verlag, 1984.

[Orło84b] E. Orłowska. Modal logics in the theory of information systems. *Zeitschr. f. Math. Logik und Grundlagen d. Math.*, 30(1):213–222, 1984.

[Paw81] Z. Pawlak. Information systems theoretical foundations. *Information Systems*, 6(3):205–218, 1981.

[PST97] L. Pacholski, W. Szwast, and L. Tendera. Complexity of two-variable logic with counting. In *LICS*, pages 318–327. IEEE, July 1997.

[PT91] S. Passy and T. Tinchev. An essay in combinatory dynamic logic. *Information and Computation*, 93:263–332, 1991.

[Vak91a] D. Vakarelov. Logical analysis of positive and negative similarity relations in property systems. In M. de Glas and D. Gabbay, editors, *First World Conference on the Fundamentals of Artificial Intelligence*, 1991.

[Vak91b] D. Vakarelov. A modal logic for similarity relations in Pawlak knowledge representation systems. *Fundamenta Informaticae*, 15:61–79, 1991.

[Vak91c] D. Vakarelov. Modal logics for knowledge representation systems. *Theoretical Computer Science*, 90:433–456, 1991.

[Var97] M. Vardi. Why is modal logic so robustly decidable? In *Descriptive complexity and finite models, A.M.S.*, 1997.

A Conditional Logic for Belief Revision

Laura Giordano, Valentina Gliozzi and Nicola Olivetti

Dipartimento di Informatica, Università di Torino, Italy
{laura, gliozzi, olivetti}@di.unito.it

Abstract. In this paper we introduce a conditional logic BC to represent belief revision. Logic BC has a standard semantics in terms of possible worlds structures with a selection function and has strong similarities with Stalnaker's logic C2. Moreover, Gärdenfors' Triviality Result does not apply to BC. We provide a representation result, which shows that each belief revision system corresponds to a BC-model and every BC model satisfying the *covering condition* determines a belief revision system.

1 Introduction

One of the most important ideas in the analysis of conditional propositions comes from a proposal of F.P.Ramsey who proposed in [15] that in order to decide whether to accept a conditional proposition $A > B$ (whose meaning is: "if A were true then B would be true") we should add the antecedent A to our belief set, changing it as little as possible, and then consider whether the consequent B follows. Stalnaker's logic [16] stems from this intuition, even if Stalnaker was interested in analysing the truth conditions of conditional propositions, whereas Ramsey maintained that conditional propositions do not have any truth value.

The acceptability criterion proposed by Ramsey has received a renewed interest ten years ago by Gärdenfors [3, 5], who developed together with Alchourrón and Makinson [1] a theory of epistemic change. In [3] it is proposed the following version of Ramsey's acceptance criterion:

Ramsey Test: $A > B \in K$ iff $B \in K * A$,

where K represents a belief set (that is, a deductively closed set of sentences) and * represents a *Belief Revision operator*. The operator * transforms ("revises") a belief set K by adding a formula A in such a way that the resulting belief set, denoted by $K*A$, is consistent if so is A; moreover, $K*A$ is obtained by minimally changing K. Gärdenfors, Alchourrón and Makinson have expressed this *minimal change requirement* and other natural conditions on revision operators by a set of *rationality* postulates (called the AGM postulates) that we will recall in the next section.

In spite of the similarities between the semantics of belief revision and the evaluation of conditionals, the above very intuitive acceptance principle leads to the well known *Triviality Result* by Gärdenfors, [3], which claims that there are no significant belief revision systems that are compatible with the Ramsey test. The core of Gärdenfors' Triviality Result is the inconsistency between the *Preservation Principle*, which governs belief revision, and the *Monotonicity Principle*, which follows from the Ramsey test. By the former principle it is intended that, if a formula B is accepted in a given belief set K and A is consistent with the beliefs in K, then B is still accepted in the minimal change of K needed to accept A. By the latter principle it is intended that $K \subseteq K'$ implies $K * A \subseteq K' * A$.

Since Gärdenfors' Triviality Result appeared, many authors have proposed a solution to the problem. Several authors have studied alternative notions of belief change, like "belief update" [7, 8, 6], which do not enforce the Preservation Principle.

Rather than considering alternative notions of belief change, we follow another line of research (see [10, 8, 12, 2]) which has focused on the problem of capturing belief revision itself within a conditional system in a less stringent way than the original formulation of RT. The dependency between conditionals and belief sets should be less strict, in the sense that if we accept a conditional proposition with respect to a belief set K, this does not entail that we are willing to accept it with respect to every larger belief set. In this sense, the acceptability of conditionals is nonmonotonic in K.

We adopt a weaker formulation of the Ramsey test. Namely:

$$\text{(RT) } A > B \text{ is "accepted" in } K \text{ iff } B \in K * A,$$

where the notion of "acceptability" of a conditional $A > B$ in K is, following the spirit of Levi's proposal, a weaker condition than "$A > B \in K$". In this work, we interpret the acceptability of a conditional $A > B$ as: $A > B$ is true in a world which satisfies the conditional theory Th_K associated to each belief set K. We get nonmonotonicity in the sense stated above from the fact that from $K \subseteq K'$ it does not follow that $Th_K \subseteq Th'_K$.

A semantical reconstruction of revision in terms of conditional logic has to depart from the standard semantics of conditionals. This aspect has been pointed out in [12] and in [2]. Semantically, to each belief set K it is associated the set of its models (or worlds) $[[K]]$ and a preorder relation $\leq_K$. To $K * A$ it is associated the set of models closest to $[[K]]$ with respect to $\leq_K$ which satisfy A. This global dependency on K (because of $\leq_K$) is the basic semantical difference with the case of the above mentioned update operator and is also the main difficulty to give a semantical reconstruction of revision in terms of conditional logics. One possible solution, pursued by Friedman and Halpern in [2] is to introduce another level of semantic objects, called epistemic states, to account for this dependency; each K corresponds to an epistemic state and each epistemic state has associated a set of worlds. In our approach, we do not introduce any further semantic object to account for this global dependency on belief sets; as we will see, each world carries with itself, so to say, the information about what is believed in it.

In this paper, we define a conditional logic which formalizes the idea of verifying acceptability of conditionals with respect to a belief set. However, we want to stay as close as possible to standard conditional logic and to define the truth conditions of formulas relatively to worlds, rather than to belief states, no matter how they are represented. To this purpose we need a belief operator which allows each world to be associated with a set of beliefs. As we will see, it is possible to define such a belief operator through the conditional implication itself. The resulting logical system is a conditional logic which allows one to represent belief sets and their revision. We will also see that our logic has strong connections with Stalnaker's logic.

In Section 2 we briefly recall the AGM postulates and the triviality result. In section 3, we introduce our conditional logic, and in section 4 we provide a representation theorem which maps belief revision systems and conditional models. Finally, section 5 concludes the paper and provides comparisons with other proposals.

2 Belief revision

We have mentioned in the previous section that Ramsey's criterion of acceptability for conditionals is to add the antecedent of the conditional to the belief set, *changing the belief set as little as possible*, and see whether the consequent belongs to the belief set so changed. But Ramsey did not worry about defining any operation to change belief sets.

In [1, 5, 4] two operations on belief sets are introduced, namely *expansion* and *revision*. Let $Cn(A)$ denote the deductive closure of A in classical propositional logic. We define a *belief set* K as a deductively closed set of propositional formulas, that is, $K = Cn(K)$. *Expansion* is the simple addition of a formula A to a belief set K, and it is defined by: $K + A = Cn(K \cup \{A\})$. *Revision* is the consistent addition of a formula A to a belief set K, denoted by $K * A$.

Alchourrón, Gärdenfors and Makinson in [1] have proposed some rationality postulates that any belief change operator must satisfy. AGM postulates enforce the *Preservation Principle*. [5]:

Revision postulates

$(K1)$ $(K * A)$ is a belief set;
$(K2)$ $A \in K * A$;
$(K3)$ $(K * A) \subseteq (K + A)$;
$(K4)$ If $\neg A \notin K, K + A \subseteq K * A$;
$(K5)$ $(K * A) = K_{\bot}$ only if $\vdash \neg A$;
$(K6)$ if $\vdash A \leftrightarrow B$ then $K * A = K * B$;
$(K7)$ $K * (A \wedge B) \subseteq (K * A) + B$;
$(K8)$ if $\neg B \notin (K * A)$, then $(K * A) + B \subseteq K * (A \wedge B)$

Postulate $(K2)$ says that revision is always successful; postulate $(K3)$ says that the revision of a belief set with a formula A does not lead to conclude more than what can be concluded by the simple expansion of K with A; $(K4)$ is equivalent to the Preservation Principle and says that when we make the revision of K with a formula consistent with K, no information of K has to be rejected. Taken together, $(K3)$ and $(K4)$ say that, if A is consistent with K, then a revision of K with A is just an expansion of K with A. Postulate $(K5)$ says that the revision is consistent unless the added formula is inconsistent by itself. Postulate $(K6)$ says that the result of revision does not depend on the syntactic form of the added information. Postulates $(K7)$ and $(K8)$ can be regarded as a generalization of $(K3)$ and $(K4)$ to deal with conjunctions.

In [3], Gärdenfors wonders whether Ramsey's proposal can be formalized using the notion of revision operator as determined by his postulates. As we have already said, the above postulates are not compatible with the Ramsey test. To see this fact, let us define a belief revision system as a pair $< \mathbf{K}, * >$, where $*$ is a revision operator and $\mathbf{K}$ is a set of belief sets closed under $*$. Moreover, we say that a belief revision system is non-trivial if there are at least three disjoint propositions A, B, C, (such that $\vdash \neg(A \wedge B)$ and $\vdash \neg(A \wedge C)$ and $\vdash \neg(B \wedge C)$) and a belief set $K \in \mathbf{K}$ such that K is consistent with A, B, C. We can roughly say that a trivial belief revision system contains only complete belief sets. Such belief revision systems represent a sort of degenerate case.

The Triviality result claims that there are no non-trivial belief revision systems that satisfy the Ramsey test. As we have already seen, the problem is that a direct consequence of the Ramsey test, the Monotonicity Principle, is inconsistent with the Preservation Principle. First of all, it is easily shown that the Monotonicity Principle follows directly from the Ramsey test: if $B \in K * A$ then by the first half of the Ramsey test $A > B \in K$. But if $K \subseteq K_1$, then $A > B \in K_1$ and, by the second half of the Ramsey test $B \in K_1 * A$. Therefore we can conclude that for each K, K_1 such that if $K \subseteq K_1$, $K * A \subseteq K_1 * A$. On the other hand the Preservation Principle makes the revision operator nonmonotonic in the sense that, given two belief sets K and K_1 such that $K \subseteq K_1$, we cannot conclude that $K * A \subseteq K_1 * A$. As an example, let us consider three belief sets K, K_1, K_2 such that K is the deductive closure of $K_1 \cup K_2$ and such that, for some formula A, $K_1 + A$ is consistent, $K_2 + A$ is consistent but $K + A$ is not consistent. From $(K4)$ we can conclude that $K_1 + A \subseteq K_1 * A$ and $K_2 + A \subseteq K_2 * A$. From the Monotonicity Principle we should conclude that $K_1 * A \subseteq K * A$ and hence that $K_1 + A \subseteq K * A$ and we should similarly conclude that $K_2 + A \subseteq K * A$. Therefore $(K_1 \cup K_2) + A \subseteq K * A$ and therefore $K * A$ would be inconsistent, which contradicts the revision postulate $(K5)$.

3 The Conditional Logic BC

Definition 1. *The language $\mathcal{L}_>$ of logic BC is an extension of the language $\mathcal{L}$ of classical propositional logic obtained by adding the conditional operator $>$. Let us define the following modalities:*

$$\Box A \equiv \neg A > \bot$$
$$\Diamond A \equiv \neg(A > \bot).$$

We call a modal formula *a formula A of the form $\bigcirc_1 \ldots \bigcirc_m B$, where $m \geq 0$ and for $i = 1, \ldots, m$, $\bigcirc_i$ is either $\Box$ or $\Diamond$, and $B \in \mathcal{L}$. The logic BC contains the following axioms and inference rules:*

(*CLASS*) *All classical axioms and inference rules;*
(*ID*) $A > A$;
(*RCEA*) *if* $\vdash A \leftrightarrow B$, *then* $\vdash (A > C) \leftrightarrow (B > C)$;
(*RCK*) *if* $\vdash A \rightarrow B$, *then* $\vdash (C > A) \rightarrow (C > B)$;

(*DT*) $(A \wedge C) > B) \rightarrow (A > (C \rightarrow B))$, *for* $A, B, C \in \mathcal{L}$;
(*CV*) $\neg(A > \neg C) \wedge (A > B) \rightarrow ((A \wedge C) > B)$, *for* $A, B, C \in \mathcal{L}$;

(*REFL*) $(\top > A) \rightarrow A$;
(*EUC*) $\neg(A > B) \rightarrow A > \neg(\top > B)$;
(*TRANS*) $(A > B) \rightarrow A > (\top > B)$;
(*BEL*) $(A > B) \rightarrow \top > (A > B)$;

(*MOD*) $\Box A \rightarrow B > A$, *where A is a modal formula;*
(*U4*) $\Box A \rightarrow \Box\Box A$, *where A is a modal formula;*
(*U5*) $\Diamond A \rightarrow \Box\Diamond A$, *where A is a modal formula.*

Note that a conditional formula $\top > A$ can be regarded as a belief operator meaning that "A is believed". Moreover, Axioms (REFL), (EUC) and (TRANS) (the last two ones for $A = \top$) make the belief operator an S5 modality.

Our logic has strong similarities with Stalnaker's logic. First, from (DT) and (REFL) we can derive (MP)($(A > B) \rightarrow (A \rightarrow B)$) restricted to the case in which $A, B \in \mathcal{L}$. Moreover, if we assume the axiom $A \rightarrow (\top > A)$, from (CV) we derive (CS)($(A \wedge B) \rightarrow (A > B)$) again restricted to the case in which $A, B \in \mathcal{L}$, from (EUC) we derive the usual (CEM) ($(A > B) \vee (A > \neg B)$) and axioms (TRANS) and (BEL) become tautological. Note that all axioms (ID), (CV), (MOD), (MP), (CS) and (CEM) belong to the axiomatization of Stalnaker's logic $C2$ (see [14]).

Our logic intends to model belief revision, in the sense of the representation theorem given in section 4. Before establishing a correspondence between the axioms of our logic and Gärdenfors' postulates, let us describe the model theory of our logic.

We develop a semantical interpretation for the logic BC in the style of standard Kripke-like semantics for conditional logics. Our structures are possible world structures equipped with a selection function.

Definition 2. *A BC-structure M has the form $\langle W, f, [[]]\rangle$, where W is a set of possible worlds, $f : \mathcal{L}_> \times W \rightarrow 2^W$ is a selection function, $[[]] : \mathcal{L}_> \rightarrow P(W)$ is a valuation function satisfying the following conditions:*

(1) $[[A \wedge B]] = [[A]] \cap [[B]]$
(2) $[[\neg A]] = W - [[A]]$
(3) $[[A \rightarrow B]] = (W - [[A]]) \cup [[B]]$
(4) $[[A > B]] = \{w : f(A, w) \subseteq [[B]]\}$;

Moreover, let Prop(S) $= \{A \in \mathcal{L}\colon \exists w \in S such that w \in [[A]]\}$ *We assume that the selection function f satisfies the following properties:*

(*ID*) $f(A,w) \subseteq [[A]]$;
(*RCEA*) *if* $[[A]] = [[B]]$ *then* $f(A,w) = f(B,w)$
(*DT*) *Prop*$(f(A,w) \cap [[C]]) \subseteq$ *Prop*$(f(A \wedge C, w))$, *for* $A, C \in \mathcal{L}$;
(*CV*) $f(A,w) \cap [[C]] \neq \emptyset \rightarrow$ *Prop*$(f(A \wedge C, w)) \subseteq$ *Prop*$(f(A,w) \cap [[C]])$, *for* $A, C \in \mathcal{L}$;
(*REFL*) $w \in f(\top, w)$;
(*TRANS*) $x \in f(A,w) \wedge y \in f(\top, x) \rightarrow y \in f(A,w)$;
(*EUC*) $x, y \in f(A,w) \rightarrow x \in f(\top, y)$
(*BEL*) $w \in f(\top, y) \rightarrow f(A,w) = f(A,y)$
(*MOD*) *If* $f(B,w) \cap [[A]] \neq \emptyset$, *then* $f(A,w) \neq \emptyset$ *where A is a modal formula*
(*UNIV*) *if* $[[A]] \neq \emptyset$, $\exists B$ *such that* $f(B,w) \cap [[A]] \neq \emptyset$, *where A is a modal formula.*

We say that a formula A is true in a BC-structure $M = \langle W, f, [[]]\rangle$ *if* $[[A]] = W$. *We say that a formula is BC-valid if it is true in every BC-structure. We also introduce the following notation* $S \models_M A$ *to say that, given a BC-structure M, a set of formulas S and a formula A, for all* $w \in M$ *if* $w \in [[B]]$ *for all* $B \in S$, *then* $w \in [[A]]$.

Note that, in a given BC-structure M, we can define through the selection function f an *equivalence relation* R on the set of worlds W as follows: for all $w, w' \in W$,

$$(w, w') \in R \text{ iff } w' \in f(\top, w).$$

The properties of R being reflexive, transitive and euclidean come from the semantic conditions (REFL), (TRANS) and (EUC) on the selection function f (and, more precisely, from the last two conditions by taking $A = \top$). Hence, we can read $\top > A$ as " A is believed", in contrast to the meaning of A which is "A is true".

The accessibility relation R determines equivalence classes among worlds. The intuition is that we can associate to each world a belief set which is the set of formulas true in all worlds in its equivalence class. As a consequence of axiom (BEL), evaluating a conditional formula $A > B$ in a world is exactly the same as evaluating that formula in a different world in the same class. This means that the value of $A > B$ in one world only depends on the beliefs true in that world and not on the objective facts true in that world.

When a conditional $A > B$ is evaluated in a world w, the selection function selects the set $f(A,w)$ of the most preferred A-worlds with respect to w, and B is evaluated in such a set of worlds. Axioms (EUC) and (TRANS) make $f(A,w)$ an equivalence class to which we can associate a belief set. Notice that, since $(A > C) \vee \neg(A > C)$ is a tautology, from (EUC) and (TRANS) we can conclude $(A > (\top > C)) \vee (A > \neg(\top > C))$, that is, C is either believed or non believed in the most preferred A-worlds. This is the conditional excluded middle, (CEM), restricted to belief formulas. While the presence of (CEM) in Stalnaker's logic causes the selection function to select a single world (i.e. $f(A,w) = \{j\}$ for all A and w), when (CEM) is restricted to belief formulas (as in our logic), it determines the unicity of the belief set associated to all worlds belonging to $f(A,w)$. When we evaluate a conditional, we want to move from one world, with an associated belief set, to other worlds with a different belief set. This ability to explicitly represent the new belief set obtained from the initial one is especially important if we want to model iterated revision by nested conditionals. We will come back to this point in section 4.

The restrictions we have imposed on our axioms are motivated by the fact that our logic is intended to model belief revision. Concerning (CV) and (DT), we observe that the revision postulates do not deal with conditional formulas. In particular, the Preservation Principle says something only about the classical formulas belonging to the revised belief set, whereas, as we have seen in the Introduction, the conditional formulas accepted in the original belief set might no longer be accepted in the revised one, since they globally depend on the whole belief set.

From the semantical condition (UNIV), which corresponds to (U4) and (U5), and from (MOD) we get the property:

$$\text{if } [[A]] \neq \emptyset, \text{ then } f(A,w) \neq \emptyset,$$

which, as we will see, is needed to model the revision postulate (K5). The restrictions we have put on (MOD), (U4), (U5) are needed since we cannot accept that the above property holds for all formulas $A \in L_>$. Having this property for arbitrary A would correspond to being able to reach any belief set from any other. This cannot be done by means of the revision operator. In general, given two belief sets K_1 and K_2, there may not exist a formula A such that $K_2 = K_1 * A$ (for instance when $B \in K_1$ but neither $B \in K_2$ nor $\neg B \in K_2$).

The axiomatization of BC is sound and complete with respect to semantic introduced above.

In the following, for readability, we use the notation $x \models A$ rather than $x \in [[A]]$.

Theorem 1 (Soundness). *If a formula A is a theorem of BC then is BC-valid.*

Proof. (Sketch) One checks each axiom and then shows that rules (RCEA) and (RCK) preserve validity. As an example, we give a proof of the validity of (CV) and (U5). Let $M = \langle W, f, [[]]^M \rangle$ be a BC structure. For (CV), let $x \in W$ and let $A, B, C \in \mathcal{L}$, $x \models \neg(A > \neg C)$ and $x \models A > B$. Let $y \in f(A \wedge C, x)$, we must show that $y \models B$. Let $Atom(B)$ be the set of propositional variables occurring in B and let

$$\psi_{B,y} = \bigwedge\{p \in Atom(B) \mid y \models p\} \wedge \bigwedge\{\neg p \in Atom(B) \mid y \not\models p\}.$$

Since $B \in \mathcal{L}$, we clearly have $y \models B$ iff $\psi_{B,y} \rightarrow B$ is a classical tautology. Moreover, $\psi_{B,y} \in Prop(f(A \wedge C, x))$. By hypothesis, $x \models \neg(A > \neg C)$ and this implies that $f(A,x) \cap [[C]] \neq \emptyset$. By condition (CV) we can conclude that $\psi_{B,y} \in Prop(f(A,x) \cap [[C]])$, thus for some $z \in f(A,x) \cap [[C]]$ we have $z \models \psi_{B,y}$

and $z \models B$, for $z \in f(A,x)$ and $x \models A > B$. Let $\psi_{B,z}$ be defined in a similar way to $\psi_{B,y}$. Since $z \models \psi_{B,y}$, it must be that $\psi_{B,z} \leftrightarrow \psi_{B,y}$ holds; on the other hand from $z \models B$, we have $\psi_{B,z} \rightarrow B$ is a classical tautology. We can conclude that $\psi_{B,y} \rightarrow B$, whence $y \models B$.

For (U5) let $x \in W$ and let (1) $x \models \Diamond A$, where A is a modal formula. Suppose that (2) $x \not\models \Box\Diamond A$. By (1) we have $f(A,x) \neq \emptyset$ (we recall that $\Diamond A \equiv \neg(A > \bot)$), whence $[[A]] \neq \emptyset$. Observe that $\Box\Diamond A \equiv (A > \bot) > \bot$, thus by (2) $x \not\models (A > \bot) > \bot$ and there is $z \in f(A > \bot, x)$. We have that $z \models A > \bot$, i.e. $f(A,z) = \emptyset$, whence $[[A]] = \emptyset$ by (MOD) and (UNIV). We have a contradiction.

□

Theorem 2 (Completeness). *If A is BC-valid then it is a theorem of BC.*

Proof. (Sketch) By contraposition, we show that if $\not\vdash A$ then there is a BC structure M in which A is not true. Let us fix the language $\mathcal{L}_>$. As usual we can prove that if $\not\vdash A$, then there is a maximal consistent set of formulas X_0 which does not contain A. We assume that the usual properties of maximal consistent sets are known (e.g. if X is maximally consistent, then $D \in X$ or $\neg D \in X$). We define $M = \langle W, f, [[]]^M \rangle$, as follows

$W = \{X \subseteq \mathcal{L}_> \mid X \not\vdash A$ and X is maximally consistent$\}$,
$f(B,X) = \{Y \in W \mid S_{B,X} \subseteq Y\}$,
where $S_{B,X} = \{C \in \mathcal{L}_> \mid B > C \in X\}$;
$[[p]]^M = \{X \in W \mid p \in X\}$.

One can prove the following facts.

Fact 1 for every formula $B \in \mathcal{L}_>$ and $X \in W$, $B \in X$ iff $X \in [[B]]^M$.

Fact 2 the structure M satisfies all conditions of definition 2, except (possibly) the condition (UNIV), namely, (ID), (RCEA), (DT), (CV), (REFL), (TRANS), (EUC),(BEL), and (MOD). As an example, we prove condition (CV) and (BEL) the other are similar and left to the reader. For (CV), let

$$f(D,X) \cap [[C]] \neq \emptyset \text{ and } \psi \in Prop(f(D \wedge C, X)),$$

where $D, C, \psi \in \mathcal{L}$. Suppose that $\psi \notin Prop(f(D,X) \cap [[C]])$, then for all $U \in f(D,X) \cap [[C]]$, we have $\neg\psi \in U$; From the fact that if $C \notin U$, then $\neg C \in U$ and the fact that the U's are deductively closed, we easily have that for all $U \in f(D,X)$, $C \rightarrow \neg\psi \in U$, this implies that $D > (C \rightarrow \neg\psi) \in X$. By (CV), we get that

$$(*)\ (D \wedge C) > (C \rightarrow \neg\psi) \in X.$$

Since $\psi \in Prop(f(D \wedge C, X))$, there exists $Z \in f(D \wedge C, X)$, such that $\psi \in Z$, on the other hand by (*), $C \rightarrow \neg\psi \in Z$ and since $C \in Z$, we get $\neg\psi \in Z$, we have a contradiction.

For (BEL) let $X \in f(T,Y)$, we show that $f(D,X) = f(D,Y)$. By definition of f, it suffices to show that $S_{D,X} = S_{D,Y}$. If $B \in S_{D,Y}$ then $D > B \in Y$, whence $T > (D > B) \in Y$ by (BEL). Thus $D > B \in X$ and $B \in S_{D,X}$. Conversely, if

$B \notin S_{D,Y}$, then $D > B \notin Y$,$T > (D > B) \notin Y$ by (REFL). This implies that $\neg(T > (D > B)) \in Y$, whence $T > \neg(T > (D > B)) \in Y$ by (EUC). Thus, $\neg(T > (D > B)) \in X$ and also $\neg(D > B) \in X$ by (BEL), whence $D > B \notin X$, i.e. $B \notin S_{D,X}$.

The structure M does not necessarily satisfy the condition (UNIV). Our plan is to define a substructure M_0 of M which is still a BC-structure, falsifies A and satisfies the universality condition. In order to define M_0, let $X_0 \in W$ such that $A \notin X_0$ (whence $\neg A \in X_0$). We define a binary relation on W, for $X, Y \in W$, let

$RXY \equiv \forall D$ modal formula $(\Box D \in X \to D \in Y)$ and then we let
$W_0 = \{Y \in W \mid RX_0Y\}$.

We first show that W_0 is closed with respect to f,i.e.

(i) if $Z \in W_0$ and $Y \in f(B, Z)$, then $Y \in W_0$,
(ii) for all $Y, Z \in W_0$, RYZ holds.

For (i) let $\Box D \in X$, then $\Box\Box D \in X$ by (U4), then $\Box D \in Z$; by (MOD) we obtain $B > D \in Z$, whence $D \in Y$.

For (ii), let RX_0Y and RX_0Z, we show that RYZ holds. Suppose $D \notin Z$, then $\Box D \notin X_0$, then $\neg\Box D \in X_0$ then $\Diamond\neg D \in X_0$, so that $\Box\Diamond\neg D \in X_0$ by (U5). Then $\Diamond\neg D \in Y$, and this implies that $\Box D \notin Y$.

Finally one, we can show that

(iii) $X_0 \in W_0$.

To this regard, if $\Box D \in X_0$, then $T > D \in X_0$ by (Mod) and $D \in X_0$ by (REFL).

We can now define a structure $M_0 = \langle W_0, f_0, [[]]^{M_0}\rangle$, where

$f_0(B, Z) = f(B, Z)$ and $[[p]]^{M_0} = [[p]]^M \cap W_0$,

in particular the definition of f is correct by virtue of (i).

Fact 3 M_0 satisfies all conditions of definition 2, in particular M_0 satisfies the condition (UNIV). In order to check (UNIV), let D be a modal formula and suppose that for all formulas B and $Z \in W_0$ $f(B, Z) \cap [[D]]^{M_0} = \emptyset$, in particular we have $f(D, Z) \cap [[D]]^{M_0} = \emptyset$; this implies $f(D, Z) = \emptyset$, that is $D > \bot \in Z$, whence $\Box\neg D \in Z$. By (ii) we have that for every $Y \in W_0$, RZY holds, thus $\neg D \in Y$, i.e. $[[D]]^{M_0} = \emptyset$.

Fact 4 For each formula C, $[[C]]^{M_0} = [[C]]^M \bigcap W_0$. This is proved by induction on the form of C, the details are left to the reader.

We can now conclude the completeness proof. If A is not a theorem of BC, then $X_0 \notin [[A]]^M$. Since $[[A]]^{M_0} = [[A]]^M \bigcap W_0$ and $X_0 \in W_0$, we have that $W_0 - [[A]]^{M_0} \neq \emptyset$, which shows that A is not true in M_0.

$\square$

4 Conditionals and Revision

The capability of defining a belief operator through conditional implication is central to our way of modelling belief revision. Given a belief set K, we will represent it by a set of belief formulas Th_K: all the formulas C in K are believed while all formulas not in K are disbelieved in Th_K. More precisely, we define

$$Th_K = \{\top > C : C \in K\} \cup \{\neg(\top > C) : C \notin K\}.$$

Checking if B belongs to the revised belief set $K * A$ will correspond, in our logic, to evaluate a conditional $A > B$ at all worlds satisfying the theory Th_K, that is, at all worlds whose corresponding belief set is K.

Before providing a representation theorem which establishes a precise correspondence among belief revision systems and our BC structures, let us give an intuitive idea of the relationship between AGM postulates and the axioms of our logic (or, equivalently, the semantic properties of BC structures).

Let us consider a single world w_K at which Th_K holds in a given BC structure. Roughly speaking, the worlds in the equivalence class of w_K are the classical interpretations of K. If we want to check if B belongs to the revised belief set $K * A$, we can evaluate the conditional $A > B$ at w_K. Then, the new belief set $K * A$ will be represented by the set of worlds $f(A, w_K)$, the set of the most preferred A-worlds with respect to the world w_K. Moreover, we can represent the belief set $K + A$, obtained by an expansion of K with A, as the subset of $f(\top, w_K)$ satisfying A, namely $f(\top, w_K) \cap [[A]]$. More precisely, given a belief set K and some BC-structure M, and a world w_K such that $w_K \models_M Th_K$, we can define

$K * A = \{B : w_K \models_M A > B\}$ and
$K + A = \{B : w_K \models_M T > (A \rightarrow B)\}$.

The below representation theorem will show that the revision operator $*$ defined above satisfies the AGM postulates. Here we provide some examples to show how BC-structures relate to the AGM postulates.

Let us consider postulate (K4): if $\neg A \notin K, K + A \subseteq K * A$. From $\neg A \notin K$, by definition of Th_K, we have that $Th_K \models_M \neg(\top > \neg A)$, and hence $w_K \models_M \neg(\top > \neg A)$. Moreover, from $B \in K + A$ we get $w_K \models_M \top > (A \rightarrow B)$. From the following consequence of axiom (CV) $\neg(\top > \neg A) \wedge (\top > (A \rightarrow B)) \rightarrow ((\top \wedge A) > (A \rightarrow B))$ we get, by (RCK), $w_K \models_M A > (A \rightarrow B)$. Hence, by using (ID), (RCK) and propositional reasoning, we can conclude $w_K \models_M A > B$, and therefore $B \in K * A$.

Let us consider postulate (K5): $(K * A) = K_\perp$ only if $\vdash \neg A$. Assume that $(K * A) = K_\perp$, then $\perp \in (K * A)$ and $w_K \models_M A > \perp$. Therefore, $f(A, w_K) = \emptyset$. As a consequence of (MOD) and (UNIV), we can conclude that $[[A]] = \emptyset$. From this we can conclude that $\vdash \neg A$ only if the model M under consideration has the property that for every satisfiable formula A there is a world in the model which satisfies it. Hence, we will make use of such a condition in the representation theorem below.

We say that a BC-structure $M = \langle W, f, [[]] \rangle$ satisfies the *covering condition* if, for any formula $A \in \mathcal{L}$ satisfiable in PC, $[[A]] \neq \emptyset$ (i.e., there is some world satisfying A in M). The following representation theorem describes the relationship between our logic BC and belief revision.

Theorem 3 (Representation theorem).

(1) Given a belief revision system $\langle \mathbf{K}, * \rangle$ such that the revision operator $*$ satisfies postulates K1-K8, there exists a BC-structure M_* such that for each consistent belief set K in $\mathbf{K}$, and $A,B \in \mathcal{L}$,

$$B \in K * A \quad \textit{if an only if} \quad Th_K \models_{M_*} A > B.$$

(2) Given a BC-structure $M = \langle W, f, [[]] \rangle$ which satisfies the covering condition, there is a belief revision system $\langle \mathbf{K_M}, *_M \rangle$ such that $\mathbf{K_M} = \{K \subseteq \mathcal{L} : K = Cn(K)$ and $[[Th_K]] \neq \emptyset\}$ and, for each belief set K of K_M, and $A,B \in \mathcal{L}$,

$$B \in K *_M A \quad \textit{if an only if} \quad w_K \models_M A > B,$$

for some w_K such that $w_K \models_M Th_K$.

Proof. (Sketch)
To prove part (1), we define a BC-structure $M_* = \langle W, f, [[]] \rangle$ as follows:

$W = \{(K, w) : w \in 2^{Prop}, K \in \mathbf{K}$ and $w \models_{PC} K\}$;
$C_K = \{(K', w) \in W : K' = K\}$ [1];
$[[p]] = \{(K, w) \in W : w \models_{PC} p\}$, for all propositional letters $p \in \mathcal{L}$;

$f(A, (K, w))$ and $[[A]]$ can be defined by double induction on the structure of the formula A. At each induction step, for each connective $\circ$, $[[A \circ B]]$ is defined by making use of the valuation of the subformulas ($[[A]]$ and $[[B]]$) and of the selection function for subformulas (for instance, $f(A, w)$); moreover, $f(A \circ B, w)$ is defined by possibly making use of the valuation of the formula $A \circ B$ itself. In particular we let:

$f(A, (K, w)) = C_{K*A}$, if $A \in \mathcal{L}$;
$f(A, (K, w)) = C_{K*\Phi_A}$, if $A \notin \mathcal{L}$ and there exists a formula $\Phi_A \in \mathcal{L}$ such that $[[A]] = [[\Phi_A]]$;
$f(A, (K, w)) = \emptyset$, otherwise.

By making use of the properties of the revision operator $*$, we can show that M_* is a BC-structure, Furthermore, it can be easily shown that the model M_* satisfies the condition: $B \in K * A$ if an only if $Th_K \models_{M_*} A > B$, by making use of the crucial property that: if $(K', w') \models_M Th_K$, then $K' = K$.

[1] In particular, for the inconsistent belief set $K_\perp$, we take $C_{K_\perp} = \emptyset$.

To prove part (2), we define a belief revision system $\langle \mathbf{K_M}, *_M \rangle$ with $\mathbf{K_M} = \{K : K = Cn(K) \text{ and } [[Th_K]] \neq \emptyset\}$ and the revision operator $*_M$ as follows:

$$K * A = \{B \in \mathcal{L} : w_K \models_M A > B\},$$

for some w_K such that $w_K \models_M Th_K$. It can be easily shown that the revision operator $*_M$ satisfies postulates (K1)–(K8). For postulates (K4) and (K5) we refer to the intuitive explanation above. We consider the remaining cases.

Postulate (K1): we have to show that $K * A$ is deductively closed. Let $C \in \mathcal{L}$ be a logical consequence of $K * A$. Then, by the compactness of classical propositional logic, there is a finite set of formulas $\beta_1, \dots, \beta_n \in K * A$ such that $\beta_1 \wedge \dots \wedge \beta_n \rightarrow C$ is a tautology. From the definition of $K * A$ we get that $w_K \models_M A > \beta_i$ for all $i = 1, \dots, n$ and, by propositional reasoning and (RCK), $w_K \models_M A > C$. Hence, $C \in K * A$.

Postulate (K2): we have to show that $A \in K * A$. This follows from the fact that $w_K \models_M A > A$ holds, by axiom (ID).

Postulate (K3): we have to show that $(K * A) \subseteq (K + A)$. Let us assume that $B \in (K * A)$. Then $w_K \models_M A > B$ and, by propositional reasoning, $w_K \models_M (\top \wedge A) > B$. Hence, by applying (DT), we conclude $w_K \models_M \top > (A \rightarrow B)$, which means that $B \in (K + A)$.

Postulate (K6): we have to show that if $\vdash A \leftrightarrow B$ then $K * A = K * B$. Assume that A and B are two equivalent formulas. Then, by (RCEA), $A > C \leftrightarrow B > C$ is valid. Hence, from $w_K \models_M A > C$ we can conclude $w_K \models_M B > C$, and vice versa. Therefore $C \in K * A$ if and only if $C \in K * B$.

Postulate (K7) can be shown to hold by making use of axiom (DT) in a way similar to case (K3), while postulate (K8) can be proved by making use of axiom (CV) in a way similar to case (K4).

□

Notice that the requirement of consistency of a belief set in the Representation Theorem, part (1), is needed since an inconsistent belief set $K_\perp$ cannot be represented by a world in a model, due to the presence of the semantic property (REFL).

Following [6] we say that a logic is *non-trivial* if there are at least four formulas A, B, C and D, such that the formulas $A \wedge B$, $B \wedge C$, $C \wedge A$ are inconsistent, and the formulas $A \wedge D$, $B \wedge D$, $C \wedge D$ are consistent. Otherwise the logic is trivial.

Theorem 4. *The logic BC is consistent and non-trivial.*

Intuitively, the reason why Gärdenfors' Triviality Result does not apply to our logic is that we have adopted a weaker formulation of the Ramsey test which does not enforce monotonicity. We can notice, indeed, that the theory Th_K depends nonmonotonically on K.

In this logic it is also possible to model iterated revision by nested conditionals. For instance, to check if B belongs to $(K * A_1) * A_2$ we will evaluate the conditional $A_1 > (A_2 > B)$ at a world w_K associated with the belief set K. In order to evaluate a formula such as $A_1 > (A_2 > B)$ in w_K, we have to evaluate $A_2 > B$ at worlds whose belief set is given by $K * A$. Hence, the belief set associated to every world in $f(A_1, w_K)$ (which are the most preferred A_1-worlds for w_K) must be $K * A_1$.

Our logic may be used to model iterated revision by making use of nested conditionals. However, we can only treat the case of revision of a consistent belief set K with a sequence of consistent formulas. The assumption that the formulas added by successive revisions are consistent is common to other proposals, as for instance [9]. In our logic, the revision of a belief set K with an inconsistent formula results in an empty set of models, that is, if w_K is a world whose associated belief set is given by K and A is an inconsistent formula, then $f(A, w_K)$ has to be empty. This means that for any formula B, the formula $A > B$ is true at w_K. In particular, for every B and C, $A > C > B$ is true at w_K. This essentially means that, once we enter in an inconsistent state we cannot get out of it. If we want to lift the assumption of consistency, a possible solution requires to modify the logic by removing axiom (REFL).

5 Conclusions and Related Work

In this paper we have introduced a conditional logic BC, which is well suited to model belief revision. Belief sets can be given a representation in this logic and can be associated with worlds by making use of the conditional operator itself. In this way we can make the evaluation of conditional formulas in one world dependent on the belief set holding at that world.

Our belief operator has some similarities with the necessity operator $\Box_K$ introduced in [12], which, however, is parametric with respect to a knowledge base K. In [12] the satisfiability of a formula is defined with respect to the model associated with a given belief set K, whereas the revision function is external to models and applies to models. As a difference, since the aim of our proposal is to depart as little as possible from standard conditional logics and their model theory, we incorporate the revision function in the models of our logic, namely in the selection function.

As a difference with conditional logic and also with our approach, Friedman and Halpern consider states rather than worlds as their primitive objects. The conditional language $\mathcal{L}^{>}$ they define is built up using a conditional operator $>$ and a belief operator $\mathbf{B}$, and it contains only subjective formulas, that is those formulas formed out by boolean combinations of conditional formulas and belief formulas. Such language $\mathcal{L}^{>}$ is completely disjoint from the language $\mathcal{L}$ containing the objective formulas (that is formulas which contain neither conditionals, nor belief operators), so that, for instance, a formula as $A \wedge (A > B)$, with $A \in \mathcal{L}$, is not allowed. Whereas objective formulas provide the belief assignment at a state, only subjective formulas are evaluated at a given state. Moreover, only objective formulas are allowed in the left hand side of a conditional, that is, in $\Phi > \Psi$ the antecedent Φ has to be an objective formula.

In contrast, we do not need to take as primitive a notion of epistemic state, nor an additional belief modality. Moreover, we do not put syntactic restrictions on the language and, in particular, we allow free occurrences of conditionals. We see that a syntactic restriction on the language is not needed, and it is sufficient to impose restrictions on some axioms: namely we require that (DT), (CV) only hold for the formulas in $\mathcal{L}$. On the contrary, we make use of the conditional operator itself in order to associate a world with a belief set, and we impose conditions on the selection function in such a way that the the selection function is defined in a world according to the belief set associated with that world.

In [8] Katsuno and Satoh present a unifying view of nonmonotonic reasoning, belief revision and conditional logic based on the notion of minimality. More precisely, they introduce ordered structures and families of ordered structures as a common ingredient. *Ordered structures* are triples $(W, \leq, V)$ containing a set W of worlds, a preorder $\leq$ and a valuation function V. They provide a semantic model to evaluate those conditional formulas that contain no nesting of $>$. *Families of ordered structures* are defined as collections of ordered structures, and their axiomatization corresponds to well known conditional logics, as VW, VC, and SS. While families of ordered structures are used to give a semantic characterization to update, ordered structures are used to give a semantic characterization to revision. In particular, Katsuno and Satoh show that, given a revision operator $*$, for each belief set K, there is an ordered structure O_K (satisfying the covering condition) such that the formulas in K are true in all minimal worlds of O_K (written $O_K \models K$), and

$$K * A = \{B : \ O_K \models A > B\}.$$

Since an ordered model O_K contains a *single* ordering relation $\leq$, it can only represent a single belief set K and its revisions. Moreover, since ordered structures do not handle nested conditionals, iterated revision cannot be captured in this formalization. As a difference, we are able to represent different belief sets and their revisions within a single structure, by associating worlds with belief sets. To this purpose, we make use of more complex semantic structures, which might be seen as a particular type of families of ordered structures.

We argue that our logic is not only well suited for modelling iterated revision, but that it also can provide a suitable framework in which to capture other forms of belief change, as belief update. This possibility is suggested by the fact that, when $A \dashrightarrow (T > A)$ is added to our axiomatization, from (REFL) we obtain the equivalence $A \leftrightarrow (T > A)$, and the belief set associated with a world becomes equal to the set of formulas true in that world. This is what we expect to happen with updates. In such a case, some of our axioms become tautological while other axioms coincide with the axioms of the conditional logic presented in [6] to deal with updates. Obviously, to deal with updates some of the axioms of BC, which is tailored for revision should be dropped, since they are not required for belief update.

References

1. C.E. Alchourrón, P. Gärdenfors, D. Makinson, On the logic of theory change: partial meet contraction and revision functions, *in* Journal of Symbolic Logic, 50:510–530, 1985.
2. N. Friedman, J.Y. Halpern, Conditional Logics of Belief Change, *in* Proceedings of the National Conference on Artificial Intelligence (AAAI 94):915-921, 1994.
3. P. Gärdenfors, Belief Revision and the Ramsey Test for Conditionals, *in* The Philosophical Review, 1996.
4. P. Gärdenfors, Belief Revision, *in* D. Gabbay (ed.), Handbook of Logic in Artificial Intelligence, 1995.
5. P. Gärdenfors, Knowledge in flux: modeling the dynamics of epistemic states, MIT Press, Cambridge, Massachussets, 1988.
6. G. Grahne, Updates and Counterfactuals, *in* Proceedings of the Second International Conference on Principles of Knowledge Representation and Reasoning (KR'91), pp. 269–276.
7. H. Katsuno, A.O. Mendelzon, On the Difference between Updating a Knowledge Base and Revising it, *in* Proceedings of the Second International Conference on Principles of Knowledge Representation and Reasoning, 1991
8. H. Katsuno, K. Satoh, A unified view of consequence relation, belief revision and conditional logic, *in* Proc. 12th International Joint Conference on Artificial Intelligence (IJCAI'91), pp. 406–412.
9. D. Lehmann, Belief revision revised, *in* Proc. 14th International Joint Conference on Artificial Intelligence (IJCAI'95), pp. 1534–1540.
10. I. Levi, Iteration of Conditionals and the Ramsey Test, *in* Synthese: 49-81, 1988.
11. D. Lewis, *Counterfactuals*, Blackwell, 1973.
12. W. Nejdl, M. Banagl, Asking About Possibilities- Revision and Update semantics for Subjunctive Queries, *in* Lakemayer, Nebel, Lecture Notes in Artificial Intelligence: 250-274, 1994.
13. D. Nute, *Topics in Conditional Logic*, Reidel, Dordrecht, 1980.
14. D. Nute, Conditional Logic, *in* Handbook of Philosophical Logic, Vol. II, 387-439, 1984.
15. F.P. Ramsey, *in* A. Mellor (editor), Philosophical Papers, Cambridge University Press, Cambridge, 1990.
16. R. Stalnaker, A Theory of Conditional, *in* N. Rescher (ed.), Studies in Logical Theory, American Philosophical Quarterly, Monograph Series no.2, Blackwell, Oxford: 98-112.

Implicates and Reduction Techniques for Temporal Logics*

Inman P. de Guzmán, Manuel Ojeda-Aciego, and Agustín Valverde

Dept. Matemática Aplicada, Universidad de Málaga, Spain
{guzman,aciego,a_valverde}@ctima.uma.es

Abstract. Reduction strategies are introduced for the future fragment of a temporal propositional logic on linear discrete time, named FNext. These reductions are based in the information collected from the syntactic structure of the formula, which allow the development of efficient strategies to decrease the size of temporal propositional formulas, viz. new criteria to detect the validity or unsatisfiability of subformulas, and a strong generalisation of the pure literal rule. These results, used as a preprocessing step, allow to improve the performance of any automated theorem prover.

1 Introduction

The temporal dimension of information, the change of information over time and knowledge about how it changes has to be considered by many AI systems. There is obvious interest in designing computationally efficient temporal formalisms, specially when intelligent tasks are considered, such as planning relational actions in a changing environment, building common sense reasoning into a moving robot, in supervision of industrial processes, ...

Temporal logics are widely accepted and frequently used for specifying concurrent and reactive agents (which can be either physical devices or software processes), and in the verification of temporal properties of programs. To verify a program, one specifies the desired properties of the program by a formula in temporal logic. The program is correct if all its computations satisfy the formula. However, in its generality, an algorithmic solution to the verification problem is hopeless. For *propositional* temporal logic, checking the satisfiability of a formula can be done algorithmically, and theoretical work on the complexity of program verification is being done [3]. The complexity of satisfiability and determination of truth in a particular finite structure are considered for different propositional linear temporal logics in [7].

Linear-time temporal logics have proven [5] to be a successful formalism for the specification and verification of concurrent systems; but have a much wider range of applications, for instance, in [2] a generalisation of the temporal propositional logic of linear time is presented, which is useful for stating and

* Partially supported by Spanish CICYT project TIC97-0579-C02-02 and EC action COST-15: *Many-valued logics for computer science applications.*

proving properties of the generic execution sequence of a parallel program. On the other hand, relatively complete deductive systems for proving *branching* time temporal properties of reactive systems [4] have been recently developed.

In recent years, several fully automatic methods for verifying temporal specifications have been introduced, in [6] a tableaux calculus is treated at length; a first introduction to the tableaux method for temporal logic can be seen in [8]. However, the scope of these methods is still very limited. Theorem proving procedures for temporal logics have been traditionally based on syntactic manipulations of the formula A to be proved but, in general, do not incorporate the substitution of subformulas in A like in a rewrite system in which the rewrite relation preserves satisfiability. One source of interest of these strategies is that can be easily included into any prover, specifically into those which are non-clausal.

In this work we focus on the development of a set of reduction strategies which, through the efficient determination and manipulation of lists of unitary implicant and implicates, *investigates exhaustively* the possibility of decreasing the size of the formula being analysed. The interest of such a set of reduction techniques is that the performance of a given prover for linear-time temporal logic can be improved because the size of a formula can be decreased, at a polynomial cost, as much as possible *before* branching.

Lists of unitary models, so-called Δ-lists, are associated to each node in the syntactic tree of the formula and used to study whether the structure of the syntactic tree has or has not direct information about the validity of the formula. This way, either the method ends giving this information or, otherwise, it decreases the size of the problem before applying the next transformation. So, it is possible to decrease the number of branchings or, even, to avoid them all.

The ideas in this paper generalise the results in [1], in a self-contained way, by explicitly extending the reduction strategy to linear-time temporal logic and, what is more important, by complementing the information in the Δ-lists by means of the so-called $\widehat{\Delta}$-sets. The former allow derivation of an equivalent and smaller formula; the latter also allow derivation of a smaller formula, not equivalent to the previous one, but equisatisfiable.

The paper is organised as follows:

- Firstly, preliminary concepts, notation and basic definitions are introduced: specifically, it is worth to note the definition of literal and the way some of them will be denoted.
- Secondly, Δ-lists, our basic tool, are introduced; its definition integrates some reductions into the calculation of the Δ-lists. The required theorems to show how to use the information collected in those lists are stated.
- Later, the $\widehat{\Delta}$-sets are defined and results that use the information in these sets are stated. One of these is a generalisation of the pure literal rule.

2 Preliminary Concepts and Definitions

In this paper, our object language is the future fragment of the Temporal Propositional Logic FNext with linear and discrete flow of time, and connectives $\neg$

(negation), $\wedge$ (conjunction), $\vee$ (disjunction), F (sometime in the future), G (always in the future), and $\oplus$ (tomorrow); $\mathcal{V}$ denotes the set of propositional variables p, q, r, ... (possibly subscripted) which is assumed to be completely ordered with the lexicographical order, e.g. $p_n \leq q_m$ for all n, m, and $p_n \leq p_m$ if and only if $n \leq m$. Given $p \in \mathcal{V}$, the formulas p and $\neg p$ are the classical literals on p.

Definition 1. *Given a classical propositional literal ℓ, the* temporal literals[1] *on ℓ, denoted* $\mathtt{Lit}(\ell)$, *are those wff of the form $\oplus^n\ell$, $F\oplus^n\ell$, $G\oplus^n\ell$, $FG\ell$, $GF\ell$ for all $n \in \mathbb{N}$.*

The notion of temporal negation normal formula, *denoted tnnf, is recursively defined as follows:*

1. *Any literal is a tnnf.*
2. *If A and B are tnnf, then $A \vee B$ and $A \wedge B$ are tnnf, which are called disjunctive and conjunctive tnnf, respectively.*
3. *If A is a disjunctive tnnf, then GA is a tnnf.*
4. *If A is a conjunctive tnnf, then FA is a tnnf.*
5. *A formula is a tnnf if and only if it can be constructed by the previous rules.*

For formulas in tnnf, we will write $\overline{p}$ for the classical negated literal $\neg p$.

As usual, a clause is a disjunction of literals and a cube is a conjunction of literals. In addition, a G-clause is a formula GB where B is a classical clause, and a F-cube is a formula FB in which B is a classical cube.

We denote $\vartheta\ell$ to mean a temporal literal on ℓ, where ϑ is said to be its *temporal prefix*; if $\vartheta\ell$ is a temporal literal, then $|\vartheta|$ denotes the number of temporal connectives in ϑ, and $\overline{\vartheta\ell}$ denotes its opposite literal, where $\overline{F} = G$, $\overline{G} = F$, $\overline{FG} = GF$, $\overline{GF} = FG$ and $\overline{\oplus} = \oplus$

The transformation of any wff into tnnf is linear by recursively applying the transformations induced by the double negation, the de Morgan laws and the equivalences in Fig. 1.

$$
\begin{array}{lll}
\neg\oplus A \equiv \oplus\neg A & \oplus FA \equiv F\oplus A & \oplus GA \equiv G\oplus A \\
FFA \equiv F\oplus A & GGA \equiv G\oplus A & FGFA \equiv GFA \\
GFGA \equiv FGA & FG\oplus A \equiv FGA & GF\oplus A \equiv GFA \\
\oplus\bigvee A_i \equiv \bigvee\oplus A_i & \oplus\bigwedge A_i \equiv \bigwedge\oplus A_i & \neg FA \equiv G\neg A \\
\neg GA \equiv F\neg A & F(\bigvee_{i\in J} A_i) \equiv \bigvee_{i\in J} FA_i & G(\bigwedge_{i\in J} A_i) \equiv \bigwedge_{i\in J} GA_i
\end{array}
$$

Fig. 1.

In addition, by using the associative laws we will consider expressions like $A_1 \vee \cdots \vee A_n$ or $A_1 \wedge \cdots \wedge A_n$ as formulas.

[1] As we will be concerned only on temporal literals, in the rest of the paper we will drop the adjective *temporal*. In addition, we will use the notation $\oplus^n$ to denote the n-folded application of the connective $\oplus$.

We will use the standard notion of tree and address of a node in a tree. Given a tnnf A, the *syntactic tree* of A, denoted by T_A, is defined as usual. An address η in T_A will mean, when no confusion arises, the subformula of A corresponding to the node of address η in T_A; the address of the root node will be denoted ε.

If T_C is a subtree of T_A, then the *temporal order of* T_C *in* T_A, denoted $ord_A(C)$, is the number of temporal ancestors of T_C in T_A.

We will also use lists with its standard notation, `nil`, for the empty list. Elements in a list will be written in juxtaposition.

If α and β are lists of literals and $\vartheta\ell$ is a literal, $\vartheta\ell \in \alpha$ denotes that $\vartheta\ell$ is an element of α; and $\alpha \subseteq \beta$ means that all elements of α are elements of β. If $\alpha = \vartheta_1\ell_1\vartheta_2\ell_2\dots\vartheta_n\ell_n$, then $\overline{\alpha} = \overline{\vartheta_1\ell_1}\,\overline{\vartheta_2\ell_2}\dots\overline{\vartheta_n\ell_n}$.

Definition 2. *A* temporal structure *is a tuple* $S = (\mathbb{N}, <, h)$, *where* $\mathbb{N}$ *is the set of natural numbers,* $<$ *is the standard strict ordering on* $\mathbb{N}$, *and* h *is a* temporal interpretation, *which is a function* $h : \mathcal{L} \longrightarrow 2^{\mathbb{N}}$, *where* $\mathcal{L}$ *is the language of the logic, satisfying:*

1. $h(\neg A) = \mathbb{N} \smallsetminus h(A)$; $\quad h(A \vee B) = h(A) \cup h(B)$
2. $h(A \to B) = (\mathbb{N} \smallsetminus h(A)) \cup h(B)$; $\quad h(A \wedge B) = h(A) \cap h(B)$
3. $t \in h(FA)$ *iff* t' *exists with* $t < t'$ *and* $t' \in h(A)$
4. $t \in h(GA)$ *iff for all* t' *with* $t < t'$ *we have* $t' \in h(A)$
5. $t \in h(\oplus A)$ *iff we have* $t + 1 \in h(A)$

A formula A is said to be *satisfiable* if there exists a temporal structure $S = (\mathbb{N}, <, h)$ such that $h(A) \neq \varnothing$; if $t \in h(A)$, then h is said to be a *model of* A in t; if $h(A) = \mathbb{N}$, then A is said to be *true in the temporal structure* S; if A is true in every temporal structure, then A is said to be *valid*, and we denote it $\models A$.

Formulas A and B are said to be *equisatisfiable* if A is satisfiable iff B is satisfiable; $\equiv$ denotes the semantic equality, i.e. $A \equiv B$ if and only if for every temporal structure $S = (\mathbb{N}, <, h)$ we have that $h(A) = h(B)$; finally, the symbols $\top$ and $\bot$ mean truth and falsity, i.e. $h(\top) = \mathbb{N}$ and $h(\bot) = \varnothing$ for every temporal structure $S = (\mathbb{N}, <, h)$.

If Γ_1 and Γ_2 are sets of subformulas in A and X and Y are subformulas, then the expression $A[\Gamma_1/X, \Gamma_2/Y]$ denotes the formula obtained after substituting in A every occurrence of elements in Γ_1 by X and every occurrence of elements in Γ_2 by Y.

If η is an address in T_A and X, then the expression $A[\eta/X]$ is the formula obtained after substituting in A the subtree rooted in η by X.

3 Adding Information to the Tree: Δ-lists

The idea underlying the reduction strategy we are going to introduce is the use of information given by partial assignments. We associate to each tnnf A two lists of literals denoted $\Delta_0(A)$ and $\Delta_1(A)$ (the associated Δ-lists of A)[2] and two

[2] It can be shown that either A is equivalent to a literal, or at most one of these lists is non-empty.

sets of lists, denoted $\widehat{\Delta_0}(A)$ and $\widehat{\Delta_1}(A)$, whose elements are obtained out of the associated Δ-lists of the subformulas of A.

The Δ-lists and the $\widehat{\Delta}$-sets are the key tools of our method to reduce the size of the formula being analysed. These reductions allow to study its satisfiability with as few branching as possible.

In a nutshell, $\Delta_0(A)$ and $\Delta_1(A)$ are, respectively, lists of temporal implicates and temporal implicants of A. The purpose of these lists is two-fold:

1. To transform the formula A into an equivalent and smaller-sized one (see Sect. 3.3).
2. To be used in the definition the $\widehat{\Delta_b}$ sets (see Sect. 4), which will be used to transform the formula A into an equisatisfiable and smaller-sized one. Furthermore, information to build a countermodel (if it exists) is provided.

The sense in which we mean temporal implicant/implicate is the following:

Definition 3.

- *A literal* $\vartheta\ell$ *is a* temporal implicant *of* A *if* $\models \vartheta\ell \to A$.
- *A literal* $\vartheta\ell$ *is a* temporal implicate *of* A *if* $\models A \to \vartheta\ell$.

3.1 The Lattices of Literals

Definition 4. *For each classical propositional literal* ℓ *we define an ordering in* $\mathtt{Lit}(\ell) \cup \{\bot, \top\}$ *as follows:*

1. $\vartheta\ell \leq \varrho\ell$ *if and only if* $\models \vartheta\ell \to \varrho\ell$
2. $\vartheta\ell \leq \top$ *for all (possibly empty)* ϑ.
3. $\vartheta\ell \geq \bot$ *for all (possibly empty)* ϑ.

Each set $\mathtt{Lit}(\ell) \cup \{\bot, \top\}$ provided with this ordering is a lattice, depicted in Figure 2. For each literal $\vartheta\ell$ we will consider its upward and downward closures, denoted $\vartheta\ell\!\uparrow$ and $\vartheta\ell\!\downarrow$.

3.2 Definition of the Δ-lists

Definition 5. *Given a tnnf* A, *we define* $\Delta_0(A)$ *and* $\Delta_1(A)$ *to be the lists of literals recursively defined below*

$$\begin{aligned}
&\Delta_0(\vartheta\ell) = \Delta_1(\vartheta\ell) = \vartheta\ell \\
&\Delta_0\left(\textstyle\bigwedge_{i=1}^n A_i\right) = \mathtt{Union}_\wedge(\Delta_0(A_1), \dots, \Delta_0(A_n)) \\
&\Delta_0\left(\textstyle\bigvee_{i=1}^n A_i\right) = \mathtt{Intersection}(\Delta_0(A_1), \dots, \Delta_0(A_n)) \\
&\Delta_1\left(\textstyle\bigwedge_{i=1}^n A_i\right) = \mathtt{Intersection}(\Delta_1(A_1), \dots, \Delta_1(A_n)) \\
&\Delta_1\left(\textstyle\bigvee_{i=1}^n A_i\right) = \mathtt{Union}_\vee(\Delta_1(A_1), \dots, \Delta_1(A_n)) \\
&\Delta_b\,(FA) = \mathtt{Add}_\mathtt{F}(\Delta_b(A)) \\
&\Delta_b\,(GA) = \mathtt{Add}_\mathtt{G}(\Delta_b(A))
\end{aligned}$$

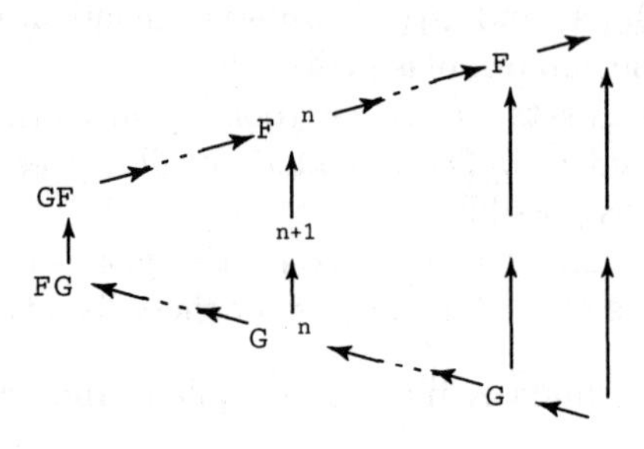

Fig. 2. Lattice $\texttt{Lit}(\ell) \cup \{\bot, \top\}$

The description of the operators involved in the definition above is the following:

1. The operators Add add a temporal connective to each element of a list of literals and simplify the results to a tnnf according to the rules in Fig. 1.
2. The two versions of Union arise because of the intended interpretation of these sets:
   (a) Elements in Δ_0 are considered as conjunctively connected, so we use $\texttt{Union}_\wedge$. This way, we obtain *minimal* implicates.
   (b) Elements in Δ_1 are considered as disjunctively connected, so we use $\texttt{Union}_\vee$. This way, we obtain *maximal* implicants.

Remark 1. By conjunctively connected, we mean that two literals in Δ_0 are substituted by its conjunction if it is either a literal or $\top$ or $\bot$, i.e. $\vartheta\ell \wedge \vartheta\ell\!\uparrow = \vartheta\ell$, $\vartheta\ell \wedge \overline{\vartheta\ell}\!\downarrow = \bot$ and the pair of literals $G\oplus^{n+1}\ell$ and $\oplus^{n+1}\ell$ is simplified to $G\oplus^n\ell$, for all n.

Similarly, the disjunctive connection in Δ_1 means the application of the following rules $\vartheta\ell \vee \vartheta\ell\!\downarrow = \vartheta\ell$, $\vartheta\ell \vee \overline{\vartheta\ell}\!\uparrow = \top$, and the pair of literals $F\oplus^{n+1}\ell$ and $\oplus^{n+1}\ell$ is simplified to $F\oplus^n\ell$ in Δ_1, for all n.

It is easy to see that, for all ℓ, we have that $\Delta_b(A) \cap \texttt{Lit}(\ell)$ contains at most one literal in the set $\{F\oplus^k\ell, G\oplus^k\ell, FG\ell, GF\ell\}$ and, possibly, several of the type $\oplus^k\ell$.

Definition 6. *If a A is a tnnf, then to Δ-label A means to label each node η in A with the ordered pair $(\Delta_0(\eta), \Delta_1(\eta))$.*

Example 1. Consider the formula $A = (\neg p \vee \neg Gq \vee r \vee G(\neg s \vee \neg q \vee u)) \wedge \neg(\neg p \vee \neg Gq \vee r \vee G(\neg s \vee u))$; the Δ-labelled tree of A is[3]

[3] For the sake of clarity, the Δ-labels of the leaves are not written.

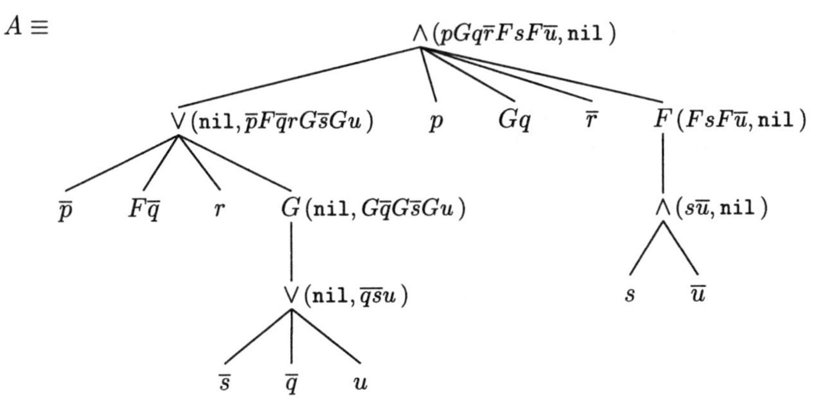

Note that in node 1, literals $F\overline{q}$ and $G\overline{q}$ are collapsed into $F\overline{q}$, because of the disjunctive connection in Δ_1.

Example 2. Let us study the validity of $A = G(\neg p \to p) \to (\neg Gp \to Gp)$. The Δ-labelled tree equivalent to $\neg A$ is

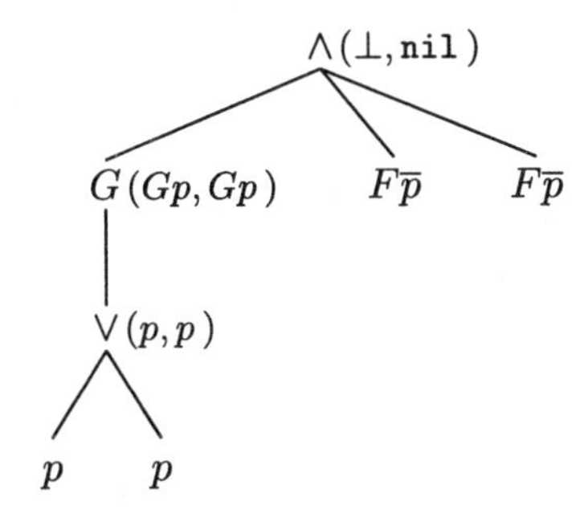

In this case, $\Delta_0(\varepsilon) = \perp$, because of the simplification of Gp and $F\overline{p}$ due to the conjunctive nature of the Δ_0-sets. We will see later that $\Delta_0(\varepsilon) = \perp$ implies that the input formula, that is $\neg A$, is unsatisfiable, therefore A is valid.

3.3 Information in the Δ-lists

As indicated above, the purpose of defining Δ_0 and Δ_1 is to collect implicants and implicates of A, as shown in the following theorem.

Theorem 1. *Let A be a tnnf,*

1. *If $\vartheta\ell \in \Delta_0(A)$, then $\models A \to \vartheta\ell$.*
2. *If $\vartheta\ell \in \Delta_1(A)$, then $\models \vartheta\ell \to A$.*

The theorem above will be used in the following equivalent form:

1. If $\vartheta\ell \in \Delta_0(A)$, then $A \equiv A \wedge \vartheta\ell$.
2. If $\vartheta\ell \in \Delta_1(A)$, then $A \equiv A \vee \vartheta\ell$.

As a literal is satisfiable, by Theorem 1 item 2, we have the following result:

Corollary 1. *If $\Delta_1(A) \neq \mathtt{nil}$, then A is satisfiable.*

3.4 Strong Meaning-Preserving Reductions

A lot of information can be extracted from the Δ-lists as corollaries of Theorem 1. The first result is a structural one, for it says that either one of the Δ-lists is empty, or both are equal and singletons.

Corollary 2. *If A is not a literal and $\Delta_1(A) \neq \mathtt{nil} \neq \Delta_0(A)$, then there exists $\vartheta\ell$ such that $\Delta_1(A) = \Delta_0(A) = \vartheta\ell$. Such tnnf A is said to be $\vartheta\ell$-simple.*

The corollary below states conditions on the Δ-lists which allow to determine the validity or unsatisfiability of the formula we are studying.

Corollary 3. *Let A be a tnnf, then*

1. *(a) If $\Delta_0(A) = \bot$, then $A \equiv \bot$.*
   (b) If $A = \bigwedge_{i=1}^{n} A_i$ in which a conjunct A_{i_0} is a clause such that $\overline{\Delta_1(A_{i_0})} \subseteq \Delta_0(A)\!\uparrow$, then $A \equiv \bot$.
   (c) If $A = \bigwedge_{i=1}^{n} A_i$ in which a conjunct A_{i_0} is a G-clause GB such that $\overline{\mathtt{Add}_{\oplus}(\Delta_1(B))} \subseteq \Delta_0(A)\!\uparrow$, then $A \equiv \bot$.
2. *(a) If $\Delta_1(A) = \top$, then $A \equiv \top$.*
   (b) If $A = \bigvee_{i=1}^{n} A_i$ in which a disjunct A_{i_0} is a cube such that $\overline{\Delta_0(A_{i_0})} \subseteq \Delta_1(A)\!\downarrow$, then $A \equiv \top$.
   (c) If $A = \bigvee_{i=1}^{n} A_i$ in which a disjunct A_{i_0} is an F-cube FB such that $\overline{\mathtt{Add}_{\oplus}(\Delta_0(B))} \subseteq \Delta_1(A)\!\downarrow$, then $A \equiv \top$.

The following definition gives a name to those formulas which have been simplified by using the information in the Δ-lists.

Definition 7. *Let A be an tnnf then it is said that A is:*

1. finalizable *if either $A = \top$, or $A = \bot$ or $\Delta_1(A) \neq \mathtt{nil}$.*
2. *A tnnf verifying either (a) or (b) or (c) of item 1 in Corollary 3 is said to be Δ_0-conclusive.*
3. *A tnnf verifying either (a) or (b) or (c) of item 2 in Corollary 3 is said to be Δ_1-conclusive.*
4. *A tnnf A is said to be* Δ-restricted *if it has no subtree which is either Δ_0-conclusive, or Δ_1-conclusive, or $\vartheta\ell$-simple.*
5. *To* Δ-restrict *a tnnf A means to substitute each Δ_1-conclusive formula by $\top$, each Δ_0-conclusive formula by $\bot$, and each $\vartheta\ell$-simple formula by $\vartheta\ell$; and then eliminate the constants $\top$ and $\bot$ by applying the 0-1 laws.*
   Note that Δ-restricting is a meaning-preserving transformation.

Example 3. Given the transitivity axiom $A = FFp \to Fp$; the tnnf equivalent to $\neg A$ is $F\oplus p \wedge G\overline{p}$; since $\Delta_0(F\oplus p \wedge G\overline{p}) = \bot$, we have that $\neg A$ is Δ_0-conclusive, therefore $\neg A$ is unsatisfiable and A is valid.

Example 4. Given the formula $A = \oplus p \wedge \oplus F\overline{p} \wedge G(p \to Fp)$, its Δ-labelled tree is

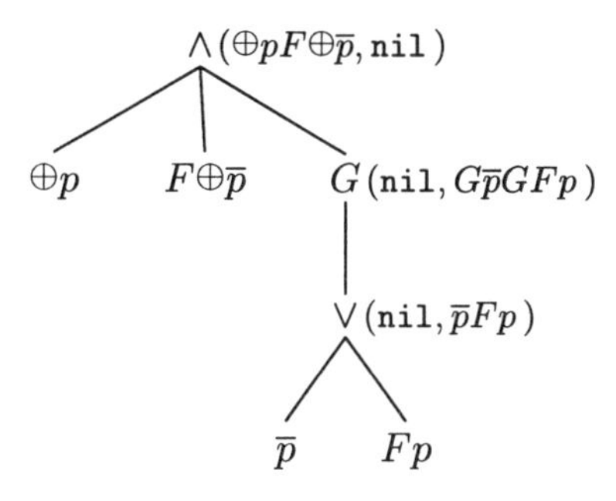

This tree is Δ_0-conclusive, since $\overline{\texttt{Add}_{\oplus}(\Delta_1(31))} = \oplus p F \oplus \overline{p} \subseteq \Delta_0(\varepsilon){\uparrow}$. In fact, what we have in this example is $\texttt{Add}_{\oplus}(\Delta_1(31)) = \Delta_0(\varepsilon)$

3.5 Weak Meaning-Preserving Reductions

The aim of this section is to give more general conditions allowing to use the information in the Δ-lists which has not been able to be used by the strong reductions. Specifically, a strong reduction uses the information in the Δ-lists in a *strong sense*, that is, to substitute a whole subformula by either $\top$, or $\bot$, or a literal. As in the propositional case, sometimes this is not possible and we can only use the information in a *weak sense*, that is, to decrease the size of the formula by eliminating literals depending on the elements of the Δ-lists.

The following notation is used in the statement of some results hereafter:

- If $\mathcal{S}$ is a set of literals in a tnnf A, then $\mathcal{S}^0$ denotes the set of all the occurrences of literals $\vartheta\ell \in \mathcal{S}$ of temporal order 0 in A
- $\texttt{Lit}(\ell, n) = \{\eta \mid \eta = \vartheta\ell \text{ and } |\vartheta| + ord_A(\eta) \geq n+1\}$
- $\texttt{Lit}(\overline{\ell}, n) = \{\eta \mid \eta = \vartheta\overline{\ell} \text{ and } |\vartheta| + ord_A(\eta) \geq n+1\}$

Theorem 2. *Let A be a tnnf and $\vartheta\ell$ a literal in A:*

1. *If $\vartheta\ell \in \Delta_0(A)$, then $A \equiv \vartheta\ell \wedge A[(\vartheta\ell{\uparrow})^0/\top, (\overline{\vartheta\ell}{\downarrow})^0/\bot]$*
2. *If $\vartheta\ell \in \Delta_1(A)$, then $A \equiv \vartheta\ell \vee A[(\vartheta\ell{\downarrow})^0/\bot, (\overline{\vartheta\ell}{\uparrow})^0/\top]$*

This theorem cannot be improved for an arbitrary literal $\vartheta\ell$; although, for some particular cases, it is possible to get more literals reduced, as shown by the following theorem, which generalises the result in Theorem 2, by dropping the restriction of order 0 for all the literals in the upward/downward closures.

Theorem 3. *Let A be a tnnf,*

1. *If $\vartheta\ell \in \Delta_0(A)$ with $\vartheta\ell \in \{FG\ell, GF\ell\} \cup \{G\oplus^n\ell \mid n \in \mathbb{N}\}$, then*
$$A \equiv \vartheta\ell \wedge A[\vartheta\ell \uparrow /\top, \overline{\vartheta\ell} \downarrow /\bot]$$
2. *If $\vartheta\ell \in \Delta_1(A)$ with $\vartheta\ell \in \{FG\ell, GF\ell\} \cup \{F\oplus^n\ell \mid n \in \mathbb{N}\}$, then*
$$A \equiv \vartheta\ell \vee A[\vartheta\ell \downarrow /\bot, \overline{\vartheta\ell} \uparrow /\top]$$

Finally, the theorem below states a number of additional reductions that can be applied when $\vartheta\ell$ equals either $G\oplus^n\ell$ or $F\oplus^n\ell$.

Theorem 4. *Let A be a tnnf and $\vartheta\ell$ a literal in A:*

1. *If $G\oplus^n\ell \in \Delta_0(A)$, then $A \equiv G\oplus^n\ell \wedge A[\texttt{Lit}(\ell,n)/\top, \texttt{Lit}(\bar{\ell},n)/\bot]$*
2. *If $F\oplus^n\ell \in \Delta_1(A)$, then $A \equiv F\oplus^n\ell \vee A[\texttt{Lit}(\ell,n)/\bot, \texttt{Lit}(\bar{\ell},n)/\top]$*

4 Adding Information to the Tree: $\widehat{\Delta}$-sets

In the previous sections, the information in the Δ-lists has been used *locally*, that is, the information in $\Delta_b(\eta)$ has been used to reduce η. The purpose of defining a new structure, the $\widehat{\Delta}$-sets, is to allow the globalisation of the information, in that the information in $\Delta_b(\eta)$ can be *refined* by the information in its ancestors.

Given a Δ-restricted tnnf A, we define the sets $\widehat{\Delta_0}(A)$ and $\widehat{\Delta_1}(A)$, whose elements are pairs (α,η) where α is a *reduced* Δ-list (to be defined below) associated to a subformula B of A, and η is the address of B in A. These sets allow to transform the formula A into an equisatisfiable and smaller sized one, as seen in Section 4.1.

The following result uses those cases in Theorems 2, 3 and 4 which allow to delete a whole subformula. The rest of possibilities only allow to delete literals; these literals will be called *reducible*.

Theorem 5. *Let A be a tnnf, B a subformula of A, and η the address in the tree of A of a subformula of B:*

1. *(a) If $\vartheta\ell$ is any literal satisfying $\vartheta\ell \in \Delta_0(\eta){\uparrow} \cap (\Delta_1(B) \cup \overline{\Delta_0(B)})$ and $ord_B(\eta) = 0$, then $A \equiv A[\eta/\bot]$.*
   (b) If $\vartheta\ell \in \{FG\ell, GF\ell\} \cup \{G\oplus^n\ell \mid n \in \mathbb{N}\}$ and satisfies and $\vartheta\ell \in \Delta_0(\eta){\uparrow} \cap (\Delta_1(B) \cup \overline{\Delta_0(B)})$, then $A \equiv A[\eta/\bot]$.
   (c) If $\vartheta\ell \in \Delta_0(\eta){\uparrow}$, and $F\oplus^n\ell \in \Delta_1(B) \cup \overline{\Delta_0(B)}$, and $|\vartheta| + ord_B(\eta) \geq n+1$, then $A \equiv A[\eta/\bot]$.
2. *(a) If $\vartheta\ell$ is any literal satisfying $\vartheta\ell \in \Delta_1(\eta){\downarrow} \cap (\Delta_0(B) \cup \overline{\Delta_1(B)})$ and $ord_B(\eta) = 0$, then $A \equiv A[\eta/\top]$.*
   (b) If $\vartheta\ell \in \{FG\ell, GF\ell\} \cup \{G\oplus^n\ell \mid n \in \mathbb{N}\}$ and satisfies and $\vartheta\ell \in \Delta_1(\eta){\downarrow} \cap (\Delta_0(B) \cup \overline{\Delta_1(B)})$, then $A \equiv A[\eta/\top]$.
   (c) If $\vartheta\ell \in \Delta_1(\eta){\downarrow}$, and $G\oplus^n\ell \in \Delta_0(B) \cup \overline{\Delta_1(B)}$, and $|\vartheta| + ord_B(\eta) \geq n+1$, then $A \equiv A[\eta/\top]$

This theorem can be seen as a generalisation of Corollary 3, in which a subformula B can be substituted by a constant even when that subformula *is not equivalent to* that constant.

The subformula at address η in A is said to be 0-*conclusive in A (resp.* 1-*conclusive in A)* if it verifies some of the conditions in item 1 (resp. item 2) above.

Definition 8. *Given a tnnf A and an address η, the* reduced *Δ-lists for A, $\Delta_b^A(\eta)$ for $b \in \{0,1\}$, are defined below,*

1. *If η is 0-conclusive in A, then $\Delta_0^A(\eta) = \bot$.*
2. *If η is 1-conclusive in A, then $\Delta_1^A(\eta) = \top$.*
3. *Otherwise, $\Delta_b^A(\eta)$ is the list $\Delta_b(\eta)$ in which the reducible literals have been deleted.*

We define the sets $\widehat{\Delta_b}(A)$ as follows

$$\widehat{\Delta_b}(A) = \{(\Delta_b^A(\eta), \eta) \mid \eta \text{ is a non-leaf address in } T_A \text{ with } \Delta_b(\eta) \neq \texttt{nil}\}$$

If A is a tnnf, to label *A means Δ-label A and to associate to the root of A the ordered pair $\left(\widehat{\Delta_0}(A), \widehat{\Delta_1}(A)\right)$.*

Example 5. From Example 1 we had the following tree

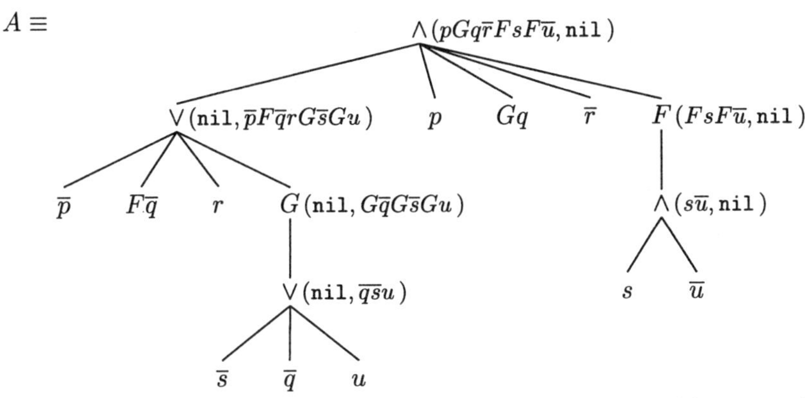

Note that literals $\overline{p}, F\overline{q}$ and r in Δ_1 of node 1 are reducible in A because of the occurrence of its duals in Δ_0 of the root. Similarly $G\overline{q}$ is also reducible in node 14, and $\overline{q}$ is reducible in 141. Therefore, the calculation of the $\widehat{\Delta}$-sets leads to

$$\widehat{\Delta}_0(A) = \{(pGq\overline{r}FsF\overline{u}, \varepsilon), (FsF\overline{u}, 5), (s\overline{u}, 51)\}$$
$$\widehat{\Delta}_1(A) = \{(G\overline{s}Gu, 1), (G\overline{s}Gu, 14), (\overline{s}u, 141)\}$$

4.1 Satisfiability-Preserving Results

In this section we study the information which can be extracted from the $\widehat{\Delta}$-sets.

Definition 9. *Let A be a tnnf then it is said that A is* restricted *if it is Δ-restricted and satisfies the following:*

- *There are not elements $(\bot, \eta)$ in $\widehat{\Delta_0}(A)$.*
- *There are not elements $(\top, \eta)$ in $\widehat{\Delta_1}(A)$.*

Remark 2. A restricted and equivalent tnnf can be obtained by using the 0-1 laws in conjunction with the elimination of conclusive subformulas in A, according to Theorem 5.

The following results will allow, by using the information in the $\widehat{\Delta}$-sets, to substitute a tnnf A by an equisatisfiable and smaller sized A'.

Complete Reduction

This section is named after Theorem 6, because after its application on a literal $G\oplus^n\ell$, gives an equisatisfiable formula whose only literals in ℓ are of the form $\oplus^n\ell$.

Definition 10. *A tnnf A is said to be $G\oplus^n\ell$-*completely reducible *if $G\oplus^n\ell \in \alpha$ for $(\alpha,\varepsilon) \in \widehat{\Delta_0}(A)$.*

Theorem 6. *If A be a $G\oplus^n\ell$-completely reducible tnnf, then A is satisfiable if and only if*

$$B[G\oplus^k\ell/\oplus^{k+1}\ell \wedge \ldots \wedge \oplus^n\ell, F\oplus^k\overline{\ell}/\oplus^{k+1}\overline{\ell} \vee \ldots \vee \oplus^n\overline{\ell}]$$

where $B = A[\mathtt{Lit}(\ell,n) \cup G\oplus^n\ell\!\uparrow\!/\top, \mathtt{Lit}(\overline{\ell},n) \cup F\oplus^n\overline{\ell}\!\downarrow\!/\bot]$.

Furthermore, if h is a model of B in t, then the interpretation h' such that $h'(q) = h(q)$ if $q \neq p$ and $h'(p) = h(p) \cup [t+n+2,\infty)$ is a model of A in t.

Example 6. Given the density axiom $A = Fp \to FFp$; the formula $\neg A$ is equivalent to the tnnf $Fp \wedge G\oplus\overline{p}$.

We have that $\Delta_0(Fp \wedge G\oplus\overline{p}) = FpG\oplus\overline{p}$. Note that, as the conjunction of Fp and $G\oplus\overline{p}$ is not a literal, no simplification can be applied. In addition, its $\widehat{\Delta_0}$-set is $\{(FpG\oplus\overline{p},\varepsilon)\}$, thus $\neg A$ is completely reducible.

Now applying Theorem 6, we get that $\neg A$ is satisfiable if and only if $\oplus p$ is satisfiable. Therefore $\neg A$ is satisfiable, a model being $h(\overline{p}) = [2,\infty)$, $h(p) = \{1\}$.

Example 7. Given the formula $A = (Gp \wedge Fq) \to F(p \wedge q)$, we have $\neg A \equiv Gp \wedge Fq \wedge G(\overline{p} \vee \overline{q})$; its Δ-restricted form is

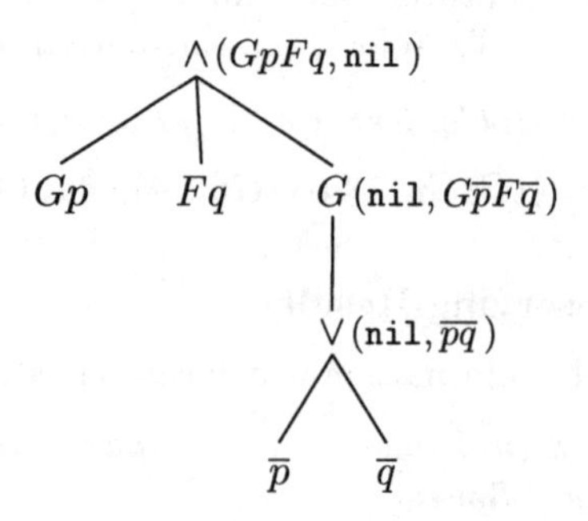

and its $\widehat{\Delta}$-sets are:

$$\widehat{\Delta}_0(A) = \{(GpFq,\varepsilon)\} \qquad \widehat{\Delta}_1(A) = \{(G\overline{p}F\overline{q},3),(\overline{pq},31)\}$$

This formula is completely reducible, by an application of Theorem 6, the leaf in node 1 is deleted, and node 3 is substituted by $G\overline{q}$.

The resulting formula is $Fq \wedge G\overline{q}$, which is 0-conclusive and, therefore, unsatisfiable.

The Pure Literal Rule

The result introduced here is an extension of the well known *pure literal rule* for Classical Propositional Logic. Existing results in the bibliography allow a straightforward extension of the concept of pure literal. Our definition makes use of the $\widehat{\Delta}$-sets, which allow to focus only on those literals which are essential parts of the formula; this is because reducible literals *are not included* in the $\widehat{\Delta}$-sets.

Definition 11. *Let A be a tnnf.*

1. *A classical literal ℓ is said to be $\widehat{\Delta}$-pure in A if a literal $\vartheta\ell$ occurs in $\widehat{\Delta_0}(A) \cup \widehat{\Delta_1}(A)$ and no literal on $\vartheta'\overline{\ell}$ occurs in $\widehat{\Delta_0}(A) \cup \widehat{\Delta_1}(A)$.*
2. *A classical literal ℓ is said to be $\widehat{\Delta}$-k-pure in A if $\oplus^k\ell$ occurs in an $(\alpha,\eta) \in \widehat{\Delta_0}(A)\cup\widehat{\Delta_1}(A)$ with $ord_A(\eta) = 0$, $\oplus^k\overline{\ell}$ does not occur in any $(\alpha,\eta) \in \widehat{\Delta_0}(A)\cup \widehat{\Delta_1}(A)$ with $ord_A(\eta) = 0$, and for any other literal $\vartheta\ell$ or $\vartheta'\ell$, occurring in some element $(\alpha,\eta) \in \widehat{\Delta_0}(A) \cup \widehat{\Delta_1}(A)$, we have $|\vartheta| + ord_A(\eta) > k$.*

Theorem 7. *Let A be a tnnf, ℓ a $\widehat{\Delta}$-pure literal in A, and B the formula obtained from A by the following substitutions*

1. *If $(\alpha,\eta) \in \widehat{\Delta_0}(A)$ with $\vartheta\ell \in \alpha$, then η is substituted by*

$$\begin{cases} \eta[\mathtt{Lit}(\ell,n) \cup G\oplus^n\ell\!\uparrow\!/\top, \mathtt{Lit}(\overline{\ell},n) \cup F\oplus^n\overline{\ell}\!\downarrow\!/\bot] & \text{if } \vartheta\ell = G\oplus^n\ell \\ \eta[\vartheta\ell\!\uparrow\!/\top, \overline{\vartheta\ell}\!\downarrow\!/\bot] & \text{if } \vartheta\ell \in \{GF\ell, FG\ell\} \\ \eta[(\vartheta\ell\!\uparrow)^0/\top, (\overline{\vartheta\ell}\!\downarrow)^0/\bot] & \text{otherwise} \end{cases}$$

2. *If $(\alpha,\eta) \in \widehat{\Delta_1}(A)$ with $\vartheta\ell \in \alpha$, then η is substituted by $\top$.*

Then, A is satisfiable if and only if B is satisfiable. Furthermore, if h is a model of B in t, then the interpretation h' such that $h'(\ell') = h(\ell')$ if $\ell' \neq \ell$ and $h'(\ell) = [t,\infty)$ is a model of A in t.

Theorem 8. *Let A be a tnnf, ℓ a $\widehat{\Delta}$-k-pure literal in A, and B the formula obtained from A by the following substitutions*

1. *If $(\alpha,\eta) \in \widehat{\Delta_0}(A)$ with $\oplus^k\ell \in \alpha$ and $ord_A(\eta) = 0$, then η is substituted by $\eta[(\oplus^k\ell\!\uparrow)^0/\top, (\oplus^k\overline{\ell}\!\downarrow)^0/\bot]$*
2. *If $(\alpha,\eta) \in \widehat{\Delta_1}(A)$ with $\oplus^k\ell \in \alpha$, then η is substituted by $\top$*

Then, A is satisfiable if and only if B is satisfiable. Furthermore, if h is a model of B in t, then the interpretation h' such that $h'(\ell') = h(\ell')$ if $\ell' \neq \ell$ and $h'(\ell) = h(\ell) \cup \{t+k\}$ is a model of A in t.

Example 8. Following with the formula in Example 5, we had

$$\widehat{\Delta}_0(A) = \{(pGq\overline{r},\varepsilon), (FsF\overline{u},5), (s\overline{u},51)\}$$
$$\widehat{\Delta}_1(A) = \{(G\overline{s}Gu,1)(G\overline{s}Gu,14), (\overline{s}u,141)\}$$

therefore

1. It is completely reducible: $Gq \in \alpha$ with $(\alpha, \varepsilon) \in \widehat{\Delta}_0(A)$.
2. literals p and $\overline{r}$ are 0-pure.

When applying the corresponding substitutions we get

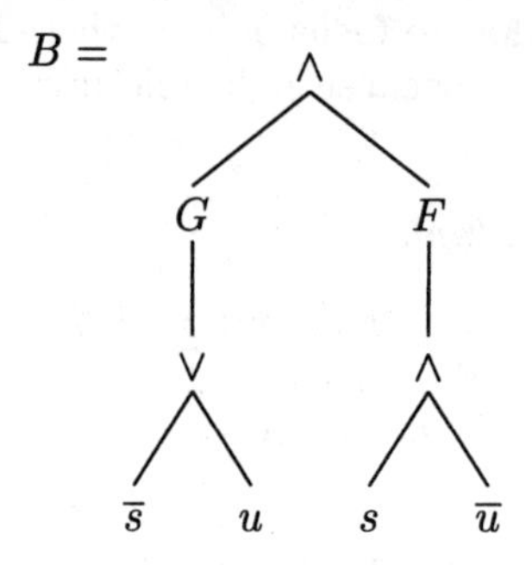

This formula cannot be reduced any longer. By applying a branching rule[4] we obtain

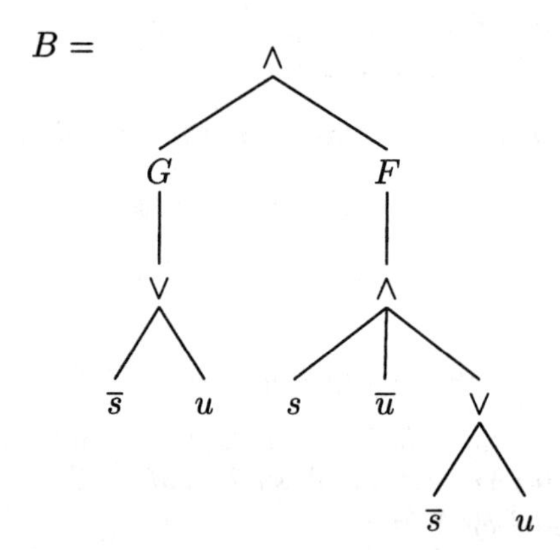

It is easy to check that node 21 is Δ_0-conclusive, by substituting this node by $\perp$ we get $\perp$ as a final result. Therefore the formula is unsatisfiable.

5 Conclusions and Future Work

We have introduced techniques for defining and manipulating lists of unitary implicants/implicates which can improve the performance of a given prover for temporal propositional logics by decreasing the size of the formulas to be branched. These strategies are interesting because can be used in *any* existing theorem prover, specially in non-clausal ones.

As future work, the information in the Δ-lists can be increased by refining the process of generation of temporal implicants/implicates. In addition, current work on G-clauses and F-cubes appears to be a new source of reduction results.

[4] Every prover for linear-time temporal logic has such rules, in the example we use just one of those in the literature.

References

1. G. Aguilera, I. P. de Guzmán, and M. Ojeda. Increasing the efficiency of automated theorem proving. *Journal of Applied Non-Classical Logics*, 5(1):9–29, 1995.
2. R. Ben-Eliyahu and M. Magidor. A temporal logic for proving properties of topologically general executions. *Information and Computation* 124(2):127–144, 1996.
3. C. Courcoubetis and M. Yannakakis. The complexity of probabilistic verification. *Journal of the ACM* 42(4):857–907, 1995.
4. L. Fix and O. Grumberg. Verification of temporal properties. *Journal of Logic and Computation* 6(3):343–362, 1996.
5. Z. Manna and A. Pnueli. *The temporal logic of reactive and concurrent systems: specifications.* Springer-Verlag, 1992.
6. Z. Manna and A. Pnueli. *Temporal verification of reactive systems: Safety.* Springer-Verlag, 1995.
7. A. P. Sistla and E. M. Clarke. The complexity of propositional linear temporal logics. *Journal of the ACM*, 32(3):733-749, 1985.
8. P. Wolper. The tableaux method for temporal logic: an overview. *Logique et Analyse* 28 année, 110-111:119–136, 1985.

A Logic for Anytime Deduction and Anytime Compilation

Frédéric Koriche

LIRMM, UMR 5506, Université Montpellier II CNRS, France
koriche@lirmm.fr

Abstract. One of the main characteristics of reasoning in knowledge based systems is its high computational complexity. Anytime deduction and anytime compilation are two attractive approaches that have been proposed for addressing such a difficulty. The first one offers a compromise between the time complexity needed to compute approximate answers and the quality of these answers. The second one proposes a trade-off between the space complexity of the compiled theory and the number of possible answers it can efficiently process. The purpose of our study is to define a logic which handles these two approaches by incorporating several major features. First, the logic is semantically founded on the notion of resource which captures both the accuracy and the cost of approximation. Second, a stepwise procedure is included for improving approximate answers. Third, both sound approximations and complete ones are covered. Fourth and finally, the reasoning task may be done off-line and compiled theories can be used for answering many queries.

1 Introduction

During these past decades, the problem of reasoning about commonsense knowledge has received a great deal of attention in the artificial intelligence community. A widely accepted framework for studying this issue is the knowledge based system approach [14]. Knowledge is described in some logical formalism, called the *representation language*, and stored in a knowledge base. This component is combined with a reasoning mechanism, which is used to determine whether a given sentence, assumed to capture the query, is entailed from the knowledge base. However, it is well known that deduction is very much demanding from a computational point of view. In particular, if the knowledge base and the query are represented in propositional logic, then checking whether the query is entailed from the knowledge base or not is a coNP-complete problem, that is, a problem which probably requires exponential time to be solved.

Anytime computation is a technique which is used in many areas of artificial intelligence to deal with the computational intractability of problems (see [24] for a recent survey). This paradigm extends the traditional notion of reasoning mechanism by allowing it to return many possible approximate answers to any given query. In the setting of propositional logic, two attractive approaches have recently appeared in the literature: *anytime deduction* and *anytime compilation*.

The goal of the first approach is to define a family of entailment relations that "approximate" classical entailment, by relaxing soundness or completeness of reasoning. The knowledge based system can provide partial solutions even if stopped prematurely; the accuracy of the solution improves with the time used in computing the solution. Hence, anytime deduction offers a compromise between the time complexity needed to compute answers by means of approximate entailment relations and the quality of these answers. Following this idea, Dalal in [5] presents a general technique for approximating deduction problems. The starting point of its framework relies on the entailment relation $\vdash^{\mathrm{BCP}}$ defined using boolean constraint propagation. Based on this relation, the author defines a family of entailment relations $\vdash^{\mathrm{BCP}}_k$, which extend $\vdash^{\mathrm{BCP}}$ by allowing chaining on sentences of size k. Each relation $\vdash^{\mathrm{BCP}}_k$ is sound but incomplete with respect to classical entailment. A different method has been proposed by Cadoli and Schaerf in [2, 17]. Their framework includes a parameter S, a set of atomic propositions, which captures the quality of approximation. Based on this parameter, the authors define two families of entailment relations, named $\vdash^S_1$ and $\vdash^S_3$, which are respectively unsound but complete and sound but incomplete with respect to classical entailment. Several extensions of this framework have been proposed in the domains of non-monotonic logics [3] diagnostic reasoning [22], and reasoning in presence of inconsistency [9, 10].

The second approach is concerned by preprocessing a knowledge base into an appropriate data structure which is used for query answering. The goal here is to invest computational resources in the preprocessing effort which will later substantially speed up query answering, in the expectation that the cost of compilation will be amortized over many queries. Compilation is called "exact" if the data structure is logically equivalent to the initial knowledge base, thus guaranteeing answers to all possible queries (see e.g. [16, 23]). However, in exact compilation it has been observed that the compiled knowledge base often occupies space exponential in the size of the initial source [19]. This undesirable effect has lead several researchers to explore the possibility of compiling the knowledge base into a family of data structures that "approximate" the initial knowledge base, giving up soundness or completeness of reasoning. The system attempts to compile a knowledge base exactly until a given resource limit is reached, and may answer queries before the completion of compilation. One can view anytime compilation as a technique which offers a trade-off between the space complexity of the compiled knowledge base and the number of queries that can be efficiently processed by this data structure. For example, several authors present anytime methods based on prime implicates generation which are sound but incomplete with respect to exact compilation [7, 13, 15]. Dually, Schrag in [18] proposes a prime implicants generation algorithm which is unsound but complete with respect to exact compilation. An analogous strategy has been proposed by Selman and Kautz in the context of "Horn approximation" for computing all the greatest lower bounds (GLB) of a clausal knowledge base [20].

The purpose of the paper is to introduce a unifying, logic oriented framework which captures the main ideas of these two approaches. Our investigation generalizes and expands in several directions previous results by Cadoli and Schaerf in [2, 17]. The framework is based on multi-modal logic which contains a well-founded semantics and a correct and complete axiomatization. Moreover, the framework integrates the following major features :

- The logic is semantically founded on the notion of *resource* which reflects both the accuracy and the computational cost of the approximations.
- The framework enables *incremental reasoning*: the quality of approximations is a nondecreasing function of the resources that have been spent. Hence, approximate answers can be improved and may converge to the right answer.
- The framework covers *dual reasoning*: both sound but incomplete and complete but unsound answers are returned at any step; they respectively correspond to the lower and upper bounds of the range of possible conclusions that approximate the right answer.
- The framework allows *off-line reasoning* : the knowledge base can be compiled and the resulting data structure may be used for efficiently processing a large set of queries.

The formalism we propose is flexible enough to be applied to several anytime reasoning methods. In this study, we concentrate on the specifications of *anytime deducers* and *anytime compilers* which extend the traditional notion of knowledge based systems. Anytime deducers incorporate the first three properties of our framework ; they approximate the reasoning task by iteratively increasing their inference capabilities. Anytime compilers also exploit off-line reasoning by iteratively computing better and better approximations of their knowledge base.

The rest of the paper is organized as follows. Section 2 formally defines the syntax, the semantics, and a sound and complete axiomatization for the logic. Sections 3 and 4 are devoted to the formal specifications of anytime deducers and anytime compilers. Finally, section 5 suggests some topics for future research. The proof of soundness and completeness of the logic is left in Appendix A.

2 The Logic

In this section, we present a propositional logic, named **ARL**, for anytime reasoning. We insist on the fact that the logic is being used here as a specification tool to describe an anytime reasoner rather than as a calculus to be used by one. We begin to define the syntax of the logic, next we examine its semantics in detail, and then we present a sound and complete axiomatization for **ARL**.

2.1 Syntax

In this study, we consider a propositional language constructed from a finite set of atomic propositions (atoms for short) P. In order to formalize the reasoning capabilities of a knowledge based system, we model the notion of deductive inference as an exploration in a space of possibilities. In a propositional setting, this space is defined by the collection of all the interpretations defined from the atoms of P. Following [17], the notion of *resource* is captured by a parameter S, a subset of P, which corresponds to a limited exploration in this space.

The main contribution of this logic relies on two families of modalities $\Box_S$ and $\Diamond_S$, defined for each subset S of P. The operator $\Box_S$ is to capture sound but incomplete inference and $\Diamond_S$ to capture complete but unsound inference. Based on these considerations, the language of **ARL** is defined by the smallest set of *sentences* built from the following rules: if p is an atom of P then p is a sentence, if α is a sentence, then $\neg\alpha$ is a sentence, if α and β are sentences then $\alpha \wedge \beta$ and $\alpha \vee \beta$ are sentences, and finally, if α is a sentence that does not contain any occurrence of the modalities $\Box_S$ and $\Diamond_S$, then $\Box_S\, \alpha$ and $\Diamond_S\, \alpha$ are sentences. We remark that the syntax does not allow nested modal operators. A sentence such as $\Box_S\, \alpha$ is read "the system necessarily infers α given the resources S"; dually $\Diamond_S\, \alpha$ is read "the system possibly infers α given the resources S".

Other connectives $\supset$ and $\equiv$ are defined in terms of $\neg$, $\wedge$ and $\vee$; that is, $\alpha \supset \beta$ is an abbreviation of $\neg\alpha \vee \beta$ and $\alpha \equiv \beta$ is an abbreviation of $(\alpha \supset \beta) \wedge (\beta \supset \alpha)$. A *declaration* is a sentence without any occurrence of modalities $\Box_S$ and $\Diamond_S$, and a *knowledge base* A is a finite conjunction of declarations. When there is no risk of confusion, we shall model knowledge bases as sets of declarations.

2.2 Semantics

The basic building block of the semantics is a domain T of truth values which determines the interpretation of sentences and the properties of logical consequence. In the context of limited reasoning, the four valued semantics first proposed by Belnap [1] and Dunn [8], and notably studied in [12] meets our needs. It is a simple modification of classical interpretation in which sentences take as truth-values subsets of $\{0, 1\}$, instead simply either 0 or 1 alone. So, in the logic **ARL**, sentences can be valued to be true, false, both, or neither.

Based on this structure, we define a *valuation* as a total function v form P to T. The space of valuations generated from P is denoted $\mathcal{V}_P$. A *possible world* is a valuation which maps every atom p of P into $\{1\}$ or $\{0\}$. The space of possible worlds generated from P is denoted $\mathcal{W}_P$. We say that a valuation v is more *specific* than v' and write $v \subseteq v'$, if for any atom $p \in P$, $v(p) \subseteq v(p')$ holds.

The concept of approximation is semantically represented by an equivalence relation between valuations. Given a resource parameter S, we say that two valuations v and v' are *S-equivalent* and write $v \sim_S v'$, if and only if for every atom $p \in P$, if $p \in S$ then $v(p) = v'(p)$. It is easy to prove that $\sim_S$ is indeed a reflexive, symmetric and transitive relation. Intuitively, a relation of S-equivalence induces a partition of the set $\mathcal{V}_P$ into equivalence classes whose granularity captures the accuracy of approximation. When the resource parameter increases, the partition becomes "finer" and the approximation more precise. The "coarsest" partition is obtained when S is the empty set; in this case, $\sim_S$ is the total relation over $\mathcal{V}_P$. Conversely, the "finest" partition is given when S is the set P; in this case $\sim_S$ is the identity relation over $\mathcal{V}_P$.

The figure 1 illustrates a space of valuations and a relation of S-equivalence defined for $P = \{p, q\}$ and $S = \{p\}$. The nodes and the edges represent the valuations and the induced inclusion relation defined from T. The sets of nodes connected by bold edges represent the equivalence classes.

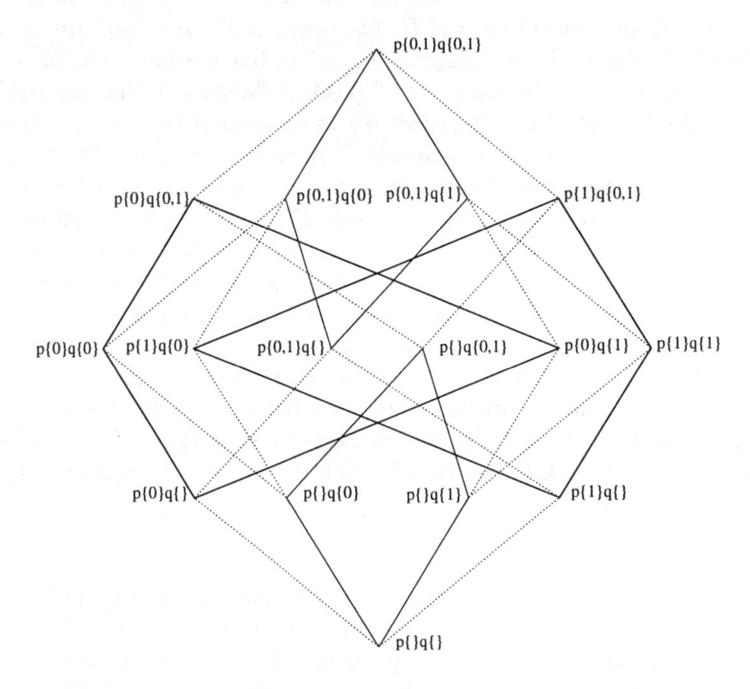

Fig. 1. A space of valuations and a relation of S-equivalence

With these notions in hand, we can now define the semantics of our logic. An *interpretation* of **ARL** consists of a *truth support relation* $\models_1$ and a *falsity support relation* $\models_0$ inductively defined by the following conditions:

$$v \models_1 p \text{ iff } 1 \in v(p), \quad v \models_0 p \text{ iff } 0 \in v(p), \tag{1}$$

$$v \models_1 \neg\alpha \text{ iff } v \models_0 \alpha, \quad v \models_0 \neg\alpha \text{ iff } v \models_1 \alpha, \tag{2}$$

$$v \models_1 \alpha \wedge \beta \text{ iff } v \models_1 \alpha \text{ and } v \models_1 \beta, \quad v \models_0 \alpha \wedge \beta \text{ iff } v \models_0 \alpha \text{ or } v \models_0 \beta, \tag{3}$$

$$v \models_1 \alpha \vee \beta \text{ iff } v \models_1 \alpha \text{ or } v \models_1 \beta, \quad v \models_0 \alpha \vee \beta \text{ iff } v \models_0 \alpha \text{ and } v \models_0 \beta, \tag{4}$$

$$v \models_1 \Box_S\, \alpha \text{ iff } \forall v' \in \mathcal{V}_P,\ \text{if } v \sim_S v' \text{ then } v' \models_1 \alpha, \quad v \models_0 \Box_S\, \alpha \text{ iff } v \not\models_1 \Box_S\, \alpha, \tag{5}$$

$$v \models_1 \Diamond_S\, \alpha \text{ iff } \exists v' \in \mathcal{V}_P \text{ such that } v \sim_S v' \text{ and } v' \models_1 \alpha, \quad v \models_0 \Diamond_S\, \alpha \text{ iff } v \not\models_1 \Diamond_S\, \alpha. \tag{6}$$

A sentence α is called *satisfiable* if and only if there exists a possible world $w \in \mathcal{W}_P$ such that $w \models_1 \alpha$. We say that a sentence α is *valid* and write $\models \alpha$, if and only if, for every possible world $w \in \mathcal{W}_P$, $w \models_1 \alpha$ holds. Finally, given two sentences α and β, we say that β is a *logical consequence* of α if and only if $\models \alpha \supset \beta$ holds. The following lemma captures an important structural property of the support relations. It will be frequently used in the remaining sections.

Lemma 1. *For any declaration α and any pair of valuations v, v' such that $v \subseteq v'$, if $v \models_1 \alpha$ then $v' \models_1 \alpha$, and if $v \models_0 \alpha$ then $v' \models_0 \alpha$.*

Proof. Straightforward by induction on the structure of α.

2.3 Axiomatization

We now focus on obtaining a sound and complete axiomatization for our logic. An *axiom system* consists of a collection of *axioms* and *inferences rules*. A *proof* in an axiom system is a finite sequence of sentences, each of which is either an instance of an axiom or follows by an application of an inference rule. Finally, we say that a sentence α is a *theorem* of the axiom system and write $\vdash \alpha$ if there exists a proof of α in the system. The axiom system of **ARL** is the following :

Axioms:

All tautologies of propositional logic. (A1)

$$\Box_S \neg\neg\alpha \equiv \Box_S \alpha \tag{A2}$$

$$\begin{aligned} &\Box_S (\alpha \wedge \beta) \equiv \Box_S (\beta \wedge \alpha) \\ &\Box_S (\alpha \vee \beta) \equiv \Box_S (\beta \vee \alpha) \end{aligned} \tag{A3}$$

$$\begin{aligned} &\Box_S (\alpha \wedge (\beta \wedge \gamma)) \equiv \Box_S ((\alpha \wedge \beta) \wedge \gamma) \\ &\Box_S (\alpha \vee (\beta \vee \gamma)) \equiv \Box_S ((\alpha \vee \beta) \vee \gamma) \end{aligned} \tag{A4}$$

$$\begin{aligned} &\Box_S (\alpha \wedge (\beta \vee \gamma)) \equiv \Box_S ((\alpha \wedge \beta) \vee (\alpha \wedge \gamma)) \\ &\Box_S (\alpha \vee (\beta \wedge \gamma)) \equiv \Box_S ((\alpha \vee \beta) \wedge (\alpha \vee \gamma)) \end{aligned} \tag{A5}$$

$$\begin{aligned} &\Box_S \neg(\alpha \wedge \beta) \equiv \Box_S (\neg\alpha \vee \neg\beta) \\ &\Box_S \neg(\alpha \vee \beta) \equiv \Box_S (\neg\alpha \wedge \neg\beta) \end{aligned} \tag{A6}$$

$$\begin{aligned} &\Box_S \alpha \;\wedge\; \Box_S \beta \equiv \Box_S (\alpha \wedge \beta) \\ &\Box_S \alpha \;\vee\; \Box_S \beta \equiv \Box_S (\alpha \vee \beta) \end{aligned} \tag{A7}$$

$$\begin{aligned} &\Box_{S \cup \{p\}} (p \vee \neg p) \\ &\Diamond_{S - \{p\}} (p \wedge \neg p) \end{aligned} \tag{A8}$$

$$\Box_S \alpha \supset \alpha \tag{A9}$$

$$\Box_S \alpha \equiv \neg\Diamond_S \neg\alpha \tag{A10}$$

Inference rule:

From $\vdash \alpha$ and $\vdash \alpha \supset \beta$ infer $\vdash \beta$ (R1)

We remark that the axiom (A1) and the inference rule (R1) come from propositional logic; hence the propositional subset of **ARL** is correctly handled. The other axioms capture the properties of the two families of modal operators. From an intuitive point of view, these axioms may be classified into two categories. The first one is concerned by the standards axioms (A2)-(A6) which capture the properties of double negation, commutativity, associativity, distributivity and DeMorgan's laws. The specificity of the logic **ARL** lies in the second category. Axiom (A7) captures the conjunctive and disjunctive properties of operators $\Box_S$. Axiom (A8) is the key point of approximate reasoning. More precisely, if the parameter S is expanded by the atom p, then the system necessarily infers the tautology $p \vee \neg p$. Dually, if S is contracted by p, then the system can infer the antilogy $p \wedge \neg p$. Axiom (A9), often called T, demonstrates that reasoning under the scope of the modality $\Box_S$ is sound. Finally, axiom (A10) captures the duality property between the modal operators $\Box_S$ and $\Diamond_S$.

The following result gives soundness and completeness for the axiom system.

Theorem 1. *For every sentence α of the logic* **ARL**,

$$\vdash \alpha \text{ iff } \models \alpha.$$

Proof. The proof is presented in Appendix A.

3 Anytime Deducers

After an excursion into the logic **ARL**, we now apply our results to the formal specifications of *anytime deducers*. In the context suggested by our approach, we define an anytime deducer as a knowledge based system that approximates the inference process by using an increasing sequence of resource parameters. Intuitively, the more time the system has to evaluate the query, the more resources it can spend. However, anytime deducers are *on-line reasoners*; there is no notion of directly processing the original knowledge base to obtain an approximation to it. This last requirement will be incorporated in the next section.

Following Levesque [11], we specify an anytime deducer as an "abstract type" that interacts with the user through a given set of service routines. To this end, we focus on two core operations, ASK and TELL, which allow a user to query the knowledge base and to add a new information to it. In the following definition, L_P denotes the set of all declarations of the logic **ARL**.

Definition 1. *An* anytime deducer *consists of an operation* TELL *from* $L_P \times L_P$ *to* L_P*, and an operation* ASK *from* $L_P \times 2^P \times L_P$ *to* $\{\texttt{YES}, \texttt{NO}, *\}$ *respectively defined as follows:*

$$\texttt{TELL}(A, \alpha) = A \wedge \alpha,$$

$$\texttt{ASK}(A, S, \alpha) = \begin{cases} \texttt{YES}, & \textit{if } \models \Box_S (A \supset \alpha), \\ \texttt{NO}, & \textit{if } \not\models \Diamond_S (A \supset \alpha), \\ *, & \textit{otherwise.} \end{cases}$$

The basic difference with the standard approach to knowledge representation relies on ASK operation that explicitly includes the notion of resource in order to capture the inference capabilities of the system.

Based on these considerations, the anytime deduction process is defined by an increasing sequence of resource parameters $\langle S_0 = \varnothing \cdots \subset S_i \cdots \subset S_n = P \rangle$ that approximate the set $\mathcal{A} = \{\alpha : \models A \supset \alpha\}$, by means of two dual families of sets $\mathcal{A}_i^{\Box} = \{\alpha : \models \Box_{S_i}(A \supset \alpha)\}$ and $\mathcal{A}_i^{\Diamond} = \{\alpha : \models \Diamond_{S_i}(A \supset \alpha)\}$. If we prove membership in any $\mathcal{A}_i^{\Box}$ then we have proved membership in $\mathcal{A}$. Dually, if we disprove membership in any $\mathcal{A}_i^{\Diamond}$ then we have disproved membership in $\mathcal{A}$. This stepwise process has the important advantage that the iteration may be stopped when a confirming answer is already obtained for a small index i. This yields a potentially drastic reduction of the computational costs. The following properties clarify the interest of our logic in the setting of anytime deduction.

Theorem 2 (Monotonicity). *For any declaration α and any resource parameters S and S' such that $S \subseteq S'$,*

$$\text{if } \models \Box_S \alpha \text{ then } \models \Box_{S'} \alpha, \tag{1}$$

$$\text{if } \not\models \Diamond_S \alpha \text{ then } \not\models \Diamond_{S'} \alpha. \tag{2}$$

Proof. Let us examine part (1). Assume that $\models \Box_S \alpha$ and $\not\models \Box_{S'} \alpha$. In the first case, for any possible world w and any valuation v such that $w \sim_S v$, we have $v \models_1 \alpha$. In the second case, there exists a possible world w' and a valuation v' such that $w' \sim_{S'} v'$ and $v' \not\models_1 \alpha$. Let us define a new valuation v'' such that $\forall p \in S$, $v''(p) = w'(p)$ and $\forall q \notin S$, $v''(q) = \varnothing$. It is clear that $w' \sim_S v''$. Moreover, we have $v'' \subseteq v'$. By application of lemma 1, it follows that $v'' \not\models_1 \alpha$. Therefore, we obtain $w' \not\models_1 \Box_S \alpha$, but this contradicts the hypothesis that $\models \Box_S \alpha$. A dual argument applies to part (2).

Corollary 1 (Convergence). *For any declaration α,*

$$\text{if } \models \alpha \text{ then there exists a resource parameter } S \text{ such that } \models \Box_S \alpha, \tag{1}$$

$$\text{if } \not\models \alpha \text{ then there exists a resource parameter } S \text{ such that } \not\models \Diamond_S \alpha. \tag{2}$$

Theorem 3 (Duality). *For any declaration α,*

$$\models \Box_S \alpha \text{ iff } \Diamond_S \neg\alpha \text{ is unsatisfiable}, \tag{1}$$

$$\not\models \Diamond_S \alpha \text{ iff } \Box_S \neg\alpha \text{ is satisfiable}. \tag{2}$$

Proof. Let us examine part (1). Assume that $\Box_S \alpha$ is valid and that $\Diamond_S \neg\alpha$ is satisfiable. By application of axioms (A10) and (A2), it follows that $\neg\Box_S \alpha$ is satisfiable. Therefore, there exists a possible world w such that $w \models_1 \neg\Box_S \alpha$. So, it follows that $w \models_0 \Box_S \alpha$ and $w \not\models_1 \Box_S \alpha$, but this contradicts the hypothesis that $\Box_S \alpha$ is valid. Dual considerations hold for part (2).

Theorem 4 (Complexity). *For any declaration α and any resource parameter S, there exists an algorithm for deciding whether $\Box_S \alpha$ is satisfiable and $\Diamond_S \alpha$ is satisfiable which runs in $O(|\alpha| \cdot 2^{|S|})$.*

Proof. We begin to show that in a S-equivalence relation, at most one valuation of each equivalence class is needed for the satisfiability test. The sentence $\Box_S \alpha$ is satisfiable if there exists a possible world w such that $w \models_1 \Box_S \alpha$. Therefore, for any valuation v such that $v \sim_S w$, we have $v \models_1 \alpha$. Let us define the valuation $v_\perp$ such that $\forall p \in S$, $v_\perp(p) = w(p)$ and $\forall q \notin S$, $v_\perp(q) = \varnothing$. It is clear that $w \sim_S v_\perp$. By application of lemma 1, if $v_\perp \models_1 \alpha$ then $v \models_1 \alpha$ pour any v such that $w \sim_S v$. Therefore, $w \models_1 \Box_S \alpha$ iff $v_\perp \models_1 \alpha$. Now we turn to the satisfiability test. For each valuation $v_\perp$, the truth value $v_\perp \models_1 \alpha$ can be determined in $O(|\alpha|)$ time. Since there exists $2^{|S|}$ valuations $v_\perp$, checking whether $\Box_S \alpha$ is satisfiable can be done in $O(|\alpha| \cdot 2^{|S|})$ time. A dual argument applies to the sentence $\Diamond_S \alpha$.

Notice that the above complexity result is just the worst case upper bound of an enumeration algorithm. Although such an analysis is important, we do not claim the "brute force" method is the most feasible one. Actually, in the case of clausal theories, we can use a resolution based algorithm which computes the satisfiability of $\Box_{S_i} A$ and $\Diamond_{S_i} A$ by means of an increasing sequence of parameters S_i. For $S_0 = \varnothing$, this respectively corresponds to checking whether A is the empty base and A contains the empty clause. For a given theory A_i which is not empty and does not contain the empty clause, the procedure starts to resolve all clauses in A_i upon the literals p_{i+1} and $\neg p_{i+1}$, next eliminates all clauses in A_i containing these literals, and then checks whether the resulting theory is the empty base or contains the empty clause. It is interesting to remark that such an algorithm is indeed *incremental*; the procedure is able to exploit information gained in previous steps and does not require to perform all computations from scratch.

The correct choice of S is crucial for the usefulness of deduction. Taking to the extreme, when S is chosen incorrectly, anytime deduction may end up as expensive as classical deduction. From this perspective, several heuristics have been proposed in the literature [6, 9, 10, 22]. For example, in a resolution based algorithm, the atoms of S may be dynamically chosen using the *minimal diversity heuristic* advocated in [6]. The diversity of an atom p is the product of the number of positive occurrences by the number of negative occurrences of p in the theory. This notion is based on the observation that an atom can be resolved upon only when it appears both positively and negatively in different clauses. Hence, choosing an atom p whose diversity is minimal will minimize the number of resolvents that can be generated upon p. This heuristic may be augmented with others strategies such as *boolean constraint propagation* and the *minimal width heuristic*. These considerations are illustrated in the following example.

Example 1. Suppose we are given $A = \{(\neg a \vee b \vee c), (a \vee b \vee \neg d), (a \vee \neg b \vee d), (\neg a \vee \neg b \vee c)\}$. We want to prove that A is satisfiable. Hence, we need to find a subset S of $\{a, b, c, d\}$ such that $\Box_S A$ is satisfiable. Starting with $S = \varnothing$ and using the minimal diversity heuristic, we gradually add the atoms b and d to S. This is sufficient for proving that A is satisfiable. Now, we want to show that $a \supset c$ is entailed by A. Hence, we need to find a subset S such that $\Diamond_S (A \wedge a \wedge \neg c)$ is unsatisfiable. Using boolean constraint propagation, we iteratively add to S the atoms a, c and b. This is sufficient for proving that $(A \wedge a \wedge \neg c)$ is unsatisfiable. Therefore, $a \supset c$ is indeed a logical consequence of A.

4 Anytime Compilers

In this section, we extend the concepts developed so far to the formal specifications of *anytime compilers*. Such systems can perform *off-line reasoning*: they approximate the original knowledge base by allowing a sequence of more and more powerful data structures. The quality of compilation depends on the computational resources that have been spent. However, the computational cost of compilation is amortized over a potentially very large set of queries and the resulting data structures can be used in processing each query.

In the setting of knowledge compilation, it is well-known that every knowledge base has two specific normal forms, namely a conjunctive normal form (CNF) and a disjunctive normal form (DNF), from which queries can be efficiently answered. These normal forms are computed by means of the so-called *prime implicates* and *prime implicants*. An attractive property of this approach stems from the fact that the program used to generate the normal forms can be stopped before completion. More specifically, an interruptible process for generating prime implicates is sound but incomplete with respect to exact compilation. On the other hand, an interruptible prime implicant generation process is unsound but complete with respect to exact compilation. Based on these considerations, we present a method for generating prime implicates and prime implicants defined in terms of the logic **ARL**.

To that end, we introduce some additional definitions. A *literal* is an atom or its negation, a *clause* is a finite conjunction of literals and a *term* is a finite disjunction of literals. When there is no risk of confusion, we shall model clauses and terms as sets of literals. A clause γ is called a *S-implicate* of a knowledge base A, if $\models \Box_S (A \supset \gamma)$ and γ does not contain two complementary literals. Dually, a term τ is a *S-implicant* of a A, if $\models \Box_S (\tau \supset A)$ and τ does not contain two complementary literals. A clause γ is called a *prime S-implicate* of A, if γ is a S-implicate of A and for every other S-implicate γ' of A, we have $\gamma' \not\subset \gamma$. In a similar way, a term τ is called a *prime S-implicant* of A, if τ is a S-implicant of A, and for every other S-implicant τ' of A, we have $\tau' \not\subset \tau$.

In the remaining paper, the conjunction of all the prime S-implicates of a knowledge base A is denoted $\text{PIC}(A, S)$ and the disjunction of all the prime S-implicants of A is denoted $\text{PID}(A, S)$. When clear from the context, such sentences will be respectively modeled as sets of clauses and sets of terms. Now, we have the formal tools for specifying anytime compilers.

Definition 2. *An* anytime compiler *consists of an operation* TELL *from* $L_P \times 2^P \times L_P$ *to* $L_P \times L_P$, *and an operation* ASK *from* $L_P \times L_P \times L_P$ *to* $\{\texttt{YES}, \texttt{NO}, *\}$ *respectively defined as follows:*

$$\texttt{TELL}(A, S, \alpha) = \Big(\text{PIC}(A \wedge \alpha, S), \text{PID}(A \wedge \alpha, S)\Big),$$

$$\texttt{ASK}(A_{\text{PIC}}, A_{\text{PID}}, \alpha) = \begin{cases} \texttt{YES}, & \textit{if } \models A_{\text{PIC}} \supset \alpha, \\ \texttt{NO}, & \textit{if } \not\models A_{\text{PID}} \supset \alpha, \\ *, & \textit{otherwise.} \end{cases}$$

The basic idea underlying the above definition is to invoke compilation during the TELL operation. Hence, the resulting data structures A_{PIC} and A_{PID} may be used in the ASK operation in order to answer many queries.

As for deduction, the compilation process may be modeled by an increasing sequence of parameters $\langle S_0 = \varnothing \cdots \subset S_i \cdots \subset S_n = P\rangle$ that approximate the deductive closure of A, denoted $\mathcal{A}$, by means of two dual families of sets $\mathcal{A}_i^{\text{PIC}} = \{\alpha : \models \text{PIC}(A, S_i) \supset \alpha\}$ and $\mathcal{A}_i^{\text{PID}} = \{\alpha : \models \text{PID}(A, S_i) \supset \alpha\}$. For a given index i, if we prove membership in $\mathcal{A}_i^{\text{PIC}}$, then we have also proved membership in $\mathcal{A}$. On the other hand, if we disprove membership in $\mathcal{A}_i^{\text{PID}}$, then we have also disproved membership in $\mathcal{A}$. The reader might wonder at this point if there exists a close relationship between the two previous families $\mathcal{A}_i^{\Box}$ and $\mathcal{A}_i^{\Diamond}$ generated during anytime deduction, and the two families $\mathcal{A}_i^{\text{PIC}}$ and $\mathcal{A}_i^{\text{PID}}$ defined during anytime compilation. In fact, as stated in the following theorem, there is a one to one correspondence between these families.

Theorem 5 (Correspondence). *For any knowledge base A, any declaration α, and any resource parameter S,*

$$\models \text{PIC}(A, S) \supset \alpha \;\text{ iff }\; \models \Box_S (A \supset \alpha), \tag{1}$$

$$\models \text{PID}(A, S) \supset \alpha \;\text{ iff }\; \models \Diamond_S (A \supset \alpha). \tag{2}$$

Proof. We begin to introduce some useful definitions. We denote $\mathcal{V}_S^{\top}$ the set of valuations v such that for every atom $p \in P$, $v(p) = \{0\}$ or $v(p) = \{1\}$ if $p \in S$, and $v(p) = \{0, 1\}$ otherwise. Dually, $\mathcal{V}_S^{\perp}$ denotes the set of valuations v such that for every $p \in P$, $v(p) = \{0\}$ or $v(p) = \{1\}$ if $p \in S$, and $v(p) = \varnothing$ otherwise.

Let us examine part (1). A sufficient condition for proving (1) is to state that for any $v \in \mathcal{V}_S^{\top}$, $v \models_1 A$ iff $v \models_1 \text{PIC}(A, S)$.

- Suppose that there is a $v \in \mathcal{V}_S^{\top}$ such that $v \models_1 A$ and $v \not\models_1 \text{PIC}(A, S)$. In this case, there must exist at least one clause γ in $\text{PIC}(A, S)$ such that $v \not\models_1 \gamma$. Since γ is a prime S-implicate, the sentence $\Diamond_S (A \wedge \neg\gamma)$ is unsatisfiable. So, either $v \not\models_1 \neg\gamma$ holds or $v \not\models_1 A$ holds. In the first case, we would obtain $v \not\models_1 \gamma \vee \neg\gamma$, but this is impossible since from definition of v there must exist at least one possible world $w \subseteq v$ such that $w \models_1 \gamma \vee \neg\gamma$ and by lemma 1, $v \models_1 \gamma \vee \neg\gamma$. So, $v \not\models_1 A$ holds, hence contradiction.
- Now, assume that there is a $v \in \mathcal{V}_S^{\top}$ such that $v \not\models_1 A$ and $v \models_1 \text{PIC}(A, S)$. Thus, for every prime S-implicate γ in $\text{PIC}(A, S)$, we must have $v \models_1 \gamma$. Suppose that A is unsatisfiable. Then it is easy to prove that $\text{PIC}(A, S)$ is either empty or contains the empty clause. In both cases, it follows that $v \not\models_1 \text{PIC}(A, S)$, hence contradiction. Now, suppose that A is satisfiable. Since γ is a prime S-implicate of A, then by theorem 2, it follows that γ is a prime implicate of A. Therefore, for every possible world w such that $w \models_1 A$ there exists at least one literal $l \in \gamma$ such that $w \models_1 l$. If for every $l \in \gamma$, we have $v \models_1 l$, then there exists at least one world $w \subseteq v$ such that $w \models_1 l$ and $w \models_1 A$. By lemma 1, it follows that $v \models_1 A$, hence contradiction. If there exists a $\gamma' \subset \gamma$ such that $v \not\models_1 \gamma'$, then for every possible world w such that $w \models_1 A$, we have $w \models_1 \gamma'$. Since $\Diamond_S (A \wedge \neg\gamma')$ is unsatisfiable, then γ' is a S-implicate of A. Therefore $\gamma \notin \text{PIC}(A, S)$, hence contradiction.

We now turn to part (2). In a dual way, a sufficient condition for proving (2) is to show that for any $v \in \mathcal{V}_S^{\perp}$, $v \models_1 A$ iff $v \models_1 \mathrm{PID}(A, S)$.

- Suppose that there exists a $v \in \mathcal{V}_S^{\perp}$ such that $v \models_1 A$ and $v \not\models_1 \mathrm{PID}(A, S)$. If $v \models_1 A$, then viewing v as a set of literals, we must have $\models v \supset A$. Morevover, from definition of v, it follows that $\models \Box_S (v \supset A)$. Hence, v is a S-implicant of A. It is clear that v cannot be a prime S-implicant, because otherwise we would have $v \models_1 \mathrm{PID}(A, S)$. Therefore, there exists a term $\tau \in \mathrm{PID}(A, S)$ such that $\tau \subset v$ and $\tau \not\models_1 A$. However, by lemma 1, it follows that $v \not\models_1 A$, hence contradiction.
- Now, suppose that there exists a $v \in \mathcal{V}_S^{\perp}$ such that $v \not\models_1 A$ and $v \models_1 \mathrm{PID}(A, S)$. In this case, there exists a term $\tau \in \mathrm{PID}(A, S)$ such that $v \models_1 \tau$, that is, $\tau \subseteq v$. Since $\Box_S (\tau \supset A)$ is valid, we must also have $\tau \models_1 A$. By lemma 1 it follows that $v \models_1 A$, hence contradiction.

The correspondence result above is very interesting because most of the properties stated for anytime deduction also hold in the setting of anytime compilation. As an example, for any parameters S and S' such that $S \subseteq S'$, we can state that $\mathrm{PIC}(A, S')$ is as complete as $\mathrm{PIC}(A, S)$ and that $\mathrm{PID}(A, S')$ is as sound as $\mathrm{PID}(A, S)$. The next result is even stronger than monotonicity; we show that anytime compilation is *incremental*, using information gained in previous steps.

Theorem 6 (Incrementality). *For any knowledge base A, and any resource parameters S and S' such that $S \subseteq S'$,*

$$\mathrm{PIC}(A, S) \subseteq \mathrm{PIC}(A, S'), \tag{1}$$

$$\mathrm{PID}(A, S) \subseteq \mathrm{PID}(A, S'). \tag{2}$$

Proof. Let us demonstrate part (1). Suppose that $\gamma \in \mathrm{PIC}(A, S)$ and $\gamma \notin \mathrm{PIC}(A, S')$. Since $\models \Box_S (A \supset \gamma)$ holds, then by theorem 2, $\models \Box_{S'} (A \supset \gamma)$ also holds. Hence, there must exists a clause γ' such that $\gamma' \subset \gamma$ and $\models \Box_{S'} (A \supset \gamma')$. However, in this case it is clear that $\models \Box_S (A \supset \gamma')$ holds. So $\gamma \notin \mathrm{PIC}(A, S)$, hence contradiction. An analogous strategy applies to part (2).

The complexity result presented below clarifies the interest of the compilation process from a computational point of view. More precisely, we demonstrate that entailment of CNF queries can be computed in time polynomial of the size of resulting data structure plus the size of the query.

Theorem 7 (Complexity). *For any knowledge base A, any resource parameter S and any declaration α in CNF, there exists an algorithm for deciding whether* $\mathrm{PIC}(A, S) \supset \alpha$ *is valid (resp.* $\mathrm{PID}(A, S) \supset \alpha$ *is valid) which runs in* $O(|\mathrm{PIC}(A, S)| + |\alpha|)$ *(resp.* $O(|\mathrm{PID}(A, S)| + |\alpha|)$*).*

Proof. $\models \mathrm{PIC}(A, S) \supset \alpha$ holds iff every non tautological clause γ' of α there is a prime S-implicate γ of $\mathrm{PIC}(A, S)$ such that $\gamma \subseteq \gamma'$. This can be done in $O(|\mathrm{PIC}(A, S)| + |\alpha|)$. On the other hand, $\models \mathrm{PID}(A, S) \supset \alpha$ holds iff every clause γ has a non-empty intersection with every S-implicant τ of $\mathrm{PID}(A, S)$. This can be done in $O(|\mathrm{PID}(A, S)| + |\alpha|)$.

Clearly, the effectiveness of compilation for subsequent query processing depends on the size of its resulting data structures. In the setting suggested by our approach, a knowledge base can have at most $3^{|S|}$ S-implicates and $3^{|S|}$ S-implicants. Using the results by Chandra and Markowsky in [4], the same knowledge base may have on average $3^{|S|}/|S|$ prime S-implicates and $3^{|S|}/\sqrt{|S|}$ prime S-implicants. From this point of view, the interest of anytime compilation is that it has potential to greatly decrease the inconvenience of using exact compilation: since off-line reasoning may be too space demanding, it is clearly desirable to be able to process queries before completion.

Several algorithms can be used to compute prime S-implicates and prime S-implicants. In the first case, we may conceive a stepwise procedure which starts by computing $\mathrm{PIC}(A, S)$ for $S = \varnothing$ and that iteratively increases the value of S. For $S = \varnothing$, this corresponds to checking whether the knowledge base A contains the empty clause or not. For a theory A which does not contain the empty clause, we check each possible implicate γ in turn by adding its negation to the base and then testing $\Diamond_S (A \wedge \neg\gamma)$ for satisfiability. If the sentence is refuted, then γ is a S-implicate of A. By allowing an increasing sequence of parameters S, we can notice already subsumed implicates and not count these in prime S-implicate generation. As far as $\mathrm{PID}(A, S)$ is concerned, dual considerations hold.

As for anytime deduction, the correct choice of the parameter S is important for the usefulness of anytime compilation. This choice may heuristic; in this case, the atoms of S are iteratively selected to minimize the predicted number of generated prime S-implicates and prime S-implicants, using strategies suggested in [6, 18, 19]. Alternatively, the choice of S may be guided by query answering considerations. The letters selected during deduction are used in turn for compilation. In other words, the work done for one query is saved for use in answering the next query. These considerations are illustrated in the following example.

Example 2. We are given $A = \{(a \vee b \vee c), (a \vee b \vee \neg c), (b \vee d \vee \neg e), (b \vee \neg d \vee e)\}$. Suppose, we want to generate at least one prime S-implicate of A. Starting with $S = \varnothing$ and using the minimal diversity heuristic, we gradually add to S the atoms b, a and c. This is sufficient to obtain $\mathrm{PIC}(A, S) = \{a \vee b\}$. Now, suppose that the system is frequently asked CNF queries containing the atom b. In this case, we iteratively add to S the atoms b, d and e. Hence, we obtain $\mathrm{PIC}(A, S) = \{b\}$. Notice that in both scenarios we have $\mathrm{PID}(A, S) = \{b\}$.

5 Conclusion

In this paper, we have dealt with the problem of reasoning in propositional knowledge bases focusing on two attractive approaches, namely, anytime deduction and anytime compilation. The first one offers a compromise between the time complexity needed to compute approximate answers and the quality of these answers. The second one proposes a trade-off between the space complexity of the compiled knowledge base and the number of possible answers that can be efficiently processed by the resulting data structure. Our aim was to provide a unifying, logic oriented framework which handles these two approaches and that enables us to specify anytime reasoners. We have stressed on a sound and complete multi-modal logic, named **ARL**, which generalizes and expands in several directions previous methods concerning approximate deduction [2, 5, 17] and anytime compilation [7, 15, 18]. Based on this logic, we have illustrated that the framework integrates several major features: resource-bounded reasoning, improvability, dual reasoning and off-line processing.

We believe that the results reported here are interesting and worth of further investigations. We outline some of them. A first extension is concerned by the empirical analysis of the parameter S. In particular, it has been found in [21] that random "3-SAT" knowledge bases can be classified in three categories, namely *under-constrained*, *over-constrained* and *critically-constrained*, according to the ratio clauses-to-atoms. An important issue here is to examine the relationship between this ratio and the resource parameter S. This will give an estimation of the number of computational resources (i.e. atoms) required to perform deduction and compilation tasks in each category of knowledge bases. A second extension is to study anytime reasoning in the setting of *pseudo-first-order logic*, which has received a great deal of interest in database theory. These representation languages are defined from a finite domain of discourse without function symbols. However, although every first order knowledge base can be replaced by an equivalent propositional theory, the size of the theory may be exponentially larger than the initial base. Hence, formal extensions of our logic should be done in order to control such a source of complexity. A third possible extension is to consider anytime reasoning in a nonmonotonic setting. As an example, in a multi-agent system, a reasoner is often confronted with uncertain and inconsistent information [9, 10]. In such circumstances, it is necessary to incorporate conflict resolution methods and preference orderings which involve additional sources of complexity. Important issues such as *anytime nonmonotonic reasoning* and *anytime recompilation* should play a key role in this setting.

References

1. N. D. Belnap. A useful four-valued logic. In J. M. Dunn and G. Epstein, editors, *Modern Uses of Multiple-Valued Logic*, pages 8–37. Reidel, Dordrecht, 1977.
2. M. Cadoli. *Tractable reasoning in artificial intelligence*, volume 941 of *Lecture Notes in Artificial Intelligence*. Springer-Verlag Inc., New York, NY, USA, 1995.
3. M. Cadoli and M. Schaerf. Approximate inference in default reasoning and circumscription. In B. Neumann, editor, *Proceedings 10th European Conference on Artificial Intelligence*, pages 319–323, Vienna, Austria, 1992. John Wiley & Sons.
4. A. Chandra and G. Markowsky. On the number of prime implicants. *Discrete Mathematics*, 24:7–11, 1978.
5. M. Dalal. Semantics of an anytime family of reasoners. In W. Wahlster, editor, *Proceedings of the 12th European Conference on Artificial Intelligence (ECAI-96)*, pages 360–364. Wiley & Sons, 1996.
6. R. Dechter and I. Rish. Directional resolution: The davis-putnam procedure, revisited. In P. Torasso, J. Doyle, and E. Sandewall, editors, *Proceedings of the 4th International Conference on Principles of Knowledge Representation and Reasoning*, pages 134–145, Bonn, Germany, 1994. Morgan Kaufmann.

7. A. del Val. Tractable databases: How to make propositional unit resolution complete through compilation. In J. Doyle, E. Sandewall, and P. Torasso, editors, *Proceedings of the 4th International Conference on Principles of Knowledge Representation and Reasoning*, pages 551–561, Bonn, 1994. Morgan Kaufmann.
8. J. M. Dunn. Intuitive semantics for first-degree entailments and coupled trees. *Philosophical Studies*, 29:149–168, 1976.
9. F. Koriche. Approximate reasoning about combined knowledge. In M. P. Singh, A. S. Rao, and M. J. Wooldridge, editors, *Intelligent Agents Volume IV*, volume 1365 of *Lecture Notes in Artificial Intelligence*, pages 259–273. Springer Verlag, Berlin, Germany, 1998.
10. F. Koriche. *Approximate Reasoning in Cooperating Knowledge Based Systems.* PhD thesis, Université de Monptellier II, Montpellier, France, January 1998.
11. H. J. Levesque. Foundations of a Functional Approach to Knowledge Representation. *Artificial Intelligence*, 23:125–212, 1984.
12. H. J. Levesque. A logic of implicit and explicit belief. In R. J. Brachman, editor, *Proceedings of the 6th National Conference on Artificial Intelligence*, pages 198–202, Austin, Texas, 1984. William Kaufmann.
13. P. Marquis. Knowledge compilation using theory prime implicates. In *Proceedings of the 14th International Joint Conference on Artificial Intelligence*, volume 2, pages 837–843, Toronto, Canada, 1995. Morgan Kaufmann.
14. J. McCarthy. Programs with common sense. In *Proceedings of the Symposium on Mechanisation of Thought Processes*, volume 1, pages 77–84, London, 1958. Her Majesty's Stationery Office.
15. T. Ngair. A new algorithm for incremental prime implicate generation. In *Proceedings 13th International Joint Conference on Artificial Intelligence*, volume 1, pages 46–51, Chambéry, France, aug 1993. Morgan Kaufmann.
16. R. Reiter and J. de Kleer. Foundations of assumption-based truth maintenance systems: Preliminary report. In K. Forbus and H. Shrobe, editors, *Proceedings of the 6th National Conference on Artificial Intelligence*, pages 183–189, Seattle, (WA) USA, July 1987. Morgan Kaufmann.
17. M. Schaerf and M. Cadoli. Tractable reasoning via approximation. *Artificial Intelligence*, 74:249–310, 1995.
18. R. Schrag. Compilation for critically constrained knowledge bases. In *Proceedings of the Thirteenth National Conference on Artificial Intelligence and the Eighth Innovative Applications of Artificial Intelligence Conference*, pages 510–516, Menlo Park (CA) USA, August 1996. AAAI Press / MIT Press.
19. R. Schrag and J. Crawford. Implicates and prime implicates in random 3-SAT. *Artificial Intelligence*, 81(1-2):199–222, 1996.
20. B. Selman and H. Kautz. Knowledge compilation and theory approximation. *Journal of the ACM*, 43(2):193–224, 1996.
21. B. Selman, D. Mitchell, and H. Levesque. Generating hard satisfiability problems. *Artificial Intelligence*, 81:17–30, 1996.
22. A. ten Teije and F. van Harmelen. Computing approximate diagnoses by using approximate entailment. In L. C. Aiello, J. Doyle, and S. Shapiro, editors, *Proceedings of the Fifth International Conference on Principles of Knowledge Representation and Reasoning*, pages 256–267, San Francisco CA, USA, 1996. Morgan Kaufmann.
23. P. Tison. Generalized consensus theory and application to the minimization of boolean functions. *IEEE Transactions on electronic computers*, 4:446–456, 1967.
24. S. Zilberstein. Using anytime algorithms in intelligent systems. *AI Magazine*, 17(3):73–83, 1996.

Appendix A. Proof of Soundness and Completeness

The following results give soundness and completeness for the axiom system.

Theorem 8 (Soundness). *For every sentence* α *of the logic* **ARL**,

$$\text{if } \vdash \alpha \text{ then } \models \alpha.$$

Proof. It is easy to see that axioms (A1)-(A6) are sound, and that the inference rule (R1) preserves validity. Let us examine the other axioms.

- Axiom (A7). The only nontrivial case is the soundness of $\Box_S(\alpha \vee \beta) \supset \Box_S \alpha \vee \Box_S \beta$. Suppose that this sentence is not valid. Then there exists a possible world w such that $w \models_1 \Box_S(\alpha \vee \beta)$ and $w \not\models_1 \Box_S \alpha \vee \Box_S \beta$. In the first case, for every valuation v such that $v \sim_S w$, we have $v \models_1 \alpha \vee \beta$. In the second case, there exists two valuations v', v'' such that $v' \sim_S w$, $v'' \sim_S w$, $v' \not\models_1 \alpha$, and $v'' \not\models_1 \beta$. Let us define a new valuation u, such that $u(p) = w(p)$ for every $p \in S$, and $u(q) = \varnothing$ for every $q \notin S$. It is clear that w, v, v', v'' and u belong to the same equivalence class defined for $\sim_S$. Moreover, $u \subseteq v'$ and $u \subseteq v''$. By lemma 1, we obtain $u \not\models_1 \alpha$ and $u \not\models_1 \beta$. So $u \not\models_1 \alpha \vee \beta$ and therefore $w \not\models_1 \Box_S(\alpha \vee \beta)$, hence contradiction.
- Axiom (A8). Let us examine the first sentence. Suppose that $\Box_{S\cup\{p\}}(p \vee \neg p)$ is not valid. Then there must exist a possible world w such that $w \not\models_1 \Box_{S\cup\{p\}}(p \vee \neg p)$. In this case, there exists a valuation v such that $w \sim_{S\cup\{p\}} v$ and $v \not\models_1 p \vee \neg p$. However, since w is a possible world, we have $1 \in w(p)$ or $0 \in w(p)$. Moreover, since $p \in (S \cup \{p\})$ it follows that $1 \in v(p)$ or $0 \in v(p)$. So $v \models_1 p \vee \neg p$, hence contradiction. We now turn to the second sentence. $\Diamond_{S-\{p\}}(p \wedge \neg p)$ is valid iff for every possible world w there exists a valuation v such that $w \sim_{S-\{p\}} v$ and $v \models_1 p \wedge \neg p$. Let us define a new valuation v' such that $v'(p) = \{0,1\}$ and for every $q \in (S - \{p\})$, $v'(q) = w(q)$. It is clear that $v' \models_1 p \wedge \neg p$. Moreover, for every possible world w, we have $w \sim_{S-\{p\}} v$. Therefore, $\Diamond_{S-\{p\}}(p \wedge \neg p)$ is valid.
- Axiom (A9). Suppose that $\Box_S \alpha \supset \alpha$ is not valid. Then there exists a possible world w such that $w \models_1 \Box_S \alpha$ and $w \not\models_1 \alpha$. If $w \models_1 \Box_S \alpha$ then for every valuation v such that $w \sim_S v$, we have $v \models_1 \alpha$. Since $\sim_S$ is reflexive, it follows that $w \models_1 \alpha$, hence contradiction.
- Axiom (A10). Let us examine the first implication (if). Suppose that $\Box_S \alpha \supset \neg\Diamond_S \neg\alpha$ is not valid. Then there exists a possible world w such that $w \models_1 \Box_S \alpha$ and $w \not\models_1 \neg\Diamond_S \neg\alpha$. In the first case, for any valuation v such that $v \sim_S w$, we have $v \models_1 \alpha$, while in the second case there exists a valuation v' such that $v' \sim_S w$ and $v' \models_0 \alpha$. It follows that $v' \models_1 \alpha \wedge \neg\alpha$. Let us define a new valuation v'' such that for every $p \in S$, $v''(p) = v'(p)$ if $p \in S$, and for every $q \notin S$, if $v'(q) = \varnothing$ then $v''(q) = \{0,1\}$ and if $v'(q) = \{0,1\}$ then $v''(q) = \varnothing$. It is easy to show by induction on the structure of α that if $v' \models_1 \alpha \wedge \neg\alpha$ then $v'' \not\models_1 \alpha \vee \neg\alpha$. Moreover, it is clear that $v' \sim_S v''$. Therefore, it follows that $w \not\models_1 \Box_S \alpha$, hence contradiction. An analogous strategy applies to the second implication (only if).

Theorem 9 (Completeness). *For every sentence α of the logic* **ARL**,

$$\text{if } \models \alpha \text{ then } \vdash \alpha.$$

Proof. We divide the proof into three steps. In the first step, we introduce the notion of *maximal consistent set* and we examine several properties of these structures. In the second step, we show that every sentence of the language has two important normal forms, namely an *extended conjunctive normal form* (ECNF) and an *extended disjunctive normal form* (EDNF) which are useful for the satisfiability test. Finally, in the third step, we prove that every sentence in a maximal consistent set is satisfiable. We conlude the proof by showing that this result is sufficient to state completeness.

We start the first step by to giving some definitions. A sentence α is *consistent* if its negation is not theorem (i.e. $\not\vdash \neg\alpha$). A finite set of sentences is consistent exactly if the conjunction of its sentences is consistent, and an infinite set of sentences is consistent exactly if all of its finite subsets are consistent. A sentence or a set of sentences is said to be inconsistent exactly if it is not consistent. Finally, a set A of sentences is called *maximal consistent* if it is consistent and for any sentence α of **ARL**, if $\alpha \notin A$ then $A \cup \{\alpha\}$ is inconsistent.

Lemma 2. *In the logic* **ARL**, *Every consistent set of sentences can be extended to a maximal consistent set of sentences. In addition, if A is a maximal consistent set, then it satisfies the following properties:*

$$\text{for every sentence } \alpha, \text{ exactly one of } \alpha \text{ and } \neg\alpha \text{ is in } A, \quad (1)$$
$$\alpha \wedge \beta \in A \text{ iff } \alpha \in A \text{ and } \beta \in A, \quad (2)$$
$$\alpha \vee \beta \in A \text{ iff } \alpha \in A \text{ or } \beta \in A, \quad (3)$$
$$\text{if} \vdash \alpha \text{ then } \alpha \in A, \quad (4)$$
$$\text{if } \alpha \in A \text{ and} \vdash \alpha \supset \beta \text{ then } \beta \in A. \quad (5)$$

Proof. Straightforward, by using standard techniques of propositional logic.

We now turn to the second step. We first recall that a *literal* is an atom or its negation, a *clause* is a finite conjunction of literals and a *term* is a finite disjunction of literals. A declaration in *conjunctive normal form* (CNF) is a conjunction of clauses and a declaration in *disjunctive normal form* (DNF) is a disjunction of terms. A sentence α is said to be in *extended conjunctive normal form* (ECNF) if, treating subformulas of the form $\Box_S \beta$ and $\Diamond_S \beta$ as atoms, α is in CNF, and for each subformula $\Box_S \beta$ and $\Diamond_S \beta$ of α, β is in CNF. The notion of *extended disjunctive normal form* (EDNF) is defined in a dual way.

Lemma 3. *For every sentence α of the logic* **ARL**,

$$\vdash \alpha \equiv \alpha_{\text{ECNF}} \text{ and } \vdash \alpha \equiv \alpha_{\text{EDNF}}.$$

Proof. By repeated applications of axioms (A2)-(A6) and rule (R1).

Now we have proved the two preparatory lemmas, we turn to the third step. We demonstrate that every sentence in a maximal consistent set is satisfiable in the semantics of our logic. This property may be formalized as follows.

Lemma 4. *If A is maximal consistent set of sentences of* **ARL**, *then there exists a possible world $w_A \in \mathcal{W}_P$ such that:*

$$\forall \alpha \in \mathbf{ARL},\ \alpha \in A \text{ iff } w_A \models_1 \alpha.$$

Proof. Let w_A be defined in the following way: for every atom $p \in P$, $w_A \models_1 p$ iff $p \in A$. We prove the lemma by induction on the structure of α. If α is an atom p, this is immediate from the definition of w_A. The cases where α is a negation, a conjunction or a disjunction follow easily from parts (1), (2) and (3) of lemma 2. In the proof below, we concentrate on the case where α is of the form $\Box_S \beta$. The final case where α is of the form $\Diamond_S \beta$ directly follows from axiom (A10).

- (if direction). Suppose that $\Box_S \beta \in A$ and that $w_A \not\models_1 \Box_S \beta$. From the first assumption and by lemma 3, we have $\Box_S \beta_{\mathrm{ECNF}} \in A$. So, by axiom (A7) and part (2) of lemma 2, for every clause γ of β_{ECNF}, we must have $\Box_S \gamma \in A$. Moreover, by axiom (A7) and part (3) of lemma 2, there exists at least one literal l of γ such that $\Box_S l \in A$. From the second assumption, we obtain $w_A \not\models_1 \Box_S \beta_{\mathrm{ECNF}}$. Therefore, there must exist a clause γ' of β_{ECNF}, such that for every $l' \in \gamma'$, we have $w_A \not\models_1 \Box_S l'$. If $l' \in S$, then $w_A \not\models_1 l'$ and by the induction hypothesis we have $l' \notin A$. By axiom (A9) we obtain $\Box_S l' \notin A$, hence contradiction. If $l' \notin S$, by axiom (A8) we obtain $\Diamond_S (l' \wedge \neg l') \in A$, and by axiom (A10) we have $\neg\Box_S \neg(l' \wedge \neg l') \in A$. By application of axioms (A2),(A3) and (A6), it follows that $\neg\Box_S (l' \vee \neg l') \in A$. From part (1) of lemma 2 we first obtain $\Box_S (l' \vee \neg l') \notin A$ and by axiom (A7) it follows that $\Box_S l' \vee \Box_S \neg l' \notin A$. Finally, from part (3) of lemma 2, we obtain $\Box_S l' \notin A$, hence contradiction.
- (only if direction). Suppose that $w_A \models_1 \Box_S \beta$ and that $\Box_S \beta \notin A$. From the first assumption, we obtain $w_A \models_1 \Box_S \beta_{\mathrm{ECNF}}$. So, for every clause γ of β_{ECNF}, there exists a literal l such that its atom is an element of S and $w_A \models_1 l$. However, by axiom (A8) we have $\Box_S (l \vee \neg l) \in A$. By axiom (A7), it follows that $\Box_S l \vee \Box_S \neg l \in A$, and from part (3) of lemma 3 we have $\Box_S l \in A$ or $\Box_S \neg l \in A$. Since $w_A \models_1 l$, by the induction hypothesis we have $l \in A$. By axiom (A9) we must obtain $\Box_S \neg l \notin A$, because otherwise we would have $\neg l \in A$. Therefore we have $\Box_S l \in A$, and by axiom (A7) it follows that $\Box_S \gamma \in A$. A similar method applies to each clause of β_{ECNF}. Therefore, by axiom (A7), it follows that $\Box_S \beta_{\mathrm{ECNF}} \in A$. Finally, by lemma 3 we obtain $\Box_S \beta \in A$, hence contradiction.

Now, we have all the results in hand for concluding the proof. By lemma 2, we know that if a sentence is consistent, then it is a member of a maximal consistent set. But in lemma 4, we have shown that if a sentence belongs to such a set, then it is satisfiable. So, we have proved that every consistent sentence is satisfiable. This is sufficient to conclude that every valid sentence is provable.

On Knowledge, Strings, and Paradoxes

Manfred Kerber

School of Computer Science, The University of Birmingham, England
M.Kerber@cs.bham.ac.uk

Abstract. A powerful syntactic theory as well as expressive modal logics have to deal with self-referentiality. Self-referentiality and paradoxes seem to be close neighbours and depending on the logical system, they have devastating consequences, since they introduce contradictions and trivialise the logical system. There is a large amount of different attempts to tackle these problems. Some of them are compared in this paper, futhermore a simple approach based on a three-valued logic is advocated. In this approach paradoxes may occur and are treated formally. However, it is necessary to be very careful, otherwise a system built on such an attempt trivialises as well. In order to be able to formally deal with such a system, the reason for self-referential paradoxes is studied in more detail and a semantical condition on the connectives is given such that paradoxes are excluded.

Keywords: Knowledge representation, self-referentiality, paradoxe, Kleene logic

Ich habe manche Zeit damit verloren,
Denn ein vollkommner Widerspruch
Bleibt gleich geheimnisvoll für Kluge wie für Toren. ...
Gewöhnlich glaubt der Mensch, wenn er nur Worte hört,
Es müsse sich dabei doch auch was denken lassen.

Johann Wolfgang von Goethe, Faust I

1 Introduction

The symbolic representation of knowledge and belief plays a crucial rôle in artificial intelligence, in particular in multi-agent systems. For their representation two principally different formal systems, modal logic and meta-systems have been developed. Both of them come with a detailed ramification of realisations depending on what is actually adequate and what is not.

The predominant formalism for representing knowledge and belief in agent systems is based on modal logics (there is a vast amount of literature on modal logic. I just point to Fitting's general introduction, where further pointers can be found, see [4, p.365–448]). However, meta-systems have some crucial advantages over modal logic, although some problems as well. As Davis points out in [2, p.77] the difference between the two approaches is pretty much the same as the difference between direct quotation ("John said, 'I am hungry.'"), corresponding to meta-systems, and indirect quotation ("John said that he was hungry."), corresponding to modal logic. Davis stresses that direct speech is more powerful, since in particular non-syntactic expressions can be expressed directly ("John said 'Arglebargle glumph'"), but not indirectly. A similar property holds for modal logic versus meta-logic. In modal logic an agent knows/believes all tautologies and as well as the deductive closure of all its knowledge/beliefs. In meta-systems, however, it is easily possible to do away with this unrealistic property. On the other hand, meta-systems allow paradoxical sentences such as the liar sentence ("This sentence is false") to be represented. This seems to be a crucial disadvantage. However, powerful modal logics are faced with the same problem [15].

In this paper, we shall see in more detail, why paradoxes occur in meta-systems and what different ways out of the problems exist. Essentially I shall advocate a system, in which paradoxes are taken seriously and formulae like the liar sentence are admitted. They are neither true nor false, but paradoxical. In order to do so, a third truth value is added. As we shall see, just adding such a truth value does not solve the problem, but the system has to fulfill an additional constraint.

2 Meta-Logic and Syntactic Theory

A reflective meta-systems is a system in which it is possible to speak about the system itself. Frege's original system (see [18]) was of that kind. Russell's discovery in the beginning of the 20th century that self-referentiality, in form of the "set of all sets that do not contain themselves", leads to paradoxes, jeopardised the whole enterprise of formalising logic. After a few years Russell himself came up with a solution – type theory – that bypasses the unwanted phenomena by syntactically forbidding them [16]. This approach is generally accepted in mathematics (and adequate for mathematics). Similar paradoxes occur in syntactic theory as well (see Grelling and Nelson [8]). If we define, for instance, a predicate long, then there is a difference in stating long(John) and long("John"), the first expression refers to a person called John, the second to a string, consisting of four letters. We can define now a new predicate heterogeneous as $\forall x.\,\mathsf{heterogeneous}(``x") \leftrightarrow \neg x(``x")$. For instance, if we define long on strings that a string has to have at least eight characters in order to be long, we have heterogeneous("long"), since ¬long("long") holds. The paradox occurs when we check, whether heterogeneous is heterogeneous or not. By definition we get: $\mathsf{heterogeneous}(``\mathsf{heterogeneous}") \leftrightarrow \neg\mathsf{heterogeneous}(``\mathsf{heterogeneous}")$.

Of course, we can argue whether the definition of heterogeneous forms really a proper definition, since it has the form $\forall x.\,\mathsf{heterogeneous}(``x") \leftrightarrow \ldots$ rather than $\forall x.\,\mathsf{heterogeneous}(x) \leftrightarrow \ldots$. Excluding this extended form of definitions (and excluding expressions of the form $x(x)$ by some kind of types) does away with self-referentiality and the paradoxes disappear. It is still possible to define predicates such as long (e.g., as $\forall x.\,\mathsf{long}(x) \leftrightarrow \mathsf{length}(x) \geq 8$), but not to define problematic predicates like heterogeneous.

However, while the predicate heterogeneous is not crucial for a meta-system, the True predicate is, and this is excluded as well. Tarski [17] gave a famous definition of truth:

$$\mathsf{True}(\text{``}\mathbf{A}\text{''}) \leftrightarrow \mathbf{A}$$

as axiom schema for arbitrary formulae **A**. While this definition seems to be obvious and to form a minimal requirement to relate strings and the objects they stand for, it already is a source of problems: The famous liar sentence "This sentence is false" can be expressed in syntactic theory as $\mathbf{L} :\equiv \neg\mathsf{True}(\text{``}\mathbf{L}\text{''})$. Note that there are no quotes around the definiendum, hence unlike the definition of heterogeneous, the definition of **L** is much more of the form we would accept as a proper definition. With Tarski's definition of truth, this definition unfolds to $\mathbf{L} \leftrightarrow \neg\mathsf{True}(\text{``}\mathbf{L}\text{''}) \leftrightarrow \neg\mathbf{L}$, that is, $\mathbf{L} \leftrightarrow \neg\mathbf{L}$. This expression is inconsistent with interpreting **L** as true as well as interpreting it as false. There are different ways out of this fundamental problem. Three of them are discussed in more detail in Section 4. First we shall take a closer look at the formalisation of the logic.

3 The Language

The language we want to investigate is a standard first-order language plus strings as terms. We assume *variables* (usually denoted by x, y, and z), *connectives* $\neg$, $\wedge$, $\vee$, $\rightarrow$, and $\leftrightarrow$, the *quantifiers* $\forall$ and $\exists$ as well as *auxiliary symbols*), (and , as usually and a signature that consists of a set *predicate symbols* (denoted by P, Q, and R, e.g.) and a set of *function symbols* (denoted by f, g, and h, e.g.), all with arities. Nullary function symbols are called constants. Terms are recursively defined as variables, constants or the application of n-ary function symbols to n terms. Formulae are either atomic formulae built by the application of an n-ary predicate symbol to n terms or formulae composed by connectives and quantifiers.

In addition to this standard setting, we use strings in the language. Strings are tuples of characters. Following the description in [2] we prefix characters by a colon, e.g., :C is the character C, the string "Cat" is syntactic sugar for tuple(:C, :a, :t). It is possible to define *syntactic* predicate symbols such as is_term and is_formula and syntactic function symbols such as conc for concatenation. Details of such a construction can be found in [5, Chapter 10]. In addition there are *semantic* predicates such as True. We shall discuss this predicate symbol in detail in the sequel. Self-referentiality occurs in this language, since it is possible to formally define the liar sentence as $\mathbf{L} :\equiv \neg\mathsf{True}(\text{``}\mathbf{L}\text{''})$. Strings as "L" are considered as ordinary ground terms in the language. They do not cause any problem as long as we do not look into them and relate them to the objects they stand for.

4 Ways Out?

In this section we distinguish three different ways how to deal with self-referential paradoxes: Firstly, forbid self-referentiality, secondly abandon or weaken the definition of truth, and thirdly change the underlying logical system.

First approach

This approach corresponds pretty much to Russell's solution of the problem, namely to syntactically exclude any self-referentiality, may it be paradoxical or not. Russell introduced types such that in any expression of the form $P(Q)$, P has to have a higher type than Q. In particular P can't express anything about itself or about anything that can make statements about P. This approach has been adapted by Tarski to meta-systems. It is formalised and implemented in the multi-level meta-systems FOL [19] and Getfol [7][1]. These systems consist of different instances of first-order logic, an object system (standard classical first-order logic) and a meta-system in which the terms stand for syntactical expressions (like terms and formulae) of the object systems[2]. The relation between the meta-level and the object level is established by so-called bridge rules. They state, whenever in the meta-system $\mathsf{True}(\text{“}\mathbf{A}\text{”})$ has been derived, it is possible to derive $\mathbf{A}$ on the object level by a rule called reflection-down, and vice versa, when $\mathbf{A}$ or $\neg\mathbf{A}$ is derived on the object level, it is possible to assume $\mathsf{True}(\text{“}\mathbf{A}\text{”})$ or $\neg\mathsf{True}(\text{“}\mathbf{A}\text{”})$ on the meta-level.

Let us exemplify this in showing how a simple meta-theorem, namely

$$\forall \mathbf{A}.\, \mathsf{True}(\mathbf{A}) \to \neg\mathsf{True}(\mathsf{conc}(\text{“}\neg\text{”}, \mathbf{A}))$$

can be proved in such a system.
conc stands for concatenation of strings. We use O to indicate the object level and M for the meta-level.

Proof:

$M\,1\;1 \vdash \mathsf{True}(\text{“}\mathbf{A}_0\text{”})$	(Ass)
$O\,2\;1 \vdash \mathbf{A}_0$	$(\mathcal{R}_{dn}\ 1)$
$O\,3\;1 \vdash \neg\neg\mathbf{A}_0$	$(\neg\neg\ 2)$
$M\,4\;1 \vdash \neg\mathsf{True}(\text{“}\neg\mathbf{A}_0\text{”})$	$(\neg\mathcal{R}_{up}\ 3)$
$M\,5\;\;\; \vdash \mathsf{True}(\text{“}\mathbf{A}_0\text{”}) \to \neg\mathsf{True}(\text{“}\neg\mathbf{A}_0\text{”})$	$(\to I\ 1,4)$
$M\,6\;\;\; \vdash \forall \mathbf{A}.\,\mathsf{True}(\mathbf{A}) \to \neg\mathsf{True}(\mathsf{conc}(\text{“}\neg\text{”}, \mathbf{A}))$	$(\forall I\ 5)$

That is, from the assumption $\mathsf{True}(\text{“}\mathbf{A}_0\text{”})$ on the meta-level, it is possible to reflect down to $\mathbf{A}_0$ on the object level. $\neg\neg\mathbf{A}_0$ is derived on the object level and reflected up to the meta-level formula $\neg\mathsf{True}(\text{“}\neg\mathbf{A}_0\text{”})$. The rest is standard reasoning on the meta-level, that is, implication introduction (from lines 1 and 4) and forall introduction (from line 5).

[1] An alternative to totally forbidding self-reference is restricting it, e.g., restrict expression like $\{x \mid x \notin x\}$ to $\{x \in A \mid x \notin x\}$. This approach is discussed in detail in [1].

[2] Actually the systems are extended to allow different meta-levels, in particular a meta-meta-level and so on.

Self-referentiality is not possible, the meta-system speaks about the object system, but never about itself. Tarski's definition of truth is built in the reflection rules ($\mathcal{R}_{up}$,$\mathcal{R}_{dn}$,and $\neg\mathcal{R}_{up}$). The liar sentence as such is not representable in that system, useful sentences are, however, excluded as well. It is not possible to extend the system in a way that two agents can make utterances about each other. For instance, A says "Everything B says is right." and B says "Everything A says is right." is not expressible. If you allow extensions of this kind, you introduce the paradoxes again in sentences such as: A says "Everything B says is wrong." and B says "Everything A says is right."

In order to show that such situations are not artificial Kripke [13] gave an interesting example that did actually occur at the time of the Watergate affair:

- Nixon says "Everything Jones says about Watergate is true."
- Jones says "Most of Nixon's assertions about Watergate are false."

You get a paradox when you assume Nixon made $2n + 1$ assertions about Watergate, n of those are definitively true, n are definitively false, plus the one above.

Second approach

In the second approach Tarski's definition of truth is changed. If you give it fully up, the liar sentence does not produce any problem any more. However, this solution is far too radical, since in such a case meta-level and object level co-existed without any relation between them. Perlis defined for that reason a weakened form of the definition of truth [14], $\mathsf{True}(``\mathbf{A}") \leftrightarrow (\mathbf{A})*$, where the $*$-operator replaces each connective occurrence of the form $\neg\mathsf{True}(``\ldots")$ in $\mathbf{A}$ by $\mathsf{True}(``\neg(\ldots)")$. Using Kripke's fixpoint semantics he can build a system in which the liar sentence, $\mathbf{L} :\equiv \neg\mathsf{True}(``\mathbf{L}")$, is not paradoxical. However, since $\mathbf{L}$ is the liar sentence, $\mathsf{True}(``\mathbf{L}")$ can't hold. Hence in the system, $\mathbf{L}$ and $\neg\mathsf{True}(``\mathbf{L}")$ have to hold at the same time. In particular $\mathbf{L}$ is true, this certainly interprets the liars sentence in a non-intuitive way: intuitively, the sentence can be neither true nor false. In Perlis' formal system it is true, however.

Compared to the second approach, a general formula like $\forall\mathbf{A}.\mathsf{True}(\mathbf{A}) \rightarrow \neg\mathsf{True}(\mathsf{conc}(``\neg", \mathbf{A}))$ cannot so easily be proved anymore, but requires an inductive argument on the construction of the formulae $\mathbf{A}$. Assume, for instance, $\mathbf{A} :\equiv \neg\mathsf{True}(``P")$ with P a nullary predicate constant. The corresponding instance of the formula above is:
$\mathsf{True}(``\neg\mathsf{True}(``P")") \rightarrow \neg\mathsf{True}(``\neg\neg\mathsf{True}(``P")")$. With the new definition of truth and elimination of double negation, this formula can be rewritten to:
$\mathsf{True}(``\neg P") \rightarrow \neg\mathsf{True}(``P")$, a formula which reduces to $\neg P \rightarrow \neg P$ in turn.

As in the first approach, the liar sentence does not cause any formal problem anymore in this approach, although the solution is not intuitive. General problems about mutually related statements about each others beliefs (A says "Everything B says is wrong." and B says "Everything A says is right.") persist, however. By the way, as Perlis showed [15], this problem is not specific to meta-systems, but occurs in modal logics as well (unless there are serious restrictions on the expressive power of the modal logic).

Third approach

Kripke [13] and others attacked the problem by changing the semantic construction of the truth values of formulae. They go beyond two-valued logic in a way that they assume a third truth value, which stands for the paradoxical expressions. There are different ways, either to assume truth value gaps [3] or a third truth value, which states undefinedness explicitly [12, p.332–340]. In both systems, Tarski's original definition of truth can be incorporated, that is, $\mathsf{True}(\text{“}\mathbf{A}\text{”}) \equiv \mathbf{A}$ and the liar sentence[3], $\mathbf{L} :\equiv \neg\mathsf{True}(\text{“}\mathbf{L}\text{”})$ can be dealt with as well.

In the first system with truth value gaps, no truth value is assigned to the liar sentence $\mathbf{L}$. In the second, $\mathbf{L}$ is evaluated to the truth value undef, in particular $\mathsf{True}(\text{“}\mathbf{L}\text{”})$ is undefined as well. The meta-theorem above $\forall \mathbf{A}. \mathsf{True}(\mathbf{A}) \rightarrow \neg\mathsf{True}(\mathsf{conc}(\text{“}\neg\text{”}, \mathbf{A})$ does not hold anymore, since with the instance “L”, we get $\mathsf{True}(\text{“}\mathbf{L}\text{”}) \rightarrow \neg\mathsf{True}(\text{“}\neg\mathbf{L}\text{”})$. This expression reduces to $\mathbf{L} \rightarrow \neg\neg\mathbf{L}$ and this in turn to $\mathbf{L} \rightarrow \mathbf{L}$. The last expression does not have any truth value in the first system and the truth value undefined in the second.

The question arises how to determine the truth values in such expressions. Kripke proposes to this end a fixpoint iteration, a process in which in each iteration truth values are assigned to more and more formulae. In the following I shall present an approach that is based on a three-valued logic as well, but does away with the complicated fix-point construction and is based on a standard calculus for three-valued Kleene logic as developed by Kohlhase and myself, for instance, see [10].

5 Kleene Logic – a Simple System for Dealing with Paradoxes

Let us now assume the following two unary connectives $\neg$ (*not*) and $\mathbf{D}$ (*defined*) in the system with the following truth tables

$\neg$	
false	true
undef	undef
true	false

$\mathbf{D}$	
false	true
undef	false
true	true

and the binary connectives $\vee$ (*or*) and $\equiv$ (*strongly equivalent*)

$\vee$	false	undef	true
false	false	undef	true
undef	undef	undef	true
true	true	true	true

$\equiv$	false	undef	true
false	true	false	false
undef	false	true	false
true	false	false	true

[3] We use $\equiv$ to denote strong equivalence, which allows for full substitutivity. That is, $A \equiv B$ is true if and only if the two truth value for A and B agree; it is false else, cp. the truth table on this page.

Other connectives as $\wedge$, $\rightarrow$, and $\leftrightarrow$ can be defined in $\vee$ and $\neg$ just as in classical two-valued logic, e.g., $\mathbf{A} \wedge \mathbf{B} :\equiv \neg(\neg\mathbf{A} \vee \neg\mathbf{B})$, and the corresponding truth tables can be calculated from these definitions.

When we look at the sentence "This sentence is undefined or false", we can define it in this system as $\mathbf{F} :\equiv \neg\mathbf{D}\mathsf{True}(\text{“}\mathbf{F}\text{”}) \vee \neg\mathsf{True}(\text{“}\mathbf{F}\text{”})$. With the definition of truth we can simplify the expression to:
$\mathbf{F} \equiv \neg\mathbf{DF} \vee \neg\mathbf{F}$. If we look at the possible truth values for $\mathbf{F}$, that is, false, undef, and true, we get the following evaluation, which shows that $\mathbf{F}$ cannot be consistently assigned to any of the three truth values that are available:

F	≡	¬	D	F	∨	¬	F
false	false	false	true	false	true	true	false
undef	false	true	false	undef	true	undef	undef
true	false	false	true	true	false	false	true

In the same line, the expression

$$\mathbf{G} :\equiv (\mathsf{True}(\text{“}\mathbf{G}\text{”}) \equiv \neg\mathsf{True}(\text{“}\mathbf{G}\text{”})) \equiv \mathsf{True}(\text{“}\mathbf{G}\text{”})$$

causes the same problem.

That means, the malign paradoxes we tried to ban by the third truth value are back. As Davis says [2, p.85]: "*Having a fundamental flaw like this in a logic is worrisome, like carrying a loaded grenade; you never know when it might go off.*" On the other hand to exclude self-referentiality altogether seems to drastically limit the expressive power of a system. As Kripke states [13, p.698, footnote 13]: "*... some writers still seem to think that some kind of general ban on self-reference is helpful in treating semantic paradoxes. In the case of self-referential sentences, such a position seems to me to be hopeless.*"

Hence there arises the question: What is the very reason for the presence of paradoxes in the first place and how can we get rid of them? The next section is devoted to this question.

How to get and how to avoid self-referential paradoxes

The source of any paradoxes of self-referentiality relies on the fact that it is possible to define a formula to which no truth value can be assigned. This can only be done if it is possible to define a function in the connectives that is *fixpoint free* on the truth values. In a two-valued system such a function is easily defined, it is just negation, that is, as soon as negation is part of a self-referential system, paradoxes are not far away. In our setting above $\lambda x. \neg\mathbf{D}x \vee \neg x$ and $\lambda x. (x \equiv \neg x) \equiv x$ are of that type as well. If we apply one of these functions to any of the three truth values false, undef, or true, the result will never be the same as the input of the function. From this the paradoxes are easily constructed.

However, if we restrict the language to a system, in which each function that can be built from the connectives has at least one fixpoint (conveniently, undef would play that rôle), it is not possible to construct self-referential paradoxes.

As the two counterexamples above show, in a three-valued setting, no language that contains the connectives $\neg$, $\vee$, and $\mathbf{D}$ and no language with unrestricted use of $\neg$ and $\equiv$ guarantees the existence of fixpoints. However, since $\neg$ has undef as fixpoint, $\lambda x.\, x \vee x$ has fixpoint undef as well, and since the fixpoint property is conserved under composition, we get:

Theorem 1. *Any function built from the connectives $\neg$ and $\vee$ contains the fixpoint* undef.

Since the connectives $\wedge$, $\rightarrow$, and $\leftrightarrow$ can be defined in $\neg$ and $\vee$ the property holds of course, when we add these connectives to the system.

Corollary 1. *Self-referential paradoxes are not constructible in the three-valued system described above that contains just the connectives $\neg$, $\vee$, $\wedge$, $\rightarrow$, and $\leftrightarrow$.*

Note that even in the case of complicated expressions in which different agents make assertion with mutual relations, the fixpoint property allows us to consider everything as undefined. Take, for example, the situation in which agent A says "What B says is wrong." and B says "What A says is right." This can be formalised as:
$\mathbf{A} :\equiv \neg\mathsf{True}(\text{“}\mathbf{B}\text{”})$ and $\mathbf{B} :\equiv \mathsf{True}(\text{“}\mathbf{A}\text{”})$. With the definition of truth, $\mathbf{A}$ reduces to $\neg\mathbf{B}$ and $\mathbf{B}$ to $\mathbf{A}$, the first expression states that $\mathbf{A}$ holds if and only if $\mathbf{B}$ does not, while the second states that $\mathbf{A}$ holds if and only if $\mathbf{B}$ does too. Both can't be consistently hold with the truth values true and false. If however, we assert undef to $\mathbf{A}$ and $\mathbf{B}$, the problem is gone, since undef is a fixpoint of $\neg$.

Note furthermore that the system with truth value gaps and the three-valued system differ. In both systems a true expression like $\neg\mathsf{long}(\text{“}\mathsf{long}\text{”})$ evaluates to the truth value true. However, if we add a paradoxical disjunct like the liar sentence, in the first system the truth value gap is contagious and the whole expression $\neg\mathsf{long}(\text{“}\mathsf{long}\text{”}) \vee \mathbf{L}$ does not get a truth value. In the second approach the expression is evaluated to $\mathsf{true} \vee \mathsf{undef}$, that is, to true.

In order to be able to introduce definitions (like the definition of the liar sentence, but of course of useful sentences as well), the $\equiv$ connective cannot be abandoned altogether, but may occur in definitions, that is, in the form $\mathbf{A} \equiv \ldots$ or $\forall x.\, \mathbf{A}(x) \equiv \ldots$, where $\mathbf{A}$ is a predicate constant.

Not all self-referential sentences are paradoxical. For instance, "This sentence is true" is self-referential, formally, if we call the sentence $\mathbf{T}$, it can be expressed as $\mathsf{True}(\mathbf{T})$. $\mathbf{T} :\equiv \mathsf{True}(\text{“}\mathbf{T}\text{”})$. All three truth values, true, false, and undefined, can consistently be assigned to this sentence.

Note that due to the restriction on the connectives, the system does not contain many tautologies (just the formulae $\mathsf{True}(\text{“}\mathbf{A}\text{”}) \equiv \mathbf{A}$ are tautological). This shouldn't be seen as a disadvantage, since under normal circumstances, we do not want to communicate tautologies, but substantial facts and reason then about consequences of them. In order to do so we enlarge the set of axioms by further assumptions and use these for deriving further formulae.

6 Semantics and Calculus

The question is how can we restore a non-trivial system when there are no proper tautologies in it. This is achieved by considering sequents, that is, pairs consisting of a set of formulae Φ and a formula $\mathbf{A}$ (written $\Phi \vdash \mathbf{A}$), in which all formulae in Φ are assumed to be defined. The semantic equivalent is the model relation $\Phi \models \mathbf{A}$ that stands for: in all models of Φ, that is, all interpretations that evaluate every formula in Φ to true, $\mathbf{A}$ holds as well, that is, $\mathbf{A}$ is evaluated to true as well (for details of the semantics see [10]). A proof theory that is built on the refutation principle makes use of the fact that all formulae in Φ must be true and refutes the assumptions that $\mathbf{A}$ might be false or undefined. Note the asymmetry between Φ and $\mathbf{A}$ in this approach, formulae in Φ can't be undefined, while $\mathbf{A}$ might be.[4]

In the following I give a short account of the proof theory of the system described above. In order to exemplify the approach a natural deduction system is introduced. It is based on restricted three-valued Kleene logic and has the advantage that it has not to make use of fixpoint iterations à la Kripke for self-referential expressions.

Since our setting is three-valued, the law of the excluded middle does not hold anymore. In particular paradoxical sentences are neither true nor false.

The rules roughly correspond to Gentzen's classical rules of natural deduction [6]. The main difference is that the classical introduction rule of negation, ¬I-rule

$$\frac{\begin{array}{c}[\mathbf{A}]\\ \bot\end{array}}{\neg\mathbf{A}}\ \neg\mathsf{I}$$

does not hold in the three-valued system. The liar sentence is a counter-example to this rule. In order to conclude from a contradiction under assumption $\mathbf{A}$ to the negation, i.e. $\neg\mathbf{A}$, it is necessary to know that $\mathbf{A}$ is not paradoxical. This can be ensured by showing that $\mathbf{A} \vee \neg\mathbf{A}$ is true . The ¬I-rule is adapted correspondingly. Remember the restricted use of $\equiv$. We use this connective for definitions only. Hence we have an elimination rule ("Subst"), but no introduction rule ($\equiv$ is to be read commutatively). By $\mathbf{C}[\mathbf{A}]$ we denote a formula with subformula $\mathbf{A}$, $\mathbf{C}[\mathbf{B}]$ is the formula in which $\mathbf{A}$ is replaced by $\mathbf{B}$.

A natural deduction calculus consists of the **axiom schema**:
Assumption: $\mathbf{A} \vdash \mathbf{A}$

and the **rules**:

[4] In the full logic with the $\mathbf{D}$ connective, this is reflected in the changed form of the deduction theorem: $\Phi \cup \{\mathbf{A}\} \models \mathbf{B}$ iff $\Phi \models \mathbf{A} \wedge \mathbf{D}\mathbf{A} \rightarrow \mathbf{B}$. The deduction theorem is not expressible in our restricted system, since we do not have the $\mathbf{D}$ connective.

$$\frac{\bot}{\mathbf{A}}\ \mathsf{Ex\ falso\ quodlibet} \qquad \frac{\mathbf{A} \equiv \mathbf{B} \quad \mathbf{C}[\mathbf{A}]}{\mathbf{C}[\mathbf{B}]}\ \mathsf{Subst}$$

$$\frac{\mathbf{A} \vee \neg\mathbf{A} \qquad \begin{matrix}[\mathbf{A}] \\ \bot\end{matrix}}{\neg\mathbf{A}}\ \neg\mathsf{I} \qquad \frac{\mathbf{A} \quad \neg\mathbf{A}}{\bot}\ \neg\mathsf{E}$$

$$\frac{\neg\neg\mathbf{A}}{\mathbf{A}}\ \neg\neg\mathsf{E} \qquad \frac{\mathbf{A} \quad \mathbf{B}}{\mathbf{A} \wedge \mathbf{B}}\ \wedge\mathsf{I} \qquad \frac{\mathbf{A} \wedge \mathbf{B}}{\mathbf{A}}\ \wedge\mathsf{EL}$$

$$\frac{\mathbf{A} \wedge \mathbf{B}}{\mathbf{B}}\ \wedge\mathsf{ER} \qquad \frac{\mathbf{A}}{\mathbf{A} \vee \mathbf{B}}\ \vee\mathsf{IL} \qquad \frac{\mathbf{B}}{\mathbf{A} \vee \mathbf{B}}\ \vee\mathsf{IR}$$

$$\frac{\mathbf{A}\{x_{\mathcal{D}}/a_{\mathcal{D}}\}}{\forall x_{\mathcal{D}}\mathbf{A}}\ \forall\mathsf{I} \qquad \frac{\forall x_{\mathcal{D}}\mathbf{A}}{\mathbf{A}\{x_{\mathcal{D}}/a_{\mathcal{D}}\}}\ \forall\mathsf{E}$$

$\forall$I goes with the usual variable condition. In addition all quantifier rules hold only for defined variables and terms, that is, not for term of the kind 1/0. I don't give a formal treatment here. A detailed description of a tableau calculus for three-valued Kleene logic with soundness and completeness proofs can be found in [10], a resolution calculus in [9].

Since we use a restricted language without a **D** connective it is possible to make use of the strategy for reusing two-valued theorem proving methods in dealing with Kleene logic as found in [11]. That means, any efficient standard two-valued theorem provers can made to an efficient theorem prover for the approach above just by adding a simple restriction strategy.

This gives a calculus for first-order Kleene logic in general. In order to handle quoted expressions, we have to add axiom schemata that describe the semantics of the meta-predicates defined on them. In the case of the truth predicate **T**, we have looked at more closely, this means we add the following axiom schema to the axioms (in an efficient implementation we would use a corresponding simplification rule of course).

Truth: $\vdash \mathsf{True}(\text{“}\mathbf{A}\text{”}) \equiv \mathbf{A}$

Example “Liar Sentence”:

Let us assume a knowledge base consisting of the liar sentence and its definition $\mathbf{L} \equiv \neg\mathsf{True}(\text{“}\mathbf{L}\text{”})$, that is formally, the knowledge base is

$$\Gamma := \{\mathbf{L}, \mathbf{L} \equiv \neg\mathsf{True}(\text{“}\mathbf{L}\text{”})\}$$

by axiom **Truth**, we know $\mathsf{True}(\text{“}\mathbf{L}\text{”}) \equiv \mathbf{L}$, hence by Subst, we get $\mathbf{L} \equiv \neg\mathbf{L}$. We can use this formula to derive by Subst from **L** the formula $\neg\mathbf{L}$. From these two we get $\bot$ by $\neg$E. That is, a knowledge base assuming the liar sentence is inconsistent. Note that it is not paradoxical, just as a classical knowledge base that contains **A** and $\neg\mathbf{A}$ is inconsistent but not paradoxical.

If the knowledge base just contains the definition of the liar sentence, but does not assert that it holds, that is,

$$\Gamma := \{\mathbf{L} \equiv \neg\mathsf{True}(\text{“}\mathbf{L}\text{”})\}$$

we can derive by the definition of truth and Subst, $\mathbf{L} \equiv \neg\mathbf{L}$. If there is nothing else in the knowledge base, nothing else can be derived from this formula. In case there are other formulae containing $\mathbf{L}$, we are allowed to replace $\mathbf{L}$ by $\neg\mathbf{L}$. The formula $\mathbf{L} \equiv \neg\mathbf{L}$ is true, since $\mathbf{L}$ is undef and undef $\equiv$ undef is true. That is, just defining the liar sentence does not cause any problem at all anymore. By our remarks above, we get that in general it is not possible to introduce contradictions by definitions.

Example "Self-referential Truth Sentence":
Let us now assume a knowledge base

$$\Gamma := \{\mathbf{T}, \mathbf{T} \equiv \mathsf{True}(\text{“}\mathbf{T}\text{”})\}.$$

This knowledge base is consistent (just as the knowledge base $\Gamma' := \{\neg\mathbf{T}, \mathbf{T} \equiv \mathsf{True}(\text{“}\mathbf{T}\text{”})\}$). Nothing can be derived if someone protests his sincerity. This again corresponds to classical logic, where a knowledge base that consists of an atom $\mathbf{A}$ is consistent as well as a knowledge base that just consists of $\neg\mathbf{A}$.

An Example with Different Agents:
Assume now a simple knowledge base, in which agent A says "B says the truth", B says "C says the truth" and C says "Joe is the murderer", let's furthermore assume that A speaks the truth. Formally we have the knowledge base

$$\Gamma := \{\mathsf{True}(\text{“}\mathbf{A}\text{”}), \mathbf{A} \equiv \mathsf{True}(\text{“}\mathbf{B}\text{”}), \mathbf{B} \equiv \mathsf{True}(\text{“}\mathbf{C}\text{”}), \mathbf{C} \equiv \mathsf{True}(\text{“}murderer(Joe)\text{”})\}.$$

Expanding the definition of truth, we get $\mathbf{A}$, again by the definition of truth and Subst, we get $\mathbf{B}$, in the next step $\mathbf{C}$ and finally $murderer(Joe)$.

Let us now assume a different knowledge base, where A says "B lies" and B says "A says the truth". Formally the knowledge base consists of $\Gamma := \{\mathbf{A} \equiv \neg\mathsf{True}(\text{“}\mathbf{B}\text{”}), \mathbf{B} \equiv \mathsf{True}(\text{“}\mathbf{A}\text{”})\}$. By the definition of truth and by Subst, we can derive from Γ the formula $\mathbf{A} \equiv \neg\mathbf{A}$, a formula that is perfectly true when we assume that $\mathbf{A}$ is undef. If we add to the knowledge base $\mathbf{A}$, however, the knowledge base is inconsistent. In that case, we can derive from the equivalence $\neg\mathbf{A}$ as well and from $\mathbf{A}$ and $\neg\mathbf{A}$ we can derive $\bot$.

7 Summary

We have seen in this work a three-valued approach to allow self-referential sentences in a formal system almost without any restrictions. The only restriction that we have to impose on such a system, is to guarantee that the third truth value is a fixpoint of any function that can be constructed by the connectives. On the first view this seems to trivialise the system, since by this restriction we have eliminated almost all tautologies from the system. However, when we assume a background theory, with respect to which reasoning takes place, this is not a serious restriction at all. All formulae in the background theory are assumed to be true (this is equivalent to saying, they are not false and not paradoxical). If we assume a paradoxical formula in the background theory as true, the system becomes inconsistent (and not paradoxical), just as a classical system becomes inconsistent by assuming a formula and its negation.

The natural deduction calculus can be used to derive true sentences from true sentences, but also to make and to get rid of assumptions. One complication is given in indirect proofs. If an assumption leads to a contradiction, we can't simply assume its negation. This holds only for defined sentences.

Summarising, we have presented a powerful system for stating facts about truth. By adding further axiom schemata it is easy to extend the framework to deal with knowledge and belief as well. No strange restrictions on the expressivity are made in order to avoid self-referential statements. The system is easier than other systems that are based on a fixpoint construction, standard calculi (with slight restrictions) can be used. The work also shows a way how to deal with paradoxes in higher-order logic without syntactically excluding them.

References

1. Jon Barwise and John Etchemendy. *The Liar – An Essay on Truth and Circularity.* Oxford University Press, New York, Oxford, 1986.
2. Ernest Davis. *Representation of Commonsense Knowledge.* Morgan Kaufmann, San Mateo, California, USA, 1990.
3. Bas C. van Fraassen. Singular terms, truth-value gaps, and free logic. *The Journal of Philosophy*, LXIII(17):481–495, 1966.
4. Dov M. Gabbay, C. J. Hogger, and J. A. Robinson, editors. *Handbook of Logic in Artificial Intelligence and Logic Programming – Volume 1: Logical Foundations.* Oxford University Press, Oxford, UK, 1993.
5. Michael R. Genesereth and Nils J. Nilsson. *Logical Foundations of Artificial Intelligence.* Morgan Kaufmann, San Mateo, California, USA, 1987.
6. Gerhard Gentzen. Untersuchungen über das logische Schließen I & II. *Mathematische Zeitschrift*, **39**:176–210, 572–595, 1935.
7. Fausto Giunchiglia. Getfol manual – Getfol version 2.0. IRST-Technical Report 9107-01, IRST, Povo, Trento, Italy, 1994.
8. K. Grelling and L. Nelson. Bemerkungen zu den Paradoxiien von Russell und Burali-Forti. *Abhandlungen der Fries'schen Schule, N.F. 2, Göttingen*, pages 300–334, 1907/1908.
9. Manfred Kerber and Michael Kohlhase. A mechanization of strong Kleene logic for partial functions. In Alan Bundy, editor, *Proc. of the 12th CADE*, pages 371–385, Nancy, France, 1994. Springer Verlag, Berlin, Germany, LNAI 814.
10. Manfred Kerber and Michael Kohlhase. A tableau calculus for partial functions. *Collegium Logicum – Annals of the Kurt-Gödel-Society*, **2**:21–49, 1996.
11. Manfred Kerber and Michael Kohlhase. Mechanising partiality without re-implementation. In Gerhard Brewka, Christopher Habel, and Bernhard Nebel, editors, *Proc. of the 21st Annual German Conference on Artificial Intelligence,*

KI-97, pages 123–134, Freiburg, Germany, 1997. Springer Verlag, Berlin, Germany, LNAI 1303.
12. Stephen Cole Kleene. *Introduction to Metamathematics.* Van Nostrand, Amsterdam, The Netherlands, 1952.
13. Saul Kripke. Outline of a theory of truth. *The Journal of Philosophy*, LXXII:690–716, 1975.
14. Donald Perlis. Languages with self-reference I: Foundations (or: We can have everything in first-order logic!). *Artificial Intelligence*, **25**:301–322, 1985.
15. Donald Perlis. Languages with self-reference II: Knowledge, belief, and modality. *Artificial Intelligence*, **34**:179–212, 1988.
16. Bertrand Russell. Mathematical logic as based on the theory of types. *American Journal of Mathematics*, XXX:222–262, 1908.
17. Alfred Tarski. Der Wahrheitsbegriff in den formalisierten Sprachen. *Studia philosophia*, **1**:261–405, 1936.
18. Jean van Heijenoort, editor. *From Frege to Gödel – A Source Book in Mathematical Logic, 1879-1931.* Havard University Press, Cambridge, Massachusetts, USA, 1967.
19. Richard W. Weyhrauch. Prolegomena to a theory of mechanized formal reasoning. *Artificial Intelligence*, **13**:133–170, 1980.

Propositional Lower Bounds: Generalization and Algorithms

Marco Cadoli[1], Luigi Palopoli[2], Francesco Scarcello[3]

[1] Dipartimento di Informatica e Sistemistica,
Università di Roma "La Sapienza", Italy
cadoli@dis.uniroma1.it

[2] Dipartimento di Elettronica Informatica e Sistemistica,
Università della Calabria, Italy
palopoli@si.deis.unical.it

[3] Istituto per la Sistemistica e l'Informatica (ISI-CNR),
c/o D.E.I.S., Università della Calabria, Italy
scarcello@unical.it

Abstract. Propositional greatest lower bounds (GLBs) are logically-defined approximations of a knowledge base. They were defined in the context of Knowledge Compilation, a technique developed for addressing high computational cost of logical inference. A GLB allows for polynomial-time complete on-line reasoning, although soundness is not guaranteed. In this paper we define the notion of k-GLB, which is basically the aggregate of several lower bounds that retains the property of polynomial-time on-line reasoning. We show that it compares favorably with a simple GLB, because it can be a "more sound" complete approximation. We also propose new algorithms for the generation of a GLB and a k-GLB. Finally, we give precise characterization of the computational complexity of the problem of generating such lower bounds, thus addressing in a formal way the question "how many queries are needed to amortize the overhead of compilation?"

1 Introduction

It is well known that problems in Logic, Automated Deduction and Artificial Intelligence are very much demanding from the computational point of view. Two of the techniques that have been proposed for addressing such computational hardness are *Knowledge Compilation* (KC) and *Knowledge Approximation* (KA). The central idea of the former technique is to divide in two phases the process of answering to the question whether a query is logically entailed by a knowledge base or not: In the first phase the knowledge base is preprocessed, thus obtaining an appropriate data structure (such a phase is sometimes called *off-line reasoning*); in the second phase, the query is actually answered using the output of the first phase (such a phase is sometimes called *on-line reasoning*). Typically, the output of off-line reasoning is a logical formula in an appropriate *target* language. The goal of preprocessing is to make on-line reasoning computationally easier (wrt query answering with no preprocessing.) This can be enforced, for example, by choosing a target language which guarantees polynomiality of reasoning.

In KA the central idea is to give up either soundness or completeness when answering to a logical query. Obviously, also in this case one aims to use less computational resources than in the sound and complete case. The two techniques can be used together: In such a case we have a method for approximate KC.

A method for approximate KC has been proposed by Selman and Kautz in several papers. In [SK91] they define KC of propositional formulae, the target language being the set of propositional Horn formulae (which admit well-known polynomial-time algorithms for inference [DG84]). Compiling a propositional formula Σ delivers two distinct Horn formulae Σ_{lb} and Σ_{ub} such that $\Sigma_{lb} \models \Sigma \models \Sigma_{ub}$. Σ_{lb} is called a *Horn lower bound* of Σ, while Σ_{ub} is called its *Horn upper bound*. As an example (cf. [SK91]), if Σ is $(master\_student \vee phd\_student) \wedge (master\_student \rightarrow student) \wedge (phd\_student \rightarrow student)$, then $(master\_student \wedge phd\_student \wedge student)$ is a Horn LB of Σ, and $(master\_student \rightarrow student) \wedge (phd\_student \rightarrow student)$ is a Horn UB of Σ. In general, Σ_{lb} and Σ_{ub} can be chosen in many different ways. A Horn LB Σ_{lb} such that there is no other Horn LB Σ' such that $\Sigma_{lb} \models \Sigma'$ and $\Sigma' \not\models \Sigma_{lb}$ is called a "greatest" Horn lower bound (GLB). Referring to the previous example, $(master\_student \wedge student)$ is a Horn GLB. This shows that a Horn GLB Σ_{glb} is a "complete and incorrect" approximation of Σ, since it might be the case that $\Sigma \not\models \Sigma_{glb}$, or, in other words, the set of models Σ_{glb} is a strict subset of the set of models of Σ. "Least" Horn upper bounds (LUB) are defined dually, and are "correct and incomplete" approximations.

In [SK96], the target language is generalized to other polynomial classes of propositional formulae, and two algorithms for generating, resp., a GLB and a LUB are provided. The reliability of the approximations (in terms of the percentage of right answers obtained using the two bounds) is studied in [KS94]. Other papers dealing with propositional approximate KC are [dV95,Sch96].

In the present paper, we focus on lower bounds, and our goal is to advance the state of the knowledge on two specific aspects:

1. definition of a more general target language,
2. study of computational properties of the compilation process.

As for the former aspect, we define the notion of *k-GLB*, which is basically the aggregate of several LBs that retains the property of polynomial-time on-line reasoning. We show that it compares favorably with the GLB, because it can be a "more sound" complete approximation.

As for the latter aspect, our main contributions are:

- new algorithms for the generation of a GLB and a k-GLB,
- precise characterization of the computational complexity of the problem of generating such lower bounds.

Although KC tries to shift the burden of query answering to off-line reasoning, in general one may ask "how many queries are needed to amortize the overhead of compilation?" Characterization of the computational complexity of the problem of generating a good LB gives a formal answer to such a question. So far, complexity of generation of a GLB has been partially addressed only in the context of Horn GLBs [SK96,Cad93].

2 Preliminaries

In this section we first give the notions of lower bound and greatest lower bound of propositional theories, following [SK96]. Then we introduce a suitable family of classes of propositional theories and study some general properties about such classes. Finally, we define k-GLBs.

We assume *theories* are propositional formulae given in conjunctive normal form (CNF), so they can be represented by sets of clauses. We call "source theory" a theory whose LBs we are interested in. A *clause* is a disjunction of literals (propositional letters or their negation), and can be represented by the set of literals it contains. We use either the set notation or the logical notation, when no confusion arises. Given a theory Σ, $\mathcal{M}(\Sigma)$ is the set of its models, defined in the standard way.

Definition 1 (LB and GLB of a theory [SK96]). *Let Σ be a CNF theory and θ be a class of CNF theories.*

- *A theory Σ_{lb} belonging to θ is a θ-LB (lower bound) of Σ if $\mathcal{M}(\Sigma_{lb}) \subseteq \mathcal{M}(\Sigma)$ (i.e., $\Sigma_{lb} \models \Sigma$).*
- *A theory Σ_{glb} belonging to θ is a θ-GLB (greatest lower bound) of Σ if there exists no θ-LB Σ_{lb} of Σ such that $\mathcal{M}(\Sigma_{glb}) \subset \mathcal{M}(\Sigma_{lb}) \subseteq \mathcal{M}(\Sigma)$.*

If no confusion arises, we skip the prefix *theta* when referring to LBs and GLBs. A *θ-clause* is a clause which belongs to class θ.

We list some general properties of the classes θ of CNF theories which are the object of compilation:

complexity of inference problem: how much does it cost to decide whether a clause γ logically follows from a GLB of a theory Σ? We require that such a problem is feasible in polynomial-time in the size of Σ plus the size of γ (cf. previous section). This clearly implies the size of a GLB is bounded by a polynomial in the size of Σ.

complexity of recognition problem: how much does it cost to decide whether a formula belongs to class θ? Typically, this is a polynomial-time problem. Nevertheless, such a condition will be relaxed in the algorithms we show in the next section.

syntactic restrictions:

unit clauses: if 1CNF theories (i.e., theories with only single literal clauses) belong to θ, we say θ is a *target class*.

closure under conjunction and resolution: if such properties hold for a target class θ, then we say θ is an *L-class* (because it has some "locality" properties, cf. Section 3).

Most classes of propositional theories used in Knowledge Representation, e.g., Horn, Dual Horn, and 2CNF, are L-classes. Although we always assume that θ is a target class, in the next section we show that assuming that θ is an L-class leads to a simpler compilation problem. It is interesting to notice that some target classes with polynomial-time inference and recognition problems, closed under resolution, are not closed under conjunction (cf. e.g., "balanced" theories [CC95]).

A list of general properties of the compiled theories which will be dealt with in the sequel of this paper follows.

covering: is the logical "or" of all GLBs equivalent to the source theory (i.e., does the union of the models of the GLBs cover the models of the source theory)? Covering is a weak form of soundness, as it guarantees that there are no inferences that are *a priori* lost.

satisfiability: does a satisfiable GLB of a satisfiable source theory always exist?

complexity of GLB checking: how much does it cost to know whether a formula is a GLB of a given source theory?

complexity of the generation problem: how much does it cost to compute a GLB of a given source theory?

number of GLBs: how many GLBs are there for a given source theory?

We are able to obtain for target classes results similar to those shown in [SK96] for Horn formulae. Σ_{lb} and Σ_{glb} denote respectively an LB and a GLB of a source theory Σ belonging to some fixed target class.

Proposition 1. *1. The union of the models of all possible Σ_{lb} equals $\mathcal{M}(\Sigma)$.*

2. The union of the models of all possible Σ_{glb} equals $\mathcal{M}(\Sigma)$.

3. Σ is satisfiable if and only if there is at least one satisfiable Σ_{glb}.

4. Checking if a theory Σ^* belonging to some target class is a GLB of Σ is coNP-complete.

5. Finding a GLB of Σ belonging to a target class with a polynomial inference problem is NP-hard.

Proof.

1. Let $M \in \mathcal{M}(\Sigma)$. The conjunction of all the literals assigned true by M is a Σ_{lb}.
2. Straightforward.
3. ($\Rightarrow$) From point 1. ($\Leftarrow$) Let Σ' be a GLB of Σ. Then $\mathcal{M}(\Sigma') \subseteq \mathcal{M}(\Sigma)$ and since Σ' is satisfiable $\mathcal{M}(\Sigma') \neq \emptyset$.
4. (coNP-hardness) Let l be a literal and $\Sigma^* = l \wedge \neg l$. Σ^* belongs to the target class and is not satisfiable. From point 3 it easily follows Σ^* is a GLB if and only if Σ is not satisfiable. (Membership to coNP) Since the size of a GLB of Σ is always bounded by a polynomial in $|\Sigma|$, we can use a simple guess-and-check algorithm.
5. Deciding whether Σ is satisfiable or not reduces to the problem of finding a GLB Σ_{glb} and to decide if $\Sigma_{glb} \models l \wedge \neg l$. The latter problem is polynomial. □

As for point 5, NP-hardness is just a lower bound to the complexity. In the rest of the paper, upper bounds in different cases will be provided. As for the number of Horn GLBs, in [SK96] it is proven that in general there are exponentially many (wrt the size of Σ), but we do not know whether this holds for any target class. Anyway, if they were just polynomially many, since their disjunction "covers" Σ (Proposition 1.2), and inference from a GLB is a polynomial-time task, we could "compile" the source theory and answer any query in polynomial time, which, using the techniques shown in [KS92], would imply that the polynomial hierarchy collapses at the second level.

As shown in [SK96], the key idea in generating a GLB is to choose an appropriate *strengthening* of each clause C of Σ, i.e., to choose which literals delete from C. The formal definition follows.

Definition 2 (θ-strengthening [SK96]). *Let θ be a target class. A θ-clause C_θ is a θ-strengthening of a clause C if $C_\theta \subseteq C$ and there is no θ-clause C'_θ such that $C_\theta \subset C'_\theta \subseteq C$.*

The first contribution of the present paper is to generalize Definition 1 to a k-tuple of LBs.

Definition 3 (k-LB and k-GLB of a theory). *Let Σ be a CNF theory, θ be a class of CNF theories, and k be a fixed number.*

- *A k-tuple of theories $\Sigma^1_{lb}, \ldots, \Sigma^k_{lb}$ belonging to θ is a θ-k-LB of Σ if $\mathcal{M}(\Sigma^i_{lb}) \subseteq \mathcal{M}(\Sigma)$ for each i $(1 \leq i \leq k)$.*
- *A k-tuple of theories $\Sigma^1_{lb}, \ldots, \Sigma^k_{lb}$ belonging to θ is a θ-k-GLB of Σ if there exists no θ-k-LB $\Sigma^1, \ldots, \Sigma^k$ of Σ such that $\mathcal{M}(\Sigma^1_{lb}) \cup \cdots \cup \mathcal{M}(\Sigma^k_{lb}) \subset \mathcal{M}(\Sigma^1) \cup \cdots \cup \mathcal{M}(\Sigma^k) \subseteq \mathcal{M}(\Sigma)$.*

Referring to the example in Section 1, $\Sigma^1 = (phd\_student \wedge student)$, $\Sigma^2 = (master\_student \wedge student)$ is a Horn-2-GLB of Σ.

A k-LB or k-GLB $\Sigma^1, \ldots, \Sigma^k$ can be used for on-line reasoning, since for each formula C, $\Sigma^1 \vee \cdots \vee \Sigma^k \models C$ iff $(\Sigma^1 \models C) \wedge \cdots \wedge (\Sigma^k \models C)$. If k is bounded by a polynomial in the size of Σ, then on-line reasoning using a k-LB or k-GLB is still a polynomial-time task.

It is interesting to note that in a k-GLB $\Sigma^1_{lb}, \ldots, \Sigma^k_{lb}$, a theory Σ^i_{lb} does not have to be a GLB (cf. forthcoming Example 2). Moreover, although for each k-GLB Σ^k_{glb} of Σ there always exist k distinct GLBs that capture the same set of models as Σ^k_{glb}, such k GLBs cannot, in general, be chosen arbitrarily (cf. forthcoming Example 3).

Moreover, observe that in principle, in Definition 3, $\Sigma^1_{lb}, \ldots, \Sigma^k_{lb}$ may belong to distinct classes θ.

In the rest of the paper we refer to the following complexity classes for decision problems: P, NP, coNP, $\Delta^P_2 = \mathrm{P}^{\mathrm{NP}}$, $\Sigma^P_2 = \mathrm{NP}^{\mathrm{NP}}$. We also refer to the corresponding classes for search problems, by adding the prefix "F" (e.g., FNP, FΣ^P_2) [Joh90]. Intuitively, an algorithm in FNP is a polynomial-time algorithm which uses a polynomial number of calls to an NP oracle (i.e., a procedure which is able to solve an NP problem at unary cost). In an $F\Sigma_2^P$ algorithm, the oracle is able to solve NP^{NP} problems, i.e., it has the power of an NP machine, with an NP oracle. In practice, a FNP algorithm can be implemented with a polynomial-time algorithm and a subroutine being able to solve an NP-complete problem.

3 Algorithms and complexity of generating a GLB

In this section we illustrate the following results:

- we show that the computation of a θ-GLB of a CNF theory can be done in $F\Sigma_2^P$ under very weak conditions on θ;
- we prove some properties of θ-GLBs by which, under some additional conditions on θ, the computation of a θ-GLB of a CNF theory can be done in FNP.

We begin with the first item above.

Proposition 2. *Let θ be a target class with a Δ_2^P recognition problem. Then finding a θ-GLB of Σ is in* $F\Sigma_2^P$.

Proof. (Sketch) As mentioned in Section 2, for each θ there is a polynomial p such that the dimension of a θ-GLB of Σ is less than or equal to $p(n)$, where n is the size of Σ's alphabet. If we assume no tautology and no redundancies occur in the GLB, we can compute it using at most $2np(n)$ calls to an oracle in Σ_2^P. In fact, since we cannot have more than $p(n)$ clauses in the GLB and each letter in Σ's alphabet can appear at most once in each clause, at the generic oracle call we shall have a set of clauses consisting of (1) a set of clauses Σ' we have already decided to be part of a θ-GLB we are constructing and (2) a (partially formed) clause C'. Then, for each literal l, we ask the oracle whether there exists a θ-GLB Σ_{glb}^{θ} of Σ such that $\Sigma' \cup \{C\} \subseteq \Sigma_{glb}^{\theta}$, where $C' \cup \{l\} \subseteq C$. □

In some cases, we can improve the upper bound given by Proposition 2, exhibiting a FNP algorithm.

Proposition 3. *Finding a 1CNF-GLB of Σ is in FNP.*

Proof. (Sketch.) The proof immediately follows from noting that any 1CNF theory is a 1CNF-GLB of Σ iff it is a prime implicant of Σ. □

On the other hand, 1CNF-GLBs are not able to cover the models of the source theory very well. In fact, if I is a 1CNF-GLB of Σ, then for each target class θ, I is a θ-LB of Σ.

Coming to the second goal of this section, we show sufficient conditions on θ that guarantee, for the problem of finding a θ-GLB, the same upper bound to complexity as the 1CNF case.

The basic assumption we make is that θ is an L-class. In such a case, it is possible to compute a GLB "a clause at a time", hereby obtaining an FNP method. In fact, for L-classes we can generalize Lemma 2 of [SK96] in the following way.

Lemma 1. *Let θ be an L-class and Σ^{θ}_{glb} be a θ-GLB of a CNF theory $\Sigma = \{C_1, \ldots, C_n\}$. Then there exists a set of clauses $\{C'_1, \ldots, C'_n\}$ where each C'_i is a θ-strengthening of C_i and such that $\Sigma^{\theta}_{glb} \equiv \{C'_1, \ldots, C'_n\}$,*

The Lemma 1 allows us to find a θ-GLB by choosing its clauses among its θ-strengthenings. If a Σ^{θ}_{glb} of Σ is composed only of θ-strengthenings of clauses in Σ (as in the Lemma above), we say that Σ^{θ}_{glb} is in *normal form*, short NF.

From this important property of L-classes, we derive the following technical results. For any theory Σ, let $IM(\Sigma)$ denote the set of its implicants.

Lemma 2. *Let Σ be a CNF theory, C_i one of its clauses, and θ an L-class. If there exists an implicant $p \in IM(\Sigma)$ and a literal $l \in C_i$ such that $p \cap C_i = \{l\}$, then, for each θ-GLB in normal form Σ' for Σ s.t. $p \models \Sigma'$, the θ-strengthening $C'_i \in \Sigma'$, corresponding to C_i, must include the literal l.*

Proof. Straightforward. □

Our FNP procedure for computing a Σ^{θ}_{glb} of Σ for any L-class θ is shown in Figure 1.

Lemma 3. *Let Σ be a CNF theory, θ an L-class, I a set of implicants for Σ, C a clause of Σ, and $l \in C$ a literal such that for each implicant $p \in IM(\Sigma)$ containing l, either $p \cap C \neq \{l\}$, or $GlobalComp(I \cup \{p\}, \Sigma)$ is false. Then, any θ-GLB Σ' for Σ such that $I \subseteq IM(\Sigma')$, is also a θ-GLB for $\Sigma \wedge (C \setminus \{l\})$.*

Proof. (Sketch) Assume, w.l.o.g., that Σ' is in NF and observe that $\Sigma' \models \Sigma \wedge (C \setminus \{l\})$. This is an consequence of the hypothesis on l, the fact that $I \subseteq IM(\Sigma')$, and the assumption Σ' is composed by θ-strengthenings of Σ.

Since $\Sigma \wedge (C \setminus \{l\}) \models \Sigma$, it is easy to verify that Σ' is in fact a θ-GLB of $\Sigma \wedge (C \setminus \{l\})$. □

The following theorem states the correctness of *ComputeGLB*.

Theorem 1. *Let Σ be a CNF theory and θ be an L-class of target theories. Procedure* Compute-GLB *computes (in the input/output parameter Σ) a θ-GLB of the source theory Σ.*

Proof. (Sketch) Denote by Σ^g the theory obtained at the end of *Compute-GLB*. First, note that Σ^g belongs to θ. At each step of the outer cycle of the algorithm, we select a clause $C \in \Sigma$ and compute a clause $\hat{C} \in \theta$. Indeed, θ is an L-class, then unit clauses belongs to θ, and θ is closed under resolution. As a consequence, if C' belongs to θ, any non-empty clause $C'' \subseteq C'$ belongs to θ as well. By construction, $\hat{C}$ is a subset of some "candidate" clause $C' \in \theta$, thus $\hat{C} \in \theta$. Furthermore, θ is closed under conjunction, hence Σ^g belongs to θ, and it is clearly a θ-LB of Σ. Moreover, by construction, no implicant $p \in I$ can violate any clause of Σ^g, hence, $I \subseteq IM(\Sigma^g)$.

Let the input theory Σ be the conjunction (or set) of clauses $C_1 \wedge \ldots \wedge C_n$, and assume, w.l.o.g., that clauses already belonging to θ are those with higher indexes. Denote by $\Sigma^s = C^s_1 \wedge \ldots \wedge C^s_n$ the value of the parameter Σ at the beginning of the sth execution of the outer *for* cycle, i.e., the step in which the clause C_s is selected.

We proceed by contradiction. Assume Σ^g is not a θ-GLB of the source theory Σ. Then, there exists a θ-GLB Σ' of Σ such that $\Sigma^g \models \Sigma' \models \Sigma$, and $\Sigma' \not\models \Sigma^g$. By Lemma 1, we can assume, w.l.o.g., that $\Sigma' = C'_1 \wedge \cdots \wedge C'_n$, where each clause C'_k, for $1 \leq k \leq n$ is a θ-strengthening of the corresponding clause C_k belonging to Σ.

Since $\Sigma' \not\models \Sigma^g$, there exists a clause belonging to Σ^g violated by some implicant of Σ'. Let i be the least index in $\{1, \ldots, n\}$ s.t. the clause C_i^g is violated by some implicant of Σ'. Such a clause C_i^g is computed at step i of the execution of the outer *for* cycle of procedure *Compute*-GLB. Now, consider the theory Σ^i computed before such a step is executed. By construction, for each $k < i$, we have $C_k^i = C_k^g$, and for each $j \geq i$, the clause C_j^i of Σ^i is identical to the corresponding clause C_j of the original input theory Σ. By the choice of i, no clause C_k^i, for $k < i$, can be violated by any implicant $p' \in IM(\Sigma')$, hence, $\Sigma' \models \Sigma^i$ clearly holds, and $IM(\Sigma') \subseteq IM(\Sigma^i)$. In fact, it is easy to see that Σ' must be a θ-GLB of Σ^i.

Now, let $\bar{C}_i$ and $\hat{C}_i$ be the value of variable C and $\hat{C}$, respectively, at the end of the inner *for*, where the clause C_i^i of Σ^i is selected. Thus, we have $\hat{C}_i = C_i^g$. Since $I \subseteq IM(\Sigma')$, by Lemma 3, Lemma 2, and Lemma 1, it follows that there exists a θ-GLB Σ'' of $\Sigma^i \wedge \bar{C}_i$ which is equivalent to Σ', and which is in normal form. Let C''_i be the θ-strengthening of $\bar{C}_i$ belonging to Σ'', hence, we have $C''_i \subseteq \bar{C}_i$.

If there exists a literal $l \in \hat{C}_i$ s.t. $l \notin C''_i$, then there exists an implicant $p' \in I$ s.t. $p' \cap \bar{C}_i = \{l\}$. However, such a p' would violate C''_i, which is a subset of $\bar{C}_i$. This is clearly a contradiction, since $I \subseteq IM(\Sigma'')$.

Otherwise, assume $\hat{C}_i \subset C''_i \subseteq \bar{C}_i$. Since each literal selected in the inner *for* is deleted from $\bar{C}_i$ or belongs to $\hat{C}_i$, for each literal $l' \in (C''_i \setminus \hat{C}_i)$, such a literal l' does not belong to any $C' \in Candidates_\theta(\bar{C}_i, I)$ s.t. $(\hat{C}_i \cup \{l'\}) \subseteq C'$. This contradicts the existence of C''_i, which should belong to $Candidates_\theta(\bar{C}_i, I)$, and thus the existence of the GLB Σ'', as well. □

The following result states the complexity of procedure *Compute-GLB*.

Theorem 2. *Let Σ be a CNF theory and θ be an L-class of target theories. If the set of θ-strengthenings is polynomial time computable, then finding a θ-GLB of Σ is in FNP.*

Proof. (Sketch) In each step of the outer cycle of the procedure *Compute-GLB*, a clause $C_i \in \Sigma$ is selected and a θ-clause $\hat{C}_i \subseteq C_i$ for it is computed. Such clause will be not selected any more. Moreover, any literal belonging to C_i is selected at most once, thus the inner cycle in the procedure is executed $|C_i|$ times, where $|C_i|$ denotes the number of literals in C_i. Hence, if the number of clauses in Σ is n, the procedure *Compute-GLB* terminates after at most $\sum_{i=1}^{n} |C_i|$ execution of the *if* statement appearing in the inner cycle.

Now, observe that the cardinality of the set I, which is the name of a parameter of both $Candidates_\theta$ and *GlobalComp*, is always $\leq \sum_{i=1}^{n} |C_i|$, and,

INPUT: A CNF theory Σ without tautological clauses and an L-class θ.
OUTPUT: A θ-GLB of Σ.

```
Procedure ComputeGLB (var Σ: Theory)

    Function Candidates_θ (C: Clause, I: Set of Implicants): SetOfClauses
    begin
      return each θ-strenghtening C' of C s.t., ∀p ∈ I, C' ∩ p ≠ ∅;
    end;
    Function GlobalComp(I: Set of Implicants, Σ: Theory): Boolean;
    begin
      if ∀C ∈ Σ  Candidates_θ(C, I) ≠ ∅ then return true
      else return false
    end;

begin
    I := ∅;
    for each non-θ clause C ∈ Σ do
    begin
      Ĉ := ∅;
      for each l ∈ (C \ Ĉ) s.t. ∃C' ∈ Candidates_θ(C, I) s.t. Ĉ ∪ {l} ⊆ C' do
          if ∀C' ∈ Candidates_θ(C, I) s.t. Ĉ ⊆ C' we have l ∈ C'
          then Ĉ := Ĉ ∪ {l};
          elsif ∃p ∈ IM(Σ) s.t. p ∩ C = {l} and GlobalComp(I ∪ {p}, Σ)
          then begin
                Ĉ := Ĉ ∪ {l};
                I:= I ∪ {p}
          end
          else C:= C \ {l};
      C := Ĉ;
    end;
    output Σ
end;
```

Fig. 1. The FNP algorithm for computing a θ-GLB.

hence, polynomial in the size of the input theory Σ. Moreover, by hypothesis, $Candidates_\theta$ is polynomial time computable, hence all the executions of either $Candidates_\theta$ or *GlobalComp* are polynomial time bounded. This entails that every execution of the *elsif* statement in the procedure can be performed by a unique call to an NP oracle, and thus all the procedure is in FNP. □

As a consequence, finding a GLB of a source theory which is Horn, Dual Horn, or 2CNF, can be done in FNP. Note that, for all such classes, the set of θ-strengthenings of any clause is very easy to compute.

The following example shows an application of the procedure *Compute-GLB* for the computation of a 2CNF-GLB.

Example 1. Let $\Sigma = C_1 \wedge C_2 \wedge C_3$ be the following CNF theory:

$$(b \vee a \vee c) \wedge (b \vee \bar{a}) \wedge (a \vee d \vee c)$$

We apply procedure *Compute*-GLB to find a $2CNF$-GLB of Σ. First, we consider the clause $C_1 = b \vee a \vee c$ and let $\hat{C} = \emptyset$.

We begin by selecting the literal $b \in C_1$. We find that there exists an implicant of Σ, $p_b = \{b, d\}$, such that $p_b \cap C_1 = \{b\}$. It is easy to verify that $GlobalComp(\{p_b\}, \Sigma)$ returns true, then we set $\hat{C} := \{b\}$ and $I := \{p_b\}$. Next, we select literal $a \in C_1$. For this literal, there exists no implicant $p \in IM(\Sigma)$ such that $p \cap C_1 = \{a\}$. Then, we can drop literal a and we get $C_1 := b \vee c$. Next, we select literal c. Since c belongs to every clause in the (singleton) set $Candidates_\theta(C_1, I)$, it is immediately added to $\hat{C}$.

At this point, we have computed a θ-clause $\hat{C} = b \vee c$, which thus replace C_1 in Σ. Then, we have the following theory Σ':

$$(b \vee c) \wedge (b \vee \bar{a}) \wedge (a \vee d \vee c)$$

Moreover, the compatible candidates for I are the following: $Candidates_\theta(C_1, I) = \{b \vee c\}$, $Candidates_\theta(C_2, I) = \{b \vee \bar{a}\}$, and $Candidates_\theta(C_3, I) = \{d \vee c,\ a \vee d\}$. It is easy to see that the procedure *Compute*-GLB leaves untouched the second clause. For the clause C_3, we begin by selecting literal a. For a, there exists an implicant $p_a = \{a, b\}$ such that $p_a \cap C_3 = \{a\}$. Now, $GlobalComp(\{p_b, p_a\}, \Sigma)$ returns true; therefore, we set $\hat{C} := \{a\}$ and $I := \{p_b, p_a\}$. Now, $Candidates_\theta(C_3, I) = \{a \vee d\}$. Therefore, d is immediately added to $\hat{C}$.

Thus, we finally get the following 2CNF-GLB of Σ:

$$(b \vee c) \wedge (b \vee \bar{a}) \wedge (a \vee d)$$

It is worth noting that, by our results, for any L-class θ, the complexity of finding a θ-GLB remains the same as for the simplest case of 1CNF, provided that the set of θ-strengthenings of any clause is polynomial time computable.

Compared with the procedure proposed in [SK96], the latter method represents an obvious improvement as far as the theoretical complexity is concerned, since the former, based on an enumeration technique, works in exponential time. Of course, if P $\neq$ NP, an effective implementation of our method cannot work in polynomial time as well, but it shows the problem is not intrinsically exponential and many optimizations can be made exploiting particular features of the selected target class.

Compared with our first algorithm working in $F\Sigma_2^P$, our method is more efficient but less general, because it requires θ be closed under conjunction and resolution.

4 Properties of k-GLBs

In this section we investigate the use of k-GLB as target theories for knowledge compilation. We begin by stating the complexity of checking θ-k-GLBs. Then, we show that, from a complexity theoretic point of view, finding a θ-k-GLB for a general target class θ in doable in $F\Sigma_2^P$ and, as such, is not more difficult than finding a θ-GLB. Finally, we concentrate on semantical considerations and show, also using some examples, that employing θ-k-GLBs often compares favorably with respect to using θ-GLBs.

Proposition 4. *Let Σ be a CNF theory and $\Sigma_1, \ldots, \Sigma_k$ be a k-tuple of theories belonging to some target class θ. Checking if $\overline{\Sigma} = \Sigma_1, \ldots, \Sigma_k$ is a θ-k-GLB of Σ is coNP-complete.*

Proof. (coNP-hardness) Let $a_1, \ldots, a_k$ be k letters from Σ. Then Σ is not satisfiable iff $\{a_i \wedge \overline{a_i} | 1 \leq i \leq k\}$ is a θ-k-GLB of Σ. (Membership to coNP) Since the size of the GLBs of Σ is always bounded by a polynomial in $|\Sigma|$ and k is fixed, we can use a simple guess-and-check algorithm. □

Such a proposition corresponds to Proposition 1.4 in the case of k-GLBs. It allows us to provide a simple $F\Sigma_2^P$ algorithm for generating a θ-k-GLB.

Proposition 5. *Let Σ be a CNF theory, k be a fixed number and θ be a target class with a Δ_2^P recognition problem. Then finding a k-θ-GLB of Σ is in* $F\Sigma_2^P$.

Proof. The proof is analogous to that of Theorem 2. The main difference is that, in this case, we need at most $2nkp(n)$ oracle calls. □

Thus, from a complexity-theoretic viewpoint, for general θ theories, constructing a θ-k-GLB is not more difficult than computing a θ-GLB. On the other hand, employing k-GLBs in the place of GLBs for knowledge compilation purposes can be advantageous from the semantical viewpoint. Indeed, the following proposition shows that, in general, k-GLBs allows to capture more models of the source theory than GLBs.

Proposition 6. *Let Σ be a CNF theory and $\Sigma_{glb_1}, \ldots, \Sigma_{glb_k}$ be k distinct θ-GLBs of Σ. Then there exists a k-GLB Σ_{glb}^k of Σ such that $\mathcal{M}(\Sigma_{glb_1}) \cup \ldots \cup \mathcal{M}(\Sigma_{glb_k}) \subseteq \mathcal{M}(\Sigma_{glb}^k)$.*

Proof. The proof is immediate by noting that either $\Sigma_{glb_1}, \ldots, \Sigma_{glb_k}$ is itself a k-GLB of Σ or otherwise there exists a k-LB Σ_{glb}^k of Σ such that $\Sigma_{glb_1} \vee \cdots \vee \Sigma_{glb_k} \models \Sigma_{glb}^k \models \Sigma$. □

Note that, in fact, there are cases where a k-GLB contains several LBs which are not GLBs, as shown in the following example.

Example 2. Let Σ be the theory $(\overline{c} \vee d) \wedge (a \vee \overline{b}) \wedge (\overline{a} \vee \overline{c})$ and assume the target class is that of implicants. Then, the theory $(\overline{a} \wedge \overline{c}) \vee (a \wedge b \wedge \overline{c}) \vee (\overline{a} \wedge b \wedge c)$ is a 1CNF-3-GLB for Σ and neither $(a \wedge b \wedge \overline{c})$ nor $(\overline{a} \wedge b \wedge c)$ are GLBs for Σ.

It is obvious, however, that for each θ-k-GLB Σ^k_{glb} of Σ there always exist k distinct *GLB*'s that capture the same set of models as Σ^k_{glb}. Nevertheless, these k GLBs cannot, in general, be chosen arbitrarily. This fact is proved by the following example.

Example 3. Let Σ be the theory $(a \vee \overline{b} \vee d) \wedge (b \vee c \vee \overline{d}) \wedge (\overline{a} \vee \overline{c})$. Assume our target class is that of implicants. On the one hand, consider the 1CNF-4-GLB $\Sigma^4_{glb} = (a \wedge \overline{c} \wedge \overline{d}) \vee (b \wedge \overline{c} \wedge d) \vee (\overline{a} \wedge c \wedge d) \vee (\overline{a} \wedge \overline{b} \wedge \overline{d})$. On the other hand, consider the following four GLBs: $\Sigma_{glb_1} = (a \wedge \overline{c} \wedge \overline{d})$, $\Sigma_{glb_2} = (a \wedge b \wedge \overline{c})$, $\Sigma_{glb_3} = (b \wedge \overline{c} \wedge d)$ and $\Sigma_{glb_4} = (\overline{a} \wedge b \wedge d)$. It is the easy to see that $\mathcal{M}(\Sigma) = \mathcal{M}(\Sigma^4_{glb})$ and that $\mathcal{M}(\Sigma_{glb_1}) \cup \mathcal{M}(\Sigma_{glb_2}) \cup \mathcal{M}(\Sigma_{glb_3}) \cup \mathcal{M}(\Sigma_{glb_4})$ does not include the following three models of Σ: $M_1 = \{\overline{a}, \overline{b}, \overline{c}, \overline{d}\}$, $M_2 = \{\overline{a}, \overline{b}, c, d\}$ and $M_3 = \{\overline{a}, \overline{b}, c, \overline{d}\}$. Moreover note that we need at least two further GLBs to be added to the four cited above to capture M_1, M_2 and M_3.

5 Conclusions

In this paper we have defined k-GLBs, generalizations of propositional GLBs, and we have investigated the computational complexity of the problems of finding a GLB and a k-GLB. For both of them, we have shown a general upper bound of $\mathrm{F}\Sigma^P_2$. As for GLBs, we have exhibited an algorithm that works in FNP. Such an algorithm is proven to be correct for L-classes, i.e., classes of propositional formulae which are closed under conjunction and resolution. Restriction to L-classes is important, since there are classes (e.g., "balanced" theories cf. [CC95]) which are not closed under conjunction. The question whether the problem of generating a k-GLB is intrinsically more complex than the one of generating a GLB is still open.

References

[Cad93] M. Cadoli, Semantical and computational aspects of Horn approximations. In *Proc. of IJCAI-93*, pages 39–44, 1993.

[CC95] M. Conforti and G. Cornuéjols. A class of logic problems solvable by linear programming. *J. of the ACM*, 42:1107–1113, 1995.

[DG84] W. P. Dowling and J. H. Gallier. Linear-time algorithms for testing the satisfiability of propositional Horn formulae. *J. of Logic Programming*, 1:267–284, 1984.

[dV95] A. del Val. An analysis of approximate knowledge compilation. In *Proc. of IJCAI-95*, pages 830–836, 1995.

[Joh90] D.S. Johnson, A catalog of complexity classes. In *Handbook of theoretical computer science*, Chapter 2, pages 67–161, J. van Leeuwen ed., Elsevier Sc. Pub., 1990.

[KS92] H. A. Kautz and B. Selman. Forming concepts for fast inference. In *Proc. of AAAI-92*, pages 786–793, 1992.

[KS94] H. A. Kautz and B. Selman. An empirical evaluation of knowledge compilation by theory approximation. In *Proc. of AAAI-94*, pages 155–161, 1994.

[Sch96] R. Schrag. Compilation for critically constrained knowledge bases. In *Proc. of AAAI-96*, pages 510–515, 1996.

[SK91] B. Selman and H. A. Kautz. Knowledge compilation using Horn approximations. In *Proc. of AAAI-91*, pages 904–909, 1991.

[SK96] B. Selman and H. A. Kautz. Knowledge compilation and theory approximation. *J. of the ACM*, 43:193–224, 1996.

Higher Order Generalization

Jianguo Lu[1], Masateru Harao[2], and Masami Hagiya[3]

[1] Robotics Institute, School of Computer Science, Carnegie Mellon University, USA
jglu@cs.cmu.edu

[2] Department of Artificial Intelligence, Kyushu Institute of Technology, Japan
harao@dumbo.ai.kyutech.ac.jp

[3] Department of Information Science, Graduate School of Science, University of Tokyo, Japan
hagiya@is.s.u-tokyo.ac.jp

Abstract. Generalization is a fundamental operation of inductive inference. While first order syntactic generalization (anti-unification) is well understood, its various extensions are needed in applications. This paper discusses syntactic higher order generalization in a higher order language $\lambda 2$[1]. Based on the *application ordering*, we proved the least general generalization exists and is unique up to *renaming*. An algorithm to compute the least general generalization is presented.

Keywords: Higher order logic, unification, anti-unification, generalization.

1 Introduction

The meaning of the word generalization is so general that we can find its occurrences in almost every area of study. In computer science, especially in the area of artificial intelligence, generalization serves as a foundation of inductive inference, and finds its applications in diverse areas such as inductive logic programming [10], theorem proving [12], program derivation [4][5]. In the strict sense, generalization is a dual problem of first order unification and is often called (ordinary) anti-unification [1]. More specifically, it can be formulated as: given two terms t and s, find a term r and substitutions θ_1 and θ_2, such that $r\theta_1 = t$ and $r\theta_2 = s$. Ordinary anti-unification was well understood as early as in 1970 [13][15]. Due to the fact that it is inadequate in many problems, there are extensions of ordinary anti-unification from various aspects.

One direction of extending the anti-unification problem is to take into consideration of some kinds of background information as in [10]. One typical example is the relative least general generalization under θ subsumption [14]. There are

[1] The words generalization and anti-unification are often used interchangeably. Here we will use anti-unification to denote the pure syntactic first order anti-unification, i.e., instantiation as the ordering, Robinson's formulation as the language. We use generalization to denote its various extensions.

various generalization methods in the area of inductive logic programming. More recently, there are generalizations under implication[8], and generalizations in constraint logic[11].

Another direction of extension is to promote the order of the underlying language. The problem with higher order generalization is that without some restrictions, the generalization is not well-defined. For example, the common generalizations of Aa and Bb without restriction would be: fx, fa, fb, fab, fA, fB, ..., $f(Aa, Bb)$, $f(g(A, B), g(a, b))$, ...,where f and g are variables. Actually, there are infinite number of generalizations. Obviously, some restrictions must be imposed on higher order generalization.

This paper is devoted to the study of higher order generalization. More specifically, we study the conditions under which the least higher order generalization exist and unique. The study is directly motivated by our research on analogical(inductive) programming and analogical(inductive) theorem proving. The most closely related works are [12] [3].

[12] studied generalization in a restricted form of calculus of constructions [2], where terms are higher-order patterns, i.e., free variables can only apply to distinct bound variables. One problem of the generalization in higher-order patterns is the over generalization. For example, the least generalization of Aa and Ba would be a single variable x instead of fa or fx, where we suppose A, B, a are constants, and f, x are variables. Another problem of higher-order pattern is that it is inadequate to express some problems. In particular, it can not represent recursion in its terms.

This motivated the study of generalization in $M\lambda$ [3]. In Mλ, free variables can apply to object term, which can contain constants and free variables in addition to bound variables. In this sense, $M\lambda$ extends $L\lambda$. On the other hand, it also added some restrictions. One restriction is that $M\lambda$ is situated in a simply typed λ calculus instead of calculus of constructions. Another restriction is $M\lambda$ does not have type variables, hence it can only generalize two terms of the same type. The result is not satisfactory in that the least general generalization is unique up to *substitution*. That means any two terms beginning with functional variables are considered equal.

Unlike the other approaches, which mainly put restrictions on the situated language, we mainly restrict the notion of the ordering between terms. Our discussion is situated in a restricted form of the language $\lambda2$[1]. The reason to choose $\lambda2$ is that it is a simple calculus which allows type variables. It can be used to formalise various concepts in programming languages, such as type definition, abstract data types, and polymorphism. The restriction we added is that abstractions should not occur inside arguments. In the restricted language $\lambda2$, we propose the following:

- an ordering between terms, called *application ordering*(denoted as $\succeq$), which is similar to, but not the same as the substitution (instantiation) ordering [15][12].

- A kind of restriction on orderings, called *subterm restriction* (the corresponding ordering is denoted as $\succeq_S$), which is implicit in first order languages, but usually not assumed in higher order languages.
- An extension to the ordering, called *variable freezing* (the corresponding ordering is denoted as $\succeq_{SF}$), which makes the ordering more useful while keeping the matching and generalization problems decidable.
- A generalization method based on the afore-mentioned ordering.

Based on the $\succeq_{SF}$ ordering, we have the following results similar to the first order anti-unification:

- For any two terms t and s, $t \succeq_{SF} s$ is decidable.
- The least general generalization exists.
- The least general generalization is unique up to *renaming.*

2 Preliminaries

The syntax of the restricted $\lambda 2$ can be defined as follows[1]:

Definition 1 (types and terms) *The set of types is defined as:*
$V = \{\alpha, \alpha_1, \alpha_2, ...\}$, *(type variables),*
$C = \{\gamma, \gamma_1, \gamma_2, ...\}$, *(type constants),*
$T = V|C|T \rightarrow T||[V]T$, *(types).*

The set of terms is defined as:
$X = \{x, x_1, x_2, ...\}$, *(variables),*
$A = \{a, a_1, a_2, ...\}$, *(constants),*
$\Lambda_1 = X|A|\Lambda_1\Lambda_1|\Lambda T$ *,(terms without abstraction),*
$\Lambda = \Lambda_1||[X : T]\Lambda||[V]\Lambda$ *, (terms).*

Here for the purpose of convenience, we use $[x : \sigma]$ instead of $\lambda x : \sigma$. Also, we use the same notation $[V]$ to denote ΛV (and $\forall V$), since we can distinguish among λ, Λ and $\forall$ from the context.

The assignment rules of $\lambda 2$ are listed here for ease of reference:

Definition 2 *Let σ, γ are types. $\Gamma \vdash t : \sigma$ is defined by the following axiom and rules:*

(start) $\quad \Gamma \vdash x : \sigma$, *if* $(x : \sigma) \in \Gamma$;

$$(\rightarrow E) \quad \frac{\Gamma \vdash t : (\sigma \rightarrow \tau),\ \Gamma \vdash s : \sigma}{\Gamma \vdash ts : \tau}$$

$$(\rightarrow I) \quad \frac{\Gamma, x : \sigma \vdash t : \tau}{\Gamma \vdash [x : \sigma]t : (\sigma \rightarrow \tau)}$$

$$(\forall E) \quad \frac{\Gamma \vdash t : [\alpha]\sigma}{\Gamma \vdash t\tau : \sigma[\alpha := \tau]}$$

$$(\forall I) \quad \frac{\Gamma \vdash t : \sigma}{\Gamma \vdash [\alpha]t : [\alpha]\sigma}, \text{ if } \alpha \notin FV(\Gamma).$$

We call a term t is valid (under Γ) if there is a type σ such that $\Gamma \vdash t : \sigma$. We use $Typ(t)$ to denote the type of t. *Atoms* are either constants or variables. By *closed terms* we mean the terms that do not contain occurrences of free variables. In the following discussion, if not specified otherwise, we assume all terms are closed, and in long $\beta\eta$ normal form. Given $\Delta \equiv [x_1 : \sigma_1][x_2 : \sigma_2]...[x_n : \sigma_n]$ and term t, $[\Delta]t$ denotes $[x_1 : \sigma_1][x_2 : \sigma_2]...[x_n : \sigma_n]t$. When type information is not important, $[x : \sigma]t$ is abbreviated as $[x]t$.

Following [12], we have a similar notion of *renaming*. Given natural numbers n and p, a partial permutation ϕ from n into p is an injective mapping from $\{1, 2, ..., n\}$ into $\{1, 2, ..., p\}$. A *renaming* of a term $[x_1 : \sigma_1][x_2 : \sigma_2]...[x_p : \sigma_p]t$ is a valid and closed term $[x_{\phi(1)} : \sigma_{\phi(1)}][x_{\phi(2)} : \sigma_{\phi(2)}]...[x_{\phi(n)} : \sigma_{\phi(n)}]t$. Intuitively, *renaming* is to permute and to drop some of the abstractions when allowed. For example, $[x_3, x_1 : \gamma]Ax_1x_3$ is a renaming of $[x_1, x_2, x_3 : \gamma]Ax_1x_3$.

3 Application orderings

3.1 Application ordering ($\succeq$)

Definition 3 ($\succeq$) *Given two terms t and s. t is more general than s (denoted as $t \succeq s$) if there exists a sequence of terms and types $r_1, r_2, ..., r_k$, such that $tr_1r_2...r_k$ is valid, and $tr_1r_2...r_k = s$. Here k is a natural number.*

To distinguish $\succeq$ with the usual instantiation ordering(denote it as $\geq$), we call $\succeq$ the *application ordering*. Compared with the instantiation ordering, the application ordering does not lose generality in the sense that for every two terms t and s in $\lambda 2$, if $t \geq s$, and t_1 and s_1 are the closed form of t and s, then $t_1 \succeq_F s_1$, where $\succeq_F$ is defined in section 3.3.

Example 1 *The following are some examples of the application ordering.*
$[\alpha][f : \alpha \to \alpha \to \alpha][x, y : \alpha]fxy$
$\succeq [f : \gamma \to \gamma \to \gamma][x, y : \gamma]fxy$
$\succeq [x, y : \gamma]Axy$
$\succeq [y : \gamma]Aay$
$\succeq Aab.$

$\succeq$ is reflexive and transitive:

Proposition 1 *For any terms t, t_1, t_2, t_3,*

1. $t \succeq t$;
2. *If* $t_1 \succeq t_2$, $t_2 \succeq t_3$, *then* $t_1 \succeq t_3$.

Proof:

1. Trivial.
2. Since $t_1 \succeq t_2$, $t_2 \succeq t_3$, there are a sequence of terms or types $r_{11}, r_{12}, ..., r_{1n}$, $r_{21}, r_{22}, ..., r_{2m}$, such that $t_1 r_{11} r_{12} ... r_{1n} = t_2$, and $t_2 r_{21} r_{22} ... r_{2m} = t_3$, hence $t_1 r_{11} r_{12} ... r_{1n} r_{21} r_{22} ... r_{2m} = t_3$. □

3.2 Application ordering with subterm restriction ($\succeq_S$)

Because $\succeq$ is too general to be of practical use, we restrict the relation to $\succeq_S$, called subterm restriction. First of all, we define the notion of subterms.

Definition 4 (subterm) *The set of subterms of term* t *(denoted as* $subterm(t)$*) is defined as* $decm(norm(t)) \cup \{Typ(r') | r' \in decm(norm(t))\}$.
Here $norm(t)$ *is to get the* $\beta\eta$ *normal form for the term* t. $decm(r)$ *is to decompose terms recursively into a set of its components, which is defined as:*

1. $decm(c) = \{c\}$ *(constants remain the same);*
2. $decm(z) = \{\}$; *(variables are filtered out);*
3. $decm(ts) = decm(t) \cup decm(s) \cup \{ts\}$, *if there is no variable in* ts;
$= decm(t) \cup decm(s)$, *otherwise;*
4. $decm([d]t) = decm(t)$.

Example 2 *Assume* $A : \gamma \to \gamma \to \gamma, B : \gamma \to \gamma$,
$subterm([x : \gamma]Axa) = \{[x, y : \gamma]Axy, a, \gamma, \gamma \to \gamma \to \gamma\}$
$subterm([f : \gamma \to \gamma][x : \gamma]f(Bx)) = \{[x : \gamma]Bx, \gamma \to \gamma\}$.

As we can see, the subterms do not contain free variables. Actually, there is no bound variables except the term having its η normal form (the $[x, y : \gamma]Axy$ in the above example). Here we exclude the *identity* and *projection* functions as subterms. This is essential to guarantee there exists least generalization in the application ordering. The intuitive behind this is that when we match two higher order terms, in general there are *imitation* rule and *projection* rule [6]. Here only *imitation* rule is used. We regard it is *projection* rule that brings about the unpleasant results and the complexities in higher order generalizations.

Definition 5 ($\succeq_S$) *Given two terms* t *and* s. t *is more general than* s *by subterms (denoted as* $t \succeq_S s$*), if there exists a sequence of* $r_1, r_2, ..., r_k$, *such that* $t r_1 r_2 ... r_k = s$. *Here* $r_i \in subterm(s), i \in \{1, 2, ..., k\}$, *and* k *is a natural number.*

Example 3 $[f][x]fx \succeq_S Aa$;
$[f][x]fx \succeq_S Bbc$;
$[\alpha][x : \alpha]x \succeq_S [x : \gamma]x \succeq_S Aa$;
$[f][x]fx \not\succeq_S a$, *since the only subterm of* a *is* a.

Due to the finiteness of *subset*(s), the ordering $\succeq_S$ becomes much easier to manage than $\succeq$.

Proposition 2 *For any terms* t_1, t_2, t_3,

1. *There exists a procedure to decide if* $t_1 \succeq_S t_2$.
2. *If* $t_1 \succeq_S t_2$, $t_2 \succeq_S t_3$, *then* $t_1 \succeq_S t_3$.

Proof:

1. Since the *subterm* of t_1 is finite, it is obvious.
2. Since $t_1 \succeq_S t_2$, $t_2 \succeq_S t_3$, there are a sequence of terms or types $r_{11}, r_{12}, ...,$ $r_{1n} \in subterm(t_2)$, $r_{21}, r_{22}, ..., r_{2m} \in subterm(t_3)$, such that $t_1 r_{11} r_{12} ... r_{1n} = t_2$, and $t_2 r_{21} r_{22} ... r_{2m} = t_3$. Hence $t_1 r_{11} r_{12} ... r_{1n} r_{21} r_{22} ... r_{2m} = t_3$. Besides, since we can not eliminate constants in t_2 when applying terms to it, and $r_{11}, r_{12}, ..., r_{1n}$ are constants in t_2, so $r_{11}, r_{12}, ..., r_{1n}$ must also be the subterms of t_3. Hence we have $t_1 \succeq_S t_3$. □

3.3 Application ordering with subterm restriction and variable freezing extension ($\succeq_{SF}$)

The ordering $\succeq_S$ is restrictive in that $[x][y]Axy \not\succeq_S [x]Axa$. To solve this problem, we have:

Definition 6 ($\succeq_F$) *t is a generalization of s by variable freezing, denoted as* $t \succeq_F s$, *if either*

- $t \succeq s$, *or*
- *for an arbitrary type constant or term constant c such that sc is valid,* $t \succeq_F sc$.

Intuitively, here we first freeze some variables in s as a constant, then try to do generalization. The word *freeze* comes from [7], which has the notion that when unifying two free variables, we can regard one of them as a constant.

The ordering $\succeq_F$ is too general to be managed, so we have the following restricted form:

Definition 7 ($\succeq_{SF}$) $t \succeq_{SF} s$, *if either*

- $t \succeq_S s$, *or*
- *For an arbitrary type constant or term constant c such that sc is valid,* $t \succeq_{SF} sc$.

Now we have $[x][y]Axy \succeq_{SF} [x]Axa$. The notion of $\succeq_{SF}$ not only mimics, but also extends the usual meaning of instantiation ordering. For example, we have $[x, y]Axy \succeq_{SF} [x]Axx$, which can not be obtained in the instantiation ordering.

Example 4 *The following relations holds:*

$$[\alpha][x:\alpha]x \succeq_{SF} [\alpha][f:\alpha\rightarrow\alpha][x:\alpha]fx$$
$$\succeq_S [f:\gamma\rightarrow\gamma][x:\gamma]fx$$
$$\succeq_S [x:\gamma]Ax$$
$$\succeq_S Aa;$$

$[f][z,x,y]f(Axy,z) \succeq_S [z,x,y]A(Axy,z) \succeq_S A(Aab,Aab)$;

$[f][z,x,y]f(Axy,z) \not\succeq_{FS} [x,y]A(Axy,Axy)$; *since* Axy *is not a subterm of* $[x,y]A(Axy,Axy)$.

$[\alpha][f:\alpha\rightarrow\alpha][x:\alpha]fx \not\succeq_{SF} [\alpha][x:\alpha]x$, *since identity and projection functions are not subterms.*

Proposition 3 *For any terms* t *and* s,

1. $t \succeq_{SF} s$ *iff there exists a sequence (possibly an empty sequence) of new, distinct constants* $c_1, c_2, ..., c_k$, *such that* $sc_1c_2...c_k$ *is of atomic type, and* $t \succeq_S sc_1c_2...c_k$.
2. *There exists a procedure to decide if* $t \succeq_{SF} s$.
3. *Suppose* $t = s$. *If* $t \succeq_{SF} r$, *then* $s \succeq_{SF} r$. *If* $r \succeq_{SF} t$, *then* $r \succeq_{SF} s$.

Proof:

1. ($\Rightarrow$) Suppose $t \succeq_{SF} s$. If s is of atomic type, then it is trivial. Now suppose s is of type $\sigma\rightarrow\tau$, c is a constant of type σ. If $t \succeq_S s$, then $t \succeq_S sc$. If $t \not\succeq_S s$. By definition of $\succeq_{SF}$, there exists c such that $t \succeq_{SF} sc$.
   ($\Leftarrow$) Suppose there exists a sequence of new constants $c_1, c_2, ..., c_k$, such that $sc_1c_2...c_k$ is of atomic type, and $t \succeq_S sc_1c_2...c_k$. By definition of $\succeq_{SF}$, $t \succeq_{SF} sc_1c_2...c_{k-1}$, $t \succeq_{SF} sc_1c_2...c_{k-2}$, ..., $t \succeq_{SF} s$.
2. Since $t \succeq_{SF} s$ iff $t \succeq_S sc_1c_2...c_k$, and we know $\succeq_S$ is decidable, hence $t \succeq_{SF} s$ is decidable.
3. If $t \succeq_{SF} r$, then there exists a sequence of new constants $c_1, c_2, ..., c_k$, such that $rc_1c_2...c_k$ is of atomic type, and $t \succeq_S rc_1c_2...c_k$. There exists a sequence of terms or types $r_1, ..., r_i$, such that $tr_1...r_i = rc_1c_2...c_k$. Since $t = s$, we have $sr_1...r_i = rc_1c_2...c_k$, $s \succeq_{SF} r$.
   The second proposition can be proved in the similar way. □

Proposition 4 *Suppose* $t_1 \equiv [\Delta]hs_1s_2...s_m$, $t_2 \equiv [\Delta']h's'_1s'_2...s'_n$, *and* $t_1 \succeq_{SF} t_2$, *then*

1. $m \leq n$,
2. $[\Delta]s_k \succeq_{SF} [\Delta']s'_{k+n-m}$, *for* $k \in \{1,2,...,m\}$,
3. *If* h *is a constant, then* h' *must be a constant, and* $h = h', m = n$.

Proof: Suppose $[\Delta'] = [z_1, z_2, ..., z_j]$. Since $t_1 \succeq_{SF} t_2$, we have $[\Delta]hs_1s_2...s_m \succeq_S (h's'_1s'_2...s'_m)[\bar{c}/\bar{z}]$, where $\bar{z}$ is a sequence $z_1, ..., z_j$, $\bar{c}$ is a sequence of new constant symbols $c_1, ..., c_j$. Now, suppose each variable in $h's'_1s'_2...s'_m$ is fixed as a new constants, $hs_1s_2...s_m$ should match $h's'_1s'_2...s'_m$ in the sense of [Huet78]. As we know, the complete minimal matches are generated by the imitation rule and the projection rule. Since in the projection rule, the

substitutions are $\{h \to [x_1, ..., x_m]x_i | i \in \{1, ..., m\}\}$, which do not satisfy our subterm restriction. Thus the only way to match is by using the imitation rule. By imitation rule we have substitutions
$\{h \to [x_1, ..., x_m]h'(h_1 x_1 ... x_m) ... (h_n x_1 ... x_m)\}$, where $h_1, ... h_n$ are new variables. On the other hand, the subterms of $h's'_1 s'_2 ... s'_n$ [2] whose head is h' could only be:
$[x_1, ..., x_n]h'x_1 x_2 ... x_n$,
$[x_2, ..., x_n]h's''_1 x_2 ... x_n$,
... ...
$[x_{i+1}, ..., x_n]h's''_1 s''_2 ... s''_i x_{i+1} ... x_n$,
... ...,
where each s''_j is either s'_j, or other possible terms inside the arguments if h' also occurs in the arguments. So the possible substitution must be $h \to [x_{i+1}, ..., x_n]h's''_1 s''_2 ... s''_i x_{i+1} ... x_n$, where $i + m = n$. After the substitution, we have to match the terms $h's''_1 ... s''_{n-m} s_1 s_2 ... s_m$ and $h's'_1 s'_2 ... s'_n$, i.e., $[\Delta]s_k \succeq_{SF} [\Delta']s'_{k+n-m}$, for $k \in \{1, 2, ..., m\}$.

When h is a constant, it is obvious that $h' = h$. □

It is clear that $\succeq_{SF}$ is reflexive and transitive:

Proposition 5 *For any terms* t, t_1, t_2, t_3,

1. $t \succeq_{SF} t$.
2. *If* $t_1 \succeq_{SF} t_2$, $t_2 \succeq_{SF} t_3$, *then* $t_1 \succeq_{SF} t_3$.

Proof:

1. Obvious.
2. We can suppose
   $t_1 \equiv [\Delta]hs_1 s_2 ... s_m$,
   $t_2 \equiv [\Delta']h'r_{11} ... r_{1i} s'_1 s'_2 ... s'_m$,
   $t_3 \equiv [\Delta'']h''r_{21} ... r_{2j} r_{31} ... r_{3i} s''_1 s''_2 ... s''_m$,
   Case 1: $m = 0$, then it is easy to verify $t_1 \succeq_{SF} t_3$.
   Case 2: $m > 0$. We have $[\Delta]s_k \succeq_{SF} [\Delta']s'_k \succeq_{SF} [\Delta'']s''_k$, for $k \in \{1, ..., m\}$. By inductive hypothesis, $[\Delta]s_k \succeq_{SF} [\Delta'']s''_k$. If h is a constant, we have $h = h' = h'', i = j = 0$, thus $t_1 \succeq_{SF} t_3$. If h is a variable, let h substitute $h''r_{21} ... r_{2j} r_{31} ... r_{3i}$.

□

Definition 8 ($\cong$) $t \cong s$ *is defined as* $t \succeq_{SF} s$ *and* $s \succeq_{SF} t$.

Example 5 $[x, y]Axy \cong [y, x]Axy \cong [z, x, y]Axy$.

Proposition 6 $t \cong s$ *iff* t *is a renaming of* s.

Proof: ($\Rightarrow$) Assume $t \cong s$, then $t \succeq_{SF} s$ and $s \succeq_{SF} t$. Suppose $t \equiv [\Delta]ht_1 t_2 ... t_m$, $s \equiv [\Delta']h's_1 s_2 ... s_n$. Since $t \succeq_{SF} s$, we have $m \geq n$. And similarly, we have $n \geq m$. So, $m = n$. If h is a constant, then $h' = h$. Similarly,

[2] Here the variables are frozen.

we can have if h' is a constant then $h' = h$. Hence, h and h' must be either the same constant, or a variable.
Case 1. $m = 0$. Obviously t and s only differs by renaming.
Case 2. $m > 0$. We have $[\Delta]t_k \succeq_{SF} [\Delta']s_k$, and $[\Delta']s_k \succeq_{SF} [\Delta]t_k$, for $k \in \{1, ..., m\}$. By inductive hypothesis, $[\Delta]t_k$ and $[\Delta']s_k$ only differs by variable renaming. On the other hand, h and h' are either variables or the same constant.

($\Leftarrow$) We only need to consider the following two cases:

Case 1. Suppose
$t \equiv [x_1, x_2, ..., x_i]ht_1t_2...t_m$,
$s \equiv [x_{\phi(1)}, x_{\phi(2)}, ..., x_{\phi(i)}]ht_1t_2...t_m$,
Then $tc_1...c_i = sc_{\phi(1)}...c_{\phi(i)}$, $t \cong s$.

Case 2. Suppose
$t \equiv [x][x_1, x_2, ..., x_i]ht_1t_2...t_m$,
$s \equiv [x_1, x_2, ..., x_i]ht_1t_2...t_m$,
where x does not occur in $ht_1t_2...t_m$. Then $tc = s$, $s \succeq_{SF} t$. Also, we have $t \succeq_{SF} s$, hence $t \cong s$.

Case 3. Suppose
$t \equiv [g : \gamma_1 \to \gamma_2 \to ... \to \gamma_i \to \gamma]gt_1t_2...t_m$,
$s \equiv [f : \gamma_{\phi(1)} \to \gamma_{\phi(2)} \to ... \to \gamma_{\phi(i)} \to \gamma]ft_{\phi(1)}t_{\phi(2)}...t_{\phi(m)}$,
$tA = s([x_{\phi(1)} : \gamma_{\phi(1)}, x_{\phi(2)} : \gamma_{\phi(2)}, ..., x_{\phi(i)} : \gamma_{\phi(i)}]Ax_1x_2...x_i)$, hence $t \cong s$. □

4 Generalization

If $t \succeq_{SF} s_1$ and $t \succeq_{SF} s_2$, then t is called a *common generalization* of s_1 and s_2. If t is a common generalization of s_1 and s_2, and for any common generalization t_1 of s_1 and s_2, $t_1 \succeq_{SF} t$, then t is called the *least general generalization (LGG)*. This section only concerned with $\succeq_{SF}$, hence in the following discussion the subscript SF is omitted.

The following algorithm $Gen(t, s, \{\})$ computes the least general generalization of t and s. Recall we assume t and s are closed terms. At the beginning of the procedure we suppose all the bound variables in t and s are distinct. Here an auxiliary (the third) global variable C is needed to record the previous correspondence between terms in the course of generalization, so that we can avoid to introduce unnecessary new variables. C is a bijection between pairs of terms(and types) and a set of variables. Initially, C is an empty set. Following the usual practice, it is sufficient to consider only long $\beta\eta$-normal forms. Not losing generality, suppose t and s are of the following forms:

$t \equiv [\Delta]h(t_1, t_2, ..., t_k)$,
$s \equiv [\Delta']h'(r_1, ..., r_i, s_1, s_2, ..., s_k)$, where h and h' are atoms. Suppose
$[\Delta, \Delta', \Delta_1]t'_1 = Gen([\Delta]t_1, [\Delta']s_1, C)$,
$[\Delta, \Delta', \Delta_2]t'_2 = Gen([\Delta, \Delta_1]t_2, [\Delta', \Delta_1]s_2, C)$,
...
$[\Delta, \Delta', \Delta_k]t'_k = Gen([\Delta, \Delta_{k-1}]t_k, [\Delta', \Delta_{k-1}]s_k, C)$,

$Typ(h) = \sigma_1, \sigma_2, ..., \sigma_k \rightarrow \sigma_{k+1}$,
$Typ(h'(r_1, ..., r_i)) = \tau_1, \tau_2, ..., \tau_k \rightarrow \tau_{k+1}$.
$Gen(t, s, \mathcal{C})$:
Case 1: $h = h' : Gen(t, s, \mathcal{C}) = [\Delta, \Delta', \Delta_k]h(t'_1, t'_2, ..., t'_k)$;
Case 2: $h \neq h'$:
Case 2.1: $\exists x.((h, h'(r_1, ..., r_i)), x) \in \mathcal{C}$:
$Gen(t, s, \mathcal{C}) = [\Delta, \Delta', \Delta_k]x(t'_1, t'_2, ..., t'_k)$
Case 2.2: $\neg\exists x.((h, h'(r_1, ..., r_i)), x) \in \mathcal{C}$,
Case 2.2.1: $Typ(h) = Typ(h'(r_1, ..., r_i))$:
$Gen(t, s, \mathcal{C}) = [\Delta, \Delta', \Delta_k][x : \sigma_1, \sigma_2, ..., \sigma_k \rightarrow \sigma_{k+1}]x(t'_1, t'_2, ..., t'_k)$
$\mathcal{C} := \{((h, h'(r_1, ..., r_i)), x)\} \cup \mathcal{C}$;
Case 2.2.2: $Typ(h) \neq Typ(h'(r_1, ..., r_i))$:
Not losing generality, suppose $\sigma_j \neq \tau_j, j \in \{1, 2, ..., k, k+1\}$.
Case 2.2.2.1: $\exists \alpha_j.((\sigma_j, \tau_j), \alpha_j) \in \mathcal{C}$:
$Gen(t, s, \mathcal{C}) = [\Delta, \Delta', \Delta_k][x : \sigma_1, ..., \alpha_j, ... \rightarrow \sigma_{k+1}]x(t'_1, t'_2, ..., t'_k)$;
$\mathcal{C} := \{((h, h'(r_1, ..., r_i)), x)\} \cup \mathcal{C}$;
Case 2.2.2.2: $\neg\exists\alpha.((\sigma_j, \tau_j), \alpha_j) \in \mathcal{C}$:
$Gen(t, s, \mathcal{C}) = [\Delta, \Delta', \Delta_k][\alpha_j][x : \sigma_1, ..., \alpha_j, ... \rightarrow \sigma_{k+1}]x(t'_1, t'_2, ..., t'_k)$;
$\mathcal{C} := \{((h, h'(r_1, ..., r_i)), x)\} \cup \mathcal{C}$;
$\mathcal{C} := \{((\sigma_j, \tau_j), \alpha_j)\} \cup \mathcal{C}$.
In the following, let $t \sqcup s \equiv Gen(t, s, \{\})$.

Example 6 *Some examples of least general generalization.*
$[x : \gamma]x \sqcup Aa = [x : \gamma][\alpha][y : \alpha]y \cong [\alpha][y : \alpha]y$, *if* Aa *is not of type* γ;
$[x : \gamma]x \sqcup Aa = [x : \gamma][y : \gamma]y \cong [x : \gamma]x$, *if* Aa *is of type* γ;
$[x]Axx \sqcup [x]Aax \cong [x, y]Axy$;
$Aa \sqcup Bb \cong [f][x]fx$, *if* A *and* B *is of the same type;*
$Aa \sqcup Bb \cong [\alpha][f : \alpha \rightarrow \gamma][x : \alpha]fx$, *if* $A : \gamma_1 \rightarrow \gamma$ *and* $B : \gamma_2 \rightarrow \gamma$;

Example 7 *Here is an example of generalizing segments of programs. For clarity the segments are written in usual notation. Let* $t \equiv [x]map1(cons(a, x)) = cons(succ(a), map1(x))$,
$s \equiv [x]map2(cons(a, x)) = cons(sqr(a), map2(x))$.
Suppose the types are
$map1 : List(Nat) \rightarrow Nat;\ succ : Nat \rightarrow Nat$,
$map2 : List(Nat) \rightarrow Nat;\ sqr : Nat \rightarrow Nat$.
Then
$t \sqcup s \cong [f : List(Nat) \rightarrow Nat; g : Nat \rightarrow Nat)][x]f(cons(a, x)) = cons(g(a), f(x)))$.

The termination of the algorithm is obvious, since we recursively decompose the terms to be generalized, and the size of the terms strictly decreases in each step. What we need to prove is the uniqueness of the generalization. The following can be proved by induction on the definition of terms:

Proposition 7 *1. (consistency)* $t \sqcup s \succeq t$, $t \sqcup s \succeq s$.
2. (termination) For any two term t *and* s, $Gen(t, s, \{\})$ *terminates.*

3. *(absorption) If* $t \succeq s$*, then* $t \sqcup s \cong t$.
4. *(idempotency)* $t \sqcup t \cong t$.
5. *(commutativity)* $t \sqcup s \cong s \sqcup t$.
6. *(associativity)* $(t \sqcup s) \sqcup r \cong t \sqcup (s \sqcup r)$.
7. *If* $t \cong s$*, then* $t \sqcup r \cong s \sqcup r$.
8. *(monotonicity) If* $t \succeq s$*, then for any term* r, $t \sqcup r \succeq s \sqcup r$.
9. *If* $t \cong s$*, then* $t \sqcup s \cong t \cong s$.

Proof:

1. It can be verified that for each case of the algorithm, we obtained a more general term.
2. It is obvious since we decompose the terms recursively.
3. Since $t \succeq s$, we can suppose
   $t \equiv [\Delta]hs_1s_2...s_m$,
   $s \equiv [\Delta']h'r_{11}...r_{1i}s'_1s'_2...s'_m$,and
   $[\Delta]s_k \succeq [\Delta']s'_k, k \in \{1, ..., m\}$.
   If $m = 0$, then it is easy to verify the conclusion. Now suppose $m > 0$. Not losing generality, suppose h is a variable which does not occur in s, and has a single occurrence in t. h has the same type as $h'r_{11}...r_{1i}$. Other cases can be proved in a similar way. Now we can suppose $t \sqcup s \equiv [\Delta''][f]ft_1t_2...t_m$. If s_k is a constant, then s'_k must be the same constant. Hence $t_k \equiv s_k$. If s_k is a variable, then t_k is a new variable. There are two cases: one is s_k has only one occurrence in t. Then t' and t only differs by renaming. The other case is that s_k has multiple occurrences in t. Since $t \succeq s$, all the occurrences of s_k must correspond to a same term in s. Hence due to the presence of the global variable $\mathcal{C}$, all the occurrences of s_k are generalized as a same variable. Hence $t \cong t'$. By inductive hypothesis, we have
   $[\Delta]s_k \sqcup [\Delta']s'_k \cong [\Delta]s_k$.
4. From $t \succeq t$ and the proposition 7.3 we can have the result.
5. It is obvious from the algorithm.
6. Not losing generality, we can suppose
   $t \equiv [\Delta]hs_1s_2...s_m$,
   $s \equiv [\Delta']h'r_{11}...r_{1i}s'_1s'_2...s'_m$,
   $r \equiv [\Delta'']h''r_{21}...r_{2j}r_{31}...r_{3i}s''_1s''_2...s''_m$,
   and suppose h, h', h'' are distinct constants, t, s, r are of the same type. The other cases can be proved in a similar way. By inductive hypothesis, for $k \in \{1, ..., m\}, p \in \{1, ..., i\}$, we can suppose:
   $([\Delta]s_k \sqcup [\Delta']s'_k) \sqcup [\Delta'']s''_k \cong [\Delta]s_k \sqcup ([\Delta']s''_k \sqcup [\Delta'']s''_k)$,
   $[\Delta]s_k \sqcup [\Delta']s'_k \cong [\Gamma]t_k$,
   $[\Gamma]t_k \sqcup [\Delta'']s''_k \cong [\Gamma'']t''_k$,
   $[\Delta']s'_k \sqcup [\Delta'']s''_k \cong [\Gamma']t'_k$,
   $[\Delta]s_k \sqcup [\Gamma']t'_k \cong [\Gamma'']t''_k$,
   $[\Delta']r_{1p} \sqcup [\Delta'']r_{3p} \cong [\Gamma']r_p$.
   Here we suppose each $\Gamma, \Gamma', \Gamma''$ are large enough to cover all the abstractions in $t_1, ..., t_m$, $t'_1, ..., t'_m$, and $t_1, ..., t''_m$, respectively.

We rename the variables in $[\Gamma]t_1, ..., [\Gamma]t_m$ such that there are multiple occurrences of a variable x in $[\Gamma]t_1, ..., [\Gamma]t_m$ if and only if its corresponding places in $s_1, ..., s_m$ hold a same term, and its corresponding places in $s'_1, ..., s'_m$ hold another same term. Similarly, we rename the terms $[\Gamma']t'_k, [\Gamma'']t''_k$. Then
$(t \sqcup s) \sqcup r$
$\cong [\Gamma][f]ft_1...t_m \sqcup [\Delta'']h''r_{21}...r_{2j}r_{31}...r_{3i}s''_1s''_2...s''_m$
$\cong [\Gamma''][f]ft''_1...t''_m$,
$t \sqcup (s \sqcup r)$
$\cong [\Delta]hs_1s_2...s_m \sqcup [\Gamma'][g]gr_1...r_it'_1...t'_m$
$\cong [\Gamma''][g]gt''_1...t''_m$.
Hence $(t \sqcup s) \sqcup r \cong t \sqcup (s \sqcup r)$.

7. Since $t \cong s$, t is a renaming of s. t and s must be of the forms $[\Delta]hr_1r_2...r_n$ and $[\Delta']hr_1r_2...r_n$. It is obvious that $[\Delta]hr_1r_2...r_n \sqcup r \cong [\Delta']hr_1r_2...r_n \sqcup r$.
8. Since $t \succeq s$, we have $t \sqcup s \cong t$, hence
$t \sqcup r$
$\cong (t \sqcup s) \sqcup r$ (by proposition 7.7)
$\cong t \sqcup (s \sqcup r)$ (commutativity)
$\succeq s \sqcup r$ (by proposition 7.1).
9. From $t \cong s$, we have $t \succeq s, s \succeq t$. Hence $t \sqcup s \cong t$, $t \sqcup s \cong s \sqcup t \cong s$. □

Based on the above propositions, we can have

Theorem 1 *$t \sqcup s$ is the least general generalization of t and s, i.e., for any term r, if $r \succeq t, r \succeq s$, then $r \succeq t \sqcup s$.*

Proof: Since $r \succeq t, r \succeq s$, we have $r \sqcup t \cong r$, $r \sqcup s \cong r$.
$r \sqcup (t \sqcup s)$
$\cong (r \sqcup r) \sqcup (t \sqcup s)$ (idempotency)
$\cong (r \sqcup t) \sqcup (r \sqcup s)$ (commutativity and associativity)
$\cong r \sqcup r$ (absorption)
$\cong r$ (idempotency).
Hence by proposition 7.1 we have $r \succeq t \sqcup s$. □

Higher order generalization is mainly used to find schemata of programs, proof, or program transformations. For example, given first order clauses
$multiply(s(X), Y, Z) \leftarrow multiply(X, Y, W), add(W, Y, Z)$, and
$exponent(s(X), Y, Z) \leftarrow exponent(X, Y, W), multiply(W, Y, Z)$,
we can obtain its least general generalization as
$P(s(X), Y, Z) \leftarrow P(X, Y, W), Q(W, Y, Z)$.

Higher order generalization also finds its applications in analogy analysis[9]. It is commonly recognized that a good way to obtain the concrete correspondence between two problems is to obtain the generalization of the two problem first. During the generalization process, we should preserve the structure as much as possible. By using the above higher order generalization method, we can find the analogical correspondence between two problems in the course of generalization.

5 Discussions

With the subterm restriction and the freezing extension, we defined the ordering $\succeq_{SF}$. As we have shown, this ordering and the corresponding generalization have nice properties almost the same as the first order anti-unification. Especially, the least general generalization exists and is unique.

To have a comparison with other kinds of generalizations, we have the following diagram:

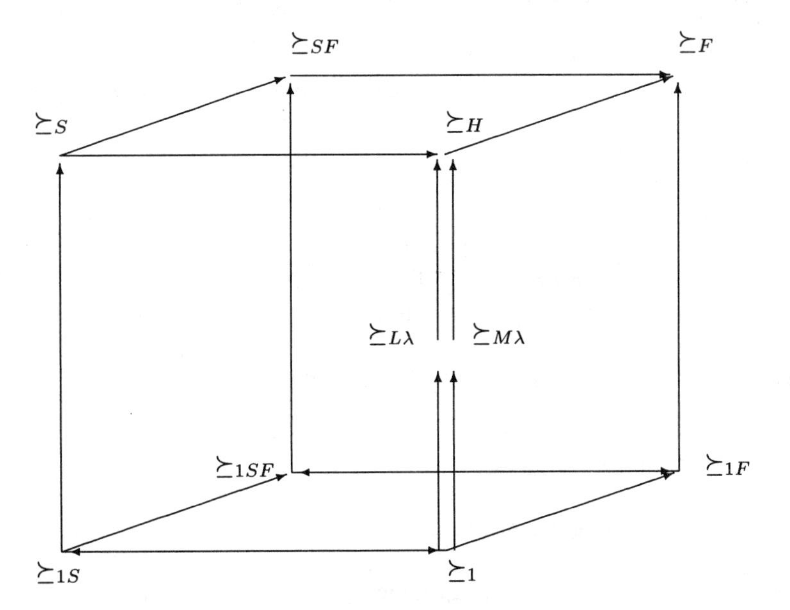

Here each vertex represents a kind of ordering. For example, $\succeq_H$ means the usual instantiate ordering in a higher order language, say $\lambda P2$ [1]. $\succeq_1$ the usual instantiation ordering in first order language, $\succeq_{M\lambda}$ the ordering in $M\lambda$, $\succeq_{L\lambda}$ the ordering in $L\lambda$ (i.e., in higher order patterns), etc.. The arrow means implication. For example, if $t \succeq_S s$, then $t \succeq_{SF} s$, and $t \succeq_H s$. It can be seen that the relations $\succeq_{SF}$ and $\succeq_H$ (also $\succeq_{L\lambda}$ and $\succeq_{M\lambda}$) are not comparable. By definition, $\succeq_{1S}$ (the ordering $\succeq_1$ with the subterm restriction) is the same as $\succeq_1$. That explains why we have good results in $\succeq_{SF}$.

Our work differs from the others in the following aspects. Firstly, we defined a new ordering $\succeq_{SF}$. In terms of this ordering, we obtain a much more specific generalization in general. For example, the terms Aab and Bab would be generalized as a single variable x in [12], or as fts in [3], where t and s are arbitrary terms. In contrast, we will have $[f]fab$ as its least general generalization. Secondly, our approach can produce a meaningful generalization of terms of different types and terms of different arities, instead a single variable x. And finally, our method is useful in applications, such as in analogical reasoning and inductive inference [5][9].

Acknowledgments

Part of the work is supported by the Japan Society of Promotion of Science, and the Education Ministry of Japan.

References

1. H. Barendregt, Introduction to generalized type systems, Journal of functional programming, Vol. 1, N0. 2, 1991. 124-154.
2. Coquand, T., Huet, G., The calculus of constructions, Information and Computation, Vol.76, No.3/4(1988), 95-120.
3. C.Feng, S.Muggleton, Towards inductive generalization in higher order logic, In D.Sleeman et al(eds.), Proceedings of the Ninth International Workshop on Machine Learning, San Mateo, California, 1992. Morgan Kaufman.
4. M. Hagiya, Generalization from partial parametrization in higher order type theory, Theoretical Computer Science, Vol.63(1989), pp.113-139.
5. R.Hasker, The replay of program derivations, Ph.D. thesis, Department of Computer Science, University of Illinois at Urbana-Champaign, 1995.
6. G.P.Huet, A unification algorithm for typed lambda calculus, Theoretical Computer Science, 1 (1975), 27-57.
7. G.Huet, Bernard Lang, Proving and applying program transformations expressed with second order patterns, Acta Informatica 11, 31-55(1978)
8. Peter Idestam-Almquist, Generalization of Horn clauses, Ph.D. dissertation, Department of Computer Science and Systems Science, Stockholm University and the Royal Institute of Technology, 1993.
9. Jianguo Lu, Jiafu Xu, Analogical Program Derivation based on Type Theory, Theoretical Computer Science, Vol.113, North Holland 1993, pp.259-272.
10. Stephen Muggleton, Inductive logic programming, New generation computing, 8(4):295-318, 1991
11. Charles David Page jr., Anti-unification in constraint logic: foundations and applications to learnability in first order logic, to speed-up learning, and to deduction, Ph.D. Dissertation, University of Illinois at Urbana-Champaign, 1993.
12. Frank Pfenning, Unification and anti-unification in the calculus of constructions, Proceedings of the 6th symposium on logic in computer science, 1991. pp.74-85.
13. Plotkin, G. D., A note on inductive generalization, Machine Intelligence 5, Edinburgh University Press 1970, pp. 153-163.
14. Plotkin, G.D., A further note on inductive generalization, Machine Intelligence 6, Edinburgh University Press 1971, pp. 101-124.
15. John C. Reynolds, Transformational systems and the algebraic structure of atomic formulas, Machine Intelligence 5, Edinburgh University Press 1970, 135-151.

The Logical Characterization of Goal-Directed Behavior in the Presence of Exogenous Events Summary

Erik Sandewall

Department of Computer and Information Science, Linkping University, Sweden
erisa@ida.liu.se

The Topic

Logics of action and change were originally conceived for the design of intellligent agents, as McCarthy proposed the situation calculus as the logic to be used by a logic-based "advice taker". In more recent research one can distinguish two motives for these logics, namely, for what is now called *cognitive robotics* and for *common-sense reasoning*. The difference is that in cognitive robotics, the logic must also account for an incoming flow of sensory data and for the actuation of actions by the robot, whereas common-sense reasoning is oriented towards natural-language communication of the premises and the conclusions of the reasoning process.

The present article addresses the case where a logic of actions and change is used for cognitive robotics purposes. An intelligent robotic agent will necessarily have a quite complex design, since it has to account for all of the following:

1. Goal-directed behavior, including the principle of *deliberated retry*, that is, if the robot takes an action in order to pursue a goal and the action fails, then it is to consider some other action or sequence of actions that is likely to achieve the same goal.

2. World modelling and the description of the effects of actions on *several levels of detail*, ranging from simple approximations such as precondition/ post-condition descriptions, to the detailed specification in control theory terms of how an action is performed.

3. The problem of *imprecise and unreliable sensors.*

4. The occurrence of *exogenous events*, including both the prediction and early reaction to those exogenous events that can reasonably be predicted, and the proper dealing with those exogenous events that come as a surprise to the agent. In fact, it is not sufficient that the agent be able to deal with exogenous events in prediction mode and synchronous mode (recognizing them at the moment they occur), but it must also sometimes be able to diagnose aberrant situations and understand them in terms of exogenous events that may have occurred earlier in time. In other words, it must be capable of postdiction as well as prediction.

In order to master the design of such complex devices, it is necessary to make use of concise and high-level specifications. It is somewhat remarkable, therefore, that most research on logics of actions and change has avoided the issues mentioned above. It has focussed on the actions performed by the agent, the restrictions that may be imposed on them, and their indirect effects due to some kind of causation. These logics have not generally been used for characterizing the behavior of the agent, goal-directed or otherwise, and there has only been a small number of contributions on reasoning about unreliable sensor data.

Research in the area of *intelligent systems* have approached some of these topics from another angle, and in a few recent cases they have also used logic as a means of characterizing the system design, complementing the traditional approach of describing the software architecture directly.

We believe that this area at the intersection of two current research traditions – reasoning about actions and change, and intelligent agent systems – will attract a lot more attention in the forthcoming years. Stringent and high-level specifications of intelligent systems will be necessary due to their inherent complexity and the severe requirements in practical applications, including the requirement of being able to assure correct behavior within a well defined envelope for what situations are assumed to arise. It will be natural to use logic for characterizing the behavior of the agent, and not only for characterizing the world in which the agent operates.

Our Approach

In the present invited lecture, we are going to describe our own recent work in the area that has now been described. In particular, [j-etai-1-105] (also available as [c-kr-98-304]) describes a method for characterizing the goal-directed behavior of deliberated retry in a first-order logic of time and action. Also, [j-aicom-9-214] and [c-hart-97-3] describe a method of relating high-level and low-level action descriptions in a similar logic. The results in those articles are complementary: the first articles address item 1 in the list of issues that was given above, whereas the latter two articles address items 2, 4, and to some extent item 3 in that list.

The common logical framework for these articles is *Time and Action Logic* (TAL), that is, the direct representation of actions and of the state of the world in first-order predicate calculus. (This is the representation that we have been using consistently since 1989, although previously without a specific label). In TAL each interpretation represents one possible history of the world, in contradistinction to e.g. the situation calculus where an interpretation contains a tree of possible developments. (The difference is actually a bit less stringent, on both sides, but that is outside the present topic). For the characterization of goal-directed agent behavior, each model will therefore represent one history of the world *in which the agent exhibits the required behavior pattern of deliberated retry* in the face of the problems that may arise.

In combining the results of those earlier articles, we obtain a logic that is able to represent the failure of actions due to exogenous events, and an agent design where the agent is able to revise its plan and make a new attempt to reach a goal after its present approach has failed.

The present lecture will review the contents of the previous articles, and show through a concrete example how the combined system works.

Related Work

As a joint reference for these and other articles, we are maintaining a webpage structure with a summary of related work by others and links to those other articles. This structure is continuously extended with new references and commentary, so it is kept up-to-date in a way that a reference list in an article can not be. Please refer to the URL of the ETAI article for links to all such further information in its up-to-date form.

References

[c-hart-97-3] Erik Sandewall. *Relating high-level and low-level action descriptions in a logic of actions and change.* Oded Maler (ed): Hybrid and Real-Time Systems, 1997, pp. 3-17. Springer Verlag.

[c-kr-98-304] Erik Sandewall. *Logic Based Modelling of Goal-Directed Behavior.* Proc. International Conference on Knowledge Representation and Reasoning, 1998, pp. 304-315.

[j-aicom-9-214] Erik Sandewall. *Towards the validation of high-level action descriptions from their low-level definitions.* AI Communications, vol. 9 (1996), pp. 214-224.

[j-etai-1-105] Erik Sandewall. *Logic-Based Modelling of Goal-Directed Behavior.* Electronic Transactions on Artificial Intelligence, vol. 1, pp. 105-128. http://www.ep.liu.se/ea/cis/1997/019/

Towards Inference and Computation Mobility: The Jinni Experiment

Paul Tarau

Department of Computer Science, University of North Texas, USA
tarau@cs.unt.edu

Abstract. We overview the design and implementation of Jinni[1] (**J**ava **IN**ference engine and **N**etworked **I**nteractor), a lightweight, multi-threaded, pure logic programming language, intended to be used as a flexible scripting tool for gluing together knowledge processing components and Java objects in networked client/server applications, as well as through applets over the Web.
Mobile threads, implemented by capturing first order continuations in a compact data structure sent over the network, allow Jinni components to interoperate with remote high performance BinProlog servers for CPU-intensive knowledge processing and with other Jinni components over the Internet.
These features make Jinni a perfect development platform for intelligent mobile agent systems.

Keywords: Java based Logic Programming languages, remote execution, Linda coordination, blackboard-based distributed logic programming, mobile code through first order continuations, intelligent mobile agents

1 The world of Jinni

Jinni is based on a simple **Things**, **Places**, **Agents** ontology, borrowed from MUDs and MOOs [14, 1, 3, 9, 18, 15].

Things are represented as Prolog terms, basically trees of embedded records containing constants and variables to be further instantiated to other trees.

Places are processes running on various computers with a server component listening on a port and a blackboard component allowing synchronized multi-user Linda [6, 10] and remote predicate call transactions.

Agents are collections of threads executing a set of goals, possibly spread over a set of different Places and usually executing remote and local transactions in coordination with other Agents. Asynchronous communication with other agents is achieved using the Linda coordination protocol.

Places and Agents are *clonable*, support inheritance/sharing of Things and are designed to be easily editable/configurable using visual tools. Agent threads

[1] Available at `http://www.cs.unt.edu/~tarau/netjinni/Jinni.html`

moving between places and agents moving as units are supported. Places are used to abstract away language differences between processors, like for instance Jinni and BinProlog. Mobile code allows very high speed processing in Jinni by delegating heavy inference processing to high-performance BinProlog components.

2 Jinni as a Logic Programming Java component

Jinni is implemented as a lightweight, thin client component, based as much as possible on fully portable, platform, version and vendor independent Java code. Its main features come from this architectural choice:

- a trimmed down, simple, operatorless syntactic subset of Prolog,
- pure Prolog (Horn Clause logic) with leftmost goal unfolding as inference rule,
- multiple asynchronous inference engines each running as a separate thread,
- a shared blackboard to communicate between engines using a simple Linda-style subscribe/publish (in/out in Linda jargon) coordination protocol based on associative search,
- high level networking operations allowing code mobility [2, 12, 11, 4, 19, 13] and remote execution,
- a straightforward Jinni-to-Java translator allows packaging of Jinni programs as Java classes to be transparently loaded over the Internet
- backtrackable assumptions [17, 8] implemented through trailed, overridable undo actions

Jinni's spartan return to (almost) pure Horn Clause logic does not mean it is necessarily a weaker language. Expressiveness of full Prolog (and beyond:-)) is easily attained in Jinni by combining multiple engines. Engines give transparent access to the underlying Java threads and are used to implement local or remote, lazy or eager findall operations, negation as failure, if-then-else, etc. at source level. Inference engines running on separate threads can cooperate through either predicate calls or through an easy to use flavor of the Linda coordination protocol. Remote or local dynamic database updates make Jinni an extremely flexible Agent programming language. Jinni is designed on top of dynamic, fully garbage collectible data structures, to take advantage of Java's automatic memory management.

3 What's new in Jinni

3.1 Engines

An engine can be seen as an abstract data-type which produces a (possibly infinite) stream of solutions as needed. To create an new engine, we use:

```
new_engine(Goal,Answer,Handle)
```

The `Handle` is a unique Java Object denoting the engine, assigned to its own thread, for further processing.

To get an answer from the engine we use:

```
ask_engine(Handle,Answer)
```

Each engine backtracks independently using its choice-point stack and trail during the computation of an answer. Once computed, an answer is copied from an engine to the master engine which initiated it.
Multiple engines enhance the expressiveness of Prolog by allowing an AND-branch of an engine to collect answers from multiple OR-branches of other engine(s). We call this design pattern *orthogonal engines.* They give to the programmer the means to see as an abstract sequence and control, the answers produced by an engine, in a way similar to Java's `Enumeration` interface. In fact, by using *orthogonal engines*, a programmer does not really need to use findall and other similar predicates anymore - why accumulate answers eagerly on a list which will get scanned and decomposed again, when answers can be produced on demand?

3.2 Coordination and remote execution mechanisms

Our networking constructs are built on top of the popular Linda [6, 10, 7] coordination framework, enhanced with unification based pattern matching, remote execution and a set of simple client-server components melted together into a scalable peer-to-peer layer, forming a 'web of interconnected worlds'. The basic operations are the following:

- `in(X)`: waits until it can take an object matching `X` from the server
- `out(X)`: puts `X` on the server and possibly wakes up a waiting in/1 operation
- `all(X,Xs)`: reads the list `Xs` matching `X` currently on the server
- `the(Pattern,Goal,Answer)`: remotely runs a thread executing `Goal` on the server and collects it's first `Answer`, of the form `the(Pattern)` if successful, and the atom `no` otherwise

The presence of the `all/2` collector compensates for the lack of non-deterministic operations. Note that the only blocking operation is `in/1`. Blocking `rd/1` is easily emulated in terms of `in/1` and `out/1`, while non-blocking `rd/1` is emulated with `all/2`.

3.3 Server-side constraint solving

A natural extension to Linda is to use constraint solving for selection matching terms, instead of plain unification. This is implemented in Jinni through the use of 2 builtins:

Wait_for(Term,Constraint): waits for a Term such that Constraint is true on the server, and when this happens, it removes the result of the match from the server with an in/1 operation. Constraint is either a single goal or a list of goals [G1,G2,..,Gn] to be executed on the server.

Notify_about(Term): notifies the server to give this term to any blocked client which waits for it with a matching constraint i.e.

```
notify_about(stock_offer(nscp,29))
```

would trigger execution of a client having issued

```
wait_for(stock_offer(nscp,Price),less(Price,30)).
```

3.4 Mobile Code: for expressiveness and for acceleration

An obvious way to accelerate slow Prolog processing for a Java based system is through use of native (C/C++) methods. The simplest way to accelerate Jinni's Prolog processing is by including BinProlog through Java's JNI.

However, a more general scenario, also usable for applets not allowing native method invocations is use of a remote accelerator. This is achieved transparently through the use of *mobile code.*

The Oz 2.0 distributed programming proposal of [19] makes *object mobility* more transparent, although the mobile entity is still the state of the objects, not "live" code.

Mobility of "live code" is called *computation mobility* [5]. It requires interrupting execution, moving the state of a runtime system (stacks, for instance) from one site to another and then resuming execution. Clearly, for some languages, this can be hard or completely impossible to achieve.

Telescript and General Magic's new Odissey [11] agent programming framework, IBM's Java based *aglets* [12] as well as Luca Cardelli's Oblique [2] have pioneered implementation technologies achieving *computation mobility.*

In the case of Jinni, computation mobility is used both as an *accelerator* and an *expressiveness lifting* device. A live thread will migrate from Jinni to a faster remote BinProlog engine, do some CPU intensive work and then come back with the results (or just sent back results, using Linda coordination). A very simple way to ensure atomicity and security of complex networked transactions is to have the agent code move to the site of the computation, follow existing security rules, access possibly large databases and come back with the results.

Jinni's mobile computation is based on the use of *first order continuations* i.e. encapsulated future computations, which can be easily suspended, moved over the network, and resumed at a different site. As continuations are first-order objects both in Jinni and BinProlog, the implementation is straightforward [16] and the two engines can interoperate transparently by simply moving computations from one to the other.

Note that a unique **move/0** operation is used to transport computation to the server. The client simply waits until computation completes, when bindings for the first solution are propagated back.

Note that mobile computation is more expressive and more efficient than remote predicate calls as such. Basically, it moves once, and executes on the server all future computations of the current AND branch.

4 Application domains

Jinni's client and server scripting abilities are intended to support platform independent Prolog-to-Java and Prolog-to-Prolog bidirectional connection over the net and to accelerate integration of the effective inference technologies developed the last 20 years in the field of Logic Programming in mainstream Internet products.

The next iteration is likely to bring a simple, plain English scripting language to be compiled to Jinni with speech recognizer/synthesizer based I/O.

Among the potential targets for Jinni based products: lightweight rule based programs assisting customers of Java enables appliances, from Web based TVs to mobile cell phones and car computers, all requiring knowledge components to adjust to increasingly sophisticated user expectations.

A stock market simulator is currently on the way to be implemented based on Jinni, featuring user programmable intelligent agents. It is planned to be connected to real world Internet based stock trade services.

5 Conclusion

The Jinni project shows that Logic Programming languages are well suited as the basic glue so much needed for elegant and cost efficient Internet programming. The ability to compress so much functionality in such a tiny package shows that building logic programming components to be integrated in emerging tools like Java might be the most practical way towards mainstream recognition and widespread use of Logic Programming technology. Jinni's emphasis on functionality and expressiveness over performance, as well as it's use of integrated multithreading and networking, hint towards the priorities we consider important for future Logic Programming language design.

Acknowledgments

We thank for support from NSERC (grants OGP0107411), the Université de Moncton, Louisiana Tech University as well as from E-COM Inc. and the Radiance Group Inc. Special thanks go to Veronica Dahl, Bart Demoen, Koen De Boschere, Ed Freeman, Don Garrett, Stephen Rochefort and Yu Zhang for fruitful interaction related to the design, implementation and testing of Jinni.

References

1. The Avalon MUD. http://www.avalon-rpg.com/.
2. K. A. Bharat and L. Cardelli. Migratory applications. In *Proceedings of the 8th Annual ACM Symposium on User Interface Software and Technology*, Nov. 1995. http://gatekeeper.dec.com/pub/DEC/SRC/research-reports/ abstracts/src-rr-138.html.
3. BlackSun. CyberGate. http://www.blaxxsun.com/.
4. L. Cardelli. Mobile ambients. Technical report, Digital, 1997. http://www.research.digital.com/ SRC/personal/Luca_Cardelli/Papers.html.
5. L. Cardelli. Mobile Computation. In J. Vitek and C. Tschudin, editors, *Mobile Object Systems - Towards the Programmable Internet*, pages 3–6. Springer-Verlag, LNCS 1228, 1997.
6. N. Carriero and D. Gelernter. Linda in context. *CACM*, 32(4):444–458, 1989.
7. S. Castellani and P. Ciancarini. Enhancing Coordination and Modularity Mechanisms for a Languag e with Objects-as-Multisets. In P. Ciancarini and C. Hankin, editors, *Proc. 1st Int. Conf. on Coordination Models and Languages*, volume 1061 of *LNCS*, pages 89–106, Cesena, Italy, April 1996. Springer.
8. V. Dahl, P. Tarau, and R. Li. Assumption Grammars for Processing Natural Language. In L. Naish, editor, *Proceedings of the Fourteenth International Conference on Logic Programming*, pages 256–270, MIT press, 1997.
9. K. De Bosschere, D. Perron, and P. Tarau. LogiMOO: Prolog Technology for Virtual Worlds. In *Proceedings of PAP'96*, pages 51–64, London, Apr. 1996.
10. K. De Bosschere and P. Tarau. Blackboard-based Extensions in Prolog. *Software — Practice and Experience*, 26(1):49–69, Jan. 1996.
11. GeneralMagicInc. Odissey. 1997. available at http://www.genmagic.com/agents.
12. IBM. Aglets. http://www.trl.ibm.co.jp/aglets.
13. E. Jul, H. Levy, N. Hutchinson, and A. Black. Fine-Grained Mobility in the Emerald System. *ACM Transactions on Computer Systems*, 6(1):109–133, February 1988.
14. T. Meyer, D. Blair, and S. Hader. WAXweb: a MOO-based collaborative hypermedia system for WWW. *Computer Networks and ISDN Systems*, 28(1/2):77–84, 1995.
15. P. Tarau. Logic Programming and Virtual Worlds. In *Proceedings of INAP96*, Tokyo, Nov. 1996.
16. P. Tarau and V. Dahl. Mobile Threads through First Order Continuations. 1997. submitted, http://clement.info.umoncton.ca/html/tmob/html.html.
17. P. Tarau, V. Dahl, and A. Fall. Backtrackable State with Linear Affine Implication and Assumption Grammars. In J. Jaffar and R. H. Yap, editors, *Concurrency and Parallelism, Programming, Networking, and Security*, Lecture Notes in Computer Science 1179, pages 53–64, Singapore, Dec. 1996. "Springer".
18. P. Tarau and K. De Bosschere. Virtual World Brokerage with BinProlog and Netscape. In P. Tarau, A. Davison, K. De Bosschere, and M. Hermenegildo, editors, *Proceedings of the 1st Workshop on Logic Programming Tools for INTERNET Applications*, JICSLP'96, Bonn, Sept. 1996. http://clement.info.umoncton.ca/~lpnet.
19. P. Van Roy, S. Haridi, and P. Brand. Using mobility to make transparent distribution practical. 1997. manuscript.

Author Index

Lecture Notes in Artificial Intelligence (LNAI)

Vol. 1374: H. Bunt, R.-J. Beun, T. Borghuis (Eds.), Multimodal Human-Computer Communication. VIII, 345 pages. 1998.

Vol. 1387: C. Lee Giles, M. Gori (Eds.), Adaptive Processing of Sequences and Data Structures. Proceedings, 1997. XII, 434 pages. 1998.

Vol. 1394: X. Wu, R. Kotagiri, K.B. Korb (Eds.), Research and Development in Knowledge Discovery and Data Mining. Proceedings, 1998. XVI, 424 pages. 1998.

Vol. 1395: H. Kitano (Ed.), RoboCup-97: Robot Soccer World Cup I. XIV, 520 pages. 1998.

Vol. 1397: H. de Swart (Ed.), Automated Reasoning with Analytic Tableaux and Related Methods. Proceedings, 1998. X, 325 pages. 1998.

Vol. 1398: C. Nédellec, C. Rouveirol (Eds.), Machine Learning: ECML-98. Proceedings, 1998. XII, 420 pages. 1998.

Vol. 1400: M. Lenz, B. Bartsch-Spörl, H.-D. Burkhard, S. Wess (Eds.), Case-Based Reasoning Technology. XVIII, 405 pages. 1998.

Vol. 1404: C. Freksa, C. Habel. K.F. Wender (Eds.), Spatial Cognition. VIII, 491 pages. 1998.

Vol. 1409: T. Schaub, The Automation of Reasoning with Incomplete Information. XI, 159 pages. 1998.

Vol. 1415: J. Mira, A.P. del Pobil, M. Ali (Eds.), Methodology and Tools in Knowledge-Based Systems. Vol. I. Proceedings, 1998. XXIV, 887 pages. 1998.

Vol. 1416: A.P. del Pobil, J. Mira, M. Ali (Eds.), Tasks and Methods in Applied Artificial Intelligence. Vol. II. Proceedings, 1998. XXIII, 943 pages. 1998.

Vol. 1418: R. Mercer, E. Neufeld (Eds.), Advances in Artificial Intelligence. Proceedings, 1998. XII, 467 pages. 1998.

Vol. 1421: C. Kirchner, H. Kirchner (Eds.), Automated Deduction – CADE-15. Proceedings, 1998. XIV, 443 pages. 1998.

Vol. 1424: L. Polkowski, A. Skowron (Eds.), Rough Sets and Current Trends in Computing. Proceedings, 1998. XIII, 626 pages. 1998.

Vol. 1433: V. Honavar, G. Slutzki (Eds.), Grammatical Inference. Proceedings, 1998. X, 271 pages. 1998.

Vol. 1434: J.-C. Heudin (Ed.), Virtual Worlds. Proceedings, 1998. XII, 412 pages. 1998.

Vol. 1435: M. Klusch, G. Weiß (Eds.), Cooperative Information Agents II. Proceedings, 1998. IX, 307 pages. 1998.

Vol. 1437: S. Albayrak, F.J. Garijo (Eds.), Intelligent Agents for Telecommunication Applications. Proceedings, 1998. XII, 251 pages. 1998.

Vol. 1441: W. Wobcke, M. Pagnucco, C. Zhang (Eds.), Agents and Multi-Agent Systems. Proceedings, 1997. XII, 241 pages. 1998.

Vol. 1446: D. Page (Ed.), Inductive Logic Programming. Proceedings, 1998. VIII, 301 pages. 1998.

Vol. 1453: M.-L. Mugnier, M. Chein (Eds.), Conceptual Structures: Theory, Tools and Applications. Proceedings, 1998. XIII, 439 pages. 1998.

Vol. 1454: I. Smith (Ed.), Artificial Intelligence in Structural Engineering.XI, 497 pages. 1998.

Vol. 1456: A. Drogoul, M. Tambe, T. Fukuda (Eds.), Collective Robotics. Proceedings, 1998. VII, 161 pages. 1998.

Vol. 1458: V.O. Mittal, H.A. Yanco, J. Aronis, R. Simpson (Eds.), Assistive Technology in Artificial Intelligence. X, 273 pages. 1998.

Vol. 1471: J. Dix, L. Moniz Pereira, T.C. Przymusinski (Eds.), Logic Programming and Knowledge Representation. Proceedings, 1997. IX, 246 pages. 1998.

Vol. 1476: J. Calmet, J. Plaza (Eds.), Artificial Intelligence and Symbolic Computation. Proceedings, 1998. XI, 309 pages. 1998.

Vol. 1480: F. Giunchiglia (Ed.), Artificial Intelligence: Methodology, Systems, and Applications. Proceedings, 1998. IX, 502 pages. 1998.

Vol. 1484: H. Coelho (Ed.), Progress in Artificial Intelligence – IBERAMIA 98. Proceedings, 1998. XIII, 421 pages. 1998.

Vol. 1488: B. Smyth, P. Cunningham (Eds.), Advances in Case-Based Reasoning. Proceedings, 1998. XI, 482 pages. 1998.

Vol. 1489: J. Dix, L. Fariñas del Cerro, U. Furbach (Eds.), Logics in Artificial Intelligence. Proceedings, 1998. X, 391 pages. 1998.

Vol. 1495: T. Andreasen, H. Christiansen, H.L. Larsen (Eds.), Flexible Query Answering Systems. Proceedings, 1998. IX, 393 pages. 1998.

Vol. 1501: M.M. Richter, C.H. Smith, R. Wiehagen, T. Zeugmann (Eds.), Algorithmic Learning Theory. Proceedings, 1998. XI, 439 pages. 1998.

Vol. 1502: G. Antoniou, J. Slaney (Eds.), Advanced Topics in Artificial Intelligence. Proceedings, 1998. XI, 333 pages. 1998.

Vol. 1504: O. Herzog, A. Günter (Eds.), KI-98: Advances in Artificial Intelligence. Proceedings, 1998. XI, 355 pages. 1998.

Vol. 1510: J.M. Zytkow, M. Quafafou (Eds.), Principles of Data Mining and Knowledge Discovery. Proceedings, 1998. XI, 482 pages. 1998.

Lecture Notes in Computer Science

Vol. 1475: W. Litwin, T. Morzy, G. Vossen (Eds.), Advances in Databases and Information Systems. Proceedings, 1998. XIV, 369 pages. 1998.

Vol. 1476: J. Calmet, J. Plaza (Eds.), Artificial Intelligence and Symbolic Computation. Proceedings, 1998. XI, 309 pages. 1998. (Subseries LNAI).

Vol. 1477: K. Rothermel, F. Hohl (Eds.), Mobile Agents. Proceedings, 1998. VIII, 285 pages. 1998.

Vol. 1478: M. Sipper, D. Mange, A. Pérez-Uribe (Eds.), Evolvable Systems: From Biology to Hardware. Proceedings, 1998. IX, 382 pages. 1998.

Vol. 1479: J. Grundy, M. Newey (Eds.), Theorem Proving in Higher Order Logics. Proceedings, 1998. VIII, 497 pages. 1998.

Vol. 1480: F. Giunchiglia (Ed.), Artificial Intelligence: Methodology, Systems, and Applications. Proceedings, 1998. IX, 502 pages. 1998. (Subseries LNAI).

Vol. 1481: E.V. Munson, C. Nicholas, D. Wood (Eds.), Principles of Digital Document Processing. Proceedings, 1998. VII, 152 pages. 1998.

Vol. 1482: R.W. Hartenstein, A. Keevallik (Eds.), Field-Programmable Logic and Applications. Proceedings, 1998. XI, 533 pages. 1998.

Vol. 1483: T. Plagemann, V. Goebel (Eds.), Interactive Distributed Multimedia Systems and Telecommunication Services. Proceedings, 1998. XV, 326 pages. 1998.

Vol. 1484: H. Coelho (Ed.), Progress in Artificial Intelligence – IBERAMIA 98. Proceedings, 1998. XIII, 421 pages. 1998. (Subseries LNAI).

Vol. 1485: J.-J. Quisquater, Y. Deswarte, C. Meadows, D. Gollmann (Eds.), Computer Security – ESORICS 98. Proceedings, 1998. X, 377 pages. 1998.

Vol. 1486: A.P. Ravn, H. Rischel (Eds.), Formal Techniques in Real-Time and Fault-Tolerant Systems. Proceedings, 1998. VIII, 339 pages. 1998.

Vol. 1487: V. Gruhn (Ed.), Software Process Technology. Proceedings, 1998. VIII, 157 pages. 1998.

Vol. 1488: B. Smyth, P. Cunningham (Eds.), Advances in Case-Based Reasoning. Proceedings, 1998. XI, 482 pages. 1998. (Subseries LNAI).

Vol. 1489: J. Dix, L. Fariñas del Cerro, U. Furbach (Eds.), Logics in Artificial Intelligence. Proceedings, 1998. X, 391 pages. 1998. (Subseries LNAI).

Vol. 1490: C. Palamidessi, H. Glaser, K. Meinke (Eds.), Principles of Declarative Programming. Proceedings, 1998. XI, 497 pages. 1998.

Vol. 1493: J.P. Bowen, A. Fett, M.G. Hinchey (Eds.), ZUM '98: The Z Formal Specification Notation. Proceedings, 1998. XV, 417 pages. 1998.

Vol. 1495: T. Andreasen, H. Christiansen, H.L. Larsen (Eds.), Flexible Query Answering Systems. IX, 393 pages. 1998. (Subseries LNAI).

Vol. 1496: W.M. Wells, A. Colchester, S. Delp (Eds.), Medical Image Computing and Computer-Assisted Intervention – MICCAI'98. Proceedings, 1998. XXII, 1256 pages. 1998.

Vol. 1497: V. Alexandrov, J. Dongarra (Eds.), Recent Advances in Parallel Virtual Machine and Message Passing Interface. Proceedings, 1998. XII, 412 pages. 1998.

Vol. 1498: A.E. Eiben, T. Bäck, M. Schoenauer, H.-P. Schwefel (Eds.), Parallel Problem Solving from Nature – PPSN V. Proceedings, 1998. XXIII, 1041 pages. 1998.

Vol. 1499: S. Kutten (Ed.), Distributed Computing. Proceedings, 1998. XII, 419 pages. 1998.

Vol. 1501: M.M. Richter, C.H. Smith, R. Wiehagen, T. Zeugmann (Eds.), Algorithmic Learning Theory. Proceedings, 1998. XI, 439 pages. 1998. (Subseries LNAI).

Vol. 1502: G. Antoniou, J. Slaney (Eds.), Advanced Topics in Artificial Intelligence. Proceedings, 1998. XI, 333 pages. 1998. (Subseries LNAI).

Vol. 1503: G. Levi (Ed.), Static Analysis. Proceedings, 1998. IX, 383 pages. 1998.

Vol. 1504: O. Herzog, A. Günter (Eds.), KI-98: Advances in Artificial Intelligence. Proceedings, 1998. XI, 355 pages. 1998. (Subseries LNAI).

Vol. 1508: S. Jajodia, M.T. Özsu, A. Dogac (Eds.), Advances in Multimedia Information Systems. Proceedings, 1998. VIII, 207 pages. 1998.

Vol. 1510: J.M. Zytkow, M. Quafafou (Eds.), Principles of Data Mining and Knowledge Discovery. Proceedings, 1998. XI, 482 pages. 1998. (Subseries LNAI).

Vol. 1511: D. O'Hallaron (Ed.), Languages, Compilers, and Run-Time Systems for Scalable Computers. Proceedings, 1998. IX, 412 pages. 1998.

Vol. 1512: E. Giménez, C. Paulin-Mohring (Eds.), Types for Proofs and Programs. Proceedings, 1996. VIII, 373 pages. 1998.

Vol. 1513: C. Nikolaou, C. Stephanidis (Eds.), Research and Advanced Technology for Digital Libraries. Proceedings, 1998. XV, 912 pages. 1998.

Vol. 1514: K. Ohta,, D. Pei (Eds.), Advances in Cryptology – ASIACRYPT'98. Proceedings, 1998. XII, 436 pages. 1998.

Vol. 1516: W. Ehrenberger (Ed.), Computer Safety, Reliability and Security. Proceedings, 1998. XVI, 392 pages. 1998.

Vol. 1518: M. Luby, J. Rolim, M. Serna (Eds.), Randomization and Approximation Techniques in Computer Science. Proceedings, 1998. IX, 385 pages. 1998.